ADVANCES IN CHEMICAL PHYSICS

VOLUME XLVIII

Advances in
CHEMICAL PHYSICS

EDITED BY

I. PRIGOGINE

University of Brussels
Brussels, Belgium
and
University of Texas
Austin, Texas

AND

STUART A. RICE

Department of Chemistry
and
The James Franck Institute
The University of Chicago
Chicago, Illinois

VOLUME XLVIII

AN INTERSCIENCE® PUBLICATION
JOHN WILEY AND SONS
NEW YORK • CHICHESTER • BRISBANE • TORONTO

AN INTERSCIENCE® PUBLICATION

Copyright © 1981 by John Wiley & Sons, Inc.

All rights reserved. Published simultaneously in Canada.

Library of Congress Catalog Card Number: 58-9935
ISBN 0-471-08294-5
ISSN 0065-2385

Printed in the United States of America
10 9 8 7 6 5 4 3 2 1

INTRODUCTION

Few of us can any longer keep up with the flood of scientific literature, even in specialized subfields. Any attempt to do more, and be broadly educated with respect to a large domain of science, has the appearance of tilting at windmills. Yet the synthesis of ideas drawn from different subjects into new, powerful, general concepts is as valuable as ever, and the desire to remain educated persists in all scientists. This series, *Advances in Chemical Physics*, is devoted to helping the reader obtain general information about a wide variety of topics in chemical physics, which field we interpret very broadly. Our intent is to have experts present comprehensive analyses of subjects of interest and to encourage the expression of individual points of view. We hope that this approach to the presentation of an overview of a subject will both stimulate new research and serve as a personalized learning text for beginners in a field.

ILYA PRIGOGINE

STUART A. RICE

CONTENTS

ADVANCES IN CHEMICAL PHYSICS

VOLUME XLVIII

ANALYSIS AND EVALUATION OF IONIZATION POTENTIALS, ELECTRON AFFINITIES, AND EXCITATION ENERGIES BY THE EQUATIONS OF MOTION— GREEN'S FUNCTION METHOD

MICHAEL F. HERMAN

Department of Chemistry
Columbia University
New York, New York

KARL F. FREED

The James Franck Institute and the Department of Chemistry
The University of Chicago
Chicago, Illinois

AND

D. L. YEAGER

Department of Chemistry
Texas A & M University
College Station, Texas

CONTENTS

1

I. INTRODUCTION

An accurate representation of the electronic structure of atoms and molecules requires the incorporation of the effects of electron correlation,[1] and this process imposes severe computational difficulties. It is, therefore, only natural to investigate the use of new and alternative formulations of the problem. Many-body theory methods[2–4] offer a wide variety of attractive approaches to the treatment of electron correlation, in part because of their great successes in treating problems in quantum field theory, the statistical mechanics of many-body systems, and the electronic properties of solids.

The pioneering work of Kelly[5] on atoms provided the first comprehensive utilization of many-body theory methods to describe electron correlation in these systems. These studies investigated the use of diagrammatic many-body perturbation theory, an approach that appeared to be quite different from the more traditional wave function methods. However, if a summation is made of the diagrams that Kelly found numerically to be the most important, the final result can then be shown to formally be equivalent to the sum-of-the-pairs wave function theories[6] that had previously been proposed by Sinanoğlu,[7] Nesbet,[8] and others. Thus, Kelly's work provided the first calculation of this sum-of-the-pairs variety. The framework of many-body perturbation theory also introduced a new vehicle for gaining physical understanding of the important processes in atomic electronic correlation. Furthermore, the work of Kelly has resulted in the introduction of a vast number of approximations and techniques that have had a wide impact on other approaches to electronic correlation.

Many-body Green's function–equations of motion methods[9, 10] appear to differ more strongly from wave function theories than does many-body perturbation theory. In wave function approaches it is necessary to evaluate energy differences, (excitation energies, ionization potentials, electron affinities, etc.) by determining the individual state energies and then evaluating their differences. On the other hand, the Green's function–equations of motion methods generate these energy differences directly.

There have also been other attempts to evaluate these energy differences directly.[11–13] These methods utilize Rayleigh-Schrödinger perturbation theory to express the energies for both states with a common orbital basis. When the perturbation series for the two state energies are subtracted, it is found that there is a considerable cancellation of identical terms from the individual series.

In all these direct energy difference methods the hope is that by a cancellation of common terms in the individual state energies, greater efficiency and accuracy can be achieved as compared with the traditional single state approaches. In addition, the equations of motion (EOM) and the many-body Green's function (MBGF) methods introduce a different operator algebra and outlook into the problem. This has the disadvantage of making the material quite incomprehensible to many practitioners of atomic and molecular quantum mechanics on one hand, but it also raises the possibility of the generation of new and useful insights into these electronic processes. These methods also introduce a new many-electron basis, to be called the many-body basis, which may be superior in some aspects, both conceptually and in terms of practical calculations, to the traditional configuration set. Throughout the discussion that follows we attempt to bridge the language gap between the many-body theory methods and the traditional wave function approaches by noting many of the strong parallels between the EOM method and traditional wave function theories, similarities that may often be obscured by the different formalism of the former.

A critical analysis of the Green's function–equations of motion method requires the resolution of the following questions: (*1*) Are these Green's function–equations of motion methods formally different from traditional wave function or many-body perturbation theory approaches? Even if they are not, these methods should still be of considerable utility because of the new insights and approaches afforded by them. (*2*) If indeed the answer to the first question is affirmative, it is of interest to determine the manner in which the many-body Green's function–equations of motion methods differ from the more traditional approaches. This is imperative if we are to be able to make meaningful comparisons of calculations that have been performed using the two types of theory. The reduction of these method types to a common language would thereby enhance our physical understanding of the important processes in determining the electronic structure of atomic and molecular systems. (*3*) It is also important to determine which types of systematic approximation can be utilized within the Green's function–equations of motion methods to provide results that are at least as accurate as those obtainable from the most sophisticated configuration interaction treatments now available.

These three questions have motivated a series of our studies of both the formal and computational aspects of the Green's function–equations of motion methods.

It is possible to provide a partial answer to question *1* without ever becoming enmeshed in the complicated details of Green's function–equation of motion theories. The simple reasoning is as follows.[14, 15] Any "black box" that produces the electronic energy levels of a many-electron system must somehow be related to the electronic Hamiltonian for the system or functions of this electronic Hamiltonian. Similarly, any theory that directly provides energy differences must be related to the only quantum mechanical operator whose eigenvalues are the energy differences. This operator is the Liouville operator L, which is defined by its action on an arbitrary operator A by

$$LA \equiv [H, A] = HA - AH$$

where H is the electronic Hamiltonian for the system and the square brackets denote the commutator as usual. Thus the equations of motion–Green's function methods must somehow differ from their wave function counterparts, which are based on the approximate solution of the eigenfunctions and eigenvalues of the electronic Hamiltonian H.

There have been a number of attempts to use Liouville operator techniques to directly evaluate energy differences. These attempts introduce the operator basis set, $\{|i\rangle\langle j|\}$, where $\{|i\rangle\}$ is a set of basis functions. The eigenfunctions of L are then represented as linear superposition of the basis operators,

$$A_\lambda = \sum_{i,j} C_{ij}^\lambda |i\rangle\langle j|$$

These attempts have considered simple problems like the anharmonic oscillator problem in a harmonic oscillator basis or the hydrogen atom in a Gaussian-type basis, generally with rather poor results. The reason for these difficulties is rather clear. Given N basis functions $\{|i\rangle\}$, there are N^2 basis operators $\{|i\rangle\langle j|\}$. Consequently, the equations for the eigenvalues and eigenvectors of L represent equations of rank $N^2 \times N^2$, as compared with the usual equations for the eigenfunctions and eigenvalues of H, which are of the much smaller dimension, $N \times N$. Hence this simple-minded Liouville operator approach merely compounds the mathematical difficulties already inherent in standard Hamiltonian methods.

The Green's function–equations of motion methods can be shown not to suffer from the N^2 problem of the naive use of Liouville operator methods. As discussed below, it is found that the corresponding Green's function–equations of motion methods are problems that generate matrices of dimension $2(N-1)$ when the original basis has been generated from all possibilities that arise from a given orbital basis set. Likewise, it can be shown that the Green's function method can be represented as particular subblocks (submatrices) of the resolvent of the Liouville operator, whereas the equations of motion methods consider the eigenvalues and eigenvectors of the Liouville operator in the same particular representation.

The one-electron Green's function has its poles at the ionization potentials and electron affinities of an atom or molecule, whereas the poles of the two-electron Green's function are located at the excitation energies.[9] Furthermore, the residues of the Green's function at these poles yield information about the transition amplitudes. Two main approaches have been followed in the evaluation of many-body Green's functions. The first involves the evaluation of a diagrammatic perturbative expansion for the Green's function,[16-35] and the latter looks for an approximate solution of the hierarchy of equations[3, 36-41] that the many-body Green's function obeys. The work of Cederbaum and co-workers[16-32] concerning the one-electron Green's function is a particularly noteworthy example of the diagrammatic technique. These investigators have developed a variety of approximations and have provided extensive numerical data concerning the importance of specific diagrams.

The propagator technique, which attempts to solve the hierarchy of equations for the many-body Green's function, has been facilitated by the use of inner projection techniques and the superoperator representation of Goscinski and Lukman.[42] Öhrn and co-workers[38-41] have applied these techniques to the evaluation of ionization potentials of atomic and molecular systems. Several authors[37, 43-48] have discussed the relationship between the EOM and MBGF approaches and have compared the various numerical schemes. Because these many-body EOM and Green's function methods are so closely related formally, the results obtained from one procedure provides information that is pertinent to all.

All calculations discussed here involve the use of a finite (therefore incomplete) set of analytical one-electron basis functions. A specific finite orbital basis set defines a finite set of N_e-electron wave functions or basis configurations that spans a finite-dimensional N_e-electron subspace of the full Hilbert space. Within this finite dimensional space any N_e-electron wave function can be expanded in terms of all the basis configurations (that have the correct symmetry). For almost all systems of interest, when reasonably accurate one-electron basis sets are used, this full N_e-electron basis expansion becomes prohibitively large, and accurate ways must be found of truncating the expansion. Much of the effort in electronic structure theory concentrates on devising better and more concise means of approximating the most important parts of the configuration space for the problem at hand. One of the central goals of this work is to systematically and critically investigate this problem for the equations of motion method.

The equations of motion method has its origins in nuclear physics, where Rowe[10] first developed it as a means of understanding nuclear energy level structure. McKoy and co-workers[49-55] refined the theory for the calculation of electron excitation energies and presented molecular calculations for a variety of different approximation schemes. Simons[56-65] and Yeager[66] independently developed the analogous EOM theory for ionization potentials and electron affinities. The present numerical work[67-72] deals mainly with the ionization potentials-electron affinity (IP-EA) variant of the EOM theory. However, because of the analogous nature of the excitation energy theory, many of the conclusions reached from the IP-EA calculations have immediate applicability to EOM excitation energy calculations. Some excitation energy calculations on simple systems are utilized here to illustrate important facets of the general theory.

Section II develops the EOM theory both for excitation energies and for ionization potentials and electron affinities. After the main EOM equations have been derived, the nature of a complete operator basis set in EOM calculations is determined and is shown to differ from the mathematically complete set. The many-body operator basis is described, and approximations introduced in practical calculations are discussed. There follows an explanation of the various divisions that are utilized to separate the IP-EA operator basis into primary and secondary subspaces. Numerical evidence, presented in Section III, indicates that the traditional division of the EOM operator space into primary and secondary subspaces (and the effectively equivalent partition in Green's function methods) for ionization potentials and electron affinities is not generally adequate. Section II also develops a more extensive IP–EA EOM theory, based on a generalized division of the operator space, which is introduced in view of the difficulties presented in Section III and of recent developments in configurational selection methods[72-75] for generating accurate approximations to the full configuration interaction matrix.

All IP–EA calculations given in Section III involve systems in which the initial state is a closed-shell state and a single determinant is used for a zeroth-order approximation to the ground-state wave function. This restriction to closed-shell ground states and single-determinant, zeroth-order wave functions has been common to nearly all EOM work, as well as to almost all the related propagator and diagrammatic Green's function calculations. In Section III.C, we present results on nitrogen that indicate the need for developing a satisfactory equation of motion—Green's function theory that allows for a multiconfigurational zeroth-order, ground-state wave function (corrected perturbatively). In Section III.F, excitation energy calculations are reported for beryllium, to compare results using a multiconfigurational reference state with the analogous calculations based on a reference wave function having a single determinant. These studies further substantiate the superiority of the multiconfigurational approach. In Section IV we briefly review current attempts[77-79] to devise an approximate theory that incorporates a multiconfigurational ground state and describe what we believe, based on our numerical evidence, to be necessary for a general, truly reliable, and accurate multiconfigurational equations of motion theory.

In Section III.A the differences between the IP–EA EOM methods of Simons[56] and Yeager[66] are analyzed numerically for nitrogen. Section III.B reports EOM ionization potentials for this gas using a series of different orbital basis sets. These results lead to the conclusion that EOM calculations using small basis sets are unreliable, much as is the case for configuration interaction and other traditional methods. This study is of interest because the early EOM results of Simons[57-65] appeared to indicate just the opposite; namely, that EOM calculations using small basis sets provided consistently accurate ionization potentials and electron affinities, presumably because of some cancellation of errors inherent in the method.

Section III.C presents results of a study of certain third-order terms in the EOM equation that had previously been neglected in IP–EA calculations. It is found that some of these terms are reasonably small but not negligible, whereas the inclusion of others in the EOM equation can cause a complete breakdown of the traditional perturbative EOM method for nitrogen when using the standard choice of the primary operator space.

Different choices of the primary space are shown to remedy the difficulties. In Section III.D the 15 to 40 eV photoelectron spectrum for nitrogen, including shake-up lines, is calculated, given the generalized definition of the EOM primary space. The peak intensities as well as peak positions are calculated.

In Section III.E, EOM ionization potentials and electron affinities are compared with accurate configuration interaction (CI) results for a number of atomic and molecular systems. The same one-electron basis sets are utilized in the EOM and CI calculations, allowing for the separation of basis set errors from errors caused by approximations made in the solution of the EOM equation. EOM results are reported for various approximations including those for the extensive EOM theory developed in Section II. Section III.F presents results of excitation energy calculations for helium and beryllium to address a number of remaining difficult questions concerning the EOM method.

Section IV summarizes the major conclusions of these investigations and outlines the extension and generalization of the EOM theory based on the results of our numerical studies.

II. THEORY

A. Derivation of the EOM Equations

Let $|0\rangle$ be the exact N_e-electron ground state of the Born-Oppenheimer Hamiltonian H for a given atomic or molecular system. Likewise, let $|\lambda\rangle$ be some exact excited state of interest for the same system with the same nuclear geometry. The corresponding state energies are denoted E_0 and E_λ, respectively. For excitation energy calculations $|\lambda\rangle$ is an excited N_e-electron state, whereas in ionization potential or electron affinity cases $|\lambda\rangle$ is an $(N_e - 1)$-electron state or an $(N_e + 1)$-electron state, respectively. The commutator of H with the operator $\tilde{O}_\lambda^\dagger = |\lambda\rangle\langle 0|$ is easily evaluated,

$$\begin{aligned}
\left[H, \tilde{O}_\lambda^\dagger \right] &= H|\lambda\rangle\langle 0| - |\lambda\rangle\langle 0| H \\
&= (E_\lambda - E_0)|\lambda\rangle\langle 0| \\
&= (E_\lambda - E_0)\tilde{O}_\lambda^\dagger
\end{aligned} \tag{1}$$

For IP (EA) calculations, it is necessary to define H to be the N_e-electron Hamiltonian when it acts on N_e-electron states and the $(N_e - 1)[(N_e + 1)]$-electron Hamiltonian when it operates on $(N_e - 1)[(N_e + 1)]$-electron states, and so on. This is simply accomplished by defining $H = \Sigma_M P_M H_M P_M$, where H_M is the M-electron Hamiltonian and P_M is the projection operator onto the space of all M-electron states. When H is expressed in second quantized notation,[2] it automatically has this property.

Equation 1 is an equation of the sort we are seeking; it yields the vertical energy difference $(E_\lambda - E_0)$ directly. One problem in calculating $E_\lambda - E_0$ from (1) is immediately obvious. $\tilde{O}_\lambda^\dagger$ involves $|\lambda\rangle$ and $|0\rangle$, and if these quantities are to be calculated separately, we have not gained anything from (1) over the traditional approach.

1. Primitive EOM Equations

One possible means of circumventing the problem of handling both $|\lambda\rangle$ and $|0\rangle$ is to expand $\tilde{O}_\lambda^\dagger$ in an appropriate set of basis operators,

$$\tilde{O}_\lambda^\dagger = \sum_i C_i^\lambda O_i^\dagger \tag{2}$$

and to determine equations governing the C_i^λ's. One such set of equations is readily obtained upon substituting (2) into (1) and multiplying from the left by the adjoint of one of the basis operators, yielding

$$\sum_j O_i \left[H, O_j^\dagger \right] C_j^\lambda = (E_\lambda - E_0) \sum_j O_i O_j^\dagger C_j^\lambda \tag{3}$$

which is an operator matrix eigenvalue equation with eigenvalues $(E_\lambda - E_0)$ and eigenvectors

$$C^\lambda = \begin{bmatrix} C_1^\lambda \\ C_2^\lambda \\ \vdots \end{bmatrix} \tag{4}$$

Equation 3 still presents problems. First, it is an operator equation. Most of the expertise that has been developed in electronic structure calculations has centered on equations involving matrix elements of operators, rather than the operators themselves. Second, and also important, the vast majority of the solutions of (3) are ones in which we have no interest. Within the limited orbital basis set approximation, there are only a finite number n of linearly independent N_e-electron states, or configuration functions, that can be formed. Within this basis, the "exact" N_e-electron energies $E_0, E_1, \ldots, E_{n-1}$ and the corresponding "exact" N_e-electron states, $|0\rangle, \ldots, |n-1\rangle$, are, respectively, the eigenvalues and eigenvectors of the $n \times n$ Hamiltonian matrix (i.e., the solutions of the complete CI problem for the finite orbital basis). For the excitation energy problem there are n^2 operators of the type $|\lambda\rangle\langle\lambda'|$ (where $|\lambda\rangle$ and $|\lambda'\rangle$ are exact N_e-electron states), and it follows that n^2 must be the dimensionality of the space that $\{O_i^\dagger\}$ spans. Hence (3) has n^2 solutions, whereas the original Schrödinger equation for the same finite orbital basis set has only n solutions. A similar difficulty with (3) persists for calculation of ionization potentials or electron affinities.

Despite these difficulties, Lasaga and Karplus[80] have discussed the calculation of excitation energies based on an operator equation related to (3). Simons and Dalgaard[81] have proposed a perturbation approach to a similar operator problem. To date, however, numerical applications have been limited to the analysis of the singlet excitation of ethylene in Pariser-Parr-Pople[82] (PPP) model, a two-level problem.[80]

Both difficulties with (3) are overcome by taking the ground-state expectation value of (3) to produce

$$\sum_j \langle 0|O_i[H, O_j^\dagger]|0\rangle C_j^\lambda = (E_\lambda - E_0)\sum_j \langle 0|O_i O_j^\dagger|0\rangle C_j^\lambda \qquad (5)$$

To show that (5) has the desired n solutions as opposed to the n^2 solutions of (3), consider the specific set of basis operators, $\{O_i^\dagger\} = \{|\lambda\rangle\langle\lambda'|\}$, where the states $|\lambda\rangle$ and $|\lambda'\rangle$ are exact N_e-electron states. In terms of this operator basis set, the matrices $\langle 0|O_i[H, O_j^\dagger]|0\rangle$ and $\langle 0|O_i O_j^\dagger|0\rangle$ are readily found to be diagonal. If $O_j^\dagger = |\lambda\rangle\langle 0|$ and $\lambda \neq 0$, then (5) yields

$$\langle 0|O_j[H, O_j^\dagger]|0\rangle C_j^\lambda = (E_\lambda - E_0)\langle 0|O_j O_j^\dagger|0\rangle C_j^\lambda$$

If $O_j^\dagger = |\lambda\rangle\langle\lambda'|$ and $\lambda' \neq 0$ ($\lambda = 0, 1, \ldots, n-1$), then (5) trivially gives $0 = 0$. The remaining case is $O_j^\dagger = |0\rangle\langle 0|$, where the matrix element $\langle 0|O_j O_j^\dagger|0\rangle = 1$ while $\langle 0|O_j[H, O_j^\dagger]|0\rangle = 0$. Thus the operator $|0\rangle\langle 0|$ corresponds to a zero eigenvalue for (5). Therefore, only n basis operators, $|\lambda\rangle\langle 0|$, $\lambda = 0, 1, \ldots, n-1$, contribute nontrivially to (5).

2. Double Commutator EOM Equations

Actual numerical calculations introduce double commutator EOM equations[10] for excitation energies and for ionization potentials and electron affinities that, respectively, are

$$\sum_j \langle 0|[O_i, [H, O_j^\dagger]]|0\rangle C_j^\lambda = \omega_\lambda \sum_j \langle 0|[O_i, O_j^\dagger]|0\rangle C_j^\lambda \qquad (6)$$

and

$$\sum_j \langle 0|\{O_i,[H,O_j^\dagger]\}|0\rangle C_j^\lambda = \omega_\lambda \sum_j \langle 0|\{O_i,O_j^\dagger\}|0\rangle C_j^\lambda \qquad (7)$$

where $\{\ ,\ \}$ is the anticommutator, $\{A,B\}=AB+BA$. Equations 6 and 7 are just simple linear matrix eigenvalue equations with $\omega_\lambda = E_\lambda - E_0$ as eigenvalues and the C_j^λ's (which give $O_\lambda^\dagger$) as eigenvectors. Equations 6 and 7 are often derived by assuming that the adjoint of $O_\lambda^\dagger$ satisfies the annihilation condition

$$O_\lambda|0\rangle = 0 \qquad (8)$$

in analogy with raising and lowering operators in the harmonic oscillator problem.[83] If (8) holds, it follows that

$$\langle 0|[H,O_\lambda^\dagger]O_i|0\rangle = \langle 0|O_\lambda^\dagger O_i|0\rangle = 0 \qquad (9)$$

Combining (5) and (9) immediately yields (6) and (7). However, Herman and Freed[72] have shown that the annihilation condition (8) is, in general, not satisfied for the excitation energy problem when $|\lambda\rangle$ is of the same symmetry as $|0\rangle$. In fact, the equation $O_\lambda^\dagger|0\rangle = |\lambda\rangle$, which is usually taken to define $O_\lambda^\dagger$, does not hold for $O_\lambda^\dagger$'s that are general solutions of (6). These conclusions result from the realization that the set of operators $\{|0\rangle\langle 0|, |\lambda'\rangle\langle\lambda''|; \lambda',\lambda''\neq 0\}$ ($|\lambda'\rangle,|\lambda''\rangle$ are eigenstates of H) give only zero matrix elements when inserted for O_i or $O_j^\dagger$ in $\langle 0|[O_i,[H,O_j^\dagger]]|0\rangle$ and $\langle 0|[O_i,O_j^\dagger]|0\rangle$. Therefore, the most general $O_\lambda^\dagger$ that satisfies

$$\langle 0|[O_i,[H,O_\lambda^\dagger]]|0\rangle = \omega_\lambda\langle 0|[O_i,O_\lambda^\dagger]|0\rangle \qquad (10)$$

is given by[72]

$$O_\lambda^\dagger = |\lambda\rangle\langle 0| + \alpha_{0,0}|0\rangle\langle 0| + \sum_{\lambda',\lambda''\neq 0} \alpha_{\lambda',\lambda''}|\lambda'\rangle\langle\lambda''| \qquad (11)$$

for arbitrary values of $\alpha_{0,0}$ and $\alpha_{\lambda',\lambda''}$. This $O_\lambda^\dagger$ does not satisfy $O_\lambda^\dagger|0\rangle = |\lambda\rangle$ and $O_\lambda|0\rangle = 0$ but rather has

$$O_\lambda^\dagger|0\rangle = |\lambda\rangle + \alpha_{0,0}|0\rangle \qquad (12)$$

and

$$O_\lambda|0\rangle = \alpha_{0,0}|0\rangle \qquad (13)$$

Since the matrices $\langle 0|[O_i,[H,O_j^\dagger]]|0\rangle$ and $\langle 0|[O_i,O_j^\dagger]|0\rangle$ are block diagonal by symmetry, $\alpha_{0,0}$ can be nonzero for excitation energy calculations only if $|\lambda\rangle$ and $|0\rangle$ are of the same symmetry. For the IP–EA variant of the theory, only operators that increase or decrease the number of electrons in the system are allowed in the basis. Therefore, there can be no $|0\rangle\langle 0|$ term in the operator expansion for IPs and EAs.

Equation 10 can be derived from (12) and (13), without introducing the inconsistent assumptions $O_\lambda^\dagger|0\rangle=|\lambda\rangle$ and $O_\lambda|0\rangle=0$, as follows.[70, 72] It follows from (12) that

$$
\begin{aligned}
\left[H,O_\lambda^\dagger\right]|0\rangle &= H(|\lambda\rangle+\alpha_{0,0}|0\rangle)-O_\lambda^\dagger H|0\rangle \\
&= E_\lambda|\lambda\rangle+E_0\alpha_{0,0}|0\rangle-E_0(|\lambda\rangle+\alpha_{0,0}|0\rangle) \\
&= (E_\lambda-E_0)|\lambda\rangle \\
&= \omega_\lambda(O_\lambda^\dagger|0\rangle-\alpha_{0,0}|0\rangle)
\end{aligned}
\tag{14}
$$

and therefore

$$
\begin{aligned}
\langle 0|O_i\left[H,O_\lambda^\dagger\right]|0\rangle &= \omega_\lambda\left[\langle 0|O_iO_\lambda^\dagger|0\rangle-\alpha_{0,0}\langle 0|O_i|0\rangle\right] \\
&= \omega_\lambda\langle 0|\left[O_i,O_\lambda^\dagger\right]|0\rangle
\end{aligned}
\tag{15}
$$

where the last step utilizes the adjoint of (13), $\langle 0|O_\lambda^\dagger=\langle 0|\alpha_{0,0}$. Similarly we may derive

$$
\begin{aligned}
\langle 0|\left[H,O_\lambda^\dagger\right]O_i|0\rangle &= E_0\langle 0|O_\lambda^\dagger O_i|0\rangle-\alpha_{0,0}\langle 0|HO_i|0\rangle \\
&= E_0\alpha_{0,0}(\langle 0|O_i|0\rangle-\langle 0|O_i|0\rangle) \\
&= 0
\end{aligned}
\tag{16}
$$

Subtracting (15) and (16) gives the desired result (10). Substitution of the operator expansion (2) for $O_\lambda^\dagger$ yields (6) for the excitation energy case. The derivation of (7) for the IP–EA variant of the theory follows in a similar fashion. It is clear that the α's in (11) are fixed by the choice of operator basis because for a given operator basis, the matrix eigenvalue equations (6) and (7) yield unique results.

The derived EOM equations (6) [(7)] are linear matrix eigenvalue equations for the exact excitation energies (ionization potentials and electron affinities). The eigenvectors give an $O_\lambda^\dagger$ satisfying conditions (12) and (13) (with $\alpha_{0,0}=0$ for IPs and EAs). The most general $O_\lambda^\dagger$ is expressed in terms of the basis operators $|\lambda'\rangle\langle\lambda''|$ by (11). In Section II.B we investigate the form of the EOM equation when represented in terms of other possible sets of basis operators, since the $\{|\lambda'\rangle\langle\lambda''|\}$ set is generally unavailable. The question of operator completeness is discussed in some detail.

3. Hermitian Equations for Approximate Calculations

In practical numerical calculations an approximate ground-state wave function is employed in (6) and (7). The matrices $\langle 0|[O_i,[H,O_j^\dagger]]|0\rangle$ and $\langle 0|\{O_i,[H,O_j^\dagger]\}|0\rangle$ are then not necessarily Hermitian[10, 50, 66, 67] (they are when $|0\rangle$ is exact). A symmetrized form of the EOM equations can readily be obtained by noticing that a similar derivation to that producing (6) and (7) yields[10, 66]

$$\langle 0|[[O_i,H],O_\lambda^\dagger]|0\rangle = \omega_\lambda\langle 0|[O_i,O_\lambda^\dagger]|0\rangle \tag{17}$$

and

$$\langle 0|\{[O_i,H],O_\lambda^\dagger\}|0\rangle = \omega_\lambda\langle 0|\{O_i,O_\lambda^\dagger\}|0\rangle \tag{18}$$

for excitation energies and ionization potentials, respectively. Introducing the expansion for $O_\lambda^\dagger$, (2), into (17) and (18) and adding (6) and (17) yields

$$\sum_j \langle 0|[O_i,H,O_j^\dagger]|0\rangle C_j^\lambda = \omega_\lambda \sum_j \langle 0|[O_i,O_j^\dagger]|0\rangle C_j^\lambda \tag{19}$$

for excitation energies, with the definition

$$[O_i,H,O_j^\dagger] \equiv \tfrac{1}{2}[O_i,[H,O_j^\dagger]] + \tfrac{1}{2}[[O_i^\dagger,H],O_j^\dagger] \tag{20}$$

The corresponding symmetrized equations for ionization potentials and electron affinities are

$$\sum_j \langle 0|\{O_i,H,O_j^\dagger\}|0\rangle C_j^\lambda = \omega_\lambda \sum_j \langle 0|\{O_i,O_j^\dagger\}|0\rangle C_j^\lambda \tag{21}$$

with

$$\{O_i,H,O_j^\dagger\} \equiv \tfrac{1}{2}\{O_i,[H,O_j^\dagger]\} + \tfrac{1}{2}\{[O_i,H],O_j^\dagger\} \tag{22}$$

The matrices in (19) and (21) are Hermitian even when an approximate ground-state wave function is employed.[10]

4. Motivation for Double Commutator Equations

So far the reasons for introducing the double commutator form of the equations in (6) and (7) and then in (19) and (21) have not been mentioned, even though this form has required a more lengthy derivation. The double commutator formalism is suggested[10] by comparison of low-order perturbation expansions of the EOM equations with the corresponding equations of many-body Green's function methods[44, 45] and time-dependent Hartree-Fock theory.[84] The practical advantage of the double commutator formulation exists because the introduction of commutators produces final equations that have a lower operator rank. The meaning of this term is most clearly illustrated by the simple example where the operators in question are expressed in second quantized notation. Consider two one-body operators in second quantized notation

$$B = \sum_{i,j} b_{ij} a_i^\dagger a_j$$

$$C = \sum_{i,j} c_{ij} a_i^\dagger a_j$$

where a_k^+ and a_k are the second quantized creation and destruction operators associated with spin orbital k, respectively, for some prescribed orbital basis set.

The commutator of these operators is

$$[B, C] = \sum_{i,j,k,l} b_{ij} c_{kl} \left[a_i^\dagger a_j a_k^\dagger a_l - a_k^\dagger a_l a_i^\dagger a_j \right]$$

$$= \sum_{j,k} \left[\sum_i \left(b_{ki} c_{ij} - c_{ki} b_{ij} \right) \right] a_k^\dagger a_j \tag{23}$$

where the usual second quantized fermion anticommutator relationships have been utilized. Thus the commutator of the two one-body operators is a one-body operator, and the simple product of the two is a two-body operator. Excitation operators conserve the number of electrons in the system, and thus must contain terms with equal numbers of creation and destruction operators. Each term therefore has a product of an even number of operators. On the other hand, ionization and electron attachment operators change the number of electrons by one; terms then have an odd number of creation and annihilation operators. The example presented in (23) can be generalized as follows; the anticommutator of two operators, each of which involves an odd number of creation/destruction operators, results in an operator with two less creation/destruction operators than the product of the original two operators. If one or both of the operators has an even number of second quantized operators, it is the commutator that lowers the operator rank by 2 from that for the product of the two operators. This can be a practical advantage in the evaluation of matrix elements.[70] More important, because of the lower hole-particle rank of the double commutators, it is generally argued[10, 50-52] that they are less sensitive to the details of the approximate ground-state wave function used in actual computations.

One manner in which this lack of sensitivity to the detailed structure of the ground-state wave function manifests itself is that the EOM–Green's function theories account for some electron correlation effects, even when an uncorrelated ground state is employed. This is evident, for instance, from the fact that an estimate of the ground state correlation energy can be obtained from time-dependent Hartree-Fock (TDHF) theory.[84] (TDHF theory is equivalent to low-order EOM excitation energy theory with an uncorrelated ground state.[10]) This TDHF correlation energy agrees in second order with the Rayleigh-Schrödinger perturbation result. Moreover, Pickup and Goscinski[36] have analyzed the second-order corrections afforded by the one-electron Green's function theory to Koopmans's theorem[85] ionization potentials and electron affinities. They demonstrate that certain terms can be ascribed to changes in correlation on removal or addition of an electron. These second-order contributions to the Green's function are present even when an uncorrelated ground state is employed.

There is a price that is incurred by the use of the double commutator. Deexcitation operators of the form

$$O_\lambda^\dagger = |0\rangle\langle\lambda| + \alpha_{0,0}|0\rangle\langle0| + \sum_{\lambda'\neq0}\sum_{\lambda''\neq0}\alpha_{\lambda',\lambda''}|\lambda'\rangle\langle\lambda''|$$

also arise as solutions of (19) with eigenvalues $E_0 - E_\lambda$. This means that the size of the operator basis set required for the excitation energy calculation has to be increased from n [for (5)] to $2(n-1)$ [for (19)]. The question of which $2(n-1)$ operators from the complete n^2-operator basis can be used for the EOM equations, a question of "EOM completeness" of operator bases is discussed more fully in Section II.B.

5. Iterative Improvement of Approximate Ground-State Wave Functions

Rowe[10] and later McKoy[50-52] suggested that the condition $O_\lambda|0\rangle = 0$ be used to improve an approximate ground-state wave function once $O_\lambda^\dagger$ has been obtained from the EOM equations (19) or (21). The EOM equations and condition $O_\lambda|0\rangle = 0$ could be solved iteratively. However, the annihilation condition is not, in general, fulfilled by an EOM excitation operator, even if the exact ground state is used and a complete operator basis is employed.[72] Instead, (13) is the appropriate condition that should be employed to iteratively improve an approximate ground-state wave function, where $\alpha_{0,0}$ can be obtained as $\langle0|O_\lambda^\dagger|0\rangle$. However, numerical evidence presented in Section III indicates that a simple self-consistent procedure based on (13) and (19) or (21) has rather poor convergence properties, and a more sophisticated (and cumbersome) procedure is necessary.[86]

Similar problems arise in the calculation of transition matrix elements using EOM excitation operators. Some previous work has been based on the equation $O_\lambda^\dagger|0\rangle = |\lambda\rangle$ for a description of the excited or ion state. Since the correct equation is $O_\lambda^\dagger|0\rangle = |\lambda\rangle + \alpha_{0,0}|0\rangle$, the component $\alpha_{0,0}|0\rangle\langle0|$ in $O_\lambda^\dagger$ must be incorporated in a correct treatment of transition matrix elements or excited-state properties. A useful way of eliminating this problem is to employ a commutator expression for the transition matrix element $\langle0|[\hat{d}, O_\lambda^\dagger]|0\rangle$, where $\hat{d}$ is the transition operator. This form has the advantage of the reduction of operator rank due to the commutator.[53]

B. Linear Independence of Basis Operators with Respect to EOM Equations

Consider the particular set of basis operators for the excitation energy case

$$\{O_i^\dagger\} = \{|\lambda\rangle\langle\lambda'|; \lambda, \lambda' = 0, 1, \ldots, N = n-1\} \tag{24}$$

with $\{|\lambda\rangle\}$ being the exact configuration interaction (CI) N_e-electron wave functions that can be constructed within the given orbital basis. The operator basis, defined by (24), is complete in the sense that any other choice of basis operators for the space can be expanded in this set of $(N+1)^2$ operators. However, it is overcomplete with respect to the EOM equation (19). Only the $2N$-dimensional subset of these operators $\{|\lambda\rangle\langle0|,|0\rangle\langle\lambda|; \lambda = 1, \ldots, N\}$ contributes nontrivially to (19). The remaining operators always give zero matrix elements when substituted into $\langle0|[O_i, H, O_j^\dagger]|0\rangle$ and $\langle0|[O_i, O_j^\dagger]|0\rangle$. With this set of $2N$ operators, (19) is diagonal, and its solution trivial.

Since the operator set described in (24) is not in general known, different operator basis sets must be chosen. Any new set of basis operators can always be expressed as a linear combination of the $(N+1)^2$ operators given in (24). It is obvious that the new operator basis set can also yield only $2N$ linearly independent equations when inserted into (19). To study this problem of linear dependence among the EOM equations in more detail, consider the following simple analysis.[72] Let (19) be expressed in terms of some set of $2N$ basis operators, $\{\Omega_i^\dagger\}$, as

$$\mathcal{C}\mathbf{C}^\lambda = \omega_\lambda \mathcal{D}\mathbf{C}^\lambda \tag{25}$$

where $\mathcal{C}$ and $\mathcal{D}$ are the matrices formed from $\langle0|[\Omega_i, H, \Omega_j^\dagger]|0\rangle$ and $\langle0|[\Omega_i, \Omega_j^\dagger]|0\rangle$, respectively. Expanding the $\Omega_i^\dagger$ in terms of the exact

$O_i^\dagger$'s of (24) gives

$$\Omega^\dagger = \mathbf{X}\mathbf{O}^{(1)^\dagger} + \mathbf{Y}\mathbf{O}^{(2)^\dagger} \tag{26}$$

where $\mathbf{O}^{(1)^\dagger}$ is the column whose elements are in the subset of relevant operators, $S_1 = \{|\lambda\rangle\langle 0|, |0\rangle\langle\lambda|; \lambda = 1, \ldots, N\}$, and $\mathbf{O}^{(2)^\dagger}$ is the column containing the remaining $N^2 + 1$ "irrelevant" $O_i^\dagger$ operators; $\mathbf{X}$ and $\mathbf{Y}$ are the corresponding matrices of expansion coefficients. Inserting (26) into (25) yields

$$\mathbf{X}\mathbf{A}'\mathbf{X}^\dagger\mathbf{C}^\lambda = \omega_\lambda \mathbf{X}\mathbf{D}'\mathbf{X}^\dagger\mathbf{C}^\lambda \tag{27}$$

The matrices $\mathbf{A}'$ and $\mathbf{D}'$ are formed from $\langle 0|[O_i^{(1)}, H, O_j^{(1)^\dagger}]|0\rangle$ and $\langle 0|[O_i^{(1)}, O_j^{(1)^\dagger}]|0\rangle$, respectively, and involve matrix elements containing only the $2N$ relevant operators. The matrix elements containing the other $(N^2 + 1)$ operators, $O_i^{(2)^\dagger}$, are identically zero; thus they do not appear in (27). Equations 25 and 27 are equivalent to the secular equation

$$|\mathbf{X}|^2 \cdot |\mathbf{A}' - \omega_\lambda \mathbf{D}'| = 0 \tag{28}$$

Therefore, the $2N$ $\Omega_i^\dagger$, chosen as the operator basis for (25), yield the same eigenvalues ω_λ when inserted into (19) as the $2N$ operators in S_1, provided the condition $|\mathbf{X}|^2 \neq 0$ is satisfied (or, equivalently, when $|\mathcal{D}| = |\mathbf{X}|^2 \cdot |\mathbf{D}'| \neq 0$).

It is to be emphasized that at most $2N$ basis operators can yield linearly independent equations when substituted into (19). A set of $2N$ basis operators for which $|\mathcal{D}| \neq 0$ forms what is often called a "complete" set of basis operators for EOM calculations in that they give the desired excitation energies. However, when expanding one set of operators in terms of a different operator basis as in (26), a basis of $(N + 1)^2$ operators is in general needed. Thus there are two entirely different senses in which an operator can be "complete." EOM completeness involves $2N$ operators, but $(N + 1)^2$ are necessary for general operator completeness.

Similar logic holds for the IP–EA case. Here the operator basis set $\{|\lambda_{m-1}\rangle\langle\lambda_m'|, m = 1, 2, \ldots\}$, with $|\lambda_{m-1}\rangle$ and $|\lambda_m'\rangle$ exact $m - 1$ and m electron states, respectively, within some orbital basis, is a complete set of basis operators in the customary sense of the term. However, only $|\lambda_{N_e-1}\rangle$ $\langle 0|$ and $|0\rangle\langle\lambda_{N_e+1}|$ contribute nontrivially to (21). Thus this smaller set is EOM complete. This EOM complete set yields the IPs and minus the EAs as its eigenvalues, as can be verified by direct substitution of the basis operators $|\lambda_{N_e-1}\rangle\langle 0|, |0\rangle\langle\lambda_{N_e+1}|$ into (21). If a general set of basis operators, reducing the number of electrons by one, is employed in the solution of the EOM equations, (21), it has a dimensionality equal to the number of IPs plus the number of EAs for the orbital basis set; and if the overlap matrix in (21) is not singular, the EOM equations yield the exact IPs and exact EAs within the finite orbital basis approximation (i.e., identical to the IPs and EAs generated by an exact CI calculation with the orbital basis).

Alternatively, the fact that (21) yields both the IPs and EAs can be viewed as a consequence of the anticommutator and symmetric double anticommutator having the symmetries

$$\{O_i, O_j^\dagger\} = \{O_j^\dagger, O_i\}$$

$$\{O_i, H, O_j^\dagger\} = -\{O_j^\dagger, H, O_i\} \tag{29}$$

The basis set appropriate for the EA case (i.e., a basis of operators that increase the number of electron by unity) is just $\{O_i\}$, the adjoint of the operator basis set employed for IPs. Equation 29 causes (21) to yield eigenvalues for the EA basis that have the same magnitude but opposite sign as when the IP basis is utilized. Thus the IP and EA calculations using (21) are essentially identical.

We proved earlier that the excitation energy equation (19) yields the exact excitation energies and the exact de-excitation energies (just the negatives of the excitation energies) when any EOM complete operator basis is used. This is, likewise, a consequence of the symmetry of the symmetric double commutator and the commutator,

$$[O_i, H, O_j^\dagger] = [O_j^\dagger, H, O_i]$$

and

$$[O_i, O_j^\dagger] = -[O_j^\dagger, O_i]$$

respectively.

C. Operator Basis Sets Used in Calculations

In the previous subsections we discussed one possible operator basis set: $\{O_i^\dagger\} = \{|\lambda\rangle\langle 0|, |0\rangle\langle\lambda|\}$, where $|0\rangle$ and $|\lambda\rangle$ are exact initial and final states within the finite space defined by the set of orbitals used. The EOM equation, (19) or (21), is diagonal in this basis. Obviously, however, this basis set is not a useful one for practical calculations, since the exact states, $|0\rangle$ and $|\lambda\rangle$, are unknowns.

The basis commonly employed in EOM calculations is expressed in terms of second quantized creation and destruction operators. To introduce this many-body basis, it is convenient to consider a reference determinant, Φ_0, approximation to $|0\rangle$. For instance, Φ_0 is usually taken to be the SCF wave function when dealing with a closed-shell system, although this is, of course, not the only possibility. Unless explicitly stated otherwise, we use the Greek letters $\mu, \nu, \ldots$, to denote orbitals that are occupied in Φ_0 (commonly referred to as hole orbitals) and the Latin letters, $m, n, \ldots$, to denote orbitals that are not occupied in Φ_0 (often called particle orbitals).

For excitation energy calculations, the customary EOM basis set has the form $\{a_m^\dagger a_\mu, a_\mu^\dagger a_m, a_m^\dagger a_n^\dagger a_\mu a_\nu, a_\mu^\dagger a_\nu^\dagger a_m a_n, \ldots\}$. The operators, $a_m^\dagger a_\mu$ and $a_m^\dagger a_n^\dagger a_\mu a_\nu$ produce approximate single and double excitations, respectively, since they remove electrons from orbitals that are occupied in Φ_0 and add an equal number of electrons to orbitals that are vacant in Φ_0, respectively. The operators $a_\mu^\dagger a_m$ and $a_\mu^\dagger a_\nu^\dagger a_m a_n$ are approximate single and double de-excitation basis operators, respectively. The full operator basis set includes up to N_e-fold excitations and de-excitations for an N_e-electron system.

For ionization potentials, the customary EOM basis has the form $\{O_i^\dagger\}$ $= \{a_\mu, a_m, a_\nu a_m^\dagger a_\mu, a_m a_\mu^\dagger a_n, \ldots\}$. These operators each annihilate one more electron than is created, thus producing an $(N_e - 1)$-electron function when operating on an N_e-electron function. The operator a_μ removes an electron from an orbital that is occupied in reference determinant, and $a_\nu a_m^\dagger a_\mu$ removes one electron from a hole orbital and excites a second electron from a hole orbital to a particle orbital. The operator $a_m^\dagger$ removes an electron from the particle orbital m and $a_m a_\mu^\dagger a_n$ removes an electron from a particle orbital and "de-excites" another electron from a particle orbital to a hole orbital. For electron affinities, the adjoints of the operators in the IP basis from the standard EOM operator basis.

Often, rather than the simple strings of second quantized operators (e.g., $a_m^\dagger a_n^\dagger a_\mu a_\nu$), spin- and/or space-symmetry-adapted linear combinations of these operators are utilized.[50, 51, 66, 67] Since the EOM matrix equations are block diagonal according to symmetry, the use of symmetry-adapted basis operators results in a reduction in the size of the resultant matrix eigenvalue problem.

When the EOM excitation basis operators $\{a_m^\dagger a_\mu, a_m^\dagger a_n^\dagger a_\mu a_\nu, \ldots\}$ or EOM ionization basis operators $\{a_\mu, a_\mu a_m^\dagger a_\nu, \ldots\}$ act on the ground-state wave function (either approximate or exact), the many-electron basis wave functions $O_i^\dagger|0\rangle$ that are formed are linear combinations of many determinants. (Since $|0\rangle$ is taken to include correlation effects, it is a linear combination of many determinants.) This is in contrast to the basis of configurations employed in CI calculations, which are generally symmetry-adapted linear combinations of only a few determinants. Since in large-scale EOM or CI calculations, it is necessary to truncate the basis set of operators or configurations employed, respectively, the EOM method often includes the effects of many more excited-state or ionized-state configurations of a given size basis than the comparable CI calculation. However, the relative weights of the determinants in $O_i^\dagger|0\rangle$ are fixed by the weights of the various determinants in $|0\rangle$. Roughly speaking, the correlation in the upper state is being crudely approximated using the information about the ground-state correlation in $|0\rangle$ based on the expectation that certain types of electron correlation are fairly invariant to changes in state. The convergence of EOM ionization potential calculations with respect to the size and constitution of the operator basis is studied in Section III.

The EOM excitation energy basis also contains the primitive de-excitation operators $\{a_\mu^\dagger a_m, a_\mu^\dagger a_\nu^\dagger a_m a_n, \ldots\}$, which have no direct counterparts in CI theory. The EOM IP–EA operator basis, likewise, has the analogous operators $\{a_m, a_m a_\mu^\dagger a_n, \ldots\}$. These operators have been shown to be important for the inclusion of effects due to changes in correlation between the initial and final states of the system.[36, 43–45]

The $O_\lambda^\dagger$ obtained from the EOM equation provide direct physical insight into the important electronic mechanisms present in the $|0\rangle \rightarrow |\lambda\rangle$ transition. By scanning the list of expansion coefficients for $O_\lambda^\dagger$, it is possible to isolate zeroth-order processes, such as, for instance, electron removal from a specific orbital accompanied by an excitation from one orbital to another, as important for a certain ionization. The same information is surely present in the results of CI calculations; however, it may not be as clearly exhibited when comparing two long lists of CI expansion coefficients.

Manne[87] has shown that the basis configurations $O_i^\dagger|0\rangle$ provide the correct number of linearly independent wave functions for the calculation of ionization potentials and electron affinities as long as $|0\rangle$ is not orthogonal the reference determinant Φ_0, which is used in the definition of the many-body basis $\{O_i^\dagger\}$. Dalgaard[88] has provided the analogous result for the excitation energy case. To prove that the operators are what we term "EOM complete," however, it remains to be proved that the EOM metrics are nonsingular for these bases. Dalgaard[88] has noted that for the IP–EA case this operator basis must yield a nonsingular metric. This can be proved as follows. Divide the operator basis into two sets

$$\{\alpha^\dagger\} = \left\{a_\mu, a_\mu a_m^\dagger a_\nu, a_\mu a_m^\dagger a_\nu a_n^\dagger a_\alpha, \ldots\right\}$$

and

$$\{\beta^\dagger\} = \left\{a_m, a_m a_\mu^\dagger a_n, a_m a_\mu^\dagger a_n a_\nu^\dagger a_p, \ldots\right\}$$

The block of the metric between the two sets vanishes, $\langle 0|\{\beta, \alpha^\dagger\}|0\rangle = 0$, since $\{a_\mu, a_\nu\} = \{a_m^\dagger, a_n^\dagger\} = 0$ whether or not $\mu = \nu$ and $m = n$. Thus it is sufficient to show that each of the blocks $\langle 0|\{\alpha, \alpha^\dagger\}|0\rangle$ and $\langle 0|\{\beta, \beta^\dagger\}|0\rangle$ is positive definite. To accomplish this, it is sufficient to show that $\Sigma_{i,j}$ $\langle 0|\{\alpha_i, \alpha_j^\dagger\}|0\rangle c_i c_j > 0$ and $\Sigma_{i,j}\langle 0|\{\beta_i, \beta_j^\dagger\}|0\rangle c_i c_j > 0$ for every choice of the c_i's such that not all $c_i = 0$. Call $\Sigma_j c_j \alpha_j^\dagger|0\rangle = |A\rangle$, $\Sigma_j c_j \alpha_j|0\rangle = |A'\rangle$, $\Sigma_j c_j \beta_j^\dagger|0\rangle = |B\rangle$, and $\Sigma_j c_j \beta_j|0\rangle = |B'\rangle$. As shown by Manne,[87] $\alpha_j^\dagger|0\rangle$ form a linearly independent set of $(N-1)$-electron wave functions and $\beta_j|0\rangle$ form a linearly independent set of $(N+1)$-electron wave functions. Therefore $\langle A|A\rangle > 0$ and $\langle B'|B'\rangle > 0$ for any set of c_i's that do not all vanish. Since $\Sigma_{i,j}\langle 0|\{\alpha_i, \alpha_j^\dagger\}|0\rangle c_i c_j = \langle A|A\rangle + \langle A'|A'\rangle \geq \langle A|A\rangle > 0$ and $\Sigma_{i,j}$ $\langle 0|\{\beta_i, \beta_j^\dagger\}|0\rangle c_i c_j = \langle B|B\rangle + \langle B'|B'\rangle \geq \langle B'|B'\rangle > 0$, this completes the proof.

A similar proof does *not* hold in the excitation energy case, since in this case the metric contains the commutator rather than the anticommutator. In fact, the above-described many-body basis for excitation energies need not be EOM complete, as the following simple example demonstrates.

Consider a system with only two levels of a given symmetry; for example, the hydrogen atom with only two s orbitals in the orbital basis or the excitations of beryllium with a $2s$ $1p$ basis in the frozen core $(1s^2)$ approximation. In the first example, the $1s \rightarrow 2s$ excitation operator and its adjoint are the only allowed basis operators, and in the second, the $2s^2 \rightarrow 2p^2(^1S)$ double excitation and its adjoint are the only basis operators of this symmetry. The two basis configurations for the system are written as $|A\rangle$ and $|B\rangle$. The normalized ground-state wave function (either exact within this basis or an approxmate one) is

$$|0\rangle = \cos\theta|A\rangle + \sin\theta|B\rangle \qquad (30)$$

The two operators in our basis are defined by

$$O_1^\dagger|A\rangle = |B\rangle \qquad (31)$$

and

$$O_2^\dagger|B\rangle = |A\rangle \qquad (32)$$

In our beryllium example $|A\rangle$ can be $|2s^2\rangle$ and $|B\rangle$ is $|2p^2(^1S)\rangle$. Then $O_1^\dagger$ is given by

$$O_1^\dagger = \frac{1}{\sqrt{3}}\left(a_{2p_x,\alpha}^\dagger a_{2p_x,\beta}^\dagger a_{2s\beta} a_{2s\alpha} + a_{2p_y,\alpha}^\dagger a_{2p_y,\beta}^\dagger a_{2s\beta} a_{2s\alpha} + a_{2p_z,\alpha}^\dagger a_{2p_z,\beta}^\dagger a_{2s\beta} a_{2s\alpha}\right)$$

and $O_2^\dagger$ is the adjoint of this. The overlap matrix is given by

$$\mathbf{D} = \langle 0 | [\mathbf{O}, \mathbf{O}^\dagger] | 0 \rangle$$
$$= \begin{pmatrix} \cos^2\theta - \sin^2\theta & 0 \\ 0 & -\cos^2\theta + \sin^2\theta \end{pmatrix} \qquad (33)$$

This matrix is singular when $\theta = \pi/4$. Otherwise, this basis is EOM complete.

Although the many-body basis is not guaranteed to be complete in the excitation case, to our knowledge no difficulties with linear dependences have emerged in numerical calculations.

Other possible operators have not yet been considered. There are operators $a_\mu^\dagger a_\nu$ that transfer an electron between two orbitals that are occupied in the reference state Φ_0. The analogous operators $a_m^\dagger a_n$ shift an electron between two orbitals that are vacant in Φ_0. There are also products of these operators with themselves, with each other, and with the ordinary operators of the many-body basis. Similar considerations apply to the IP–EA case. These unorthodox operators can improve the EOM calculations if they are added to an incomplete operator basis set and if the resulting basis yields a nonsingular metric. It is conceivable that in some cases the lower rank of these "moving" operators will make them more convenient to use than the higher excitations and de-excitations of the usual many-body basis. The effect of adding these unorthodox basis operators to an incomplete basis is investigated numerically in Section III.F.

D. Approximations

1. The Ground-State Wave Function

In general, the exact ground-state wave function $|0\rangle$ is unknown. It has been customary in EOM calculations and, in effect, for Green's function methods, to approximate $|0\rangle$ by perturbation theory. As mentioned above, because the double commutators in the EOM equations are of lower operator rank than simple products of operators, it has been widely believed that the EOM method (as well as the related Green's function and propagator theories) should be fairly insensitive to small errors in the ground-state wave function.[10, 50–52] The ground-state wave function can be expressed as a linear combination of some reference N_e-electron state plus all possible single, double, ..., N_e-fold excitations out of this reference state. Usually in EOM work, the reference state is chosen to be a closed-shell SCF determinant,

$|HF\rangle$. Restricting the discussion to this particular reference state results in some simplification in the evaluation of matrix elements.

The use of a closed-shell reference ground-state wave function is not as great a restriction in the IP–EA theory as it might seem, since only one state, the neutral or the ion, need be closed shell. For instance, in Section III.E we compute the EA of F as the lowest ^2P ionization potential of F$^-$.

Yeager and McKoy[54] have extended the EOM theory at a low order (random phase approximation) to allow for the calculation of excitation energies of systems with one or two open shells. They have reported results for the lithium atom and the oxygen molecule. The complexities of an open-shell treatment have led Purvis and Öhrn[38] to propagator calculations for the O_2 molecule that employ unrestricted SCF orbitals, and thus utilize a single-determinant, zeroth-order description of the ground-state wave function. Cederbaum and Domcke[16] have described the formal extension of the diagrammatic Green's function theory to open shells, but to the best of our knowledge they have not presented any calculations. Recently, Jørgensen and co-workers[89, 90] have presented second-order propagator calculations for ionization from a doublet ground-state in Li, Na, and BeH and for doublet-to-doublet excitations in Li and BeH.

When a closed-shell SCF determinant is employed as a reference state, the ground-state wave function can be written as

$$|0\rangle = N_0\left\{1 + \sum_{\mu,p} C_\mu^p a_p^\dagger a_\mu + \sum_{\substack{\mu < \nu \\ m < n}} C_{\mu\nu}^{mn} a_m^\dagger a_n^\dagger a_\nu a_\mu + \dots\right\}|HF\rangle \qquad (34)$$

where N_0 is the normalization constant. Generally, the correlation coefficients $C_\mu^p, C_{\mu\nu}^{mn}, \dots$, have been evaluated by the use of Rayleigh-Schrödinger perturbation theory (RSPT).[50–52, 56, 66]

Yeager and Freed[91] considered the effect on calculated excitation energies of evaluating the correlation coefficients by Epstein-Nesbet perturbation theory (ENPT).[92] For N_2 they found a considerable difference between the two approaches, and this is symptomatic of the fact that the perturbation approach of (34) is not optimal in all cases. In Section III.C we compare EOM results for the IPs of N_2 when the ground state is approximated using RSPT and ENPT. In Section III.F the use of a ground state based on a multiconfigurational reference state, augmented by generalized perturbation theory, is investigated in a simple Be(^1S) excitation energy calculation. Recently some EOM theory has been undertaken by Simons[78, 79] (for excitation energies and IPs and EAs) and by Yeager and

Jørgensen[77] (for excitation energies) that employs a multiconfigurational reference state within an RPA-type approximation.

2. Partitioning Theory and Approximating the Contribution of Q-Space

Partitioning theory[93] is employed to reduce the apparent dimensionality of the EOM equations and subsequently to enable the introduction of useful numerical approximations. The EOM operator basis is separated into the operators that are considered to be the most important, called the primary (P) space, and the remaining operators, comprising the secondary or Q-space. The EOM equations (19) or (21) can be written in block matrix form,

$$\begin{pmatrix} A_{PP} & A_{PQ} \\ A_{QP} & A_{QQ} \end{pmatrix} \begin{pmatrix} C_P \\ C_Q \end{pmatrix} = \omega \begin{pmatrix} D_{PP} & D_{PQ} \\ D_{QP} & D_{QQ} \end{pmatrix} \begin{pmatrix} C_P \\ C_Q \end{pmatrix} \tag{35}$$

This is equivalent to the two matrix equations

$$A_{PP}C_P + A_{PQ}C_Q = \omega(D_{PP}C_P + D_{PQ}C_Q)$$

and

$$A_{QP}C_P + A_{QQ}C_Q = \omega(D_{QP}C_P + D_{QQ}C_Q) \tag{36}$$

The second equation can be solved formally for C_Q in terms of C_P. The results are then substituted into the first equation (36) to produce the familiar equation[93]

$$\left[A_{PP} - (A_{PQ} - \omega D_{PQ})(A_{QQ} - \omega D_{QQ})^{-1}(A_{QP} - \omega D_{QP}) \right] C_P = \omega D_{PP}C_P \tag{37}$$

Equation 37 is a pseudoeigenvalue problem of the form

$$L(\omega)C = \omega DC \tag{38}$$

It can be solved iteratively for ω and C. Equation 37 has the dimensionality of the P space.

Computationally (37) is no improvement over (35) because it contains the large inverse matrix $(A_{QQ} - \omega D_{QQ})^{-1}$. Since the Q-space contains only basis operators that are assumed to be of secondary importance, it is reasonable to approximate the effect of the Q-space in (37). This is done by separating $(A_{QQ} - \omega D_{QQ})$ into a zeroth-order part M_0 and the remaining part, M_1. The inverse matrix can then be expanded in a Born expansion

$$(M_0 + M_1)^{-1} = M_0^{-1} - M_0^{-1} M_1 M_0^{-1} + \cdots \tag{39}$$

Equation 39 is inserted into (37), and the expansion is truncated in low orders. In Section III.C it is shown that inclusion of the second term in the Born expansion in (39) can have very significant results if proper care is not taken in the choice of the P-space of basis operators.

Two divisions of $A_{QQ} - \omega D_{QQ}$ have been employed in our treatment of the EOM equations. In both M_0 is chosen to be diagonal (not the only possibility, of course), since this results in a great simplification in inverting M_0. The first method, the unshifted ω-dependent denominators method, chooses M_0 to be the zeroth-order parts of $A_{QQ} - \omega D_{QQ}$ in the electron-electron interaction. The other common method, called the shifted ω-dependent denominators method, includes in M_0 the first-order parts (in electron-electron interaction) of the diagonal elements in addition to the zeroth-order matrix elements. For excitation energies, McKoy and co-workers[50-52] employed unshifted ω-dependent denominators. Yeager and Freed[91] investigated the use of shifted denominators in N_2 excitation energy calculations, finding that the shifted results differ substantially from the unshifted results for this system. Simons and co-workers[56-65] obtained accurate results for IPs and EAs with shifted denominators. In Sections III.C and III.E we compare ionization potentials calculated using both these methods.

Because (37) is calculated through some prescribed order in the electron-electron interaction, the accuracy to which the different submatrices (i.e., A_{PP}, A_{PQ}, etc.) need be evaluated is limited. This restriction, in turn, limits the order to which the various correlation coefficients in (34) are needed. Furthermore, (37) is simplified by orthogonalizing the operator basis set such that

$$D_{PP} = I_{PP}$$

$$D_{QQ} = I_{QQ}$$

$$D_{QP} = D_{PQ} = 0 \tag{40}$$

In practice, the operator basis need be adjusted only so that (40) hold through certain finite orders in the electron-electron interaction, so that (37) is consistent. Orthogonalized basis operators are employed in most of the numerical calculations presented in Section III.

E. Choice of the Primary Space for Ionization Potentials

Most of the numerical work presented here employs the IP–EA variant of the theory. Since the changes in correlation and orbital relaxation are generally greater for ionization and electron attachment processes than for excitation processes, calculation of IPs and EAs might be expected to provide a more severe test of the theory than the excitation energy form of the theory.

On the other hand, in certain technical aspects the ionization potential–electron affinity form of the theory is easier to implement than the excitation energy EOM version. This is because the basis operators of the excitation energy EOM theory contain strings of two, four, ..., second quantized operators, whereas the IP–EA theory basis operators have strings of one, three, ..., creation and destruction operators. Thus if an ionization potential calculation employs an operator basis that is truncated to include only operators containing one or three second quantized operators (a common point of truncation), the operator basis is much smaller than in the excitation energy calculation (using the same set or orbitals) employing basis operators with two and four $a_i^\dagger$'s and a_i's (i can be either hole or particle here). Furthermore, because of the fewer $a_i^\dagger$'s and a_i's in the basis operators, the matrix elements are somewhat easier to evaluate in the IP–EA case.

At this juncture, it is convenient to introduce some nomenclature and notation. The basis operators, which correspond to simple electron removal, a_μ and a_m, are referred to as 1-block operators, and the operators, which are products of three, five, ..., creation and destruction operators, are called the 3-block, the 5-block, ..., operators, respectively. This nomenclature is also used when spin- and/or space-symmetry-adapted linear combinations of these operators are employed as the operator basis or when the basis operators are orthogonalized to each other. The orthogonalization of the operator basis mixes the primitive nonorthogonal operators, which have different numbers of creation and destruction operators. However, in all cases considered, this mixing is slight, and the orthogonalized set of basis operators retains predominately the character of the 1-block or 3-block and so on.

It is also convenient to divide the EOM equation into block form,

$$
\begin{bmatrix}
\mathbf{A}^{(1,1)} & \mathbf{A}^{(1,3)} & \cdots \\
\mathbf{A}^{(3,1)} & \mathbf{A}^{(3,3)} & \cdots \\
\vdots & \vdots &
\end{bmatrix}
=
\omega
\begin{bmatrix}
\mathbf{C}^{(1)} \\
\mathbf{C}^{(3)} \\
\vdots
\end{bmatrix}
\begin{bmatrix}
\mathbf{D}^{(1,1)} & \mathbf{D}^{(1,3)} & \cdots \\
\mathbf{D}^{(3,1)} & \mathbf{D}^{(3,3)} & \cdots \\
\vdots & \vdots &
\end{bmatrix}
\begin{bmatrix}
\mathbf{C}^{(1)} \\
\mathbf{C}^{(3)} \\
\vdots
\end{bmatrix}
$$

where $A_{ij}^{(1,1)} = \langle 0 | \{ O_i, H, O_j^\dagger \} | 0 \rangle$ when $O_i^\dagger$ and $O_j^\dagger$ are both in the 1-block.

For $A_{ij}^{(3,1)}$, $O_i^\dagger$ is in the 3-block and $O_j^\dagger$ is in the 1-block, and so on; $D_{ij}^{(1,1)}$, $D_{ij}^{(3,1)}$,..., are defined analogously.

Prior to the present work, the universal choice of the EOM P-space for ionization potentials[56-66] was just the simple electron removal operators a_μ and a_m, the 1-block. This is effectively also the customary choice in the propagator and diagrammatic Green's function methods.

Simons[56] and Yeager,[66] working separately, developed the IP–EA variant of the EOM method based on this limited choice of the P-space. These theories included some third-order terms but neglected others. Simons[57-65] has reported IPs of many small diatomic molecules using this approach which, apart from a few notable exceptions, are highly accurate.

Both Simons[56] and Yeager[66] employ the 3-block basis operators as the secondary operator space, retaining only portions of the diagonal matrix elements thereof. When the correlation coefficients are calculated by RSPT and the 5-block operators (i.e., $a_\mu a_m^\dagger a_\nu a_n^\dagger a_\lambda$ and $a_m a_\mu^\dagger a_n a_\nu^\dagger a_p$) are Schmidt orthogonalized to the simple electron removal operators (the 1-block), the $A^{(1,5)}$ matrix vanishes through first order.[70] Therefore, the 5-block basis operators do not contribute until fourth order [since (37) is bilinear in $A^{(1,5)}$]. Differences between the approaches of Yeager and Simons are described more fully and tested numerically in Section III.A.

Section III.C presents evidence that suggests that the foregoing choice of the P-space can lead to severe difficulties. The problems arising when this simple choice is employed are overcome by enlarging the P-space to include some operators from the 3-block.

The 3-block basis operators that are to be transferred to the P-space are chosen in most of our calculations by a numerical selection criterion, similar to the configuration selection procedures[74-76] that have proved very useful in CI work. The lowest order perturbation correction of a basis operator $O_k^\dagger$ to the diagonal EOM matrix element A_{ii} due to an operator in the P-space, $O_i^\dagger$, is given by $A_{ik}A_{ik}/(A_{kk}-A_{ii})$. Therefore, the quantity

$$\Delta_k = \sum_i \left| \frac{A_{ik}^2}{A_{kk}-A_{ii}} \right| \tag{41}$$

is employed as an estimate of the magnitude of a contribution of the various shake-up basis operators (i.e., the 3-block) to the P-space in the numerical work.[68] The summation in (41) is over some prescribed (by the user) list of P-space operators that are most important for the IP of interest. All shake-up operators with a contribution greater than some tolerance are added to the P-space. Various tolerances are used to test convergence on a number of systems in Section III.E.

Diagrammatic Green's function theory attempts to evaluate a perturbation expansion for

$$\left[\mathbf{A}_{PP} - \mathbf{A}_{PQ}(\mathbf{A}_{QQ} - \omega\mathbf{I}_{QQ})^{-1}\mathbf{A}_{QP} - \omega\mathbf{I}_{PP}\right]^{-1} \tag{42}$$

where the 1-block of operators comprises the P-space for the one-electron Green's function $G_1(\omega)$. The matrix $\mathbf{A}_{PQ}(\omega\mathbf{I}_{QQ} - \mathbf{A}_{QQ})^{-1}\mathbf{A}_{QP}$ corresponds to the ω-dependent part of the self-energy. Since (42) is essentially the inverse of the partitioned EOM equation (37), the poles of $G_1(\omega)$ lie at the eigenvalues of (37), which are the ionization potentials and electron affinities of the system.

Inclusion of basis operators from the 3-block into the primary operator space would entail, in Green's function language, the summation of certain classes of diagrams through all orders in perturbation theory. This would represent a rather tedious task of diagram summation, and the repartitioning approach accomplishes this in a conceptually simple and numerically automatic fashion. The relationship of the EOM theory, with an expanded P-space, to the diagrammatic one-electron Green's function approach is discussed more fully in Section II.F.

When some shake-up operators are introduced into the P-space, the 5-block basis operators now make second-order contributions to (37), since there are nonvanishing first-order $A^{(3,5)}$ matrix elements. In Section III.E, we investigate the effect of these terms as well as other second-order contributions to $\mathbf{A}_{PP}$ that arise from the second-order parts of $A_{ik}^{(3,3)}$ matrix elements for the $O_i^\dagger$ and $O_k^\dagger$ in the P-space. An estimate of the contribution of the 5-block operators to the P space can be obtained in similar fashion to (41),[71] and 5-block operators can be included in the P space based on their estimated contributions to $\mathbf{A}_{PP}$. Section III.E reports results of EOM calculations in which the P-space includes operators from the 1-, 3-, and 5-blocks to provide tests of the convergence of the EOM method.

In related propagator work,[40, 41] it has been conjectured that because of orbital relaxation effects, the 5-block basis operators make important contributions. While adhering to the traditional choice of the P-space, (i.e., a_μ and a_m), Öhrn and co-workers[40, 41] explicitly include the 5-block in calculations on Ne and N_2, respectively, via a continued fraction formalism. Their numerical calculations confirm our findings of the importance of these terms.

Cederbaum and co-workers[16–32] do not explicitly evaluate diagrams that include effects of the 5-block of basis operators in their diagrammatic Green's function approach. Cederbaum[19] has, however, employed a self-consistent procedure that iteratively replaces unperturbed Green's function, G_0, lines with perturbed ones, G, in self-energy diagrams. This results in the replacement of orbital energies in the denominators of the perturbation series for the self-energy of the one-electron Green's function, with an improved estimate of the corresponding ionization potential provided by the previous iteration. This method implicitly calculates some diagrams containing effects of the 5-, 7-, ..., blocks of operators. However, many other diagrams resulting from 5- and higher blocks are not incorporated by this procedure.[71] Numerical evidence indicates that this self-consistent approach causes only slight shifts in the calculated ionization potentials and electron affinities.[19] In contrast, results presented in Section III.E emphasize the significance of the more complete treatment of the 5-block in the·EOM theory.

Wendin[35] also implicitly includes part of the effect of the 5- and higher blocks by replacing hole orbital energies in the denominators of self-energy diagrams with ΔSCF values for the ionization potentials for the removal of an electron from the orbitals in question. Orbital relaxation is treated in this fashion. This approach is likely to be reasonable for holes in core levels, the situation with which Wendin is concerned, since relaxation effects tend to be greater than changes in correlation for ionization from these levels.

F. Relationship of the EOM Method with an Expanded *P*-Space to Diagrammatic Green's Function Theory

The one-electron Green's function $G_1(E)$ is the resolvent of the Liouville operator within the 1-block of basis operators. A comparison of $G_1(E)$ and the EOM method with its expanded *P*-space can be accomplished by folding the partitioned EOM equation, which has the dimensionality of the *P*-space, into the 1-1 subblock. The *P*-space repartitioned EOM pseudo-eigenvalue equation (38) is written in block matrix form

$$\begin{bmatrix} \mathbf{L}(\omega)^{(1,1)} & \mathbf{L}(\omega)^{(1,3)} \\ \mathbf{L}(\omega)^{(3,1)} & \mathbf{L}(\omega)^{(3,3)} \end{bmatrix} \begin{pmatrix} \mathbf{C}^{(1)} \\ \mathbf{C}^{(3)} \end{pmatrix} = \omega \begin{pmatrix} \mathbf{I}^{(1,1)} & 0 \\ 0 & \mathbf{I}^{(3,3)} \end{pmatrix} \begin{pmatrix} \mathbf{C}^{(1)} \\ \mathbf{C}^{(3)} \end{pmatrix} \qquad (43)$$

with an orthogonal operator basis assumed in (43). Equation 43 explictly displays the ω dependence of the $\mathbf{L}(\omega)$ matrix that arises from the $-\mathbf{A}_{PQ}(\mathbf{A}_{QQ} - \omega\mathbf{I}_{QQ})^{-1}\mathbf{A}_{QP}$ term in (37), and $(\mathbf{A}_{QQ} - \omega\mathbf{I}_{QQ})^{-1}$ is approximated by its diagonal elements (through first order) in the present formulation of the theory. Applying partitioning theory to (43) and suppressing the ω dependence of L to simplify the notation, the EOM equation becomes

$$\left[\mathbf{L}^{(1,1)} - \mathbf{L}^{(1,3)} (\mathbf{L}^{(3,3)} - \omega\mathbf{I}^{(3,3)})^{-1} \mathbf{L}^{(3,1)} \right] \mathbf{C}^{(1)} = \omega\mathbf{C}^{(1)} \qquad (44)$$

Diagonalizing (43) directly (the current EOM theory) corresponds to the exact inversion of $(\mathbf{L}^{(3,3)} - \omega\mathbf{I}^{(3,3)})$ in (44) or, in other terminology, to summing all self-energy diagrams generated by the Born expansion of this inverse matrix to infinite order. The algebraic expression for the general term in this Born series is

$$\sum_{k_1, k_2, \ldots, k_n} L_{ik_1}^{(1,3)}\big(E_{k_1} - \omega\big)^{-1} L_{k_1 k_2}^{\prime(3,3)}\big(E_{k_2} - \omega\big)^{-1} \cdots L_{k_{n-1}k_n}^{\prime(3,3)}\big(E_{k_n} - \omega\big)^{-1} L_{k_n j}^{(3,1)}$$

where $k_1, k_2, \ldots, k_n$ label shake-up operators, E_{k_i} is the zeroth-order contribution to $L_{k_i k_i}^{(3,3)}$ (a linear combination of three orbital energies), and $\mathbf{L}^{\prime(3,3)}$ is $\mathbf{L}^{(3,3)}$ minus its zeroth-order part. Hence the present formalism is equivalent to a summation through all orders of series of self-energy diagrams containing the zeroth-order shake-up poles corresponding to the 3-block operators in the P-space. The situation is actually very complicated, since $L_{ik_j}^{(1,3)}$, $L_{k_i k_j}^{(3,3)}$, and $L_{k_i j}^{(3,1)}$ matrix elements already contain extensive summation of first- and second-order terms. These include first-order electron-electron interaction matrix elements (V_{ijkl}), second-order terms that are products of electron-electron interaction matrix elements and first-order ground-state correlation coefficients (these add ω-independent denominators to the diagrams), and second-order $\mathbf{A}_{PQ}(\mathbf{A}_{QQ} - \omega\mathbf{I}_{QQ})^{-1}\mathbf{A}_{QP}$ terms. The latter type of term contains ω-dependent denominators that have the zeroth-order poles corresponding to 5-block operators. Suffice it to say that it is conceptually much clearer and numerically much more efficient to accurately account for important shake-up basis operators by simply and automatically including them in the primary operator space, as is done in the present EOM theory, rather than attempting extensive diagrammatic summations.[43]

Wendin[35] attempts to diagrammatically account for some of the collective excitation effects that have prompted the use of a generalized P-space in the present theory. Wendin utilizes a renormalized electron-electron interaction that includes the summation of a prescribed, limited set of ring diagrams. In this manner, he includes some of the interaction of certain quasi–particle shake-up processes. He also employs a V^{N-2} potential in the calculation of virtual orbitals (i.e., the virtual orbitals are generated by an SCF description of the motion of electrons in the field of a system in which two electrons have been removed). This presumably provides a better set of virtual orbitals for the description of shake-up process, which has one ejected electron and one electron excited. The excited electron "sees" only the $N-2$ remaining electrons, where N is the number of electrons in the neutral atom or molecule. We have also suggested that the use of V^{N-2} virtual orbitals might be helpful in the evaluation of shake-up spectra[69] but have not performed any calculations in this way. One possible problem arises insofar as the virtual orbitals from a V^{N-2} of V^{N-1} potential have, in general, lower orbital energies than the virtuals derived from the full V^N SCF potential. Since all the denominators in the EOM–Green's function theories contain differences between occupied and virtual orbital energies, it is possible that the use of V^{N-2} virtual orbitals could cause convergence problems for the perturbation series.[69] Yeager et al. have observed similar problems in EOM excitation energy calculations where they used V^{N-1} potential virtual orbitals.[94]

Cederbaum and co-workers[16] also employ a renormalized interaction in their calculations, obtained by approximating certain sets of diagrams as geometric series and summing the series to all orders. Cederbaum et al.[16] note, however, that this approximation is expected to be accurate only when there is adequate separation of the valence ionization potentials and the shake-up energies. Cederbaum and co-workers[24-30] obtain more accurate estimates of shake-up energies from the diagonalization of the large CI matrices, $\langle \mathrm{HF}|a_\mu^\dagger a_m a_\nu^\dagger H a_\gamma a_p^\dagger a_\delta|\mathrm{HF}\rangle$ and $\langle \mathrm{HF}|a_m a_\mu^\dagger a_n H a_p^\dagger a_\gamma a_q^\dagger|\mathrm{HF}\rangle$ when they are interested in the shake-up portion of the spectrum. The eigenvalues and eigenvectors of these matrices are then employed in the second-order Dyson equation, which couples the primitive shake-up and simple ionization processes. Thus the first-order couplings between different shake-up states are included in the evaluation of inner valence ionization potentials and shake-up energies, yet the second-order couplings, which arise from correlation effects in the ground-state wave function, are ignored. Numerical results described in Section III demonstrate that the diagonalization of the entire CI shake-up matrices is not necessary; rather, only a relatively small number of configurationally selected shake-up operators need be directly diagonalized.

III. NUMERICAL STUDIES

In this section we describe numerical investigations concerning the accuracy of various approximate solutions of the EOM equations. The work centers mainly on the ionization potentials–electron affinity variant of the equations of motion theory.

A. Differences Between Simons' and Yeager's EOM Calculations

The simplest EOM method for EAs and IPs has been developed independently by Simons and co-workers[56] and by Yeager.[66] This approximation uses a P-space consisting of the simple electron removal basis operators a_μ and a_m. Only the 3-block of basis operators is retained in the Q-space. Simons approximates the ground-state wave function by first-order RSPT as a linear combination of the SCF determinant $|\mathrm{HF}\rangle$, plus all double excitations with respect to $|HF\rangle$. Only the diagonal matrix elements of the inverse matrix in the partitioned EOM equation, (37), are retained through first order (in the electron-electron interaction). The matrix $A^{(1,1)}$ is approximately calculated through third order, and $A^{(3,1)}$ through second order. However, third-order terms in $A^{(1,1)}$ and second-order terms in $A^{(3,1)}$ involving single excitation correlation coefficients are neglected in Simons' work.[46, 48, 68, 95] (The single excitations do not enter into the ground-state wave function until second order.) Fourth-order terms in (37) that are bilinear in the second-order parts of $A^{(1,3)}$ and $A^{(3,1)}$ are discarded. The result is a $2\frac{1}{2}$-order theory. The neglected third-order terms, which either involve single excitation correlation coefficients or the off-diagonal parts of $A^{(3,3)}$ through the second term in the Born expansion of $(A_{QQ} - \omega I_{QQ})^{-1}$, are studied in detail in Section III.C. First, however, we consider the implications of the two differences between the formulation of Simons and that of Yeager.

The first difference concerns the manner in which the EOM matrix is symmetrized. Simons[56] uses the unsymmetrized form of the EOM equation, (7), whereas the present work,[67–71] as well as Yeager's,[66] employs the symmetrized form, (21). However, the unsymmetrized $A = \langle 0|\{O,[H,O^{\dagger}]\} |0\rangle$, is, in general, non-Hermitian when an approximate ground-state wave function is employed. (If an nth-order approximation to $|0\rangle$ is utilized, the unsymmetrized A is Hermitian, at least through order n.[39]) Simons[56] apparently symmetrizes his A matrix by replacing the upper right triangular block with the adjoint of the lower left triangular block. This method is called $2\frac{1}{2}$-order EOM method I, below. When the symmetrized EOM equation (21) is employed, it is termed $2\frac{1}{2}$-order EOM method II. These two methods differ by third-order contributions to the partitioned EOM equations.

The second difference between Simons's work and Yeager's development of the theory arises from the use of spin-symmetry-adapted shake-up basis operators in Yeager's method[66] and unsymmetrized ones in Simons's method.[56] The present calculations, $2\frac{1}{2}$-order EOM methods I and II, follow Yeager in utilizing spin-symmetry-adapted 3-block operators.[67] The primitive operators $a_{\mu}a_{m}^{\dagger}a_{\nu}$ and $a_{m}a_{\mu}^{\dagger}a_{n}$ are related to the spin-adapted forms by a unitary transformation. The EOM matrix is blocked by symmetry when symmetry-adapted operators are employed. The EOM equation (21) and the partitioned EOM equation (37) are invariant to unitary transformations within the operator basis. However, when the inverse matrix in (37) is approximated by a truncated form of the Born expansion, (39), this invariance to unitary transformations is lost.[68] Of course, the use of symmetry-adapted operators reduces the dimensionality of the secondary operator space. This reduction is accomplished at the expense of more complicated expressions for the $A^{(1,3)}$, $A^{(3,1)}$, and $A^{(3,3)}$ matrix elements.

Table I presents the results of EOM calculations of the three lowest IPs of nitrogen. Comparison of the first two columns of Table I demonstrates that there is a difference of 0.2 to 0.3 eV in the IPs when the EOM A matrix is symmetrized as by Simons, $2\frac{1}{2}$-order method I, and when the symmetrized form of the EOM equations, (21), $2\frac{1}{2}$-order method II, is employed. The lack of symmetry in $\langle 0|\{O,[H,O^\dagger]\}|0\rangle$ in a $2\frac{1}{2}$-order calculation arises from the inclusion of certain second-order $A^{(3,1)}$ and $A^{(1,3)}$ terms, which contain the products of electron-electron interaction matrix elements with first-order double excitation correlation coefficients, and the neglect of other second-order $A^{(3,1)}$ and $A^{(1,3)}$ terms, which involve second-order single excitation correlation coefficients multiplied by linear combinations of orbital energies. The discrepancies between the EOM $2\frac{1}{2}$-order methods I and II are a measure of the importance of the terms due to single excitations in the ground-state wave function. In Section III.C, we consider the third-order terms not included in this primitive $2\frac{1}{2}$-order EOM theory. The calculations imply although these terms are small, they are certainly not negligible.[68]

By comparing the calculations of the $2\frac{1}{2}$-order EOM method I with Simons' N_2 results[57] (using the same orbital basis set), columns one and three of Table I, respectively, the effect of spin-symmetry adapting the operators basis can be assessed, since this is the only difference between the calculations. The results exhibit a shift in the IPs of 0.30 to 0.44 V when the symmetry-adapted operators are employed. This quite large shift is really a reflection of the restricted accuracy of the diagonal approximation

TABLE I

Comparison of Primitive EOM Results[a] for N_2 Using Simons' Basis[57]

	Ionization potential (eV)			
Ion state	$2\frac{1}{2}$-Order EOM method I[b]	$2\frac{1}{2}$-Order EOM method II[b]	Simons' results[57,c]	Experiment[96]
$X^2\Sigma_g^+$	16.01	15.74	15.69	15.60
$A^2\Pi_u$	18.33	18.09	18.03[97]	16.98
$B^2\Sigma_u^+$	19.22	19.00	18.63	18.78

[a]All results neglect singly excited configurations in the ground-state wave function.

[b]Uses spin-symmetry-adapted basis operators.

[c]Uses non-spin-symmetry-adapted basis operators.

for the $(A_{QQ} - \omega I_{QQ})$ matrix. This is the first indication of another problem that dictates the course of much of the subsequent work presented below.

In all the IP–EA calculations below, we follow Yeager[66] and employ the symmetrized EOM equation (21) and symmetrized shake-up operators.

B. Basis Set Dependence of Calculated Ionization Potentials for N_2

The logic presented in Section II.A demonstrates that within the same finite orbital basis, an exact EOM and a full CI calculation yield the same energy difference (i.e., when the only approximation is due to the choice of a finite orbital basis). Evidence from CI studies has shown that rather extensive orbital basis sets are needed to provide reliable results in molecular calculations. On the other hand, the early EOM computations of Simons and co-workers[57–65] had been restricted to rather small basis sets because of computer limitations. Nevertheless, these investigators obtained very good agreement with experimental values for the IPs and EAs of diatomics, involving the atoms H through F. An accuracy of 0.1 to 0.2 eV was frequently claimed, although the basis sets (Slater-type orbitals) are generally double zeta quality without polarization functions. This high degree of accuracy certainly would not be expected of CI calculations with such primitive sets of orbitals.

Griffing and Simons[60] vary the exponents of those basis functions with large expansion coefficients in the highest occupied orbital in the SCF determinant, to maximize the ionization potential in CN^- and BO^-, claiming that this simple variation procedure leads to orbital basis sets that yield accurate IPs and EAs. Liu[98] has criticized this method of basis set "optimization," presenting results for CN showing that the major effect of this variation is an increase (worsening) of the SCF energy of CN, while the SCF energy of CN^- remains relatively unchanged. Liu contends that this procedure improves the calculated EA by actually providing a poorer basis set for the calculation of the energy levels of the neutral. Since this method is not guided by any variational bounds, its application seems rather dubious.

In this section we compare EOM results for the three lowest IPs of N_2 using four different orbital basis sets, attempting to clarify in the most direct fashion the basis set dependence of the EOM method. The basis sets used are the double zeta (no polarization functions) Slater basis set of Simons,[57] the same basis set augmented with a set of $d\sigma$ and $d\pi$ polarization functions, the partially optimized Slater basis set due to Nesbet,[99] and the modest Gaussian basis employed by Yeager and Freed[91] for their N_2 excitation energy calculations. This Gaussian basis set consists of a $[3s2p]$ Dunning contraction of a $(9s5p)$ set of basis orbitals with two diffuse p_z and two sets of diffuse $d\pi$ functions placed at the center of the N—N bond.

The nitrogen molecule N_2 is an interesting and very difficult system for the calculation of ionization potentials. Both Koopmans' theorem[85] and the ΔSCF method predict the wrong ordering of the IPs when accurate basis sets are employed. This is the case for the extensive SCF calculations of Cade, et al.[100] on the N_2 molecule and the N_2^+ ion. These calculations are believed to be near the Hartree-Fock limit. Cade's ground-state energy for N_2 at $R = 2.068$ au is -108.9928 au. The four basis sets employed in this study yield SCF ground-state N_2 energies of -108.8644, -108.9350, -108.9714, and -108.87855 au, respectively, which are 3.49, 1.57, 0.58, and 3.11 eV from Cade's value. These absolute errors are very large for all the basis sets except Nesbet's. Even for this basis set it is still 0.58 eV. However, since we are investigating energy differences, the final results ought to be less sensitive to errors in the orbital basis set employed.

Table II shows that Simons' basis set (set I) reproduces the experimental values rather accurately for the $^2\Sigma_g^+$ and $^2\Sigma_u^+$ ionization potentials, the errors being 0.14 and 0.22 eV, respectively. These results represent vast improvements over the Koopmans' theorem values of 17.58 and 21.75 eV. However, the $^2\Pi_u$ state is grossly in error, being 1.11 eV different from experiment. (Simons, at first, mistakenly[97] reported his value to be 1 eV lower than this, leading him to conclude, erroneously, that the EOM method gave accurate IPs for all three states with this basis set.[57]) Adding the polarization functions to Simons' basis set (giving basis set II) lowers the calculation of the troublesome $^2\Pi_u$ IP by 0.36 eV and leaves the other states relatively unaffected. The EOM results with Nesbet's basis (set III) are much better still: all the IPs are within 0.34 eV of experiment. The calculations employing basis set IV indicate that the EOM Gaussian basis set calculations are of comparable accuracy to the double zeta Slater basis sets with polarization functions.[67]

Cederbaum and co-workers[21, 23] have also performed basis set studies on a number of species. Their conclusions are similar to those arrived at from the calculations presented above. In choosing the basis set for EOM IP–EA

TABLE II
Ionization Potentials (eV) of N_2

Ion state	Basis set I		Basis set II		Basis set III		Basis set IV		Experiment[96]
	KT[a]	EOM	KT[a]	EOM	KT[a]	EOM	KT[a]	EOM	
$X^2\Sigma_g^+$	17.58	15.74	17.34	15.78	17.49	15.94	17.49	15.09	15.60
$A^2\Pi_u$	17.77	18.09	17.54	17.73	16.94	17.20	17.47	17.30	16.98
$B^2\Sigma_u^+$	21.75	19.00	21.54	19.09	21.38	19.01	21.03	18.71	18.78

[a] Koopmans' theorem prediction of ionization potential.

calculations, a compromise must be made between basis sets optimized for the neutral and ionic states. Basis set errors are propagated through the EOM–Green's function treatments just as they are in SCF and CI calculations. It appears that, to consistently yield accurate ionization potentials for molecules, the basis sets must be, at the very least, of double zeta accuracy with polarization functions.

C. Third-Order Results and the Breakdown of the Perturbation Expansion

The EOM ionization potential calculations discussed in Sections III.A and III.B include some numerically important third-order terms while neglecting other third-order terms whose importance is suggested by means of the various comparisons. In this section we investigate these remaining third-order contributions in EOM IP calculations.

Three types of third-order contribution are omitted in the calculations in Sections III.A and III.B. Two of these arise because of single excitations from the SCF determinant in the ground-state wave function $|0\rangle$. These singly excited configurations (SEC) in $|0\rangle$ produce third-order terms in $A^{(1,1)}$ and second-order ones in $A^{(1,3)} - \omega D.^{(1,3)68}$

Both these provide third-order contributions to the partitioned EOM equation given the restricted definition of the P-space as the set of all simple electron removal operators, a_μ and a_m. These terms are neglected in the calculations of Simons and of Sections III.A and III.B because singly excited configurations contribute to $|0\rangle$ only in second order, and only a first -order $|0\rangle$ is considered in these works for simplicity.[56] Analysis of the $A^{(1,3)}$ and $A^{(1,1)}$ matrix elements shows that the other second-order perturbation terms in the ground-state wave function, involving doubly, triply, and quadruply excited configurations with respect to the SCF determinant, $|HF\rangle$, cannot contribute to the partitioned EOM equation through third order for the present definition of the EOM P-space. (This represents yet another simplification over EOM excitation energy calculations where the second-order double excitation correlation coefficients do enter into the third-order theory with the traditional partitioning.[91]) The final third-order contribution to EOM ionization potentials involves the second term in the Born expansion (39) of the inverse matrix $(A_{QQ} - \omega I_{QQ})^{-1}$ that is present in the partitioned EOM equation, (37).[68, 95]

Table III demonstrates the effect on the ionization potentials of including single excitations from $|HF\rangle$ in the ground-state wave function for N_2. Simons' basis set is employed in these calculations. Table III exhibits the small, but nonnegligible, effect of retaining third-order terms involving singly excited configurations in $|0\rangle$; in the partitioned EOM equation it

TABLE III

Effect of Including Singly Excited Configurations (SEC) in the Ground
State Using Simons' Basis on EOM Ionization
Potentials (eV) of N_2

	$X^2\Sigma_g^+$	$A^2\Pi_u$	$B^2\Sigma_u^+$
Normal EOM[a]	15.74	18.09	19.00
Including effects of SECs in normal EOM[a]	15.58	17.96	18.87
Including effects of SECs in $A^{(1,1)}$ only[a]	15.29	17.72	18.64
Including effects of SECs in $(A^{(3,1)} - \omega D^{(3,1)})$ only[a]	16.02	18.33	19.24
Including effects of SECs using orthogonalized operator basis	15.57	17.96	18.86

[a] Calculation employs nonorthogonal basis operators.

produces shifts of 0.16, 0.13, and 0.13 eV for the $^2\Sigma_g^+$, $^2\Pi_u$, and $^2\Sigma_u^+$ symmetries, respectively. These contributions to the EOM ionization potentials are of comparable magnitude to the claimed accuracy of many early EOM calculations. Decomposing these contributions further and considering the extra third-order terms in $A^{(1,1)}$ and second-order terms in $A^{(1,3)} - \omega D^{(1,3)}$ separately reveals that each of these individual contributions is much larger, as large as 0.45 eV. However, these two types of term partially cancel each other.

The $D^{(1,3)}$ terms are present because an orthogonalized operator basis set is not employed in these calculations. The last row of Table III presents results for calculations that include both the third-order contributions to the partitioned EOM equation due to single excited configurations and use an operator basis in which the 3-block operators are Schmidt-orthogonalized to the 1-block (therefore, $D^{(1,3)} = 0$) through second order. These results differ from the results using the nonorthogonal operator basis by only 0.0 to 0.01 eV for each symmetry.

The only remaining third-order contribution to the partitioned EOM equation (37) arises from the second term in the Born expansion of $(A_{QQ} - \omega D_{QQ})^{-1}$, (39). This term has the form $A_{pq}M_0^{-1}M_1M_0^{-1}A_{qp}$, where $A_{pq}(A_{qp})$ is the first-order part of $A_{PQ}(A_{QP})$. When shifted ω-dependent denominators are employed, M_1 contains the first-order off-diagonal matrix elements of A_{QQ}. On the other hand, when unshifted ω-dependent denominators are employed, M_1 includes the first-order diagonal elements as well. In any event, the evaluation of this term involves a double summation over all the operators in the Q-space (i.e., the 3-block here). There is a contribution of this type to each element of the partitioned EOM matrix (which has the dimensionality of the P-space). Thus inclusion of these terms constitutes a substantial increase in the labor involved in solving the EOM equation, and this is, prehaps, the motivation for its neglect in the $2\frac{1}{2}$-order theories. The results, presented in Table IV using both shifted and unshifted ω-dependent denominators, are startling when the traditional partitioning of the operator basis is used. The shift upon inclusion of the second Born term is large for the $^2\Pi_u$ IP and enormous for the (shifted) $^2\Sigma_g^+$ calculation. The shifted $^2\Sigma_u^+$ calculation oscillates wildly and does not even converge. These results strongly imply that the expansion of $(A_{QQ} - \omega I_{QQ})^{-1}$ is failing for this system. The large differences in the shifted versus unshifted ω-dependent denominators, which occurs for both the zeroth- and first-order Born expansions, is a further indication of a complete breakdown of the perturbative expansions of the inverse matrix.

This breakdown can be qualitatively explained by a simple molecular orbital picture of N_2. A partial orbital energy level diagram for Simons's and for Nesbet's bases is provided in Fig. 1. The $1\pi_g$ orbitals are very tight,

TABLE IV

Effect of Different EOM Partitioning Schemes on the Stability of the Expansion for $(A_{QQ} - \omega I_{QQ})^{-1}$ Using Simons' Basis for Ionization Potentials of N_2

	ω-Dependent denominators	Order of expansion of $(A_{QQ} - \omega I_{QQ})^{-1}$	Ionization potentials (eV) $X^2\Sigma_g^+$	$A^2\Pi_u$	$B^2\Sigma_u^+$
Traditional partitioning	Unshifted	Zeroth	16.21	18.00	19.96
	Shifted	Zeroth	15.74[a]	18.09[a]	19.00[a]
	Unshifted	First	16.41	17.71	20.36
	Shifted	First	17.75	17.51	—[b]
Repartitioning scheme I	Unshifted	Zeroth	15.99	17.98	19.33
	Shifted	Zeroth	15.86	18.11	19.18
	Unshifted	First	15.82	17.70	19.20
	Shifted	First	15.81	17.51	19.30
Repartitioning scheme II	Unshifted	Zeroth	16.00	17.82	19.33
	Shifted	Zeroth	15.85	17.70	19.17
	Unshifted	First	15.79	17.76	19.19
	Shifted	First	15.81	17.76	19.16
Diagonalization of unpartitioned EOM matrix[c]			15.61	17.56	18.97

[a] "Standard EOM."

[b] Calculation does not converge.

[c] Includes effects of single excitation correlation in the ground state.

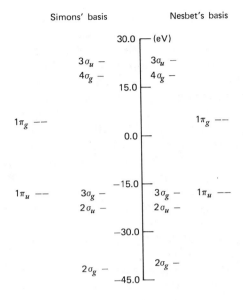

Fig. 1.   Partial orbital energy diagram for N_2.

low-lying virtual orbitals, reflecting the almost bound nature of N_2^-, as witnessed experimentally by the shape resonance in low-energy N_2-electron-scattering profiles.[101] Thus the presence of this low-lying, tight π_g orbital is not dependent on a peculiar choice of orbital basis. Some primitive shake-up ionization operators, involving removal of one electron form a valence orbital $(2\sigma_u, 3\sigma_g, 1\pi_u)$ and excitation of a second valence electron to $1\pi_g$, mix strongly with the simple ionization process of removing a single electron from a valence orbital. These shake-up basis operators have been relegated to the Q-space, supposedly of minor importance. Numerically, this mixing of valence ionization and shake-up processes is reflected in relatively small diagonal $(\mathbf{A}_{QQ} - \omega\mathbf{I}_{QQ})^{-1}$ matrix elements and in large $\mathbf{A}^{(1,3)}$ coupling elements. A check of the corresponding matrix elements, indeed, confirms the molecular orbital picture. Furthermore, many of the off-diagonal $A_{ij}^{(3,3)}$ couplings between these important shake-up basis operators are large and comparable in magnitude to the splittings between their diagonal elements, $A_{ii}^{(3,3)} - A_{jj}^{(3,3)}$. Hence the criterion for convergence of the Born series, an expansion in powers of $A_{ij}^{(3,3)}/(A_{ii}^{(3,3)} - \omega)$, is violated in this case.

This example provides strong evidence that the traditional many-body choice of the single electron operators, a_μ and a_m, as the primary space cannot be justified in general.

A comparison of different partitionings of the EOM operator basis is given in Table IV. Included in the P-space in repartitioning scheme I are shake-up operators, corresponding to removal of an electron from a $2\sigma_u$, $1\pi_u$, or $3\sigma_g$ orbital and excitation of a second electron from one of these orbitals to the $1\pi_g$ orbitals. The expansion of the inverse matrix is then well behaved for the $^2\Sigma_g^+$ and $^2\Sigma_u^+$ states, in contrast to the situation when the traditional definition of the P-space is employed. The $^2\Pi_u$ IP still changes as much as 0.6 eV when the first-order Born term is added for the shifted ω-dependent denominators case. This instability in the Born expansion for the $^2\Pi_u$ symmetry is removed when the P-space is expanded further to include all 3-block basis operators involving just $2\sigma_g$, $2\sigma_u$, $1\pi_u$, $3\sigma_g$, and $1\pi_g$ orbitals (repartitioning scheme II). This partitioning of the EOM P-space includes some particle removal and de-excitation operators, and these operators are, in fact, important for an accurate description of the $^2\Pi_u$ ionization. This illustrates the need for a more general means of choosing the EOM P-space, such as the one based on an operator selection criterion described in Section II.E, since the de-excitation basis operators would not necessarily have been anticipated to belong in the P-space a priori.

Because of the smallness of Simons' basis, it is possible to completely diagonalize the unpartitioned EOM matrix. The calculations also appear in Table IV. They agree with the IPs from repartitioning scheme II to about 0.2 eV, indicating the validity of our approach of adding 3-block basis operators to the EOM P-space. Furthermore, the IPs calculated by direct diagonalization of the (unpartitioned) EOM matrix include the effect of the singly excited configurations in $|0\rangle$ in $A^{(1,1)}$ and $A,^{(1,3)}$ whereas the calculations using repartitioning scheme II do not. The inclusion of singly excited configurations in the ground state is expected to lower the partitioned EOM results by 0.1 to 0.2 eV, based on the results of Table III discussed above. This lowering of the results of repartitioning scheme II would bring them into even closer agreement with the unpartitioned solutions within the 1- and 3-blocks for the basis set.

By analogy with the shifted and unshifted ω-dependent denominator methods, there are the corresponding approaches for the perturbation calculation of the ground-state wave function. Above, unshifted ω-independent denominators, that is, Rayleigh-Schrödinger perturbation theory (RSPT), have been used in all calculations. The alternate, shifted ω-independent denominators method, that is Epstein–Nesbet perturbation theory (ENPT),[92] treats the complete (through first order in the electron-electron interaction) diagonal Hamiltonian (CI) matrix as the zeroth-order Hamiltonian. The off-diagonal elements, which arise in the configuration mixing, give the perturbation. RSPT includes only the lowest (zeroth) order diagonal elements of H in the zeroth-order Hamiltonian. The situation is exactly analogous to the expansion of the inverse matrix in the partitioned EOM equation. The unshifted method just has sums and differences of orbital energies in the denominators, whereas the shifted method has electron-electron interaction terms added to this.

The typical situation in EOM calculations for ionization potentials has been to employ shifted ω-dependent and unshifted ω-independent (RSPT) denominators. This practice seems somewhat inconsistent. Table V compares results of N_2 calculations that employed unshifted ω-dependent and ω-independent denominators with corresponding calculations using only shifted denominators. There are -0.13, -0.99, and 0.30 eV differences for the $^2\Sigma_g^+$, $^2\Pi_u$, and $^2\Sigma_u^+$ symmetries, respectively. Comparison of these shifted results with calculations employing shifted ω-dependent denominators, but calculating $|0\rangle$ by RSPT shows that the mode of calculating $|0\rangle$ yields an 0.99 eV difference in the $^2\Pi_u$ IP and about 0.2 eV differences for the $^2\Sigma_g^+$ and $^2\Sigma_u^+$ IPs. These large changes can again be explained by considering a simple molecular orbital picture of N_2. Configurations involving the $1\pi_g$ orbitals provides important correlation corrections to the SCF configuration. The weighting coefficients of these configurations are as large as 0.1 in magnitude for Simons' basis when RSPT is used. The weights of these configurations nearly double in size when ENPT is employed,[68] indicating deficiencies in the assumption of a ground-state wave function for N_2 based on a single configuration zeroth-order approximation. A more general approach is to use a multiconfigurational zeroth-order ground state, although the technical problems in developing an efficient EOM method based on a multiconfigurational zeroth-order ground state are large. We return to this point again in Sections III.F and IV.

TABLE V

Shifted Versus Unshifted Denominators in EOM Ionization Potential Calculations for N_2 Using Simons' Basis[a]

ω-dependent denominators	ω-independent denominators	Ionization potentials (eV)		
		$X^2\Sigma_g^+$	$A^2\Pi_u$	$B^2\Sigma_u^+$
Unshifted	Unshifted	15.60	17.51	19.06
Shifted	Shifted	15.47	16.52	19.36
Shifted	Unshifted	15.65	17.36	19.17

[a] Repartitioning scheme I is employed with a first-order Born expansion of the inverse matrix. Third-order terms arising from singly excited configurations in the ground-state wave function are included.

D. Calculation of the Shake-Up X-Ray Photoelectron Spectroscopy Spectrum of N_2

A notable feature of X-ray photoelectron spectroscopy (XPS)[102, 103] and ultraviolet photoelectron spectroscopy (UPS)[96] is the presence of satellite lines that appear along with the lines attributable mainly to simple ionization from a given molecular orbital. N_2 provides a good example of this type of spectrum. Electron spectroscopy for chemical analysis (ESCA) yields the XPS spectrum of N_2 due to Siegbahn et al.[102] reproduced in Fig. 2. The three peaks in the 15–20 eV range arise mainly from ionization from the $3\sigma_g$, $1\pi_u$, and $2\sigma_u$ orbitals (in order of increasing energy). The remaining peaks are primarily due to "two-electron processes" involving the ionization of one valence electron and the excitation of a second. Hence the shake-up lines give additional information about electronic correlations. An understanding of the assignment of these shake-up transitions may be of additional value in the use of photoelectron spectroscopy as an analytical tool.

Section III.C describes how the inclusion of some of the basis operators from the 3-block in the EOM P-space is important for the calculation of the main peaks in the N_2 spectrum. Once this is done, the increased dimensionality of the partitioned EOM matrix leads to the appearance of additional roots of the EOM pseudoeigenvalue problem. If the 3-block basis operators, which are moved to the P-space, correspond to the major components of the ionization operators for the shake-up lines, the new roots of the EOM matrix give the positions of these shake-up peaks. Thus the shake-up energies are naturally produced by the EOM method given our generalized definition of the EOM P space

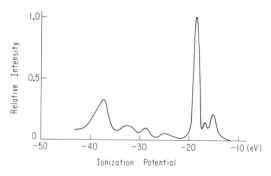

Fig. 2. Experimental X-ray photoelectron spectrum of N_2 measured by Siegbahn et al.[102] Incident photons have an energy of 1254 eV (MgK_α).

For the sake of comparison with the ESCA spectrum of N_2, and to provide a more critical test of the EOM method, we calculate the peak intensities as well as the peak positions. The details of the intensity calculations are left to the Appendix. The dipole approximation is invoked, as the calculation is based on Fermi's golden rule.[104] A plane wave approximation is employed for the outgoing electron. The cross section for the ejected electron is averaged over all possible molecular orientations and polarizations of the incident photon as described by Ellison.[105]

The final expression for the total cross section for λth ionization peak is given by[69]

$$\sigma_T(\lambda) = \frac{2e^2 k^3 L^3}{3\mu\omega c} \sum_{l,l'} \overline{\langle\phi_l|\mathbf{k}\rangle^* \langle\phi_{l'}|\mathbf{k}\rangle} B_l^*(\lambda) B_{l'}(\lambda) \tag{45}$$

In (45) ϕ_l is the lth molecular orbital (l and l' are summed over all basis orbitals) and $|\mathbf{k}\rangle$ is the plane wave representing the outgoing electron. The magnitude of the wave vector $\mathbf{k}$ is determined by energy conservation; e and μ are the charge and mass of the electron, c is the speed of light, and L^3 is the volume of the "box" in which the experiment is enclosed. The L^3 factor, which arises from the plane wave density of states, cancels a factor of L^{-3} from the box normalization of the plane wave, $|\mathbf{k}\rangle$. The bar over $\overline{\langle\phi_l|\mathbf{k}\rangle^*\langle\phi_{l'}|\mathbf{k}\rangle}$ denotes an average over all molecular orientations, and $B_l(\lambda)$ is given by

$$B_l(\lambda) = \langle 0|a_l^\dagger O_\lambda^\dagger|0\rangle \tag{46}$$

where $O_\lambda^\dagger$ is the EOM ionization operator; $O_\lambda^\dagger$ is approximated by its $P-$space projection in the numerical calculations presented below.

The term $\sigma_T(\lambda)$ contains the overlap between the orbital of the outgoing electron $|\mathbf{k}\rangle$ and the molecular orbitals ϕ_l. Since a plane wave approximation for $|\mathbf{k}\rangle$ fails to accurately follow the oscillations of the true continuum orbital for the ejected electron in the region near the molecule, the use of the plane wave for $|\mathbf{k}\rangle$ is expected to introduce some errors into the calculations. A Coulomb orbital, which has the correct long-range behavior, would likewise be inaccurate in the vicinity of the molecule. [Equation 45 shows that it is the short-range behavior of $|\mathbf{k}\rangle$ that is relevant for the evaluation of $\sigma_T(\lambda)$, since $|\mathbf{k}\rangle$ appears only through the overlap integrals $\langle\phi_\mu|\mathbf{k}\rangle$.] Clearly, the plane wave approximation is more accurate at higher photon energies ($h\nu = 1254$ eV in the present example), since the effect of the field of the remaining molecular ion is diminished as compared to the kinetic energy of the outgoing electron (roughly speaking, the momentum of the electron is more nearly constant). More accurate work requires the development of a feasible procedure for obtaining continuum orbitals with the correct short-range behavior for molecules.

Substitution of the perturbation expansion for $|0\rangle$ into (46) leads to a perturbation expansion for $B_l(\lambda)$. This, in turn, gives an expansion for $\sigma_T(\lambda)$. The zeroth-order terms for $\sigma_T(\lambda)$ have the form (see Appendix)

$$
\sigma^{(0)}(\lambda) \propto \sum_{\mu,\nu} \overline{\langle \phi_\mu | \mathbf{k} \rangle^* \langle \phi_\nu | \mathbf{k} \rangle} \langle HF | a_\mu^\dagger a_\mu | HF \rangle \langle HF | a_\nu^\dagger a_\nu | HF \rangle C_\mu^\lambda C_\nu^\lambda
$$

$$
= \sum_{\mu,\nu} \overline{\langle \phi_\mu | \mathbf{k} \rangle^* \langle \phi_\nu | \mathbf{k} \rangle} C_\mu^\lambda C_\nu^\lambda \tag{47}
$$

where $|HF\rangle$ is the zeroth-order (SCF) ground state; μ and ν are occupied orbitals in $|HF\rangle$ and C_μ^λ is the coefficient in the operator expansion of $O_\lambda^\dagger$ for simple electron removal a_μ from the orbital μ ($O_\lambda^\dagger$ can apply to simple ionizations or shake-up processes). A single term of this sort is expected to dominate the cross section for simple ionization, since one C_μ^λ generally dominates the operator expansion. For shake-up processes, these terms represent "intensity borrowing" from the main ionization peaks of the same symmetry.

The correlation part of the wave function makes first-order contributions to the total photoionization intensities. This is in contrast to the situation for the peak positions, where ground-state correlation does not contribute until second order in electron correlation. If $\Gamma_{\mu m \mu'}^\dagger$ represents a 3-block basis operator that removes an electron from orbital μ and excites a second electron from μ' to m and $C_{\mu m \mu'}^\lambda$ is the corresponding expansion coefficient in the operator expansion of $O_\lambda^\dagger$, then one first-order term is given by

$$
\sigma^{(1)}(\lambda)^{\text{corr}} \propto \overline{\langle \phi_\mu | \mathbf{k} \rangle^* \langle \phi_n | \mathbf{k} \rangle} \langle HF | a_\mu^\dagger a_\mu | HF \rangle \langle 0 | a_n^\dagger \Gamma_{\mu m \mu'}^\dagger | 0 \rangle C_\mu^\lambda C_{\mu m \mu'}^\lambda \tag{48}
$$

where $\langle 0 | a_n^\dagger \Gamma_{\mu m \mu'}^\dagger | 0 \rangle$ is simply a linear combination of first-order ground-state correlation coefficients.[69] For shake-up peaks, a few $C_{\mu m \mu'}^\lambda$ dominate the expansion of $O_\lambda^\dagger$, and terms of the form given in (48) can be expected to make significant contributions to $\sigma_T(\lambda)$.

If there are no main peaks of a given symmetry, the zeroth-order $B_l(\lambda)$ terms, $\langle HF | a_\mu^\dagger a_\mu | HF \rangle$, are absent because there are no occupied orbital μ in $|HF\rangle$ of this symmetry (or else there would be a corresponding main peak). The lowest order contribution to $\sigma_T(\lambda)$ is then of second order for any shake-up peak of that symmetry, since all $B_l(\lambda)$ then begin in first order. Thus low intensities are expected for shake-up peaks of a symmetry different from the nearby main peaks. This is, in fact, exhibited in our calculations for the Π_g symmetry of N_2.[69]

The calculated photoionization spectrum of N_2 is presented in Figs. 3 and 4. Nesbet's basis set[99] is employed in both calculations. In Fig. 3 the EOM P-space is as in repartitioning scheme I described previously. Figure 4 displays the predicted spectrum obtained with a more extensive P-space. The configuration selection criterion, as described in Section II.E, is used to identify the shake-up basis operators that mix strongly with the more limited P-space (i.e., the P-space used in the calculation of Fig. 3). Each peak is fitted with a Gaussian whose total area is given by the calculated cross section. The widths are adjusted arbitrarily. Alternatively, the widths could be chosen from a simple estimate of the Franck-Condon overlaps for the vibrational fine structure, as is done by Cederbaum and Domcke.[16, 31, 32]

The calculated results agree qualitatively with experiment for both peak positions and relative intensities. However, the higher energy peaks in the spectrum appear at too high energy. The other salient feature is that the peak at about 40.0 eV is actually made up of two peaks of nearly equal intensity. The $2\sigma_g$ electron removal operator makes strong contributions to these peaks and also to the $^2\Sigma_g^+$ peak at 29.44 eV. This contribution of the $2\sigma_g$ ionization to a number of lines is in agreement with the Green's function results of Schirmer et al.[24]

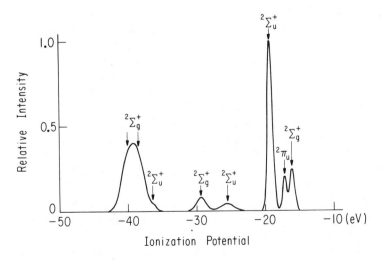

Fig. 3. Calculated N_2 X-ray photoelectron spectrum for incident photons with energy of 1254 eV. The primary operator space for the calculations contains only a minimal number of shake-up-basis operators (repartitioning scheme I).

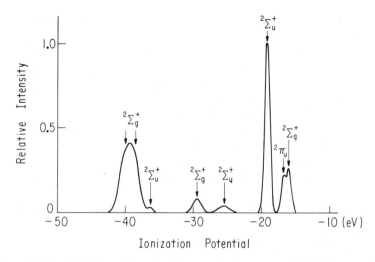

Fig. 4.   Calculated N_2 X-ray photoelectron spectrum for incident photons with energy of 1254 eV. The primary operator space is chosen by perturbation theory.

Nearly all the peaks in the calculated N_2 spectrum have a number of basis operators that contribute significantly. This indicates that the simple molecular orbital picture of the shake-up process is insufficient.[24, 69] The results emphasize the need for some selection technique, such as the perturbation theory approach employed here, in the choice of the primary operator space. The addition of several extra shake-up basis operators by the perturbation selection criterion lowers the $^2\Pi_u$ peak from 17.12 to 16.79 eV. The most important of these extra operators involves removal of a $1\pi_g$ electron and de-excitation of a second $1\pi_g$ electron to the $1\pi_u$ level. The importance of this operator, which acts only on the correlation part of $|0\rangle$, is not obvious by pure chemical intuition.

Cederbaum and his co-workers[24] have also calculated the N_2 shake-up spectrum using the diagrammatic Green's function approach. As mentioned in Section II.E, the approximate poles of the self-energy (corresponding to shake-up energies) are obtained by diagonalization of the ion CI matrices $\langle HF | a_\mu^\dagger a_m a_\nu^\dagger H a_\gamma a_n^\dagger a_\delta | HF \rangle$ and $\langle HF | a_m a_\mu^\dagger a_n H a_p^\dagger a_\gamma a_q^\dagger | HF \rangle$. This approach includes the important couplings between shake-up states, but at the expense of the diagonalization of these large-ion CI matrices. After obtaining the eigenvalues and eigenvectors of these matrices, Cederbaum and co-workers[24] then include the couplings of these eigenvectors to the simple electron removal operators, a_μ and a_m, by solving a second-order Dyson equation for the one-particle Green's function.

Purvis and Öhrn[38] have reported calculations of the N_2 shake-up spectrum using the propagator approach. They diagonalize small blocks of matrix elements between different shake-up operators, to obtain spin- and space-symmetry-adapted operators. In this manner they include what appears to be the most important couplings between the shake-up operators that are degenerate in lowest order. However, these operators are retained in the Q-space, and the couplings between important nondegenerate shake-up configurations are apparently ignored.

Iwata and co-workers[106] have performed extensive configuration interaction calculations of the shake-up energies of N_2. These results also show a large mixing of basis configurations in the shake-up states, and the lower shake-up states correspond nicely to our EOM results. The higher shake-up states are 2 to 4 eV below our EOM calculations, giving better agreement with experiment.

A number of steps can be taken to improve the EOM calculations. The coupling between shake-up operators is only evaluated through first order in the calculations displayed in Figs. 3 and 4. Ground-state correlation effects contribute to these couplings in second order. There are also second-order contributions to the partitioned EOM equations because of the 5-block basis operators, which are not included in these calculations. These contributions are described in Section II.E. Preliminary calculations of the shake-up positions including these terms have been attempted and the results are presented in Table VI. While the calculations did run into numerical difficulties, they indicate significant reductions can be expected in the calculated positions of the higher energy shake-up peaks. The numerical difficulties are similar to those encountered in Section III.C concerning the solution of the partitioned EOM equation for the simple ionizations of N_2 when the standard many-body partitioning scheme is employed. These difficulties suggest that many of the 5-block operators should be moved into the P-space to account for the greater orbital reorganization that accompanies the shake-up process as compared with the primary ionizations. Some shake-up basis operators cause instabilities in the calculation of the main IPs unless they are placed in the P-space, and the situation is expected to become more severe in the case of shake-up states.

Iwata and co-workers[106] found that configurations obtained from the SCF determinant by removal of a valence electron together with a double excitation to the $1\pi_g$ orbitals provide very important contributions to the higher shake-up states of N_2. These significant CI configurations are present in our EOM calculations, since the EOM basis operators act on every determinant in an approximately correlated ground-state wave function. However, the coefficients of these important CI configurations are not

TABLE VI
Calculated Ionization Potentials for N_2^a Using Nesbet's Basis

| Symmetry | Ionization potentials (eV) | |
	No 5-block or second-order $A^{(3,3)b}$	Includes 5-block and second-order $A^{(3,3)c,d}$
$^2\Sigma_g^+$	15.98	15.71 (15.77)
	29.40	27.72 (29.94)
	38.57	32.61 (36.27)
	39.96	41.27 (39.65)
$^2\Pi_u$	16.79	16.76 (16.84)
	29.05	27.66 (30.20)
	36.31	29.22 (33.59)
$^2\Sigma_u^+$	19.10	18.68 (18.98)
	25.43	24.66 (26.36)
	36.24	30.66 (34.04)

[a] Partitioned using perturbation criterion, P-space as for Fig. 4.
[b] Shifted ω-dependent denominators.
[c] Shifted ω-dependent denominators (unshifted ω-dependent denominators in parentheses).
[d] Underlined digits not converged.

independent in the EOM calculation, but rather they are fixed in the "EOM configurations" $O_i^\dagger|0\rangle$, by the perturbation expansion for the ground-state wave function, $|0\rangle$. Hence it seems that the accurate incorporation of the 5-block basis operators into the EOM calculations (i.e., in the P-space) is necessary if quantitative results are to be obtained for the higher shake-up states of N_2.

It is likely that in a more accurate calculation the two $^2\Sigma_g^+$ peaks, which now make up the highest energy peak ($\sim$40 eV) in Figs. 3 and 4, will split noticeably. This is indicated by the results in Table VI. Thus it seems possible that the second highest energy peak in the experimental spectrum could be a superposition of a $^2\Sigma_u^+$ peak and a $^2\Sigma_g^+$ peak. In fact, more recent XPS experiments[103] resolve the 30–35 eV region of the spectrum into two peaks rather than one, as displayed in Fig. 2.

There are a number of problems that weigh against the further enlarging the P-space to include the important 5-block operators in an attempt to obtain a more accurate calculated spectrum. There can be as many as ten 5-block operators that are degenerate in zeroth order, that is, involve the same spatial orbitals but have different arrangements of spin. Thus the dimensionality of the P-space grows very quickly when 5-block operators are added to it. This makes the EOM calculations much more cumbersome. Furthermore, the 7-block has first-order couplings to the 5-block, and these may be required in lowest order to obtain truly accurate shake-up calculations. Most important, as explained in Section III.C, it is unlikely that highly accurate results can be obtained for N_2 when a single configuration zeroth-order approximation is employed for $|0\rangle$. The same feature that produces a strong shake-up spectrum (a low-lying $1\pi_g$ level) also makes N_2 a particularly difficult example.

The present calculations indicate that the approximate EOM method can provide qualitative agreement with XPS data for both the peak positions and intensity. The calculations are reasonably simple and inexpensive, and they are capable of giving insight into the interpretation of the experimental results.

E. Extended EOM Calculations and Comparison with Configuration Interaction Results

The preceding sections consider cases in which the EOM operator basis consists of only the 1-block and 3-block basis operators. These ionization potential calculations indicate that the traditional division of the operator space into the 1-block P-space and the 3-block Q-space is generally not sufficient. The theory section, however, describes a more general division of the operator basis, which introduces additional second-order contributions to the partitioned EOM equation. These are the second-order $A^{(3,3)}$ matrix elements that are now in A_{PP} and contributions from including the 5-block in the Q space. In this section we report EOM calculations for a number of atomic and molecular systems. The effects of these additional second-order terms are specifically studied and compared with accurate configuration interaction studies. By comparison of the EOM IPs and EAs to highly converged CI results, rather than to experimental numbers, basis set errors can be eliminated. (Recall that full EOM and CI calculations yield identical results for a given orbital basis.) This focuses on the validity of various approximations introduced in solving the EOM equations.

The lowest ionization potentials of BH, HF, and Ne, and the largest electron affinities of OH and F are investigated.[71] Moderate-sized Slater basis sets are employed: $6s4p2d/3s1p$ for BH, $6s5p1d/3s1p$ for HF and OH, and $6s5p$ for Ne and F. The δ orbitals are removed in the diatomic calculations. By economizing slightly on the basis sets (these basis sets are actually larger than those employed in most EOM calculations), it is possible to carry the EOM and CI expansions further and thus perform a truly definitive comparison. The internuclear separations for the BH, HF, and OH calculations are 1.7328, 1.8342 and 2.336 au, respectively.

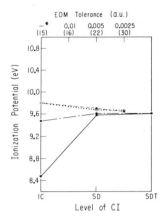

Fig. 5. EOM and CI vertical ionization potentials for BH: solid line, relaxed CI; long and short dashes, unrelaxed CI, using SCF orbitals of BH; dashed curve, extensive EOM; dotted curve, primitive repartitioned EOM. The EOM results are plotted against the tolerance for retaining shake-up-basis operators in the primary operator space, and the dimension of the primary operator space is given in parentheses for each tolerance. The CI values are presented at the one configuration level (1C), for single and double excitations CI (SD), and for single, double, and triple excitations CI (SDT). EOM calculations are not performed at tolerance of 0.01 au because this tolerance does not result in an appreciable increase in the dimensionality of the P-space. Experimental value is 9.77 eV. Asterisk: EOM primary operator space restricted to simple ionization operators.

Two types of CI calculation are presented. The relaxed CI calculations employ the neutral SCF orbitals in the calculation of the ground-state energy of the neutral, and the ion SCF orbitals are used for the ion ground-state energy. The unrelaxed CI calculations use a common set of orbitals for both the neutral and ion calculations. The orbitals are taken from a ground-state SCF calculation for whichever of the two is the closed-shell system. The unrelaxed CI calculations are similar in this respect to the EOM calculations, which utilize only one set of closed-shell SCF orbitals throughout. Comparison of the relaxed and unrelaxed CI calculations also affords a check of the CI convergence. A number of different relaxed and unrelaxed CI calculations are made for each system. The simplest involves

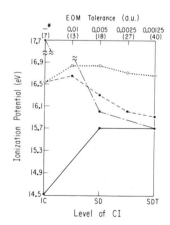

Fig. 6. EOM and CI vertical ionization potentials for HF: solid curve, relaxed CI; long and short dashes, unrelaxed CI, using SCF orbitals of HF; dashed curve, extensive EOM; dotted curve, primitive repartitioned EOM. Meaning of the x-axis is the same as in Fig. 5. Experimental value is 16.01 eV. Asterisk EOM primary operator space restricted to simple ionization operators.

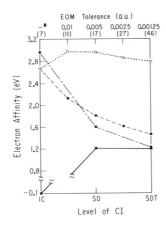

Fig. 7. EOM and CI values for the vertical electron affinity of OH: solid curve, relaxed CI; long and short dashes, unrelaxed CI, using SCF orbitals of OH⁻; dashed curve, extensive EOM; dotted curve, primitive repartitioned EOM. Meaning of the x-axis is the same as in Fig. 5. Experimental value is 1.83 eV. Asterisk: EOM primary operator space restricted to simple ionization operators.

a single configuration calculation (1C); one with all single and double excitations out of |HF⟩(SD) is also performed; and finally one with the singly, doubly, and triply excited configurations (SDT) is performed. Natural orbitals[1] are utilized, and to make the calculation tractable, the list of triple excitations is truncated[71, 107, 108] for HF and OH.

The results for the first $^2\Sigma^+$ IP of BH are displayed in Fig. 5, the HF calculations are given in Fig. 6, the EA of OH is shown in Fig. 7, the IP of Ne in Fig. 8, and the EAs of F in Fig. 9. In all cases the CI treatment appears to be very well converged. The relaxed and unrelaxed cases differ by no more than 0.03 eV in all the SDT cases. Furthermore, the effect of the triple excitations is never more than 0.07 eV for the relaxed calculations. Estimates[109] of the effect of quadruple excitations suggest that 0.1 to 0.2 eV

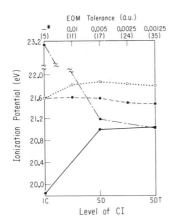

Fig. 8. EOM and CI values for the vertical ionization potential of Ne: solid curve, relaxed CI; long and short dashes, unrelaxed CI, using SCF orbitals of Ne; dashed curve, extensive EOM; dotted curve; primitive repartitioned EOM. Meaning of the x-axis is the same as in Fig. 5. Experimental value is 21.56 eV. Asterisk: EOM primary operator space restricted to simple ionization operators.

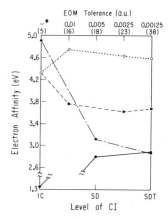

Fig. 9. EOM and CI vertical electron affinities for F: solid curve, relaxed CI; long and short dashes, unrelaxed CI, using SCF orbitals for F^-; dashed curve, extensive EOM; dotted curve, primitive repartitioned EOM. Meaning of the x-axis is the same as in Fig. 5. EOM calculations not performed at a tolerance of 0.05 au because this tolerance does not result in an appreciable increase in the dimensionality of the P-space. Experimental value is 3.339 eV. Asterisk: EOM primary operator space restricted to simple ionization operators.

shifts in the CI results might be expected because of the quadruples. Thus 0.1 to 0.2 eV is about the magnitude of the "error bars" on our best CI calculations.[71]

1. Ionization Potential of BH

The EOM results for BH are well converged even when the primary operator space is restricted to simple ionization basis operators (the dimension of A_{PP} is 15), yielding an IP that is 0.2 eV above the best CI case (SDT, relaxed). Since there are no shake-up basis operators in the P-space, this calculation does not contain any effects due to the 5-block basis opeators or any off-diagonal $A^{(3,3)}$ couplings. (Only the diagonal terms in A_{QQ} through first order are retained.)

As shake-up operators are added to the EOM P-space, the agreement with the CI calculation improves further. With a selection criterion tolerance of 0.0025 au, corresponding to 15 shake-up basis operators in the P-space, the discrepancy between the EOM and CI results is only 0.05 eV. The inclusion of the 5-block opeators in the Q space and the second-order $A^{(3,3)}$ terms within A_{PP} has very little effect on the EOM result. The difference between what we call the extensive EOM method, which includes these terms, and the primitive repartitioned EOM method, which neglects these terms, is on the order of 0.01 eV for each EOM tolerance.

It is interesting to note that the Koopmàns's theorem prediction for BH (i.e., the 1C, unrelaxed CI result) is also very accurate for this basis set. This indicates that the effects of orbital relaxation and the changes in correlation approximately cancel for this example. This cancellation may be at least partially responsible for the excellent accuracy of the EOM ionization potential for BH for all systematic approximation schemes.

2. Ionization Potential of HF and Electron Affinity of OH

The calculations of the EA of OH and the IP of HF bear strong similarities to each other. This is not surprising, since electron removal from HF and OH $^-$ are isoelectronic processes. When the EOM P-space for HF is restricted to the 1-block, the IP is about 0.8 eV above the best CI results. If the P-space is enlarged, the tendency of the primitive repartitioned EOM (without the 5-block and second-order 3-3 couplings) is to further increase the IP somewhat, and the EOM calculation appears to be converging to a number about 1 eV above the CI value. The difference between the EOM and converged CI IPs for OH $^-$ is even larger, being about 1.5 eV when the 1-block comprises the P-space. The inclusion of the additional second-order terms due to the 5-block operators and the $A^{(3,3)}$ elements in A_{PP} produces a dramatic improvement in the EOM results for both HF and OH. At the lowest tolerance utilized, 0.00125 au, the EOM IP for HF differs from the CI value by only 0.21 eV. The best EOM and CI electron affinities of OH differ by 0.27 eV. If the extensive EOM results are extrapolated to a tolerance of 0.0 au (corresponding to a P space of all 1-block and 3-block operators, and a Q space of all 5-block basis operators), the EOM results agree with the best CI values to about 0.08 eV for the IP of HF and 0.15 eV for the EA of OH.[71] This is certainly within the accuracy of the CI calculations, indicating that these EOM calculations are highly converged for this type of approximation. If the contributions of each of the additional terms are considered separately, it is found that the second-order $A^{(3,3)}$ elements lower the IP of HF by 0.20 eV and the EA of OH by 0.26 eV. The inclusion of the 5-block in the Q space accounts for the remaining improvement in the EOM results (0.55 eV for HF and 1.07 eV for OH).

3. Ionization Potential of Ne and Electron Affinity of F

The situation for the IP of Ne and the EA of F is somewhat different, however. It appears that the extensive EOM, though a definite improvement over the primitive repartitioned EOM, is not converging to the same limit as the CI calculations. At the lowest selection tolerance of 0.00125 au, the extensive EOM results are 0.43 eV above our best CI calculation for the ionization potential of Ne and 0.79 eV above the SDT relaxed CI EA for F. These extensive EOM treatments produce a 0.33 eV lowering of the IP of Ne and a 0.91 eV lowering of the EA of F over the primitive repartitioned EOM values at the same tolerances. In both cases, this lowering is due almost entirely to the inclusion of the 5-block basis operators in the Q-space. The inclusion of the second-order $A^{(3,3)}$ elements in A_{PP} results in only about a 0.01 eV shift in both cases.

4. Analysis of Calculations

Why is the agreement between the EOM and CI results worse for the atomic processes of the ionization of Ne and the electron attachment to F than for the related processes in the diatomics, HF and OH, respectively? One case is just the united atom limit of the other. The atomic systems, of course, differ from the isoelectronic diatomics by their higher symmetry and a concomitant increased local electron density. In HF (OH), an electron is being removed from (added to) a doubly degenerate π orbital. In Ne (F), the electron is being removed from (added to) the triply degenerate p orbitals. In BH, with its singly degenerate σ orbital, the 5-block of basis operators has very little effect. The results are apparently converged on inclusion of the 1-block (simple ionization) and the 3-block (shake-up) basis operators. In HF and OH with their π systems, the 5-block of basis operators has a significant effect, and when these operators are included in the space of basis operators, the EOM method yields results in good agreement with the CI method. It may be that the 7-block is necessary to describe the high degree of intrashell reorganization and to obtain accurate IPs and EAs involving triply degenerate p orbitals. Among the other higher order matrix elements that are ignored in the present calculations are second-order $A^{(5,1)}$, $A^{(5,3)}$, and $A^{(5,5)}$ matrix elements. Since the lowest order effects of the 5-block do play a significant role in these systems, these higher order contributions from the 5-block basis operators may also be important. However, there seems to be no obvious reason, apart from their role in describing a higher degree of intrashell reorganization, for these terms to make important contributions to the atomic systems but not to the related diatomic systems. (Notice that the effects due to the first-order $A^{(5,3)}$ and diagonal $A^{(5,5)}$ matrix elements are actually greater in the HF and OH systems than in Ne and F.)

An alternative interpretation of the data is that the discrepancy in the Ne calculations between the best CI and best EOM values is not all that bad, 0.43 eV, and the quadruple excitations may have some nonnegligible effect on the CI values. Adopting this viewpoint, the agreement between CI and EOM is poor only for F, and F is generally admitted to be a difficult system on which to obtain reliable calculations (although in this case we have the advantage that errors due to an incomplete orbital basis set are not a factor).

Shifted ω-dependent denominators are employed in all the foregoing EOM calculations. This has been the common choice in EOM IP–EA calculations. Comparison between the shifted and unshifted ω-dependent denominator EOM methods for the largest P-space employed in each example shows that the unshifted calculations provide IPs that are 0.19, 0.18, and 0.10 eV lower for BH, HF, and Ne, respectively. The corresponding reductions in the EOM EAs of OH and F are somewhat larger (0.46 and 0.36 eV, respectively). In every case except BH, the unshifted results provide closer agreement with the CI results. However, when operators from the 5-block are added to the P-space in F, the shifted results stay almost constant. The unshifted results monotonically increase as 5-block operators are included in the P-space until the difference between the shifted and unshifted results is 0.09 eV when eighteen 5-block operators are included in the P-space.[71] These calcultions indicate that the shifted ω-dependent denominators method provides a better approximation for the expansion of the inverse matrix in the partitioned EOM equation for IPs and EAs. Comparison of shifted and unshifted ω-dependent denominator calculations as a function of the number of shake-up basis operators in the P-space for these five systems also indicates the general superiority of the shifted method.[71]

5. Comparisons with Experiment

The experimental values for the ionization potentials of BH, HF, and Ne are 9.77, 16.01, and 21.56 eV, respectively. The experimental electron affinities of OH and F are 1.83 and 3.339 eV, respectively. For each system, the agreement between the most converged EOM results and experiment is closer than the agreement between the CI and experiment. However, this comparison with experiment is partially misleading. In the cases of HF and OH, for example, as the EOM results are systematically improved (by reducing the tolerance for the operator selection criterion), they tend toward the CI values and away from the experimental ones. This illustrates the well-known fallacy in estimating the accuracy of theoretical method through comparison with experiment of approximate results obtained with a particular finite orbital basis set. This also demonstrates the need for studies such as this one to provide a reliable understanding of the accuracy of the EOM–Green's function methods.

F. Consideration of a Number of Lingering Difficulties

Notwithstanding the preceding description of converged EOM calculations that agree, as required, with CI ones, a number of questions concerning the EOM method remain to be resolved. Problems involving the addition of still more basis operators or different types of ground-state wavefunctions, become very expensive and require extensive additional computer programming when they are attacked within the framework of the large calculations in Section III.E. It is, therefore, useful to study these questions with rather small basis set problems, where the exact solutions within the basis is obtainable. Hence, we now consider calculations of the lowest ^1S excitation energy of He and Be within very small orbital basis sets ($3s$ for He and $2s1p$ for Be), where complete EOM and CI calculations can be easily performed. Approximate calculations can then be utilized to allow identification of the ramifications of a given approximation for the simple example studied. It is hoped that these conclusions can be extrapolated to larger scale calculations for larger systems.

1. Unorthodox Basis Operators

The first of the three topics discussed in this section concerns the role of operators that are not present in the standard EOM operator basis. These are operators that involve moving an electron from one hole orbital to another, $a_\mu^\dagger a_\nu$ ($\mu \neq \nu$), or from one particle orbital to another, $a_m^\dagger a_n$ ($m \neq n$). As discussed in Section II.C, the standard EOM basis can be shown to be complete for the calculation of IPs and EAs. Although it is not necessarily complete in the excitation energy variant of the theory, it has the correct number of operators for a complete operator basis for any finite set of orbitals. However, the choice of a complete operator basis is not unique, and the calculated excitation operator can depend on the choice. The only requirement is that the basis have the correct number of operators and the overlap matrix $\mathbf{D}$ in the EOM equation be nonsingular (see Section II.B). Thus it may be possible to improve a truncated operator basis by including in it these hole-hole moving or particle-particle electron-shifting operators. It might even be simpler in some cases to add these unorthodox operators to the basis than to include the higher excitation and de-excitation operators of the standard basis, because the hole-particle rank of the unorthodox operators is lower and the resultant matrix elements are simpler.

Various operator basis sets used in the calculations are described in Table VII. Table VIII shows that very accurate results emerge for He when the complete standard EOM basis is employed and when first-order perturbation theory, based on a one-configuration reference (zeroth-order) state, is used to calculate an approximate ground-state wave function. The calculations differ by only 0.06 eV from the exact EOM values. The overlap between the approximate ground state, $|\tilde{0}\rangle$, and the exact one within this orbital basis, $|0\rangle$, differs from unity by only 0.002%. Second-order perturbation theory for $|\tilde{0}\rangle$ produces almost exact agreement with the exact EOM calculation. Operator basis set II is obtained by removing the double excitation operators from the complete standard EOM basis. The calculated excitation energy is in error by 0.82 eV for basis set II. Clearly this basis set of single excitation and single de-excitation operators is unable to provide an accurate description of the excitation under consideration, even though that excitation is basically a one-electron $1s^2 \rightarrow 1s2s$ transition (a $1s \rightarrow 2s$ excitation). Adding the particle$\rightarrow$particle operator for the $2s \rightarrow 3s$

TABLE VII

Basis Operators Used for EOM Calculations on Be and He[a,b]

Be

Set I[c] $O^\dagger_{2s^2 \to 2p^2}, O^\dagger_{2p^2 \to 2s^2}, O^\dagger_{1s2s \to 2p^2}, O^\dagger_{2p^2 \to 1s2s},$
$O^\dagger_{1s^2 \to 2p^2}, O^\dagger_{2p^2 \to 1s^2}, O^\dagger_{1s^2 2s^2 \to 2p^4}, O^\dagger_{2p^4 \to 1s^2 2s^2}$

He

Set I[c] $O^\dagger_{1s \to 2s}, O^\dagger_{2s \to 1s}, O^\dagger_{1s \to 3s}, O^\dagger_{3s \to 1s}, O^\dagger_{1s^2 \to 2s^2},$
$O^\dagger_{2s^2 \to 1s^2}, O^\dagger_{1s^2 \to 2s3s}, O^\dagger_{2s3s \to 1s^2}, O^\dagger_{1s^2 \to 3s^2}, O^\dagger_{3s^2 \to 1s^2}$

Set II $O^\dagger_{1s \to 2s}, O^\dagger_{2s \to 1s}, O^\dagger_{1s \to 3s}, O^\dagger_{3s \to 1s}$

Set III $O^\dagger_{1s \to 2s}, O^\dagger_{2s \to 1s}, O^\dagger_{1s \to 3s}, O^\dagger_{3s \to 1s}, O^\dagger_{2s \to 3s}, O^\dagger_{3s \to 2s}$

Set IV $O^\dagger_{1s \to 2s}, O^\dagger_{2s \to 1s}, O^\dagger_{1s \to 3s}, O^\dagger_{3s \to 1s}, O^\dagger_{1s^2 \to 2s^2}, O^\dagger_{2s^2 \to 1s^2}$

Set V $O^\dagger_{1s \to 2s}, O^\dagger_{2s \to 1s}, O^\dagger_{1s \to 3s}, O^\dagger_{3s \to 1s}, O^\dagger_{2s \to 3s}, O^\dagger_{3s \to 2s},$
$O^\dagger_{1s^2 \to 2s^2}, O^\dagger_{2s^2 \to 1s^2}$

[a] The operators are space- and spin-symmetry adapted.

[b] The operators represented in this table as $O^\dagger_{A \to B}$ are operators that "move" electrons from the orbitals A to the orbitals B. For example, $O^\dagger_{1s^2 \to 2s^2} = a^\dagger_{2s\alpha} a^\dagger_{2s\beta} a_{1s\beta} a_{1s\alpha}$, where α and β are spin indices.

[c] The complete set of standard EOM operators.

transition and its adjoint to basis II (giving basis set III), reduces the error in the excitation energy substantially, to 0.33 eV. However, the addition of the standard $1s^2 \to 2s^2$ double excitation operator and its adjoint to basis set II (giving basis set IV) is still more effective in reducing the error. The error due to basis set IV is only 0.05 eV, which is just about as good as that obtained using the larger set (set V) formed by the union of sets III and IV

TABLE VIII

Results of EOM Calculations on ^1S He

Number of configurations in zeroth-order ground state	Order of perturbation expansion for ground-state wave function	$\langle 0|\tilde{0}\rangle$	Operator basis	Lowest ^1S excitation energy (eV)
—	—	1.0	Set I	38.6280
1	0	0.99775	Set I	37.8379
1	1	0.99998	Set I	38.5683
1	2	$1.0 - (1.5 \times 10^{-7})$	Set I	38.6223
1	1	0.99998	Set II	39.4486
1	1	0.99998	Set III	38.9628
1	1	0.99998	Set IV	38.6731
1	1	0.99998	Set V	38.5886

(error of 0.04 eV). Thus, in this simple example, the addition of the particle -particle operators to the incomplete basis produces a sizable improvement in the calculated excitation energy. However, this improvement is not as great as that obtained when equal numbers of standard double excitation and double de-excitation operators are added.

2. Employing a Multiconfigurational Zeroth-Order Ground State

The second problem centers about the use of an approximate ground-state wave function that eminates from a multiconfigurational zeroth-order approximation. The N_2 calculations in Section III.C suggest that the restriction to a single configuration zeroth-order ground state imposes a fundamental limitation on the quality of the calculated EOM ionization potentials for that system.

The Be data in Table IX clearly demonstrate the need for an accurate ground-state wave function in the EOM equation. A ground-state wave function, based on a single configuration reference state, does not appear to be accurate enough to give a good excitation energy for this example. Even with a second-order ground state, the excitation energy is in error by 0.38 eV. However, when a two-configuration reference state is employed, the error is reduced to less than 0.002 eV because of the strong mixing of the $1s^2 2p^2$ (^1S) configuration into the ground state. This reduced error is one reason for choosing the Be system for study, and it reinforces the conclusions indicated by the earlier N_2 calculations with respect to the de-

TABLE IX
Effect of Solving for New Ground-State Wave Function

System	$|0\rangle$[a]	$\langle 0|\tilde{0}\rangle$	Lowest ^1S excitation (eV)	$\alpha_{0,0} =$ $\langle \tilde{0}|O_t^{\dagger}|\tilde{0}\rangle$	$\langle 0|\tilde{0}\rangle$ Iterated
Be	1C, first order	0.99912	11.4227	0.3357	0.99977
Be	1C, second order	0.99913	11.4189	0.3355	0.99978
Be	2C, zeroth order	$1.0-10^{-5}$	11.0355	0.3468	$1.0-(2.5\times10^{-7})$
Be	2C, first order	$1.0-(4\times10^{-8})$	11.0395	0.3468	$1.0-(4\times10^{-8})$[c]
Be	—[b]	1.0	11.0372	0.3465	1.0
He	1C, first order	0.999976	38.5683	0.0082	0.999984
He	1C, second order	$1.0-(1.5\times10^{-7})$	38.6223	0.0140	$1.9-10^{-7}$
He	—[b]	1.0	38.6280	0.0144	1.0

[a] nC, ith-order means that $|\tilde{0}\rangle$ is calculated by ith-order perturbation theory beginning with an n-configuration zeroth-order approximation.

[b] Exact EOM calculation within finite one-particle orbital basis.

[c] Calculation showed slight improvement in $|\tilde{0}\rangle$.

ficiency of using a single configuration reference wave function in EOM calculations on systems with near degeneracies that should be described by multiconfigurational zeroth-order reference ground-state wave functions.

Yeager and Jørgensen,[77] using a two-configuration Be MCSCF ground-state, wave function have reported EOM-like calculations. Their approach, which is called the multiconfigurational time-dependent Hartree-Fock (MCTDHF) method or the multiconfigurational random phase approximation (MCRPA), uses an expansion for $O_\lambda^\dagger$ that involves all one-body excitation operations for which $\langle 0|[O_i, O_i^\dagger]|0\rangle \neq 0$. In addition, the other multiconfigurational states orthogonal to the multiconfigurational reference state are included in the basis set they employ for the expansion for $O_\lambda^\dagger$. Using an orbital basis of 48 Slater functions, they calculate the lowest 16 excitation energies, generating results that deviate from experiment by a maximum of 0.31 eV, with an average deviation of 0.18 eV.

3. Iterative Procedure to Improve an Approximate Ground State

We now consider the feasibility of employing the condition (13), $O_\lambda|0\rangle = \alpha_{0,0}|0\rangle$, where $\alpha_{0,0} = \langle 0|O_\lambda^\dagger|0\rangle$, along with $O_\lambda^\dagger$ from the EOM equation to determine an improved ground state $|0\rangle$, and then, in turn, to solve the EOM equation for a new $O_\lambda^\dagger$, and so on, iteratively.

Table IX contains the overlaps between the exact ground-state wave function (within this orbital basis) and an iterated ground-state wave function obtained using the condition, (13), $O_\lambda|0\rangle = \alpha_{0,0}|0\rangle$, for the Be and He calculations. Table IX shows that although in every case the iterated ground state is an improvement over the initial one, the improvement is variable. Only for one case, Be with a two-configuration reference state, does the improvement exceed an order in magnitude (and in this case the initial approximate ground-state wave function is very good). Another Be calculation shows negligible improvement in the ground state.

An indication of the cause of this rather slow convergence can be obtained by considering the simple two-level system. This system has the advantage that the normalized ground-state wave function depends on only one parameter. If we express the ground state as $|0\rangle = \cos\theta|\tilde{1}\rangle + \sin\theta|\tilde{2}\rangle$, where $|\tilde{1}\rangle$ and $|\tilde{2}\rangle$ are two independent basis configurations, then θ can be taken as this parameter. Arbitrarily choosing the Hamiltonian matrix within this configuration basis to be

$$\mathbf{H} = \begin{pmatrix} -0.5 & 0.1 \\ 0.1 & 0.3 \end{pmatrix}$$

yields an exact excitation energy of 0.82462 and an exact θ of -0.12249.

Using a trial ground-state wave function calculated by first-order perturbation theory ($\tilde{\theta} = -0.12435$), the results presented in Table X show that $\tilde{\theta} - \theta$ improves by only about a factor of 2 on each iteration, when the simple iteration scheme, employed in the Be and He examples above, is utilized.

Since $O_\lambda^\dagger$ is calculated from the EOM equation, it is a complicated function of θ, that is, the approximate ground state $|\tilde{0}\rangle$ employed in the solution of the EOM equation. When $O_\lambda |\tilde{0}\rangle = \alpha_{0,0} |\tilde{0}\rangle$ is solved to produce a new $|\tilde{0}\rangle$, this dependence of $O_\lambda^\dagger$ on θ is not accounted for in the foregoing iteration scheme. The situation can be improved as follows.[72] For a two-level system, a complete set of basis operators consists of $2(N-1) = 2(2-1) = 2$ operators (see Sections II.A and II.B). Call these $O_1^\dagger$ and $O_2^\dagger$; then $O_\lambda^\dagger = c_1 O_1^\dagger + c_2 O_2^\dagger$. The θ dependence of c_1 and c_2 can be treated in a linear approximation as

$$c_i = c_i^0 + \frac{dc_i}{d\theta} \Delta\theta, \quad i = 1, 2$$

with $\Delta\theta = \theta - \theta_0$; θ_0 and c_i^0 are the values of θ and c_i for the present iteration, and $dc/d\theta$ can be approximated as the ratio of the changes of c_i and θ on the previous iteration ($dc/d\theta$ is set equal to zero on the initial iteration). In this manner, the θ dependence of $O_\lambda^\dagger$ is crudely accounted for

TABLE X

Results of Different Methods for Iteratively Improving the
Ground-State Wave Function for the Two-Level System[a]

Iteration	Method A[b]		Method B[c]	
	$\tilde{\theta}$	ΔE	$\tilde{\theta}$	ΔE
0[d]	−0.12435	0.82500	−0.12435	0.82500
1	−0.12342	0.82481	−0.12342	0.82481
2	−0.12295	0.82472	−0.12250	0.82462
3	−0.12272	0.82467		
4	−0.12260	0.82464		
5	−0.12255	0.82463		
6	−0.12252	0.82463		
7	−0.12250	0.82462		

[a] Exact value of θ is -0.12249 and exact value of excitation energy ΔE is 0.82462.
[b] Iterative method ignores the dependence of $O_\lambda^\dagger$ on $|\tilde{0}\rangle$.
[c] Iterative method uses estimate of $dO_\lambda^\dagger/d\theta$ to account for approximate dependence of $O_\lambda^\dagger$ on $|\tilde{0}\rangle$.
[d] Initial value of θ chosen by perturbation theory.

when the equation $O_\lambda^\dagger|0\rangle = \alpha_{0,0}|0\rangle$ is solved for $\Delta\theta$ (i.e., a new θ and $|0\rangle$). When $dc/d\theta$ is set to zero, the previous scheme, which ignores any θ dependence of $O_\lambda^\dagger$, is recovered. When the EOM equation and the condition on $O_\lambda|\tilde{0}\rangle$ are iterated in this fashion, the results are converged in the second iteration, which is the first iteration for which an estimate of $dc_i/d\theta$ is available (see Table X). Unfortunately, extension of this procedure to full-scale calculations is very difficult, since the ground state is a function of a large number of parameters. To make this iterative approach tractable, therefore, requires an efficient means of estimating the matrix whose elements are $dc_i/d\theta_j$, where the c_i are the expansion coefficient of $O_\lambda^\dagger$ and θ_j are the parameters on which $|\tilde{0}\rangle$ depends.

IV. CONCLUSIONS

Most of the numerical data presented in Section III involve the single determinant SCF approximation for the lowest order description of the ground-state wave function. However, analysis of the EOM ionization potentials computed for N_2 provides evidence that a single determinant, zeroth-order ground state may not be sufficient for accurate EOM calculations. This is the result of the pseudo–open shell nature of the ground state due to the low-lying $1\pi_g$ orbital. A ground-state wave function, based on a multiconfigurational zeroth-order approximation that is then perturbatively corrected, should provide a much more reliable and flexible theory. If the zeroth-order wave function contains all dominant configurations, first order perturbation theory will likely provide an accurate treatment. Excitation energy calculations on Be support this outlook.

Work in this general direction by Yeager and Jørgensen[77] has been presented recently for excitation energies. Their present work is limited to an RPA level with a multiconfigurational ground state. Simons and coworkers[78, 79] have described a propagator theory that employes a multiconfigurational SCF ground-state wave function. However, Simons's work is based on an unusual and untested definition of orders of perturbation theory.

Our formal and numerical analysis of the EOM–Green's function methods has indicated a number of generalizations of these methods that are necessary for an accurate description of the electronic processes that accompany excitation, ionization, and electron attachment. Although the bulk of the numerical work has centered on the IP–EA variant of the EOM theory, similar behavior should be expected in the excitation energy version of the theory. These generalizations necessarily introduce complications in the implementation of the EOM approach.

The accurate description of the electronic structure of atoms and molecules is a fundamentally difficult problem, and it is not surprising that there are no "quick and dirty" means of obtaining accurate answers. In our discussion we have attempted to relate these many-body methods to the traditional wave function techniques of bound-state quantum mechanics. We have indicated some similarities between these approaches, and at the same time have demonstrated how the EOM–Green's function theories have features not found in the conventional methods, leading to new and useful insights.

We have demonstrated that within a finite orbital basis, the EOM–Green's function and conventional wave function theories yield the same values for ionization potentials, electron affinities, and excitation energies. This important theorem enables the careful study of approximations to the complete solution of the EOM equations through comparison with accurate results obtained via the more fully understood wave function techniques.

The EOM–Green's function theories differ fundamentally from conventional methods in that they are based on a Liouville operator formalism, whereas the wave function theories deal with the Hamiltonian operator. We have discussed in detail what constitutes a complete basis set for these many-body methods and have proved that they do not suffer from the N^2 problem encountered by more naive Liouville operator formalisms.

These many-body theories utilize an altogether different operator basis, the many-body basis. These basis operators account for correlation in an approximate way, since they act on the correlation part of the ground state as well as the SCF term. Hence, the many-body basis operators have interesting physical interpretations as primitive ionization or excitation operators. In addition to the excitation operators, the complete many-body basis set for excitation energies includes primitive de-excitation operators, which have no analogs in traditional configuration interaction theory. The many-body basis for ionization processes includes operators that remove electrons from particle orbitals. These operators are also without simple counterparts in CI theory. The various terms in the expression for photo-ionization cross sections have been analyzed in light of the physical content of the many-body basis set.

Based on the present formal and numerical investigations, it is possible to draw some conclusions about what is necessary for a systematic, accurate, and reliable EOM theory. The ground-state wave function should contain in zeroth order all configurations that are necessary to account for near-degeneracy effects, as well as any other physically important valence configurations. This zeroth-order wave function can then be perturbatively corrected. Our numerical data indicate that a flexible procedure for the selection of the primary operator space has important advantages over a strictly perturbative approach based on some fixed choice of the P-space. This primary operator space must at least contain operators from the 1- and 3-blocks of basis operators for ionization potentials. It seems likely that a P-space containing parts of the 2- and 4-blocks will be required for excitation energy calculations. It is necessary to treat all couplings between important operators accurately, and all operators that couple directly to the primary operator space in low orders must be included in the secondary space. The formal and programming difficulties to implement such a theory are great, but the recent multiconfigurational approaches of Yeager and Jørgensen[77] and of Simons and co-workers[78, 79] represent a beginning.

The relative calculational efficiency of EOM–Green's function methods and conventional configuration interaction methods is a difficult matter to assess, since it is intimately bound to the question of optimization of computer codes. Our major emphasis has been on determining the requirements for an accurate and reliable EOM theory. Of necessity, the program optimization has to an extent taken a back seat to the constant changes introduced in the theory in the course of this work. However, the demonstrated ability to obtain accurate results for the simple ionization potentials of small molecules with very small primary operator spaces bodes well for the EOM method.

Hence a clear and theoretically sound picture of the EOM–Green's function methods is emerging. More numerical work is still needed to resolve some remaining questions, but our increased understanding of these theories should render them useful and important complements to the traditional wave function approaches.

APPENDIX

Here we describe briefly the evaluation of the peak intensities for the calculated X-ray photoelectron spectra presented in Section III.D. Details of the calculations can be found in Ref. 69. The state of the outgoing electron is approximated by a plane wave, $|\mathbf{k}_0\uparrow\rangle$, which is orthogonalized to the SCF molecular orbitals,

$$|\mathbf{k}_0\uparrow\rangle = a^\dagger_{\mathbf{k}_0\uparrow}|-\rangle = N_0\left\{|\mathbf{k}\uparrow\rangle - \sum_{\phi_j \in \{MO\}} |\phi_j\uparrow\rangle\langle\phi_j\uparrow|\mathbf{k}\uparrow\rangle\right\} \qquad \text{(A.1)}$$

where $a^\dagger_{\mathbf{k}_0\uparrow}$ is the second quantized creation operator associate with $|\mathbf{k}_0\uparrow\rangle$, $|-\rangle$ is the vacuum state, $|\mathbf{k}\uparrow\rangle$ is the pure plane wave $L^{-3/2}\exp\{-i\mathbf{k}\cdot\mathbf{r}\}$

multiplied by the spin function $|\uparrow\rangle$. Box normalization is employed with box length L, and the summation in (A.1) is over all SCF molecular orbitals. The magnitude of the wave vector $\mathbf{k}$ is fixed by energy conservation for a given incident photon energy $\hbar\omega$. Within this plane wave approximation, the final electronic state of the ionized molecule plus ejected electron is described by

$$|\lambda\rangle = (2)^{-1/2}\left[a_{\mathbf{k}_0\downarrow}^\dagger O_{\lambda,\uparrow}^\dagger|0\rangle + a_{\mathbf{k}_0\uparrow}^\dagger O_{\lambda,\downarrow}^\dagger|0\rangle \right] \qquad (A.2)$$

In (A.2), $|0\rangle$ is the ground-state wave function of the molecule prior to ionization, and $O_\lambda^\dagger$ is the ionization operator obtained from the EOM calculation using (37); $O_\lambda^\dagger$ is approximated by the part of its expansion in the primary operator space.

Fermi's golden rule for the differential ionization cross section[104] states:

$$\frac{d\sigma}{d\Omega} = \frac{\pi e^2}{\mu^2\omega c}|\hat{\mathbf{u}}\cdot\langle 0|\sum_n \mathbf{p}_n|\lambda\rangle|^2\rho(E) \qquad (A.3)$$

where $\hat{\mathbf{u}}$ is the unit polarization vector for the incident photon, μ and e are the electronic mass and charge, respectively, and c is the speed of light; $\rho(E)$ is the density of continuum states for the ejected electron. In the plane wave approximation we have

$$\rho(E) = \frac{\mu k L^3}{2\pi^2\hbar} \qquad (A.4)$$

In (A.3) the velocity form of the dipole approximation is used. The factor of L^3 in $\rho(E)$ cancels with the normalization for the plane wave, thus providing the correct continuum limit ($L\to\infty$). If it is assumed that $|0\rangle$ is a closed-shell state, the two terms on the right-hand side of (A.2) yield identical results in (A.3). Therefore, we simplify the notation by combining the two terms and suppress the spin designations. The electronic momentum operator for our system, expressed in second quantized notation, is given by

$$\sum_n \hat{\mathbf{p}}_n = -i\hbar \oint_j \oint_{j'} \langle j|\nabla|j'\rangle a_j^\dagger a_j \qquad (A.5)$$

where $\oint$ includes a summation over all the SCF molecular orbitals and over all orthogonalized plane wave states (or integration over all $|\mathbf{k}\rangle$ in the continuum limit). Using the anticommutation relationship for the second quantized operators and the property that $\langle 0|a_{\mathbf{k}_0}^\dagger = 0$, the vector $\mathbf{P}_{0\lambda} = \langle 0|\Sigma_n \mathbf{p}_n|\lambda\rangle$ is obtained[69] as:

$$\mathbf{P}_{0\lambda} = -(2)^{1/2} i\hbar \sum_l \langle\phi_l|\nabla|\mathbf{k}_0\rangle\langle 0|a_l^\dagger a_{\mathbf{k}_0} a_{\mathbf{k}_0}^\dagger O_\lambda^\dagger|0\rangle$$

$$= -(2)^{1/2} i\hbar \sum_l \langle\phi_l|\nabla|\mathbf{k}_0\rangle\langle 0|a_l^\dagger O_\lambda^\dagger|0\rangle \qquad (A.6)$$

Equation A.6 results because only the terms in (A.5) that survive when (A.5) is substituted into $\langle 0|\Sigma_n p_n|\lambda\rangle$ are those for which $j' = \mathbf{k}_0$ and $j = l$ (where l designates any SCF molecular orbital). Substituting for $|\mathbf{k}_0\rangle$ from (A.1) yields

$$\mathbf{P}_{0\lambda} = -(2)^{1/2} N_0 i\hbar \left\{ \sum_l \langle\phi_l|\nabla|\mathbf{k}\rangle\langle 0|a_l^\dagger O_\lambda^\dagger|0\rangle \right.$$

$$\left. - \langle\phi_l|\nabla|\phi_m\rangle\langle\phi_m|\mathbf{k}\rangle\langle 0|a_l^\dagger O_\lambda^\dagger|0\rangle \right\} \qquad (A.7)$$

Since $\langle\phi_l|\nabla|\mathbf{k}\rangle = -i\mathbf{k}\langle\phi_l|\mathbf{k}\rangle$, the first term in (A.7) is expected to dominate for large k.[110] Because the present calculations are of X-ray photoelectron spectra, we retain only the first summation on the right-hand side of (A.7). There (A.3) gives

$$\frac{d\sigma(\lambda)}{d\Omega} = \frac{e^2 k L^3}{\mu\pi\omega c} |\hat{\mathbf{u}}\cdot\mathbf{k}|^2 \sum_{l,\,m} \langle\phi_l|\mathbf{k}\rangle^* \langle\phi_m|\mathbf{k}\rangle B_l^*(\lambda) B_m(\lambda) \qquad (A.8)$$

where the amplitudes $B_l(\lambda)$ are

$$B_l(\lambda) = \langle 0|a_l^\dagger O_\lambda^\dagger|0\rangle \qquad (A.9)$$

The factors $|\hat{\mathbf{u}}\cdot\mathbf{k}|^2\langle\phi_l|k\rangle^* \langle\phi_m|\mathbf{k}\rangle$ can be averaged over all possible molecular orientations and over all polarizations of the incident photon,[105] to yield (45) for the averaged total cross section. The term $O_\lambda^\dagger$ is approximated by

$$O_\lambda^\dagger = \sum_\mu C_\mu^\lambda a_\mu + \sum_m C_m^\lambda a_m + \sum_{\substack{\mu < \nu \\ m}} C_{\mu m\nu}^\lambda a_\mu a_m^\dagger a_\nu$$

$$+ \sum_{\substack{m < n \\ \mu}} C_{m\mu n}^\lambda a_m a_\mu^\dagger a_n \qquad (A.10)$$

and $|0\rangle$ by (34). Since only the basis operators from the P-space are included in the intensity calculations, it is reasonable to treat all the expansion coefficients in (A.10) (i.e., C_μ^λ, C_m^λ, $C_{\mu m\nu}^\lambda$, and $C_{m\mu n}^\lambda$) as zeroth order. Because (34) is a perturbation expansion for $|0\rangle$, inserting (34) and (A.10) into (A.9) yields a perturbation expansion for (45).

In the actual calculations, spin-symmetry-adapted operators[66, 67] $\Gamma_{\mu m\nu}^\dagger$ and $\Omega_{m\mu n}^\dagger$ are employed in (A.10), rather than the simple operators $a_\mu a_m^\dagger a_\nu$ and $a_m a_\mu^\dagger a_n$, respectively. The exact form of the various terms in $B_i(\lambda)$ is discussed in detail in Ref. 69.

Acknowledgments

We wish to thank Dr. R. S. Berry, Dr. J. C. Light and Dr. R. E. Stanton for many helpful comments on this manuscript. This research is supported, in part, by National Science Foundation (NSF) Grant CHE 80-23456. We are grateful to Dr. Bowen Liu for carrying out the CI calculations in collaboration with our EOM studies. One of us (MFH) is grateful for support provided by NSF graduate, McCormick, and James Franck Fellowships during the course of this work.

References

1. H. F. Schaefer, III, *The Electronic Structure of Atoms and Molecules, A Survey of Rigorous Quantum Mechanical Results*, Addison-Wesley, Reading, MA, 1972.
2. N. H. March, W. H. Young, and S. Sampanthar, *The Many-Body Problem in Quantum Mechanics*, Cambridge University Press, Cambridge, 1967.
3. J. Linderberg and Y. Ohrn, *Propagators in Quantum Chemistry*, Academic Press, New York, 1973.
4. A. L. Fetter and J. D. Walecka, *Quantum Theory of Many-Particle Systems*, McGraw-Hill, New York, 1973.
5. H. P. Kelly, *Adv. Chem. Phys.*, **14**, 129 (1969); *Adv. Theor. Phys.*, **2**, 75 (1968), and references therein.
6. K. F. Freed, *Phys. Rev.*, **173**, 1 (1968); *Chem. Phys. Lett.*, **4**, 496 (1970); *Annu. Rev. Phys. Chem.*, **22**, 313 (1971).
7. O. Sinanoğlu, *Adv. Chem. Phys.*, **6**, 315 (1964); **14**, 239 (1969).
8. R. K. Nesbet, *Adv. Chem. Phys.*, **9**, 321 (1965).
9. G. Csanak, H. S. Taylor, and R. Yaris, *Adv. At. Mol. Phys.*, **7**, 287 (1971).
10. D. J. Rowe, *Rev. Mod. Phys.*, **40**, 153 (1968).
11. D. P. Chong, F. H. Herring, and D. McWilliams, *J. Chem. Phys.*, **61**, 78 (1974).
12. D. P. Chong and Y. Takahata, *Int. J. Quantum Chem.*, **12**, 549 (1977).
13. N. Ohmichi and T. Nakajima, *J. Chem. Phys.*, **67**, 2078 (1977).
14. K. F. Freed and D. L. Yeager, *Chem. Phys.*, **22**, 401 (1977).
15. K. F. Freed, M. F. Herman, and D. L. Yeager, *Proceedings of the Nobel Symposium No. 46, Physica Scripta*, **21**, 242 (1980).
16. L. S. Cederbaum and W. Domcke, *Adv. Chem. Phys.*, **36**, 205 (1977).
17. L. S. Cederbaum, G. Hohlneicher, and S. Peyerimhoff, *Chem. Phys. Lett.*, **11**, 421 (1971).
18. L. S. Cederbaum, G. Hohlneicher, and W. von Niessen, *Chem. Phys. Lett.*, **18**, 503 (1973); *Mol. Phys.*, **26**, 1405 (1973).
19. L. S. Cederbaum, *Theor. Chim. Acta*, **31**, 239 (1973); *J. Phys. B*, **8**, 290 (1975).

EQUATIONS OF MOTION—GREEN'S FUNCTION METHOD 67

20. W. von Niessen, L. S. Cederbaum, and W. P. Kraemer, *J. Chem. Phys.*, **65**, 1378 (1976).
21. W. von Niessen, G. H. F. Diercksen, and L. S. Cederbaum, *Chem. Phys.*, **10**, 345 (1975); *J. Chem. Phys.*, **67**, 4124 (1977).
22. W. Domcke, L. S. Cederbaum, W. von Niessen, and W. P. Kraemer, *Chem. Phys. Lett.*, **43**, 258 (1976).
23. L. S. Cederbaum, W. Domcke, and W. von Niessen, *J. Phys. B*, **10**, 2963 (1977).
24. J. Schirmer, L. S. Cederbaum, W. Domcke, and W. von Niessen, *Chem. Phys.*, **26**, 149 (1977).
25. L. S. Cederbaum, J. Schirmer, W. Domcke, and W. von Niessen, *J. Phys. B*, **10**, L549 (1977).
26. L. S. Cederbaum, W. Domcke, H. Köppel, and W. von Niessen, *Chem. Phys.*, **26**, 169 (1977).
27. J. Schirmer and L. S. Cederbaum, *J. Phys. B*, **11**, 1889 (1978).
28. J. Schirmer, W. Domcke, L. S. Cederbaum, and W. von Niessen, *J. Phys. B*, **11**, 1901 (1978).
29. L. S. Cederbaum, W. Domcke, J. Schirmer, W. von Niessen, G. H. F. Diercksen, and W. P. Kraemer, *J. Chem. Phys.*, **69**, 1591 (1978).
30. L. S. Cederbaum, J. Schirmer, W. Domcke, and W. von Niessen, *Int. J. Quantum Chem.*, **14**, 593 (1978).
31. L. S. Cederbaum and W. Domcke, *J. Chem. Phys.*, **60**, 2878 (1973); **64**, 603 (1976).
32. W. Domcke and L. S. Cederbaum, *J. Chem. Phys.*, **64**, 612 (1976).
33. G. Wendin, *J. Phys. B*, **5**, 110 (1972); **6**, 42 (1973).
34. M. Ohno and D. Wendin, *Physica Scripta*, **14**, 148 (1976); *J. Phys. B*, **11**, 1557 (1978).
35. G. Wendin, "Many-body theory of hole spectra in atoms, molecules and solids: The missing $4p$ ESCA line in Xe and related problems," preprint, 1979.
36. B. T. Pickup and O. Goscinski, *Mol. Phys.*, **26**, 1013 (1973).
37. J. Oddershede and P. Jørgensen, *J. Chem. Phys.*, **66**, 1541 (1977).
38. G. D. Purvis and Y. Öhrn, *J. Chem. Phys.*, **60**, 4063 (1974); **62**, 2045 (1975); **65**, 917 (1976).
39. C. Nehrkorn, G. D. Purvis, and Y. Öhrn, *J. Chem. Phys.*, **64**, 1752 (1976).
40. L. T. Redmon, G. D. Purvis, and Y. Öhrn, *J. Chem. Phys.*, **63**, 5011 (1975).
41. G. D. Purvis and Y. Öhrn, *Int. J. Quantum Chem. Symp.*, **11**, 359 (1977).
42. O. Goscinski and B. Lukman, *Chem. Phys. Lett.*, **7**, 573 (1970).
43. F. S. M. Tsui and K. F. Freed, *Chem. Phys.*, **5**, 337 (1974).
44. F. S. M. Tsui and K. F. Freed, *Chem. Phys. Lett.*, **32**, 345 (1975).
45. F. S. M. Tsui and K. F. Freed, *Chem. Phys.*, **14**, 27 (1976).
46. P. Jørgensen and J. Simons, *J. Chem. Phys.*, **63**, 5302 (1975).
47. J. Simons and P. Jørgensen, *J. Chem. Phys.*, **64**, 1413 (1976).
48. T. T. Chen, J. Simons, and K. D. Jordan, *Chem. Phys.*, **14**, 145 (1976).
49. T. H. Dunning and V. McKoy, *J. Chem. Phys.*, **47**, 1735 (1967); **48**, 5263 (1968).
50. T. Shibuya and V. McKoy, *Phys. Rev. A*, **2**, 2208 (1970); *J. Chem. Phys.*, **53**, 3308 (1970); **54**, 1738 (1971).
51. J. Rose, T. Shibuya, and V. McKoy, *J. Chem Phys.*, **58**, 74 (1973).
52. T. Shibuya, J. Rose, and V. McKoy, *J. Chem. Phys.*, **58**, 500 (1973).
53. D. L. Yeager, M. Nascimento, and V. McKoy, *Phys. Rev. A*, **11**, 1168 (1975).
54. D. L. Yeager and V. McKoy, *J. Chem. Phys.*, **63**, 4861 (1975).
55. C. W. McCurdy, T. N. Rescigno, D. L. Yeager, and V. McKoy, in H. F. Schaeffer, III, Ed., *Modern Theoretical Chemistry*, Vol. III, *Methods of Electronic Structure Theory*, Plenum Press, New York, 1977.
56. J. Simons and W. D. Smith, *J. Chem. Phys.*, **58**, 4899 (1973).

57. T. T. Chen, W. D. Smith, and J. Simons, *Chem. Phys. Lett.*, **26**, 296 (1974).
58. W. D. Smith, T. T. Chen, and J. Simons, *Chem. Phys. Lett.*, **27**, 499 (1974); *J. Chem. Phys.*, **61**, 2670 (1974).
59. J. Kenney and J. Simons, *J. Chem. Phys.*, **62**, 592 (1975).
60. K. M. Griffing and J. Simons, *J. Chem. Phys.*, **62**, 535 (1975); **64**, 3610 (1976).
61. K. M. Griffing, J. Kenney, J. Simons, and K. D. Jordan, *J. Chem. Phys.*, **63**, 4073 (1975).
62. K. D. Jordan, K. M. Griffing, J. Kenney, E. Anderson, and J. Simons, *J. Chem. Phys.*, **64**, 4730 (1976).
63. E. Anderson and J. Simons, *J. Chem. Phys.*, **64**, 4548 (1976).
64. K. D. Jordan and J. Simons, *J. Chem. Phys.*, **65**, 1601 (1976).
65. J. Simons, *Annu. Rev. Phys. Chem.*, **28**, 15 (1977).
66. D. L. Yeager, Ph.D. thesis, California Institute of Technology, 1975.
67. M. F. Herman, D. L. Yeager, K. F. Freed, and V. McKoy, *Chem. Phys. Lett.*, **46**, 1 (1977).
68. M. F. Herman, D. L. Yeager, and K. F. Freed, *Chem. Phys.*, **29**, 77 (1978).
69. M. F. Herman, K. F. Freed, and D. L. Yeager, *Chem. Phys.*, **32**, 437 (1978).
70. M. F. Herman, K. F. Freed, and D. L. Yeager, *J. Chem. Phys.*, **72**, 602 (1980).
71. M. F. Herman, K. F. Freed, D. L. Yeager, and B. Liu, *J. Chem. Phys.*, **72**, 611 (1980).
72. M. F. Herman and K. F. Freed, *Chem. Phys.*, **36**, 383 (1979).
73. For a recent review of configuration interaction methods, see I. Shavitt, H. F. Schaeffer; III, Ed., in *Modern Theoretical Chemistry*, Vol. III, *Methods of Electronic Structure Theory*, Plenum Press, New York, 1976.
74. Z. Gershogorn and I. Shavitt, *Int. J. Quantum Chem.*, **2**, 751 (1968).
75. G. A. Segal and R. W. Wetmore, *Chem. Phys. Lett.*, **32**, 556 (1975).
76. P. J. Fortune and B. J. Rosenberg, *Chem. Phys. Lett.*, **37**, 110 (1976).
77. D. L. Yeager and P. Jørgensen, *Chem. Phys. Lett.*, **65**, 77 (1979).
78. A. Banerjee, R. Shepard, and J. Simons, *Int. J. Quantum Chem. Symp.*, **12**, 389 (1978).
79. A. Banerjee, J. W. Kenney, and J. Simons, *Int. J. Quantum Chem.*, **16**, 1209 (1979).
80. A. C. Lasaga and M. Karplus, *Phys. Rev. A*, **16**, 807 (1977); *J. Chem. Phys.*, **71**, 1218 (1979).
81. E. Dalgaard and J. Simons, *J. Phys. B*, **10**, 2767 (1977).
82. R. Pariser and R. G. Parr, *J. Chem. Phys.*, **21**, 466 (1953); **21**, 767 (1953).
83. A Messiah, *Quantum Mechanics*, Vol. I, Wiley, New York, 1964, p. 436.
84. A. D. McLachlan and M. A. Ball, *Rev. Mod. Phys.*, **36**, 844 (1964).
85. T. Koopmans, *Physica*, **1**, 104 (1933).
86. Lasaga and Karplus[80] take quite a different point of view from the one presented here concerning the number of basis operators needed to solve the EOM equation and to iteratively improve an approximate ground state using the condition $O_\lambda^\dagger |0\rangle = 0$. They claim that N^2 basis operators are needed. The reason they come to this conclusion arises from their application of the annihilation condition $O_\lambda^\dagger |0\rangle = 0$. For this to hold, even for a exact $|0\rangle$, $\alpha_{0,0}$ must vanish. However, it is in general not possible to eliminate the $|0\rangle\langle 0|$ component from a linear combination of operators from an arbitrary basis without using all the N^2 bases operators present. However, it seems to be more sensible to take the $\alpha_{0,0}|0\rangle\langle 0|$ term in $O_\lambda^\dagger$ into account, as proposed here, than to attempt to solve a rank N^2 problems (actually it is rank N^2+N+1,[80] since the N correlation coefficients in $|0\rangle$ and ω are also treated as independent variables).
87. R. Manne, *Chem. Phys. Lett.*, **45**, 470 (1977).
88. E. Dalgaard, *Int. J. Quantum Chem.*, **15**, 169 (1979).
89. P. Albertsen and P. Jørgensen, *J. Chem. Phys.*, **70**, 3254 (1979).

90. P. Swanstrom and P. Jørgensen, *J. Chem. Phys.*, **71**, 4652 (1979).
91. D. L. Yeager and K. F. Freed, *Chem. Phys.*, **22**, 415 (1977).
92. R. K. Nesbet, *Proc. R. Soc. London, Ser. A*, **230**, 312 (1955); 322 (1955).
93. P. O. Lowdin, in C. H. Wilcox, Ed., *Perturbation Theory and Its Application in Quantum Mechanics*, Wiley, New York, 1966.
94. D. L. Yeager, W. McCurdy and V. McKoy, unpublished results.
95. G. D. Purvis and Y. Ohrn, *Chem. Phys. Lett.*, **33**, 396 (1975).
96. D. W. Turner, C. Baker, A. D. Baker, and C. R. Brundle, *Molecular Photoelectron Spectroscopy*, Wiley, New York, 1970.
97. J. Simons, private communication.
98. B. Liu, *J. Chem. Phys.*, **67**, 373 (1977).
99. R. K. Nesbet, *J. Chem. Phys.*, **40**, 3619 (1964).
100. P. E. Cade, K. D. Sales, and A. C. Wahl, *J. Chem. Phys.*, **44**, 1973 (1966).
101. R. S. Berry, *Annu. Rev. Phys. Chem.*, **20**, 357 (1969).
102. K. Siegbahn, C. Nordling, G. Johansson, J. Hedman, P. F. Heden, K. Hamrin, U. Gelius, T. Bergmark, L. O. Werne, and Y. Baer, *ESCA Applied to Free Molecules*, North-Holland, Amsterdam, 1969.
103. R. Nilsson, R. Nyholm, A. Berndtssson, J. Hedman, and C. Nordling, *J. Electron Spectrosc.*, **9**, 337 (1976).
104. G. Wentzel, *Z. Phys.*, **40**, 574 (1926); **41**, 828 (1927).
105. F. O. Ellison, *J. Chem. Phys.*, **61**, 507 (1974).
106. N. Kosugi, H. Kuroda, and S. Iwata, *Chem. Phys.*, **39**, 337 (1979).
107. B. Liu, *J. Chem. Phys.*, **58**, 1927 (1973).
108. P. Siegbahn and B. Liu, *J. Chem. Phys.*, **68**, 2457 (1978).
109. S. R. Langhoff and E. R. Davidson, *Int. J. Quantum Chem.*, **8**, 6 (1974).
110. R. L. Martin and D. A. Shirley, *J. Chem. Phys.*, **64**, 3685 (1976).

KINETIC THEORY OF CHEMICAL
REACTIONS IN LIQUIDS

RAYMOND KAPRAL

Department of Chemistry
University of Toronto
Toronto, Ontario, Canada

CONTENTS

I. INTRODUCTION

When a chemical reaction takes place in the condensed phase, the solvent must clearly play some role in determining the outcome of a reactive event. A complete microscopic theory for such reactions would necessarily require a description of the coupling between the dynamics of the reactive event, involving the solute molecules, and the dynamics of the solvent molecules—a very ambitious and difficult task. Some of this difficulty can be bypassed by adopting a more modest description. It is customary to assume that the approach of the solute molecules is governed by a diffusion equation and that the reactive event can be described by a boundary condition. In this way a treatment of the solvent dynamics is avoided; solvent properties enter only through the diffusion coefficient. A number of objections to this type of treatment can be raised. For example, how can a diffusion equation possibly be valid for the small internuclear separations between solute molecules that reaction entails? Here the molecular nature of the solvent must be especially evident to the solute molecules, since they may, for instance, have to displace a solvent molecule before they can react. Also, can we really expect a diffusion equation to be valid on the very short time scales that characterize the reactive event?

There is another, often easier, route to take. Transition state theory provides us with a method for calculating the reaction rate coefficient. Here, the dynamic problem is avoided and replaced by an equilibrium "one-way flux" calculation. Solvent effects enter through the free energy at the transition state. This theory is, of course, not generally valid. The reactive molecules are constantly buffeted by the solvent, and so a "molecule" that has just crossed the reaction surface may be forced because of solvent collisions to recross it rather than to form stable products. The actual rate will then be lower than the transition state theory prediction.

Nonetheless, these theories and others like them are applicable in certain circumstances, and their usefulness in correlating large amounts of experimental data cannot be overlooked. A theory of condensed-phase reactions must help us to understand why these theories work when they do and provide alternate descriptions when they do not. It is probably no surprise to find that a diffusion equation description of the coagulation of large colloidal particles in solution is adequate,[1] but should such an approach be at all applicable to the study of iodine recombination in carbon tetrachloride on picosecond time scales? It is difficult to imagine that the iodine atoms think they are moving in a continuum solvent in this case.

This chapter is intended to serve as a framework for the discussion of some of these questions. Thus we construct a kinetic theory that treats both the solute and solvent dynamics. We need to adopt such a detailed point of view if we are to attempt an answer to the questions posed above. Only the beginnings of a theory are presented, but we hope to provide some insight into how condensed-phase reactions might be described by a microscopic theory.

II. CHEMICAL RATE LAWS AND RATE COEFFICIENTS

Since a portion of this chapter is devoted to the derivation of rate laws and various microscopic expressions for the rate coefficients of condensed-phase chemical reactions, it is useful to first write down the phenomenological rate law we expect to obtain, to define the various rate coefficients and relaxation times, and to present the different points of view that we shall adopt in describing the system.

We confine our attention almost exclusively to the description of reactions among *dilute* solute species in a dense inert solvent. As an example, consider the bimolecular reaction

$$A + B \underset{k_r}{\overset{k_f}{\rightleftarrows}} C + B \qquad (2.1)$$

taking place in a solvent S. The irreversible version of this reaction scheme might, for instance, represent a fluorescence quenching reaction, where A is de-excited to C upon collision with quencher B, while the reversible case could represent an isomerization induced by collision with B or an excitation–de-excitation process, with A and C the two states of the molecule.

The macroscopic rate law describes the time evolution of the average densities of the reactive species. We let $\overline{n_\alpha(\mathbf{r}, t)}$ be the average density of species α at point $\mathbf{r}$ in the solution at time t. We initially consider reacting systems that are only slightly disturbed from complete equilibrium. The deviation of the average density from its equilibrium value n_{eq}^α is

$$\overline{\delta n_\alpha(\mathbf{r})} = \overline{n_\alpha(\mathbf{r})} - n_{eq}^\alpha \qquad (2.2)$$

The overbar does not refer to an equilibrium average; rather, it implies an average over spatial regions containing many molecules and times long compared to microscopic relaxation times. On a more formal level, we can consider it to be a nonequilibrium average where the density fields are constrained to have fixed values. The linearized macroscopic law then describes the evolution of these average fields. Taking into account the two possible mechanisms for their change, diffusion and reaction, the rate law is

$$\frac{d\delta n_A(\mathbf{r}, t)}{dt} = \left(D_A \nabla^2 - k_f n_{eq}^B\right)\overline{\delta n_A(\mathbf{r}, t)} + k_r n_{eq}^B \overline{\delta n_C(\mathbf{r}, t)} \qquad (2.3)$$

Here, D_A is the diffusion coefficient for the A species and we have dropped cross-diffusion effects. Similar equations can be written for the other species.

If we consider a spatially homogeneous closed system, the more familiar chemical rate law for the deviation of the average number of A molecules from its equilibrium value

$$\delta N_A(t) = \int_V d\mathbf{r} \overline{\delta n_A(\mathbf{r}, t)} \qquad (2.4)$$

is

$$\frac{d\delta N_A(t)}{dt} = -k_f n_{eq}^B \overline{\delta N_A(t)} + k_r n_{eq}^B \overline{\delta N_C(t)} \qquad (2.5)$$

The integration in (2.4) is over the volume of the system, V. The forward and reverse rate coefficients are related by the detailed balance condition,

$$\frac{k_f}{k_r} = K_{eq} \qquad (2.6)$$

where K_{eq} is the equilibrium constant. The rate coefficients are not obtained from one-way fluxes; rather, each coefficient contains information about microscopic forward and reverse processes.[2-4]

Introducing the progress variable $\xi(t)$ by

$$\xi(t) = \overline{\delta N_A(t)} = -\overline{\delta N_C(t)} \qquad (2.7)$$

we find that (2.5) takes the form

$$\frac{d\xi(t)}{dt} = -(k_f + k_r)n_{eq}^B \xi(t)$$

$$\equiv -\tau_{chem}^{-1}\xi(t) \qquad (2.8)$$

which also serves to define the chemical relaxation time, τ_{chem}.

A basic assumption, which is made when writing such equations, is that the chemical relaxation time is much longer than other characteristic times in the system, such as internal (vibrational, rotational) or translational relaxation times.[3] One might inquire about the generalization of the rate law when such a time-scale separation is not satisfied. From a theoretical point of view, a convenient generalization of (2.8) is[5, 6]

$$\frac{d\xi(t)}{dt} = -\int_0^t dt' \tau_{chem}^{-1}(t')\xi(t-t') \tag{2.9}$$

where the memory kernel

$$\tau_{chem}^{-1}(t) \equiv n_{eq}^B(k_f(t)+k_r(t)) \tag{2.10}$$

takes into account the finite response times of the internal or other degrees of freedom. A description of this type is also appropriate if the chemical relaxation time is determined by measuring the frequency response of the system, as, for example, in light-scattering experiments.[7] We shall call the time- (or frequency-) dependent quantities $k_f(t)$ and $k_r(t)$ rate *kernels*. Their precise relation to the usually measured rate coefficient will depend on the type of experiment. Some examples illustrate this point. If the rate coefficient is determined by examining the response for small frequencies, then

$$k_f = \int_0^\infty dt\, k_f(t) \tag{2.11}$$

This relation will also hold more generally, provided the time-scale separation holds.[5, 6] We discuss this point more fully when we describe the microscopic basis of these rate laws. As a second example, consider the fluorescence quenching reaction mentioned earlier. In this case, the quenching process is conveniently described by introducing[8] a time-dependent rate coefficient $k_q(t)$,

$$A + B \xrightarrow{k_q(t)} \text{products} \tag{2.12}$$

The rate law is

$$\frac{d\overline{N_A(t)}}{dt} = -k_q(t)n_{eq}^B \overline{N_A(t)} \tag{2.13}$$

and $k_q(t)$ is related to the rate kernel by

$$k_q(t) = \int_0^t dt'\, k_f(t') \tag{2.14}$$

Thus the rate kernel plays a central role in the description of relaxation processes in the chemically reacting system, and a good deal of our attention is devoted to this quantity in the sequel.

It is often convenient to adopt a somewhat different view of the system. Before, we imagined that a system in equilibrium was disturbed in some way and watched the decay back to equilibrium. However, even in an equilibrium system there are fluctuations and, according to the regression hypothesis,[9] the decay of these fluctuations is given by the macroscopic laws for long enough times and on large enough distance scales.

The time development of these fluctuations is conveniently described in terms of correlation functions of the form

$$C_{\alpha\beta}(\mathbf{r}, \mathbf{r}'; t) = \langle \delta n_\alpha(\mathbf{r}, t) \delta n_\beta(\mathbf{r}') \rangle$$
$$\equiv C_{\alpha\beta}(\mathbf{r} - \mathbf{r}', t) \qquad (2.15)$$

where $\delta n_\alpha(\mathbf{r}, t)$ is the *microscopic* expression for the number of molecules of species α at point $\mathbf{r}$ in the solution at time t; it depends in general on the phase point of the system at time t. The angular brackets denote a system equilibrium average in some convenient ensemble. The second line of (2.15) makes explicit the unique dependence of these correlation functions on the difference between the two spatial points in the solution; that is, they do not depend on the origin. This follows from translational symmetry.[10] Since the decay of these correlation functions, for sufficiently long times and distance scales, is given by macroscopic laws, for example,

$$\frac{dC_{AA}(\mathbf{r}, t)}{dt} = \left(D_A \nabla^2 - k_f n_{eq}^B \right) C_{AA}(\mathbf{r}, t) + k_r n_{eq}^B C_{CA}(\mathbf{r}, t) \qquad (2.16)$$

we may alternatively use this description to obtain expressions for the rate coefficients and rate kernels. This type of formulation is especially convenient for the development of microscopic theories. We shall in fact show later that (2.16) follows from the microscopic equations of motion when certain conditions apply, thus verifying the statements above. We should, however, point out that although this approach is convenient for the analysis of the validity of the macroscopic laws and the study of rate coefficients and rate kernels, it cannot describe the decay from arbitrary initial states required for the analysis of certain types of experiment, for example, atom recombination that follows a photodissociation event. We describe processes of these types in Sections XI and XII.

The previous considerations, of course, apply equally well to the more general bimolecular reaction

$$A + B \underset{k_r}{\overset{k_f}{\rightleftarrows}} C + D \qquad (2.17)$$

The linearized rate law is

$$\frac{d\overline{\delta n_\alpha(\mathbf{r}, t)}}{dt} = D_\alpha \nabla^2 \overline{\delta n_\alpha(\mathbf{r}, t)}$$

$$- \sum_{\mu=A}^{D} k_{\alpha\mu} \overline{\delta n_\mu(\mathbf{r}, t)} \tag{2.18}$$

where, for example, $k_{AA} = n_{eq}^B k_f$, $k_{AB} = n_{eq}^A k_f$, $k_{AC} = -n_{eq}^D k_r$, and $k_{AD} = -n_{eq}^C k_r$, with similar definitions for the other $k_{\alpha\beta}$'s. In terms of correlation functions, (2.18) can be written as

$$\frac{dC_{\alpha\beta}(\mathbf{r}, t)}{dt} = D_\alpha \nabla^2 C_{\alpha\beta}(\mathbf{r}, t)$$

$$- \sum_{\mu=A}^{D} k_{\alpha\mu} C_{\mu\beta}(\mathbf{r}, t) \tag{2.19}$$

We next consider the theoretical description of the rate coefficients introduced here.

III. CONVENTIONAL APPROACHES

The most widely used descriptions of reactions in liquids are formulated in terms of configuration space equations for the dynamics of the reacting molecules. Approaches of this type have a long history in the condensed-phase reaction rate literature, dating back to M. von Smoluchowski's classic studies of colloid coagulation.[1] More modern applications still frequently employ such descriptions, often with considerable success and justification.[8, 11] Nevertheless, these theories must fail on short distance and time scales—just the domains in which modern experimental techniques are capable of providing new information on reaction dynamics. Among the goals of the microscopic theories described below are to delineate the range of validity of these conventional approaches and to provide suitable extensions. Therefore we now briefly summarize the results of the diffusion equation approaches and describe how Fokker-Planck and Langevin equations may be used to approximately account for velocity relaxation effects on the reaction.

A. Simple Diffusion Equation

As a first example of the diffusion equation approach, we consider a very simple problem: the absorption of small (point) particles A, which are dilutely dispersed in a viscous continuum, by a collection of large stationary particles ("sinks") B (Fig. 3.1). The sinks are assumed to be sufficiently large that changes in their radii may be neglected during the course of the reaction, and sufficiently dilute that they act independently and competition effects are small. This corresponds to a primitive model for the growth of droplets in aerosols or the growth of crystal nuclei from solution.[12, 13]

The density field of the A species at point $\mathbf{r}$ in the fluid, $n_A(\mathbf{r})$, is measured relative to the center of a sink and is assumed to obey a simple diffusion equation:

$$\frac{\partial n_A(\mathbf{r}, t)}{\partial t} = D_A \nabla^2 n_A(\mathbf{r}, t) \tag{3.1}$$

The absorption of molecules at the surface of each sink is taken into account by a boundary condition. Perhaps one of the most useful boundary conditions is the "radiation" boundary condition introduced by Collins and Kimball,[14]

$$4\pi D_A R^2 \hat{\mathbf{r}} \cdot \nabla n_A(\mathbf{r}, t)\big|_{r=R} = k_i n_A(\hat{\mathbf{r}}R, t) \tag{3.2}$$

where R is the radius of the sink and k_i is some intrinsic rate constant for the absorption process (cf. Fig. 3.1). This boundary condition allows for the possibility of reflection as well as absorption on the sink's surface and is similar to radiation boundary conditions in heat transport problems.[14] If $k_i \gg k_D = 4\pi D_A R$, the boundary condition reduces to the complete absorption boundary condition used by Smoluchowski,[1]

$$n_A(\hat{\mathbf{r}}R, t) = 0 \tag{3.3}$$

When describing transient effects in these systems, it often proves convenient to deal with the Fourier transform of the local density field,

$$n_A(\mathbf{r}, \omega) = \int_{-\infty}^{\infty} dt\, e^{i\omega t} n_A(\mathbf{r}, t)$$

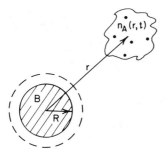

Fig. 3.1. An illustration of some features of the absorption of A molecules by the "sink" B. The density field of the A species, $n_A(\mathbf{r}, t)$, is referred to a coordinate frame centered on B. The dashed circle about B signifies a region about the particle where the diffusion equation no longer applies (a diffusion boundary layer). In the application of the radiation boundary condition, the presence of this boundary layer is approximately taken into account by the effective rate coefficient k_i [(3.2)], and its spatial extent is neglected.

which satisfies

$$-i\omega n_A(\mathbf{r}, \omega) = D_A \nabla^2 n_A(\mathbf{r}, \omega) \tag{3.4}$$

The half-sided Fourier transform of the rate kernel for the absorption of A,

$$k_f(\omega) = \int_0^\infty dt\, e^{i\omega t} k_f(t)$$

can be obtained by explicitly solving the diffusion equation with the radiation boundary condition, then calculating the flux across the surface of the sink,

$$\Phi(\omega) = k_D R \hat{\mathbf{r}} \cdot \nabla n_A(\mathbf{r}, \omega)|_{r=R}$$

$$= \frac{k_i k_D(1+\alpha R)}{k_i + k_D(1+\alpha R)} n_A^0(\omega) \equiv k_f(\omega) n_A^0(\omega) \tag{3.5}$$

where $\alpha = (-i\omega/D_A)^{1/2}$ and $n_A^0(\omega)$ is the density field far from the sink, and is assumed to be uniform in space. Thus the rate kernel is given by[15]

$$k_f(\omega)^{-1} = k_i^{-1} + k_D(\omega)^{-1} \tag{3.6}$$

where $k_D(\omega) = k_D(1+\alpha R)$.

The rate coefficient then follows by taking the $\omega = 0$ limit of the rate kernel

$$k_f(\omega = 0)^{-1} \equiv k_i^{-1} + k_D^{-1} \tag{3.7}$$

The rate coefficient may also be written in an equivalent form as the difference of the intrinsic rate constant k_i and a relaxing part

$$k_f(\omega) = k_i - \frac{k_i^2}{k_i + k_D(\omega)}$$

$$\equiv k_i + \Delta k_f(\omega) \tag{3.8}$$

The t-space expression for the rate kernel then follows after inversion of the half-sided Fourier transform

$$k_f(t) = 2k_i \delta(t) - \frac{k_i^2}{k_D} \left\{ \left(\frac{D_A}{\pi t R^2} \right)^{1/2} - \left(1 + \frac{k_i}{k_D} \right) \frac{D_A}{R^2} \right.$$

$$\left. \times \exp\left[\frac{(1+k_i/k_D)^2 t D_A}{R^2} \right] \mathrm{erfc}\left[\left(1 + \frac{k_i}{k_D} \right) \left(\frac{t D_A}{R^2} \right)^{1/2} \right] \right\} \tag{3.9}$$

Note that the rate kernel has a long time tail that decays as $t^{-3/2}$,

$$k_f(t) \underset{t \text{ large}}{\to} -\frac{D_A}{2\sqrt{\pi}\, R^2} \frac{k_i^2}{k_D}\left(1 + \frac{k_i}{k_D}\right)^{-2}\left(\frac{tD_A}{R^2}\right)^{-3/2}$$

and a singular contribution at $t=0$.

We shall discuss the results above for $k_f(t)$ and k_f in the course of the derivation of such results from a microscopic point of view.

Simple diffusion equations of this type have also been applied to much wider classes of reaction, for example, to cases of "sinks" and particles being absorbed that are both moving and similar in size (e.g., colloid coagulation and small molecule reactions). The physical background for such applications is widely discussed in the literature.[11] When the sink density is not small, competition effects come into play and it is no longer sufficient to consider reaction at a single sink.[11, 16, 17] These competition effects lead to a nonanalytic dependence of the rate coefficient on the sink density.[18, 19] Such effects are not discussed here.

In contrast to the foregoing approach, which utilizes equations for the density field of the reacting particles, it is often convenient to focus on the dynamics of a pair of particles. For dilute solutions the descriptions are equivalent. We shall make use of both types of descriptions. As an example of this pair formulation, we outline the Smoluchowski equation description for reactive pair dynamics.

B. Smoluchowski Equation

We now adopt a somewhat different point of view: suppose we have an ensemble of isolated pairs of potentially reactive A and B molecules and wish to describe the time evolution of the probability that a given pair at relative separation $\mathbf{r}$ remains unreacted at time t, $P(\mathbf{r}, t)$. In contrast to the simple diffusion equation, we now also allow for the possibility that forces act between the molecules in the pair. It is customary to assume that the dynamics for this situation may be modeled by a Smoluchowski equation[1]

$$\frac{\partial P(\mathbf{r}, t)}{\partial t} = \nabla \cdot \mathbf{D} \cdot [\nabla - \beta \mathbf{F}] P(\mathbf{r}, t) \tag{3.10}$$

where $\mathbf{D}$ is a relative diffusion tensor, which may be $\mathbf{r}$-dependent, and $\mathbf{F} = -\nabla W$ is the mean force acting between the particles in the pair, with W the corresponding mean potential. When the particles approach to within a distance σ, reaction is possible and is again accounted for by a "radiation" boundary condition, now of the following form.[20]

$$4\pi R^2 \hat{r} \cdot \mathbf{D} \cdot [\nabla - \beta \mathbf{F}] P(\mathbf{r}, t)|_{r=\sigma} = k_i P(\hat{r}\sigma, t) \tag{3.11}$$

The problem may also be formulated in an alternate but equivalent way using a Smoluchowski equation with sink terms.[21]

$$\frac{\partial P(\mathbf{r}, t)}{\partial t} = \nabla \cdot \mathbf{D} \cdot [\nabla - \beta \mathbf{F}] P(\mathbf{r}, t) - \frac{k_i}{4\pi\sigma^2} \delta(r - \sigma) P(\mathbf{r}, t) \quad (3.12)$$

This equation may be formally solved to yield an expression for the rate kernel with the general form of (3.7), (cf. Northrup and Hynes[15, 21]),

$$k_f(\omega)^{-1} = k_f^{0^{-1}} + k_D(\omega)^{-1} \quad (3.13)$$

Here, k_f^0 is the equilibrium one-way flux rate coefficient

$$k_f^0 = g(\sigma)k_i \quad (3.14)$$

with $g(\sigma) = e^{-\beta W(\sigma)}$ the pair distribution function at contact and $\beta = (k_B T)^{-1}$. The contribution $k_D(\omega)$ is given by the average over the reaction surface of the propagator for the pair motion in the absence of reaction but influenced by the forces

$$k_D(\omega) = g(\sigma) \left[(4\pi\sigma^2)^{-2} \int d\mathbf{r} \, d\mathbf{r}' \, \delta(r - \sigma) \mathcal{G}(\mathbf{r}|\mathbf{r}'; \omega) \delta(r' - \sigma) \right]^{-1} \quad (3.15)$$

with $\mathcal{G}(\mathbf{r}; \omega)$ formally given by

$$\mathcal{G}(\mathbf{r}; \omega) = \{ -i\omega - \nabla \cdot \mathbf{D} \cdot [\nabla - \beta \mathbf{F}] \}^{-1} \quad (3.16)$$

The derivation of this result is sketched in Appendix A.

In the limit of no forces other than simple excluded volume effects ($g(r) = \theta(r - \sigma)$), (3.15) reduces to the simple diffusion equation result in Section III.A. In the $\omega = 0$ limit, k_D may be written in the form[22-24]

$$k_D^{-1} = \int_\sigma^\infty dr \left[4\pi r^2 e^{-\beta W(r)} D(r) \right]^{-1} \quad (3.17)$$

where we have taken $\mathbf{D}(r) = D(r)\mathbf{1}$ for simplicity. This result clearly displays the effects of the forces and nonlocality of D on the diffusive rate coefficient.

The method we have just outlined is certainly well suited to describe the system to which it was first applied, the coagulation of large colloidal particles in solution. Here, the assumption of a continuum solvent is likely to be a very good approximation, and a configuration space description will be appropriate for the dynamics of the large, massive colloidal particles. For the reactions of small molecules in solution, these conditions no longer apply and the validity of the method must be tested.

C. Fokker-Planck and Langevin Equations

Velocity relaxation effects can be accounted for in an approximate fashion by going to a phase-space description in terms of Fokker-Planck or Langevin equations.[25] Perhaps the best known study of this type is due to Kramers,[26] who studied the escape of particles over potential barriers as a model for certain types of isomerization or dissociation reaction.

Suppose the "particle" moves in the one-dimensional potential shown schematically in Fig. 3.2. Kramers assumes that the time evolution of the phase-space distribution function $F(q, v; t)$ is given by the Fokker-Planck equation

$$\frac{\partial}{\partial t} F(q, v; t) = -v \frac{\partial}{\partial q} F(q, v; t) + \frac{1}{m} \frac{\partial W}{\partial q} \frac{\partial}{\partial v} F(q, v; t)$$

$$+ \frac{\zeta}{m} \frac{\partial}{\partial v} \left(v + \frac{k_B T}{m} \frac{\partial}{\partial v} \right) F(q, v; t) \tag{3.18}$$

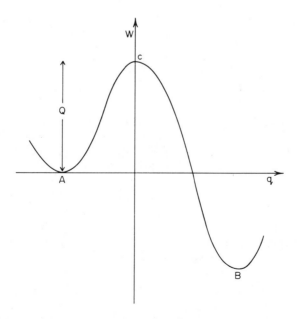

Fig. 3.2.   The double minimum potential W as a function of the reaction coordinate q. The stable A and B species are separated by a barrier located at $C(q_c)$ with height Q.

where ζ is the friction coefficient. Several situations are treated by Kramers: in the high friction limit, the Fokker-Planck equation can be reduced to a Smoluchowski equation and then solved to obtain the rate coefficient for irreversible passage across the barrier. By assuming that the potential near the barrier top C has the form

$$W = Q - \frac{1}{2} m \omega'^2 (q - q_c)^2 \tag{3.19}$$

where Q is the barrier height, and is harmonic near A,

$$W = \frac{1}{2} m \omega_0^2 q^2 \tag{3.20}$$

Kramers finds the following:

$$k_f = \frac{\omega_0 \omega' m}{2 \pi \zeta} e^{-\beta Q} \tag{3.21}$$

In the intermediate to high friction limit, the steady-state Fokker-Planck equation can be solved with the potential in (3.19) to obtain the following more general result;

$$k_f = \frac{\zeta \omega_0}{4 \pi m \omega'} e^{-\beta Q} \left[\left(1 + \frac{4\omega'^2 m^2}{\zeta^2} \right)^{1/2} - 1 \right] \tag{3.22}$$

For high friction, $\zeta/m \gg 2\omega'$, the equation reduces to (3.21), whereas for intermediate friction, $\zeta/m \ll 2\omega'$, (3.22) reduces to the transition state theory (TST) result.[27]

$$k_f^{\text{TST}} = \frac{\omega_0}{2 \pi} e^{-\beta Q} \tag{3.23}$$

The transition state theory result assumes an equilibrium distribution at the barrier top; then the equilibrium one-way flux across the barrier is calculated to obtain the rate. In the high friction limit, collisions with the solvent will cause frequent recrossings of the barrier, hence leading to the reduction in the rate coefficient given in (3.21). Neither of these results will apply in the extreme low friction limit. In this case there will be an insufficient number of collisions to maintain equilibrium throughout the well, especially near the barrier top, and collisional activation will be rate-controlling step. Kramers solved the problem in this limit by converting the Fokker-Planck equation to an equation in energy space and solving for the flux across the barrier to obtain:

$$k_f = \frac{\zeta \beta Q}{m} e^{-\beta Q} \qquad (3.24)$$

Since Kramers's paper there have been a large number of studies devoted to the problem of passage over a barrier, focusing especially on the transition between the low and high friction regimes. We discuss some of the more recent developments later (cf. Section XII).

The problem can also be approached by considering the Langevin equation rather than the Fokker-Planck equation.[25] In the Langevin description, the motion of the particle is given by the stochastic equations

$$\dot{q}(t) = v(t)$$
$$m\dot{v}(t) = -\zeta v(t) + F(t) + f(t) \qquad (3.25)$$

where $F = -\partial W/\partial q$ is the force on the "particle" and the random force $f(t)$ is assumed to be a Gaussian random process. The correlations of $f(t)$ are related to the friction by the fluctuation-dissipation theorem

$$\int_0^\infty dt \langle f(t) f \rangle = k_B T \zeta \qquad (3.26)$$

The results obtained by averaging over the stochastic trajectories computed according to (3.25) are equivalent to those obtained via the solution of the corresponding Fokker-Planck equation. In many cases it is more convenient to work with the Langevin equation. We discuss a specific example in detail in Section XII.

These are the theoretical approaches that have been customarily used to describe the rates of chemical reactions in liquids. In the sections that follow, we examine the microscopic basis of these approaches from a kinetic theory point of view, to appreciate their range of validity and to see how they might be extended to describe the reactions of small molecules in solution on short time scales.

IV. MICROSCOPIC SPECIFICATION OF SPECIES IN A REACTING SYSTEM

Any microscopic theory of chemical reactions must, at some stage, define what characterizes a particular chemical species and prescribe the interconversion of the species. A general specification of functions or operators that characterize species is difficult, since it depends on the type of chemical reaction under consideration.[28, 29] In quantum mechanical formulations, a representation in terms of number operators is often a convenient way of formulating the problem.[30, 31] This kind of description does not really simplify the species specification problem; instead, it shifts the problem to the construction of a suitable Hamiltonian, which describes how the chemical reaction takes place. This section focuses on a classical theory of chemical reactions. A formal description is given first; then specific examples are discussed.

In a formal way, we may introduce functions (operators in species space) Θ_i^α that characterize the species α of a given molecule i. In general, the Θ_i^α will be a function of whatever coordinates are needed to specify molecule i. These operators should be constructed to have the following properties:

$$\Theta_i^\alpha \Theta_i^\beta = \delta_{\alpha\beta} \tag{4.1}$$

that is, molecule i cannot be of species types α and β at the same time; also,

$$\sum_\alpha \Theta_i^\alpha = 1 \tag{4.2}$$

molecule i must be some species type.

We may then write the number of molecules of species α in terms of these operators as

$$N_\alpha = \sum_{i=1}^N \Theta_i^\alpha \tag{4.3}$$

where N is total number of molecules in the system.

The average number of molecules of species α at equilibrium is obtained by averaging N_α over an equilibrium ensemble,

$$N_{eq}^\alpha = \langle N_\alpha \rangle \tag{4.4}$$

Local density fields can also be introduced easily. The number of molecules of species α at point $\mathbf{r}$ in the fluid is given by

$$n_\alpha(\mathbf{r}) = \sum_{i=1}^N \delta(\mathbf{r} - \mathbf{R}_i) \Theta_i^\alpha \tag{4.5}$$

Similarly, the number of molecules at the phase point $(\mathbf{r}, \mathbf{v})$ is given by

$$n_\alpha(\mathbf{r}, \mathbf{v}) = \sum_{i=1}^{N} \delta(\mathbf{r} - \mathbf{R}_i)\delta(\mathbf{v} - \mathbf{V}_i)\Theta_i^\alpha \qquad (4.6)$$

In these equations $\mathbf{R}_i$ and $\mathbf{V}_i$ are the position and velocity, respectively, of the center of mass of the ith molecule. The corresponding average values of these fields are

$$\langle n_\alpha(\mathbf{r}) \rangle = \frac{N_{eq}^\alpha}{V} = n_{eq}^\alpha \qquad (4.7)$$

and

$$\langle n_\alpha(\mathbf{r}, \mathbf{v}) \rangle = n_{eq}^\alpha \phi_\alpha(v) \qquad (4.8)$$

where V is the volume and $\phi_\alpha(v)$ is the Maxwell distribution function. Although these results formally specify how species are introduced in the description, specific expressions for the Θ_i^α must be given before they can be implemented. To illustrate their use, we now consider a few examples.

Perhaps the simplest example is provided by the case of an isomerization reaction $A \rightleftarrows B$, which is modeled by the motion of the molecule in the one-dimensional double minimum potential of Fig. 3.2. If the reaction coordinate for this motion of molecule i is q_i, the usual way of specifying species is to say that the molecule is species A if $q_i < 0$, and is species B if $q_i > 0$. Thus

$$\Theta_i^A = \theta(-q_i)$$

and

$$\Theta_i^B = \theta(q_i) \qquad (4.9)$$

where $\theta(x)$ is the Heaviside function [$\theta(x) = 1$ for $x > 0$ and zero otherwise]. Several examples in the recent literature employ this type of microscopic species identification.[32, 33]

Another example is provided by the case of atom recombination in liquids. As the atoms approach and enter the reaction region, they experience a strong attractive force. This should be contrasted with the isomerization reaction just described, where a chemical barrier several $k_B T$ high must be surmounted before reaction can occur. In a liquid the force between the atoms will be modified because of the presence of the solvent.

The potential of mean force thus has maxima and minima, which reflect the solvent structural correlations. For example, the first maximum is due to the extra energy the atoms need to push beyond the solvent cage or, alternatively, the energy needed to squeeze out an intervening solvent molecule. Figure 4.1 is a schematic diagram of this situation. Since the atoms encounter only a small solvent barrier $\mathcal{O}(k_B T)$, one might anticipate that the dynamics and theoretical treatment of such reactions would be rather different from those with high chemical barriers.

For atomic recombination reactions in the gas phase, a surface separating molecules from unbound atoms is most conveniently drawn in energy space. For instance, one might adopt the convention that configurations with energies less than the dissociation energy E_d correspond to molecules, and those with energies above E_d are unbound atoms. Several examples of dividing surfaces for atomic recombination-dissociation reactions are given by Keck.[34] In a dense liquid, where energy dissipation is large, a criterion based on a configuration space prescription will probably be useful. Consider, for example, the distance r_0 denoted by the broken line in Fig. 4.1, which corresponds to the position of the first maximum in the potential of

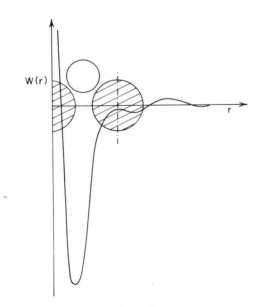

Fig. 4.1. General structure of the potential of mean force for an atomic recombination reaction. The second minimum occurs at a distance corresponding to a separation of the atoms by one solvent molecule. To illustrate the effects of solvent structure, the effective hard-sphere representations (cf. Section VI) of the solute and solvent atoms are also shown.

mean force. Configurations with internuclear separations larger than r_0 will correspond to solvent-separated pairs atoms. Since there is a shallow well in the mean potential, such solvent-separated pairs may actually have an energy less than E_d at a given time. Molecule formation takes place inside the solvent cage for separations less than r_0. At short internuclear separations the strong attractive forces will lead to rapid, stable molecule formation. We might then write for the number of AB molecules as follows:

$$N_{AB} = \sum_{i=1}^{N_A} \sum_{i=1}^{N_B} \theta(r_0 - R_{ij}) \qquad (4.10)$$

However, in the vicinity of the very small solvent barrier, there will be frequent recrossings of the r_0 surface; thus the recombination rate will differ considerably from the one-way flux across this surface. Alternatively, one might select an r_0 at shorter separations, which correspond to stable molecule formation (cf. Section XII).

Another more complex example is given by Stillinger[35] for the ionization of water.

There is one final case, which we describe very briefly here and in more detail later. The classical description can be written in a form that is quite similar to the number operator representation in quantum mechanics. An operator Θ_i^α is assigned to molecule i, which is one if i is of type α and zero otherwise. Now, however, these operators do not themselves depend on the positions and momenta; they follow a dynamics that is specified by the classical Liouville operator of the system. In particular, the Liouville operator determines the conditions under which species interconversion is possible. Hence, just as in the quantum mechanical case, the problem of the specification of the precise conditions for reaction is deferred to the Liouville operator. Section VI describes how such Liouville operators can be constructed.

In summary then, the mode of species identification is largely dictated by the particular reaction being studied and the experimental conditions. This chapter illustrates how the formal considerations above can be implemented in a theory of reactions in liquids.

V. CORRELATION FUNCTION EXPRESSIONS FOR RATE COEFFICIENTS

Before presenting a full kinetic theory description of a reacting fluid, we give a more formal microscopic treatment of reactions via linear response theory. This discussion will serve to make the conditions under which we expect the macroscopic rate law to hold more precise, and will also provide formal expressions for the rate kernels and rate coefficients, whose detailed structure we shall examine by kinetic theory.

The discussion presented below is most conveniently given in terms Fourier-Laplace transforms of the density fields

$$\delta n_\alpha(\mathbf{k}, z) = \int_0^\infty dt\, e^{-zt} \int d\mathbf{r}\, e^{-i\mathbf{k}\cdot\mathbf{r}} \delta n_\alpha(\mathbf{r}, t) \tag{5.1}$$

In this space, the rate law given in (2.19) takes the form

$$(z + k^2 D_\alpha) C_{\alpha\beta}(\mathbf{k}; z) + \sum_{\mu=A}^{D} k_{\alpha\mu} C_{\mu\beta}(\mathbf{k}; z) = C_{\alpha\beta}(\mathbf{k}) \tag{5.2}$$

We shall examine the conditions necessary to obtain this result from the microscopic theory.

The microscopic basis of the phenomenological rate law can be investigated by making use of projection operator methods.[36, 37] The time evolution of the microscopic density fields, discussed in Section IV, is given by the Liouville equation

$$\frac{d\delta n_\alpha(\mathbf{k}, t)}{dt} = i\mathcal{L}\, \delta n_\alpha(\mathbf{k}, t) \tag{5.3}$$

where $\mathcal{L}$ is the Liouville operator for the system. Using projection operator techniques, the time correlation function of the solute density fields can be shown to satisfy the exact equation of motion[36]:

$$\frac{dC_{\alpha\beta}(\mathbf{k}; t)}{dt} = -\sum_{\mu=A}^{D} \int_0^t dt'\, \phi_{\alpha\mu}(\mathbf{k}, t') C_{\mu\beta}(\mathbf{k}; t-t') \tag{5.4}$$

or

$$z C_{\alpha\beta}(\mathbf{k}; z) + \sum_{\mu=A}^{D} \phi_{\alpha\mu}(\mathbf{k}; z) C_{\mu\beta}(\mathbf{k}; z) = C_{\alpha\beta}(\mathbf{k}) \tag{5.5}$$

We make frequent use of these now standard projection operator methods. Although no details are given, the various results are easily derived by application of the operator identity $[z - A]^{-1} = [z - QA]^{-1} + [z - A]^{-1} \mathcal{P} A [z - QA]^{-1}$.

The damping matrix or memory kernel $\phi_{\alpha\mu}$ is defined by

$$\phi_{\alpha\mu}(\mathbf{k}; t) = \langle \delta \dot{n}_\mu(-\mathbf{k}) e^{Qi\mathcal{L}t} \delta \dot{n}_\alpha(\mathbf{k}) \rangle (N_{eq}^\mu)^{-1} \tag{5.6}$$

with $Q = 1 - \mathcal{P}$, where $\mathcal{P}$ projects onto the density fields

$$\mathcal{P}A(\mathbf{k}) = \sum_\alpha \langle A(\mathbf{k})\delta n_\alpha(-\mathbf{k})\rangle (N_{eq}^\alpha)^{-1}\delta n_\alpha(\mathbf{k}) \tag{5.7}$$

We have used time reversal symmetry to write

$$\langle \delta \dot{n}_\alpha(\mathbf{k})\delta n_\mu(-\mathbf{k})\rangle = 0 \tag{5.8}$$

in these equations. Also, for solute species dilute in the solvent, we used the following relation:

$$C_{\alpha\beta}(\mathbf{k}) = \langle \delta n_\alpha(\mathbf{k})\delta n_\beta(-\mathbf{k})\rangle = \delta_{\alpha\beta}N_{eq}^\alpha \tag{5.9}$$

To make connection with the macroscopic law, we must analyze the structure of the damping matrix. The local density fields can change by both motion of the center of mass of the molecules and by reactive events that change the number of molecules of a given species. Thus we have

$$\delta \dot{n}_\alpha(\mathbf{k}) = -i\mathbf{k}\cdot\sum_{i=1}^N \mathbf{V}_i e^{-i\mathbf{k}\cdot\mathbf{R}_i}\Theta_i^\alpha + \sum_{i=1}^N e^{-i\mathbf{k}\cdot\mathbf{R}_i}\dot{\Theta}_i^\alpha$$

$$\equiv -i\mathbf{k}\cdot\mathbf{j}_\alpha(\mathbf{k}) + R_\alpha(\mathbf{k}) \tag{5.10}$$

A result with the same structure as the macroscopic law follows when terms to order k^2 are retained, cross-diffusion and cross-diffusion–reaction terms are neglected, and a Markov approximation $\phi_{\alpha\mu}(t) \simeq 2\phi_{\alpha\mu}(z=0)\delta(t)$, is made. When these approximations are introduced, we may write[36]

$$\phi_{\alpha\mu}(\mathbf{k}; z) \simeq k^2 D_\alpha \delta_{\alpha\mu} + k_{\alpha\mu} \tag{5.11}$$

where we have defined the diffusion coefficient for species α as

$$D_\alpha = \lim_{z\to 0}\lim_{k\to 0} \hat{k}\cdot\langle\mathbf{j}_\alpha(\mathbf{k}, z)\mathbf{j}_\alpha(-\mathbf{k})\rangle\cdot\hat{k}(N_{eq}^\alpha)^{-1}$$

$$= (3N_{eq}^\alpha)^{-1}\int_0^\infty dt \lim_{k\to 0}\left\langle\left[\sum_{i=1}^N \mathbf{V}_i\Theta_i^\alpha\right]\cdot e^{Qi\mathcal{L}t}\left[\sum_{j=1}^N \mathbf{V}_j\Theta_j^\alpha\right]\right\rangle \tag{5.12}$$

and

$$k_{\alpha\mu} = \lim_{z\to 0}\lim_{k\to 0} \langle R_\alpha(\mathbf{k}, z)R_\mu(-\mathbf{k})\rangle (N_{eq}^\mu)^{-1}$$

$$= \int_0^\infty \lim_{k\to 0}\left\langle\left[\sum_{i=1}^N \dot{\Theta}_i^\mu\right]e^{Qi\mathcal{L}t}\left[\sum_{j=1}^N \dot{\Theta}_j^\alpha\right]\right\rangle(N_{eq}^\mu)^{-1}\,dt$$

$$\equiv \lim_{z\to 0} k_{\alpha\mu}(z) \tag{5.13}$$

The results given above are the correlation function expressions for the rate coefficient, which we wished to obtain. The last line also serves to define the quantity $k_{\alpha\mu}(z)$, the rate kernel. As noted earlier, this quantity is central to discussion of reaction rate theory. Here and in the sections that follow, we attempt to elucidate its structure.

A few comments will help to clarify the nature of these approximations. Because of the presence of reactive terms, the expression for D_α differs somewhat from the usual autocorrelation function expression for the self-diffusion coefficient of species α. To examine this situation a bit more closely, consider the case of a nonreactive fluid. Then $\delta \dot{n}_\alpha(\mathbf{k}) = -i\mathbf{k} \cdot \mathbf{j}_\alpha(\mathbf{k})$ and, using the operator identity $(A+B)^{-1} = A^{-1} - A^{-1}B(A+B)^{-1}$, one may show[36] that

$$e^{Qi\mathcal{L}t} = e^{i\mathcal{L}t} + \mathcal{O}(\mathbf{k}) \tag{5.14}$$

Thus in the $k \to 0$ limit, we find the expected result[39, 40]

$$D_\alpha = \frac{1}{3} \int_0^\infty dt \langle \mathbf{V}_1^\alpha \cdot \mathbf{V}_1^\alpha(t) \rangle \tag{5.15}$$

where we have used the fact that α is dilute in the solvent. When reactive terms are present, the projected evolution operator cannot be reduced to usual evolution operator by taking only the $k \to 0$ limit; the nonconserved pieces remain.

The procedure above has not in any sense derived the macroscopic relaxation equations; only some formal conditions have been stated under which the structures of the microscopic and macroscopic equations become the same. One crucial point, which certainly deserves further comment, is the physical basis of the Markov approximation. This approximation removes the memory effects from (5.5) so that the structures of the microscopic and macroscopic equations become similar. For this approximation to be useful, the memory kernel $\phi_{\alpha\mu}$ must decay much more rapidly than the density fields. The projected time evolution will guarantee that this is the case, provided these fields decay much more slowly than other variables in the system.

In the case of a nonreacting fluid, where one is usually interested in macroscopic equations for conserved (in the limit $k \to 0$) variables, the origin and region of validity of this approximation is clear. In the small k limit the conserved fields do decay much more slowly than other variables in the system, and the limit $z \to 0, k \to 0$ has the effect of extracting the decay on this slow time scale. (Mode coupling contributions spoil some of these arguments, but it is now known how to account for these effects.[41] We discuss this aspect of the problem in Section VII.)

In a chemically reacting fluid, where the number of molecules of a given species is no longer conserved, the small k limit is not sufficient to guarantee this time-scale separation. For the Markov approximation to be useful, we must also require that the chemical reaction be slow. We can make this discussion more precise by using the same methods that have been applied in the past to justify this point in other contexts.[6] We consider, for simplicity, the case of a spatially homogeneous system and take the $k \to 0$ limit of (5.4) at the outset to obtain

$$\frac{dC_{\alpha\beta}(t)}{dt} = -\sum_{\mu=A}^{D} \int_0^t dt' \, \phi_{\alpha\mu}(t') C_{\mu\beta}(t-t') \tag{5.16}$$

where

$$C_{\alpha\beta}(t) = \lim_{k \to 0} C_{\alpha\beta}(\mathbf{k}, t) = \langle \delta N_\alpha(t) \delta N_\beta \rangle \tag{5.17}$$

and $\phi_{\alpha\mu}(t)$ is the $k \to 0$ limit of the damping matrix given earlier. In this limit we also have

$$\delta \dot{N}_\alpha = \sum_{i=1}^{N} \dot{\Theta}_i^\alpha = R_\alpha = \gamma \bar{R}_\alpha \tag{5.18}$$

where we have introduced the parameter γ to keep track of the manner in which the reactive terms enter. It is a dimensionless parameter, which is $\mathcal{O}(\tau_0/\tau_{\text{chem}})$, where τ_0 is a relaxation time characteristic of internal, translational, or other relaxation processes in the system. Thus γ gauges how fast the chemical reaction proceeds relative to other relaxation processes. The Markov approximation can be shown to be valid in the limit of a slow chemical reaction by standard weak coupling techniques.[38] To implement these techniques, we write

$$\phi_{\alpha\mu}(t) = \gamma^2 \langle \bar{R}_\mu e^{\mathcal{Q} i \mathcal{L} t} \bar{R}_\alpha \rangle (N_{\text{eq}}^\mu)^{-1} \tag{5.19}$$

and let $\tau = \gamma^2 t$ and $C_{\alpha\beta}(t) = \mathcal{C}_{\alpha\beta}(\tau)$. We may then write (5.16) as

$$\frac{d\mathcal{C}_{\alpha\beta}(\tau)}{d\tau} = -\sum_\mu \int_0^{\tau/\gamma^2} dt' \langle \bar{R}_\mu e^{\mathcal{Q} i \mathcal{L} t'} \bar{R}_\alpha \rangle (N_{\text{eq}}^\mu)^{-1} \mathcal{C}_{\mu\beta}(\tau - \gamma^2 t') \tag{5.20}$$

The slow-time-scale τ variation can then be extracted by taking the limit

$\gamma^2 \to 0$ at fixed τ. We find

$$\frac{dC_{\alpha\beta}(t)}{dt} = -\gamma^2 \sum_\mu \left[\int_0^\infty dt' \lim_{\gamma \to 0} \langle \overline{R}_\mu e^{Qi\mathcal{L}t'} \overline{R}_\alpha \rangle (N_{eq}^\mu)^{-1} \right] C_{\mu\beta}(t)$$

$$\equiv - \sum_\mu k_{\alpha\mu} C_{\mu\beta}(t) \tag{5.21}$$

Notice that by keeping track of the γ dependence, one can show that

$$e^{Qi\mathcal{L}t} = e^{i\mathcal{L}t} + \mathcal{O}(\gamma) \tag{5.22}$$

by using the operator identity given earlier. Thus this procedure leads to the expression

$$k_{\alpha\mu} = \lim_{z \to 0} \lim_{\gamma \to 0} \lim_{k \to 0} \phi_{\alpha\mu}(\mathbf{k}, z)$$

$$= \gamma^2 \int_0^\infty dt \lim_{\gamma \to 0} \langle \overline{R}_\mu e^{i\mathcal{L}t} \overline{R}_\alpha \rangle (N_{eq}^\mu)^{-1} \tag{5.23}$$

where a double limiting procedure in k and γ, in addition to the $z \to 0$ limit, is required if the usual macroscopic law is to apply. Although the formal limiting process $\gamma \to 0$ is well defined, γ is not a parameter that is under our control. In any given physical situation, γ may be small but not zero; otherwise reaction would not be possible. Equation 5.23 simply states that if the reaction is slow enough, the rate coefficient may be calculated from the autocorrelation function of the reactive flux, with time evolution calculated via the nonreactive part of the Liouville operator. A discussion of this point in the context of a Boltzmann equation description of reactions is given in Ref. 42. For a slow reaction, the results in (5.23) and (5.13) are equivalent, since, using (5.22), they differ only by terms $\mathcal{O}(\gamma)$. For a fast chemical reaction, the rate kernel $k_{\alpha\mu}(z)$ is well defined and is an interesting quantity to study; however, the macroscopic rate law no longer applies, and memory effects, due to the competition between the reaction and other relaxation processes, will occur.

There is another variant of the correlation function expression for $k_{\alpha\mu}$ that is useful and worth noting. Thus far the analysis has stressed that if the macroscopic law is to apply, the time scale for the relaxation of $C_{\alpha\beta}(t)$, τ_{chem}, must be well separated from that of $\phi_{\alpha\mu}(t)$, say τ_0. If this is the case, we may then introduce a time t^* that satisfies the inequality $\tau_0 \ll t^* \ll \tau_{\text{chem}}$. Then we may write[43]

$$k_{\alpha\mu} = \int_0^{t^*} dt \langle R_\mu e^{i\mathcal{L}t} R_\alpha \rangle (N_{eq}^\mu)^{-1} \tag{5.24}$$

For a slow reaction, $k_{\alpha\mu}$ will be independent of the particular value of t^*, provided the inequality is satisfied. This form is often useful in numerical calculations. Using the fact that $e^{i\mathcal{L}t}R_\alpha = \delta\dot{N}_a(t)$, (5.24) can also be written[43] as

$$k_{\alpha\mu} = \langle \delta\dot{N}_\mu [\,\delta N_\alpha(t^*) - \delta N_\alpha\,]\rangle (N_{\text{eq}}^\mu)^{-1} \qquad (5.25)$$

If, for instance, species α corresponds to a diatomic molecule defined by

$$\Theta_\alpha = \sum_{i,\,j=1}^{N_A N_B} \theta(r_0 - R_{ij}),$$

then

$$k_{AA} = \left\langle \sum_{i,\,j}^{N_A N_B} (\mathbf{V}_{ij}\cdot\hat{\mathbf{R}}_{ij})\delta(R_{ij}-r_0)\theta(r_0 - R_{ij}(t^*)) \right\rangle (N_{\text{eq}}^\alpha)^{-1} \qquad (5.26)$$

This represents the average flux across the surface $R_{ij} = r_0$, given that the molecule AB is formed at time t^*.[43]

VI. MICROSCOPIC MODEL OF A REACTING LIQUID

The microscopic description of a chemically reacting fluid depends to a considerable extent on the specific chemical reaction under consideration. To present the kinetic theory results in a manner that is not purely formal, we focus primarily on a particular class of models that should be appropriate for several types of reaction. We provide some physical background for these models and then turn to their mathematical description.

A. Examples

First consider the general bimolecular reaction in (2.17) and suppose that the energy change during a reactive collision event is given schematically by Fig. 6.1. As the A and B molecules approach, they experience a strong short-range repulsion because of the chemical activation barrier, which is assumed to be of order several $k_B T$. If the molecules do not have sufficient relative translational energy to surmount the barrier, reaction will not take place and the A and B molecules will elastically or inelastically (nonreactively) scatter.

A simple model that corresponds to a reaction with these foregoing characteristics can be constructed along the following lines. We suppose that the internal states of reactant and product molecules can be treated in an average way by assigning effective one-level internal energies to them. Then, for energies below the reaction threshold, only elastic scattering occurs, whereas for higher energies the one-level reactant states convert to one-level product states. This type of model has been widely used in the gas-phase kinetics literature.[44, 45] It provides only a schematic treatment of the reactive event, since the mechanism by which reactants convert to products depends crucially on the internal constitution of the molecular species. Nevertheless, this model is useful for studying certain aspects of condensed-phase chemical reactions. An elucidation of the precise nature of the approach of the reactive molecules through the solvent is one of the most important features of liquid-state reactions. Thus although this type of reactive model does not treat the reactive event in full generality, it does permit a detailed study of molecular motion in the dense solvent and its coupling to reaction.

This type of model is not especially appropriate in all situations, however—for example, the atom recombination process. The relevant potential energy surface for this process was shown in Fig. 4.1. We noted in Section IV that there are no large chemical barriers for the recombination, and the details of the strongly attractive forces play an essential role in the reaction dynamics of this system. A theoretical treatment of this reaction must therefore include this direct chemical force, since it will play an essential role in governing the dynamics of the approach of the atoms through the solvent. We shall defer a thorough discussion of this case to Section XII, where the atom recombination problem is discussed in more detail, but the kinetic theory is formulated in a way that permits this case also to be studied.

Any fully microscopic theory of condensed-phase reactions must also specify the nature of the solute-solvent and solvent-solvent forces. Once

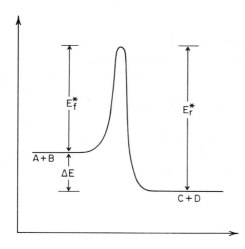

Fig. 6.1. Schematic representation of the energy changes for a bimolecular reaction with a localized high barrier to reaction.

again, these will depend on the system under investigation. For instance, solute-solvent attractive forces may be essential in the reaction mechanism, as in the case of *gas-phase* atom recombination (see Section XII), or may influence the reaction dynamics in a liquid in certain cases. In many instances, however, only the strongly repulsive forces are important in describing solvent effects on condensed-phase reactions. We shall attempt to justify this approximation for certain reactions later. Here we simply restrict our considerations to this case.

We also adopt a similar description for the solvent. This type of model requires some comment, even when applied to the simple solvents such as dense liquid argon or other noble gases. Although the static structural properties of such fluids are represented quite well by taking into account only the strongly repulsive parts of the potential,[46] the weak attractive forces do have noticeable effects on dynamic properties such as the velocity autocorrelation function.[47] However, a model that includes only the repulsive forces is not unreasonable for a description of the solvent dynamics in dense liquids, and this expedient is adopted. We focus on general features that are not expected to be especially sensitive to this approximation.

In summary, we are primarily concerned with two classes of reactions: (*1*) bimolecular reactions with a steep chemical barrier or possibly a steric constraint to reaction, and (*2*) recombination reactions in which motion of the atoms in the strongly attractive well must be treated. In both instances we assume that only the strongly repulsive solute-solvent and solvent-solvent forces need to be taken into account. We present a type of kinetic theory that is capable of handling more general cases, but the two reaction classes suffice to illustrate the use of the theory without overly elaborate calculations. Our goal is a detailed treatment of the effects of solvent dynamics on the reaction.

B. Mathematical Description

The time evolution in the N body reacting fluid is, in general, given by the Liouville operator introduced earlier. If, however, we make the additional assumption that the strongly repulsive solute-solvent and solvent-solvent forces can be approximated by effective hard-sphere interactions, the theory can be formulated in a way that greatly simplifies the calculation. This can be accomplished by the use of the pseudo-Liouville representation for the dynamics in a hard-sphere system.[48, 49] In a hard-sphere system, the time evolution of a dynamic variable is given by the pseudo-Liouville equation

$$\frac{d A(\mathbf{X}^N)}{dt} = \mathcal{L}_{\pm} A(\mathbf{X}^N),$$

(6.1)

where $A(X^N)$ is any function of the phase point $X^N = (X_1, X_2, \ldots, X_N)$ with $X_i = (V_i, R_i)$, and $\mathcal{L}_+ (\mathcal{L}_-)$ is the pseudo-Liouville operator for $t > 0$ $(t < 0)$,

$$\mathcal{L}_\pm = \mathcal{L}_0 \pm \mathcal{L}'_\pm \tag{6.2}$$

The free-streaming operator $\mathcal{L}_0$ is

$$\mathcal{L}_0 = \sum_\alpha \sum_{i=1}^N \Theta_i^\alpha V_i \cdot \nabla_i \tag{6.3}$$

and the collisional part $\mathcal{L}'_\pm$ is given in terms of the binary collision operators $T_\pm^{\alpha\beta}(ij)$ as

$$\mathcal{L}'_\pm = \sum_{\alpha, \beta} \sum_{1 = i < j}^N T_\pm^{\alpha\beta}(ij) \tag{6.4}$$

$$T_\pm^{\alpha\beta}(ij) = |V_{ij} \cdot \hat{R}_{ij}| \theta(\mp V_{ij} \cdot \hat{R}_{ij}) \delta(R_{ij} - \sigma_{\alpha\beta})(b_{ij} - 1)\Theta_i^\alpha \Theta_j^\beta \tag{6.5}$$

Here $\sigma_{\alpha\beta}$ is the mean diameter of the α and β species and the operator b_{ij} changes the velocities of particles i and j to their postcollision values V_i^* and V_j^*,

$$b_{ij} A(X^N) = A(X_1, \ldots, V_i^*, R_i, \ldots, V_j^*, R_j, \ldots, X_N) \tag{6.6}$$

where the postcollision velocities are given by

$$V_i^* = V_i - \frac{2\mu_{ij}}{m_i}(V_{ij} \cdot \hat{R}_{ij})\hat{R}_{ij}$$

and

$$V_j^* = V_j + \frac{2\mu_{ij}}{m_j}(V_{ij} \cdot \hat{R}_{ij})\hat{R}_{ij} \tag{6.7}$$

To write the sums over particles in a convenient way, following the latter part of the discussion on species identification in Section IV, we have introduced the Θ_i^α operators simply as follows:

$$\Theta_i^\alpha = \begin{cases} 1 & \text{if molecule } i \text{ is of species } \alpha \\ 0 & \text{otherwise} \end{cases}$$

We have thus far not considered a reactive system. An important point should be noticed at this stage, namely, the pseudo-Liouville equation is not time reversible. This feature enters the description of hard-sphere systems because of the need to distinguish forward streaming, where particles are going to collide, from backward streaming, where they have collided. As we shall see, this lack of time reversal symmetry has important and convenient implications for the description of reactive systems.

The pseudo-Liouville operator for hard-sphere interactions can be written directly in terms of the binary collision operator $T^{\alpha\beta}$, and this leads to considerable simplifications in the formalism. It is not difficult to treat more general interactions, but a considerable number of manipulations must be carried out to express the results in terms of generalized binary collision operators. To avoid these difficulties, we make use of hard-sphere interactions whenever no violence is done to the qualitative features of the effects we are studying.

In some of the models described above, soft attractive chemical forces are an important ingredient and we must allow for their presence. This is easily done by appending a soft force contribution to the pseudo-Liouville operator written above. Thus, for this more general case,[48]

$$\mathcal{L}_{\pm} = \mathcal{L}_0 \pm \mathcal{L}'_{\pm} + \mathcal{L}_s \tag{6.8}$$

where

$$\mathcal{L}_s = \sum_{\alpha\beta} \sum_{1=i<j}^{N} \Theta_i^{\alpha} \Theta_j^{\beta} \nabla_i V_s^{\alpha\beta}(R_{ij}) \cdot \left(\frac{\partial}{\partial m_i V_i} - \frac{\partial}{\partial m_j V_j} \right) \tag{6.9}$$

with $V_s^{\alpha\beta}(R_{ij})$ the soft potential.

The prescription above is suitable for the description of the time evolution in a nonreactive system. We next consider how reactions might be described in the context of this formalism.

For atomic recombination, the formulation above is complete. In this case a strong chemical force, described by $\mathcal{L}_s$, is assumed to act among the solute atoms, while the hard-sphere solute-solvent and solvent-solvent collisions are described by $T_{\pm}^{\alpha\beta}$ for these pairs. The description of the reactive event then hinges on the specification of a bound molecular species as described in Section IV. Here, however, we turn to the situation described in the latter part of Section IV and construct a dynamic model for the general bimolecular reaction described earlier, where more complex molecular species are described in some average sense as one-level species. It is then up to us to give a prescription for the interconversion of these molecular species. This can be done by introducing collision operators corresponding to reactive collisions.[50–53]

As an illustration of the construction of these reactive collision operators, we first consider the simplest case. Suppose that when an AB pair collides there is a probability α_R that reaction to form CD will take place. For a reaction with a barrier, $\alpha_R \sim \mathcal{O}(e^{-\beta E_j^*})$, whereas for a steric constraint α_R might be related to the probability that the molecules have proper orientation for reaction to occur. Correspondingly, there is a probability $(1 - \alpha_R)$ that the pair will elastically scatter. The collision operator for this process can be written (by analogy with the purely elastic scattering case above) as

$$T_\pm^{\alpha\beta}(ij) = (1 - \alpha_R)T_{E\pm}^{\alpha\beta}(ij) + \hat{T}_{R\pm}^{\alpha\beta}(ij) \qquad (6.10)$$

where $\alpha\beta = $ AB or CD. Although the operator $T_{E\pm}^{\alpha\beta}$ is the same as $T_\pm^{\alpha\beta}$ in (6.5), we have introduced the subscript E to stress that for an AB or CD collision, both elastic and reactive events are possible. If only elastic collisions can occur for the pair, we suppress the E notation. In this simple example, the reactive operator has the form

$$\hat{T}_{R\pm}^{AB}(ij) = \alpha_R |V_{ij} \cdot \hat{R}_{ij}| \theta(\mp V_{ij} \cdot \hat{R}_{ij}) \delta(R_{ij} - \sigma_{AB}) [b_{ij} \mathcal{P}_i^{AC} \mathcal{P}_j^{BD} - 1] \Theta_i^A \Theta_j^B$$
$$(6.11)$$

where $\mathcal{P}_i^{AC}$ changes the species label of molecule i from A to C if it was of type species A. After this labeling change the C and D molecules are given postcollision velocities as if they had elastically scattered. Thus, this is indeed a simple dynamic model for the reaction, and amounts to a "coloring" process with probability α_R. For this model $K_{eq} = 1$. One cautionary remark should be made: we exclude from consideration cases of the diameters of the species changing upon reaction. If this were allowed, the possibility of overlapping hard-sphere configurations would have to be taken into account, leading to a more complex theory.

It is also convenient to write the collision operator in (6.10) as a sum of two terms, one which is independent of α_R and one proportional to α_R,

$$T_\pm^{\alpha\beta}(ij) = T_{E\pm}^{\alpha\beta}(ij) + T_{R\pm}^{\alpha\beta}(ij) \qquad (6.12)$$

where

$$T_{R\pm}^{AB}(ij) = \alpha_R |V_{ij} \cdot \hat{R}_{ij}| \theta(\mp V_{ij} \cdot \hat{R}_{ij}) \delta(R_{ij} - \sigma_{AB}) b_{ij} [\mathcal{P}_i^{AC} \mathcal{P}_j^{BD} - 1] \Theta_i^A \Theta_j^B$$
$$(6.13)$$

As another example, we consider the slightly more complex (and realistic) case of molecules that must have their relative translational energy along the line of centers at contact greater than the activation energy before reaction can occur. Taking into account the possibility that the reaction may be exothermic or endothermic (see Fig. 6.1), we may now introduce a $b_{ij}(\Delta E)$ operator that acts as in (6.6), although now[52]

$$\mathbf{V}_i^{C^*} = \mathbf{V}_i^A - \frac{\mu_{ij}^{AB}}{m_i^A}\left(\mathbf{V}_{ij}^{AB}\cdot\hat{\mathbf{R}}_{ij}\right)\left\{1+\left[1+\frac{\Delta E}{E_{AB}}\right]^{1/2}\right\} \qquad (6.14)$$

and

$$\mathbf{V}_i^{D^*} = \mathbf{V}_j^B + \frac{\mu_{ij}^{AB}}{m_j^B}\left(\mathbf{V}_{ij}^{AB}\cdot\hat{\mathbf{R}}_{ij}\right)\left\{1+\left[1+\frac{\Delta E}{E_{AB}}\right]^{1/2}\right\}$$

where

$$E_{AB} = \tfrac{1}{2}\mu_{AB}\left(\mathbf{V}_{ij}^{AB}\cdot\hat{\mathbf{R}}_{ij}\right)^2 \qquad (6.15)$$

is the relative kinetic energy along the line of centers. The reactive collision operator for this case can again be written in the form of (6.12), with $T_{R\pm}^{AB}$ given by

$$T_{R\pm}^{AB}(ij) = |\mathbf{V}_{ij}\cdot\hat{\mathbf{R}}_{ij}|\theta\left(\mp\mathbf{V}_{ij}\cdot\hat{\mathbf{R}}_{ij}\right)\theta\left(E_{AB}-E_f^*\right)\delta\left(R_{ij}-\sigma_{AB}\right)$$
$$\times\left[b_{ij}(\Delta E)\mathscr{P}_i^{AC}\mathscr{P}_j^{BD}-b_{ij}\right]\Theta_i^A\Theta_j^B \qquad (6.16)$$

The activation energy is not precisely specified in this model, and may contain solvent effects.

Most of the results using this formalism will not depend on the specific nature of the reactive collision operators, but only on their general structure and properties. For this reason, we shall not attempt to construct more elaborate models.

There is one other notational change that will prove useful in the calculations that follow. We may split the reactive collision operators into two parts: a part that changes species,

$$T_{Rr\pm}^{AB}(ij) = \alpha_R\left(\mathbf{V}_{ij}\cdot\hat{\mathbf{R}}_{ij}\right)\theta\left(\mp\mathbf{V}_{ij}\cdot\hat{\mathbf{R}}_{ij}\right)\delta\left(R_{ij}-\sigma_{AB}\right)b_{ij}\mathscr{P}_i^{AC}\mathscr{P}_j^{BD}\Theta_i^A\Theta_j^B$$
$$(6.17)$$

and a part that does not.

$$T_{Rf\pm}^{AB}(ij) = \alpha_R|\mathbf{V}_{ij}\cdot\hat{\mathbf{R}}_{ij}|\theta\left(\mp\mathbf{V}_{ij}\cdot\hat{\mathbf{R}}_{ij}\right)\delta\left(R_{ij}-\sigma_{AB}\right)b_{ij}\Theta_i^A\Theta_j^B \qquad (6.18)$$

Thus

$$T_{R\pm}^{AB}(ij) = T_{Rr\pm}^{AB}(ij) - T_{Rf\pm}^{AB}(ij) \qquad (6.19)$$

A similar decomposition applies for any model of the reactive T operator.

In summary, the major feature of the dynamic model just described is the approximation that solute-solvent and solvent-solvent collisions can be described by hard-sphere interactions. This greatly simplifies the calculations; the formal calculations are not difficult to carry out in the more general case, but the algebra is tedious. We want to describe the effects of solute and solvent dynamics on the reactive process as simply as possible, and the model is ideal for this purpose. Specific reactive events among the solute molecules are governed by the interaction potentials that operate among these species. The particular reactive model described here allows us to examine certain features of the coupling between reaction and diffusion dynamics without recourse to heavy calculations. More realistic treatments must of course be handled via the introduction of species operators for the system under consideration.

C. Structure of the Rate Kernel

The formal linear response analysis of rate coefficient expressions that was carried out in Section V can also be applied to this model of the reacting liquid. Now, however, the starting point of the analysis is the pseudo-Liouville equation (6.1). The lack of time reversal symmetry has important implications, which we shall now discuss.

In the course of the calculation in Section V, we used the fact that $\langle \delta \dot{n}_\alpha(\mathbf{k}) \delta n_\mu(-\mathbf{k}) \rangle = 0$, which follows because $\delta \dot{n}_\alpha$ and δn_μ have opposite parity under time reversal. As a consequence of the different evolution operators for $t > 0$ and $t < 0$, the foregoing result is no longer true. Therefore we briefly sketch the modification in the results of Section V.

Starting from the pseudo-Liouville equation for the solute density fields

$$\frac{d\delta n_\alpha(\mathbf{k}, t)}{dt} = \mathcal{L}_+ \delta n_\alpha(\mathbf{k}, t) \qquad \text{for} \quad t > 0 \qquad (6.20)$$

we may rewrite this equation by using the projection operator in (5.7) to find the exact equation

$$\frac{dC_{\alpha\beta}(\mathbf{k}, t)}{dt} = - \sum_{\mu=A}^{D} \phi_{\alpha\mu}^s(\mathbf{k}) C_{\mu\beta}(\mathbf{k}, t)$$

$$- \sum_{\mu=A}^{D} \int_0^t dt' \, \phi_{\alpha\mu}^d(\mathbf{k}; t') C_{\mu\beta}(\mathbf{k}, t - t') \qquad (6.21)$$

The lack of time reversal symmetry manifests itself by the appearance of a

R. KAPRAL

singular contribution to the memory kernel, the ϕ^s-matrix, with elements

$$\phi^s_{\alpha\mu}(\mathbf{k}) = -\langle [\, \mathcal{L}_+ \delta n_\alpha(\mathbf{k}) \,] \delta n_\mu(-\mathbf{k}) \rangle (N^\mu_{eq})^{-1} \qquad (6.22)$$

The dynamic contribution to the damping matrix is also slightly modified because ϕ^s now appears. We have

$$\phi^d_{\alpha\mu}(\mathbf{k}; t) = \langle f^\dagger_\mu(\mathbf{k}) e^{\mathcal{QL}_+ t} f_\alpha(\mathbf{k}) \rangle (N^\mu_{eq})^{-1} \qquad (6.23)$$

where

$$f_\alpha(\mathbf{k}) = \mathcal{QL}_+ \delta n_\alpha(\mathbf{k}) = \mathcal{L}_+ \delta n_\alpha(\mathbf{k}) + \sum_\mu \phi^s_{\alpha\mu}(\mathbf{k}) \delta n_\mu(\mathbf{k}) \qquad (6.24)$$

The rate kernel is again most conveniently obtained by taking the Laplace transform of (6.21),

$$\sum_{\mu=A}^{D} \left\{ z\delta_{\alpha\mu} + \left[\phi^s_{\alpha\mu}(\mathbf{k}) + \phi^d_{\alpha\mu}(\mathbf{k}; z) \right] \right\} C_{\mu\beta}(\mathbf{k}, z) = C_{\alpha\beta}(\mathbf{k}) \qquad (6.25)$$

and identifying $k_{\alpha\mu}(z)$ as

$$k_{\alpha\mu}(z) \equiv \lim_{k \to 0} \left[\phi^s_{\alpha\mu}(\mathbf{k}) + \phi^d_{\alpha\mu}(\mathbf{k}; z) \right] \qquad (6.26)$$

The rate kernel can therefore be written as the sum of a z-independent part

$$\lim_{k \to 0} \phi^s_{\alpha\mu}(\mathbf{k}) \equiv k^0_{\alpha\mu} = \langle [\, \mathcal{L}_+ \delta N_\alpha \,] \delta N_\mu \rangle (N^\mu_{eq})^{-1} \qquad (6.27)$$

and a z-dependent part

$$\lim_{k \to 0} \phi^d_{\alpha\mu}(\mathbf{k}, z) \equiv \Delta k_{\alpha\mu}(z) = \lim_{k \to 0} \int_0^\infty dt\, e^{-zt} \langle f^\dagger_\mu(\mathbf{k}=0) e^{\mathcal{QL}_+ t} f_\alpha(\mathbf{k}=0) \rangle (N^\mu_{eq})^{-1} \qquad (6.28)$$

Thus,

$$k_{\alpha\mu}(z) = k^0_{\alpha\mu} + \Delta k_{\alpha\mu}(z) \qquad (6.29)$$

This result implies that $k_{\alpha\mu}(t)$ is the sum of a singular part and a relaxing part,

$$k_{\alpha\mu}(t) = 2k^0_{\alpha\mu} \delta(t) + \Delta k_{\alpha\mu}(t) \qquad (6.30)$$

Some insight into this expression can be gained by examining the structure of $k^0_{\alpha\mu}$ for the reaction models described above. Consider the calculation of k^0_{AA}. We have

$$
k^0_{AA} \equiv k^0_f n^B_{eq} = -\left\langle \delta N_A \sum_{i<j}^N T^{AB}_{R+}(ij)\delta N_A \right\rangle (N^A_{eq})^{-1}
$$

$$
= \left\langle \sum_{i,j=1}^N \theta(-\mathbf{V}_{ij}\cdot\hat{\mathbf{R}}_{ij})|\mathbf{V}_{ij}\cdot\hat{\mathbf{R}}_{ij}|\theta(E_{AB}-E^*_f)\delta(R_{ij}-\sigma_{AB})\Theta^A_i\Theta^B_j \right\rangle (N^A_{eq})^{-1}
$$

$$
(6.31)
$$

where, consistent with the condition that the solute species are dilute in the solvent, terms lowest order in the solute densities have been kept. This is just the equilibrium average one-way flux of A and B molecules with sufficient energy along the line of centers to surmount the barrier across the reaction surface located at $R_{ij}=\sigma_{AB}$, that is, the transition state theory expression for the rate coefficient.[27, 54, 55] Direct calculation of (6.31) leads to the usual transition state result

$$
k^0_f = \sigma^2_{AB}\left(\frac{8\pi k_B T}{\mu_{AB}}\right)^{1/2} e^{-Q/k_B T}
\tag{6.32}
$$

where the free energy at the barrier, Q, is given by

$$
e^{-Q/k_B T} \equiv g_H(\sigma_{AB})\exp-\left[\frac{\varepsilon_A+\varepsilon_B+E^*_f}{k_B T}\right]
$$

$$
= \exp-\left[\frac{W_H(\sigma_{AB})+\varepsilon_A+\varepsilon_B+E^*_f}{k_B T}\right]
\tag{6.33}
$$

where W_H is the hard-sphere potential of mean force,

$$
W_H(r) = -k_B T \ln g_H(r)
\tag{6.34}
$$

Solvent effects enter through the potential of mean force and the activation energy; they may cancel or nearly cancel in the expression for k^0_f (cf. Northrup and Hynes[15]). The collision frequency per unit density of B, $\sigma^2_{AB}(8\pi k_B T/\mu_{AB})^{1/2}$, in the expression for k^0_f for a bimolecular reaction, takes the place of the frequency ω_0 in (3.23) for an isomerization reaction.[21, 44] This analysis shows that the transition state expression for the rate coefficient appears in this theory as a singular contribution to the rate kernel for the hard-sphere model of the reaction.

Thus this formulation of the reaction problem provides a convenient and natural separation of the rate kernel into a direct collisional, one-way flux contribution of the transition state theory form $k^0_{\alpha\mu}$, and a relaxing part $\Delta k_{\alpha\mu}(t)$. The relaxing part contains the effects of solute and solvent dynamics on the rate process and provides corrections to $k^0_{\alpha\mu}$ arising from a nonequilibrium in configuration and velocity spaces generated by the reaction. In the simple diffusion equation theory described in Section III.A, the relaxing contribution is explicitly given by the second term in (3.9) and arises solely from a nonequilibrium in configuration space. This rate kernel is displayed in Fig. 6.2. We expect that more realistic treatments will yield a rate kernel expression with some of these features, for example, a singular, "one-way flux" part and a relaxing part that varies as $t^{-3/2}$ for long times. The short time behavior of $\Delta k_{\alpha\mu}(t)$ is of course not accurately given by the diffusion equation theory. Section X analyzes the kinetic theory expression for the rate kernel and shows what approximations to this more general theory lead to the diffusion equation result.

If the reaction can not be modeled by an impulsive collision event, as for atomic recombination or some isomerization reactions, then $\phi^s_{\alpha\mu}$ is zero by time reversal symmetry as discussed earlier. Nevertheless, the structure of the rate kernel has the same qualitative structure as described above.[32] To

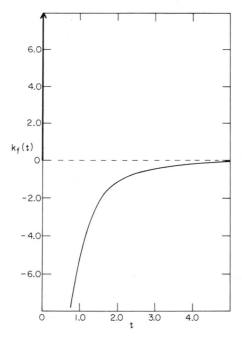

Fig. 6.2. Rate kernel in units of $k_i^2 D_A / k_D R^2$ as a function of time (tD_A/R^2) for simple diffusion equation dynamics. The ratio $k_i/k_D = 1.0$ for this graph. The heavy vertical axis indicates the singular contribution to the rate kernel. For diffusion equation dynamics, $\Delta k_f(t)$ also diverges at $t=0$ (cf. (3.9)). This will not be true in general. The breakdown of the simple diffusion model comes as no surprise. Apart from this, the gross features of this diagram illustrate the general situation: a singular part followed by a negative relaxing part, which decays as $t^{-3/2}$ for long times.

see this, one need only calculate the rate coefficient in (5.24) for a small positive time ε, which is taken to zero at the end of the calculation,

$$
\begin{aligned}
k_{\alpha\mu}^{\varepsilon} &= \int_0^{\varepsilon} dt \langle R_{\mu} e^{i\mathcal{L}t} R_{\alpha} \rangle (N_{eq}^{\mu})^{-1} \\
&= \langle \delta \dot{N}_{\mu} \delta N_{\alpha}(\varepsilon) \rangle (N_{eq}^{\mu})^{-1}
\end{aligned}
\tag{6.35}
$$

For the isomerization reaction described in Section IV, this reduces to

$$
\begin{aligned}
\lim_{\varepsilon \to 0} k_{AA}^{\varepsilon} &= \lim_{\varepsilon \to 0} \left\langle \sum_i^{N_A} v_i \delta(q_i - q_c) \theta(q_c - q_i(\varepsilon)) \right\rangle (N_{eq}^A)^{-1} \\
&= \left\langle \sum_i^{N_A} v_i \delta(q_i - q_c) \theta(v_i) \right\rangle (N_{eq}^A)^{-1}
\end{aligned}
\tag{6.36}
$$

the usual transition state theory result.[32] A similar result holds for other types of reactions. Thus, once again the rate kernel consists of a singular "one-way flux" part and a relaxing contribution.

VII. KINETIC EQUATIONS

The previous sections have attempted to provide some insight into the form of the microscopic expressions for the rate kernels and rate coefficients that characterize condensed-phase reactions. Although the "equilibrium" one-way flux rate coefficient k_f^0 is relatively easy to calculate and under certain circumstances may yield an adequate description of the rate, a variety of important dynamic effects are contained in the relaxing part of the rate kernel, $\Delta k_{\alpha\mu}(t)$. In this section, we describe a kinetic theory that provides a means of examining the collision events that enter in $\Delta k_{\alpha\mu}(t)$.

We present the formal structure of the theory and motivate its development, and then give some examples.

A. Background

The kinetic theory of condensed-phase chemical reactions is a direct outgrowth of kinetic theory and mode coupling descriptions of dense, simple fluids, which have been developed primarily in the past 10 years. This work in turn relies on an older body of literature, but we shall, when possible, draw parallels with the more recent interpretations of liquid-state dynamics. At present there are a variety of techniques available for constructing kinetic equations that are useful for describing dense, simple, nonreacting liquids. These range from approaches based on the dynamic hierarchy[56-59] to renormalized kinetic theory[60-63] and generalized Langevin equation methods.[64, 65] These methods are ultimately equivalent and occasionally differ in their predictions as a result of the introduction of approximations that are required to make the theories tractable. Although we have used all these methods to study condensed-phase reactions, to present a coherent discussion, we focus solely on the formulation based on the generalized Langevin equation.[66, 67] This approach has the added advantage of being a natural extension of the linear response theory description in the earlier sections.

Before presenting the mathematical formulation of the theory, it is useful to describe the types of dynamic event that are likely to be important for the description of liquid-state reactions; this discussion shows why current theories of the dynamics of simple liquids are relevant for this problem.

One of the primary features distinguishing liquid-state reactions from those that occur in dilute gases is the enhanced probability of reaction, which arises from the presence of solvent molecules in high concentration. Such solvent effects in condensed-phase chemical kinetics are often referred to as "cage" effects.[68] The schematic diagram in Fig. 7.1 shows a configuration that may occur in the course of a bimolecular reaction in a liquid. The solvent molecules trap (cage) the solute pair, and thus the rate coefficient will be different from the gas-phase case. There are both static and dynamic aspects to this cage effect. On the static level, we expect that the solvent will affect the interaction potential between the reactive pair of molecules. Thus the "bare" (gas-phase) intermolecular potential will be replaced by the potential of mean force, which accounts for the solvent

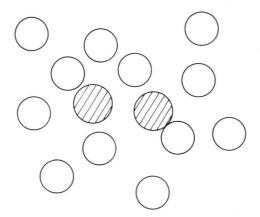

Fig. 7.1. A caged solute pair in a high-density solvent.

structure. This was discussed previously (especially Section IV). Dynamic features also enter in the description of the cage effect.

The presence of a high density of solvent molecules leads to recollisions between the potentially reactive pair of molecules. Some examples of such recollision events are shown schematically in Fig. 7.2. In Fig. 7.2a the solute molecules A and B collide elastically, and after collision of A with a solvent molecule S, the A molecule recollides with B and reacts. An event of this type is extremely unlikely in a dilute gas. The description of collision sequences of this type is outside of the scope of a Boltzmann equation, which accounts only for *uncorrelated* binary collision events. Collision sequences of this kind are expected to play an increasingly important role as the solvent density increases, and as we shall see, they are often the dominant contribution to the dynamics. A similar sequence of reactive events is shown in Fig. 7.2b.

These ring collision events are now a familiar part of the kinetic theory description of dynamic processes in simple dense fluids.[63] A brief comparison of the theory for the velocity autocorrelation function with that for the chemically reacting fluid will help motivate our description. Recent developments in the theory of the velocity autocorrelation function have arisen out of an attempt to understand the slow $t^{-3/2}$ power law decay observed by Alder and Wainwright in a computer simulation of a dense hard-sphere fluid.[69] This work also showed that the translational motion of a small hard sphere in a fluid of similar hard spheres has a significant collective (hydrodynamic) component. On the theoretical side,[70] this type of behavior was discussed from the kinetic theory point of view in terms of the ring collision events[71] described above and provided a microscopic basis for the introduction of collective effects.[72] In addition, it was shown that mode

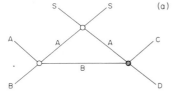

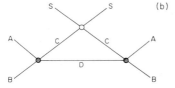

Fig. 7.2. Examples of correlated collision events. (a) Solute molecules A and B first collide elastically; after collision of A with S, A and B recollide to produce C and D. (b) Molecules A and B initially collide to produce C and D; after C collides with S, C and D collide to produce A and B.

coupling theories provided another route for the inclusion of these collective effects.[73, 74] The most recent work has focused on the detailed calculation of the velocity correlation function and corresponding self-diffusion coefficient[63, 75–77] as well as the microscopic basis of Stokes' law from kinetic theory.[78]

There is a close parallel between this development and the microscopic theory of condensed-phase chemical reactions. First, the questions one asks are very nearly the same. In Section III we summarized several configuration space approaches to this problem. These methods assume the validity of a diffusion or Smoluchowski equation, which is based on a continuum description of the solvent. Such theories will surely fail at the close encounter distance required for reaction to take place. In most situations of chemical interest, the solute and solvent molecules are comparable in size and the continuum description no longer applies. Yet we know that these simple approaches are often quite successful, even when applied to the small molecule case. Thus we again have a microscopic relaxation process exhibiting a strong hydrodynamic component. This hydrodynamic component again gives rise to a power law decay ($t^{-3/2}$) in the rate kernel[79] (cf. Section III). In addition, there is a strong parallel between the microscopic basis of Stokes' law ($\zeta = c\pi\eta R$, where ζ is the friction coefficient, η the viscosity, R the radius, and c a constant that depends on boundary conditions) for small molecule motion, and the Smoluchowski value of the rate coefficient, $k_D = 4\pi DR$ (cf. Section III).

In many respects, at a superficial level, the theory for the chemical reaction problem is much simpler than for the velocity autocorrelation function. The simplifications arise because we are now dealing with a scalar transport phenomenon, and it is the diffusive modes of the solute molecules that are coupled. In the case of the velocity autocorrelation function, the coupling of the test particle motion to the collective fluid fields (e.g., the viscous mode) must be taken into account. At a deeper level, of course, the same effects must enter into the description of the reaction problem, and one is faced with the problem of the microscopic treatment of the correlated motion of a pair of molecules that may react. In the following sections, we attempt to clarify and expand on these parallels.

B. General Theory

We now consider the construction of kinetic equations suitable for the description of the dynamics in a reacting liquid. Let $F_\alpha(1, t)$ be the distribution function for species α, and $1 = \mathbf{x}_1 = (\mathbf{r}_1, \mathbf{v}_1)$ the field point in the fluid. The distribution function may be regarded as the nonequilibrium average of the phase-space density

$$n_\alpha(\mathbf{x}_1) = \sum_{i=1}^N \delta(\mathbf{x}_1 - \mathbf{X}_i)\Theta_i^\alpha \tag{7.1}$$

Thus $F_\alpha(1, t) = \overline{n_\alpha(\mathbf{x}_1, t)}$. An exact equation of motion for the distribution function can be written in terms of a hierarchy that relates the lower order distribution functions to higher order ones. The first member of the hierarchy is well known and for hard-sphere interactions may be written in the following form[40] (cf. also Appendix B).

$$\frac{\partial F_\alpha(1, t)}{\partial t} = - \left\{ \mathbf{v}_1 \cdot \nabla_1 + \frac{\mathbf{F}_{ex}}{m_1} \cdot \nabla_{v_1} \right\} F_\alpha(1, t)$$
$$+ \sum_\beta \int d\mathbf{x}_2 \, \overline{T}_-^{\alpha\beta}(12) F_{\alpha\beta}(12, t)$$

The first term on the right-hand side gives the change in F_α due to the free streaming in the external force $\mathbf{F}_{ex}$; the second term contains the effects of collisions. [The relation between the $\overline{T}$ operator and the previously introduced T operator is given in (B.9).]

If we factor the two-particle distribution function,

$$F_{\alpha\beta}(12, t) \simeq g^{\alpha\beta}(r_{12}) F_\alpha(1, t) F_\beta(2, t)$$

the nonlinear Boltzmann-Enskog equation results;

$$\frac{\partial F_\alpha(1, t)}{\partial t} = - \left\{ \mathbf{v}_1 \cdot \nabla_1 + \frac{\mathbf{F}_{ex}}{m_1} \cdot \nabla_{v_1} \right\} F_\alpha(1, t)$$
$$+ \sum_\beta \int d\mathbf{x}_2 \, \overline{T}^{\alpha\beta}(12) g^{\alpha\beta}(r_{12}) F_\alpha(1, t) F_\beta(2, t)$$

This equation accounts only for dynamically uncorrelated collisions, and thus misses many of the important classes of correlated collision events discussed earlier. We need a new kinetic equation that does not suffer from this limitation.

If we wish to examine only the structure of the rate kernel, we know that it is sufficient to study the decay of fluctuations about equilibrium. If we let

$$\delta F_\alpha(1, t) = F_\alpha(1, t) - n_\alpha^{eq} \phi_\alpha(v_1)$$

and linearize about complete equilibrium, we obtain the linearized form of the Boltzmann-Enskog equation. The linearized collision operator is

$$\sum_\beta g^{\alpha\beta}(\sigma_{\alpha\beta}) \int dx_2 \, \bar{T}^{\alpha\beta}(12) \big[n_\alpha^{eq} \phi_\alpha(v_1) \delta F_\beta(2, t) + n_\beta^{eq} \phi_\beta(v_2) \delta F_\alpha(1, t) \big]$$

$$\equiv \sum_\beta K_{\alpha\beta}^{BE}(1, \bar{1}') \delta F_\beta(\bar{1}', t)$$

The second line defines the linearized Boltzmann-Enskog collision operator. The bar over a variable signifies that that variable is integrated over.

The form of the Boltzmann-Enskog collision operator is thus specified; out task is to find its generalization. We denote the general collision operator by $K_{\alpha\beta}(1, \bar{1}'; t')$, where we have allowed for the possibility that it may be nonlocal in time as well as space. The general kinetic equation may then be written as

$$\frac{\partial \delta F_\alpha(1, t)}{\partial t} = -\left\{ v_1 \cdot \nabla_1 + \frac{F_{ex}}{m_1} \cdot \nabla_{v_1} \right\} \delta F_\alpha(1, t)$$

$$+ \sum_\beta \int_0^t dt' \, K_{\alpha\beta}(1, \bar{1}'; t') \delta F_\beta(\bar{1}', t-t') \qquad (7.2a)$$

Using an operator notation for $K_{\alpha\beta}$, and taking the Laplace transform, we may also write

$$\left\{ z + v_1 \cdot \nabla_1 + \frac{F_{ex}}{m_1} \cdot \nabla_{v_1} - K_{\alpha\alpha}(1; z) \right\} \delta F_\alpha(1, z)$$

$$= \sum_{\alpha \neq \beta} K_{\alpha\beta}(1; z) \delta F_\beta(1, z) + \delta F_\alpha(1) \qquad (7.2b)$$

In the spirit of the discussion of Section II, we may consider the time evolution of the correlation functions of $\delta n_\alpha(x_1)$ rather than its nonequilibrium average. These correlation functions will satisfy the same kinetic equations as the $\delta F_\alpha(1, t)$. This is the approach we take here. We now outline a method for explicitly constructing $K_{\alpha\beta}(1; z)$.

The procedure for constructing kinetic equations using the generalized Langevin equation is well known[64, 66]; one uses as variables in this description the *phase*-space density fields. We could of course simply use the solute phase-space fields, (7.1), and follow the methods of Section V to obtain a formal kinetic equation for their time evolution. This procedure would not, however, lead to a useful result, since the kernel in the kinetic equation would be a complex memory function whose structure would be difficult to compute. What is needed is a systematic method for extracting from the memory kernel the dynamic processes that are important for the description of the rate kernel in a liquid. We argued above that the new feature, which must be taken into account for this case, is the (diffusive) approach of the potentially reactive molecules and the corresponding re-collision events, which occur because of the presence of a high concentration of solvent molecules. Thus, in the spirit of mode coupling methods, one might expect such effects to be contained in higher order (e.g., pair density fields), which describe the correlated motion of the solute and solvent molecules. The introduction of these fields explicitly into the formulation will then lead to a theory that clearly shows how such correlated motions enter into the description of the rate coefficient.

There are also compelling reasons for the introduction of these fields from more formal considerations. The usefulness of the linear response formulation in Section V hinges on the assumption that the linear density fields vary more slowly than other variables in the system. It is known that this is not the case even for fields that are conserved in the zero wave vector limit.[41] Consider, for instance, the solute density field $n_\alpha(\mathbf{k})$ for a nonreacting fluid. As discussed in Section V, for small $\mathbf{k}$ this field will decay slowly in time. But so will the product field $n_\alpha(\mathbf{q})n_\beta(\mathbf{k}-\mathbf{q})$, provided both $\mathbf{k}$ and $\mathbf{q}$ are small, since

$$\frac{d}{dt}\left[n_\alpha(\mathbf{q})n_\beta(\mathbf{k}-\mathbf{q}) \right] = -i\mathbf{q}\cdot\mathbf{j}_\alpha(\mathbf{q})n_\beta(\mathbf{k}-\mathbf{q})$$

$$-i(\mathbf{k}-\mathbf{q}))j_\beta(\mathbf{k}-\mathbf{q})n_\alpha(\mathbf{q}) \qquad (7.3)$$

These are, of course, just the standard arguments used to motivate the mode coupling theories, which have proved useful in the description of critical phenomena[41] and the asymptotic decay of correlation functions.[70]

With this background in mind, we may proceed to construct a formal kinetic theory for a condensed-phase, chemically reacting system along the following lines.[66, 67] We shall be interested primarily in the time development of the solute phase-space densities $n_\alpha(\mathbf{x}_1)$, (7.1). Recall that $n_\alpha(\mathbf{x}_1)$ is a function of the phase point $\mathbf{X}_i = (\mathbf{R}_i, \mathbf{V}_i)$ and depends parametrically on $\mathbf{x}_1 = (\mathbf{r}_1, \mathbf{v}_1)$, the *field point* in the fluid. Following the arguments given above, the dynamics of these singlet fields is best exposed by considering the coupling to higher order phase-space density fields, for example, the doublet field

$$n_{\alpha_1\alpha_2}(\mathbf{x}_1\mathbf{x}_2) = \sum_{i_1, i_2 = 1}^{N} \delta(\mathbf{x}_1 - \mathbf{X}_{i_1})\delta(\mathbf{x}_2 - \mathbf{X}_{i_2})\Theta_{i_1}^{\alpha_1}\Theta_{i_2}^{\alpha_2} \qquad (7.4)$$

Higher order fields may be introduced in a completely analogous fashion by defining

$$n_{\alpha_1\alpha_2\cdots\alpha_n}(\mathbf{x}_1\cdots\mathbf{x}_n)= \sum_{i_1,i_2,\ldots,i_n=1}^{N} \delta(\mathbf{x}_1-\mathbf{X}_{i_1})\cdots\delta(\mathbf{x}_n-\mathbf{X}_{i_n})\Theta_{i_1}^{\alpha_1}\Theta_{i_2}^{\alpha_2}\cdots\Theta_{i_n}^{\alpha_n}$$

(7.5)

These fields may be written in a more compact way by defining $\boldsymbol{\alpha}= \{\alpha_1,\alpha_2,\ldots\}$ and $\mathbf{x}=\{\mathbf{x}_1,\mathbf{x}_2,\ldots\}$. The general phase-space density field is then given by $n_\alpha(\mathbf{x})$. In the course of carrying out calculations using such fields, it is convenient to work with a set of orthogonal fields. In addition, in the spirit of the discussion in Section II, we shall focus on the decay of fluctuations from equilibrium, thus considering the deviation of the above-mentioned fields from their equilibrium values. We shall denote such orthogonal fields whose equilibrium averages vanish by $\delta n_\alpha(\mathbf{x})$. The phase-space density correlation functions are defined by

$$C_{\alpha,\alpha'}(\mathbf{x},\mathbf{x}';t)=\langle\delta n_\alpha(\mathbf{x},t)\delta n_{\alpha'}(\mathbf{x}')\rangle \tag{7.6}$$

The kinetic equations for these correlation functions then follow by application of standard projection operator techniques.[36, 37] We first introduce a projection operator onto these fields by

$$\mathcal{P}A_\alpha(\mathbf{x})= \sum_{\alpha'} \int d\mathbf{x}'\,d\mathbf{x}''\langle A_\alpha(\mathbf{x})\delta n_{\alpha'}(\mathbf{x}')\rangle C_{\alpha',\alpha'}^{-1}(\mathbf{x}',\mathbf{x}'')\delta n_{\alpha'}(\mathbf{x}'') \tag{7.7}$$

and use this projection operator to write the pseudo-Liouville equation

$$\frac{d\delta n_\alpha(\mathbf{x},t)}{dt} =\mathcal{L}_+\delta n_\alpha(\mathbf{x},t) \tag{7.8}$$

in the alternate form

$$\frac{d\delta n_\alpha(\mathbf{x},t)}{dt} = -\sum_{\alpha'}\int d\mathbf{x}'\,\phi_{\alpha,\alpha'}^s(\mathbf{x},\mathbf{x}')\delta n_{\alpha'}(\mathbf{x}',t)$$

$$-\int_0^t dt'\sum_{\alpha'}\int d\mathbf{x}'\,\phi_{\alpha,\alpha'}^d(\mathbf{x},\mathbf{x}';t')\delta n_{\alpha'}(\mathbf{x}',t-t')+f_\alpha(\mathbf{x},t)$$

(7.9)

The static and dynamic memory kernel matrix elements $\phi_{\alpha,\alpha'}^s$ and $\phi_{\alpha,\alpha'}^d$ have definitions similar to those in Section VI,

$$\phi_{\alpha,\,\alpha'}^{s}(\mathbf{x},\mathbf{x}') = -\int d\mathbf{x}'' \langle [\,\mathcal{L}_+ \delta n_\alpha(\mathbf{x})\,] \delta n_{\alpha'}(\mathbf{x}'') \rangle C_{\alpha',\,\alpha'}^{-1}(\mathbf{x}'',\mathbf{x}') \qquad (7.10)$$

and

$$\phi_{\alpha,\,\alpha'}^{d}(\mathbf{x},\mathbf{x}';\,t) = \int d\mathbf{x}'' \langle [\,e^{\mathcal{Q}\mathcal{L}_+ t} f_\alpha(\mathbf{x})\,] f_{\alpha'}^{+}(\mathbf{x}'') \rangle C_{\alpha',\,\alpha'}^{-1}(\mathbf{x}'',\mathbf{x}') \qquad (7.11)$$

with

$$f_\alpha(\mathbf{x}) = (1-\mathcal{P})\mathcal{L}_+ \delta n_\alpha(\mathbf{x}) \equiv \mathcal{Q}\mathcal{L}_+ \delta n_\alpha(\mathbf{x}) \qquad (7.12)$$

and

$$f_\alpha^{+}(\mathbf{x}) = -\mathcal{Q}\mathcal{L}_- \delta n_\alpha(\mathbf{x}) \qquad (7.13)$$

In view of the definition of $f_\alpha(\mathbf{x})$ in (7.12), it is clear that $\langle f_\alpha(\mathbf{x})\delta n_{\alpha'}(\mathbf{x}') \rangle = 0$; thus the correlation function $C_{\alpha,\,\alpha''}(\mathbf{x},\mathbf{x}'';\,z)$ satisfies the equation

$$zC_{\alpha,\,\alpha''}(\mathbf{x},\mathbf{x}'';\,z) + \sum_{\alpha'} \int d\mathbf{x}' [\,\phi_{\alpha,\,\alpha'}^{s}(\mathbf{x},\mathbf{x}') + \phi_{\alpha,\,\alpha'}^{d}(\mathbf{x},\mathbf{x}';\,z)\,] C_{\alpha',\,\alpha''}(\mathbf{x}',\mathbf{x}'';\,z)$$

$$= C_{\alpha,\,\alpha''}(\mathbf{x},\mathbf{x}'') \qquad (7.14)$$

The low-order phase-space density correlation functions are the quantities of primary interest to us. For example, the singlet correlation function $C_{\alpha\alpha'}(\mathbf{x}_1,\mathbf{x}_1';\,z) \equiv C_{\alpha\alpha'}(1,1';\,z)$ yields information about the configuration space correlation functions, discussed in connection with the phenomenological rate laws in Section II, when integrations over field point velocities are carried out [cf. (2.15)],

$$C_{\alpha,\,\alpha'}(\mathbf{r}_1,\mathbf{r}_1';\,z) = \int d\mathbf{v}_1 d\mathbf{v}_1' C_{\alpha,\,\alpha'}(1,1';\,z) \qquad (7.15)$$

Also [cf. (5.17)],

$$C_{\alpha,\,\alpha'}(z) = \int d\mathbf{x}_1 d\mathbf{x}_1' C_{\alpha,\,\alpha'}(1,1';\,z) \qquad (7.16)$$

Thus, once the kinetic equation for $C_{\alpha,\,\alpha'}(1,1';\,z)$ is known, the validity and microscopic basis of the phenomenological rate law can be investigated. Equation 7.14 is such a kinetic equation. Before providing explicit examples of this formalism, we discuss a few properties of the structure of these equations.

In Appendix B we show that the action of the pseudo-Liouville operator on the singlet field generates a coupling to the doublet field, and so on. One may show that if the phase-space density fields are defined as in (7.5), if density fields up to the nth order are included explicitly in the description, the random "forces" corresponding to all fields lower than the nth are zero. Hence only the damping matrix corresponding to this nth-order field is nonzero. As an example, consider the case of singlet and doublet fields that are explicitly treated. In this case, (7.14) reduces to two coupled equations of the form

$$zC_{\alpha\beta}(1,1';z) + \sum_{\mu} \phi^s_{\alpha,\mu}(1,\bar{1}'')C_{\mu\beta}(\bar{1}'',1';z)$$

$$+ \sum_{\mu_1\mu_2} \phi^s_{\alpha,\mu_1\mu_2}(1,\bar{1}''\bar{2}'')C_{\mu_1\mu_2,\beta}(\bar{1}''\bar{2}'',1';z) = C_{\alpha\beta}(1,1') \qquad (7.17)$$

and

$$zC_{\mu_1\mu_2,\beta}(12,1';z) + \sum_{\nu_1\nu_2} \left[\phi^s_{\mu_1\mu_2,\nu_1\nu_2}(12,\bar{1}''\bar{2}'') + \phi^d_{\mu_1\mu_2,\nu_1\nu_2}(12,\bar{1}''\bar{2}'';z) \right]$$

$$C_{\nu_1\nu_2,\beta}(\bar{1}''\bar{2}'',1';z) + \sum_{\mu} \phi^s_{\mu_1\mu_2,\mu}(12,\bar{1}'')C_{\mu\beta}(\bar{1}'',1';z) = 0$$

$$(7.18)$$

In these equations, as noted earlier, a bar over an argument signifies that that variable is to be integrated over. These two coupled equations may now be formally solved for $C(1,1';z)$ (we introduce a matrix notation),

$$\{z + \phi^s(1,\bar{1}'') - R(1,\bar{1}'';z)\}C(\bar{1}'',1';z) = C(1,1') \qquad (7.19)$$

with

$$R(1,1';z) = \phi^s(1,1'\bar{2}') \left[z + \phi^s(1'\bar{2}';\bar{1}''\bar{2}'') + \phi^d(1'\bar{2}',\bar{1}''\bar{2}'';z) \right]^{-1}$$

$$\times \phi^s(\bar{1}''\bar{2}'',1') \qquad (7.20)$$

This is the standard form of a mode-coupling result.[73] The "matrices"

$\phi^s(1, 1'2')$ and $\phi^s(1''2'', 1')$ couple the linear and nonlinear fields, and the inverse operator in (7.20) gives the evolution of the nonlinear field. If higher order nonlinear fields are included explicitly, the structure of $\phi^d(12, 1'2'; z)$ is elaborated and has a form very similar to that of $\mathbf{R}$ itself, but involving the propagation of these higher order fields. A phenomenological version of a mode-coupling theory of reactions has been given by Jhon and Dahler.[80]

To show that the formal theory leads to useful results, we now consider some specific examples. These examples show that $\phi^s(1, 1')$ contains free-streaming and uncorrelated collision effects, while correlated collision terms reside in $\mathbf{R}(1, 1'; z)$. Thus an explicit expression for the collision kernel $K_{\alpha\beta}$ in (7.2) can be obtained by this method.

In summary, the strategy of these calculations is to explicitly consider products of phase-space density fields in the theory. These product fields describe the correlated motion of the solute and solvent molecules. An examination of the coupling of these higher order fields to the solute fields should then lead to a precise description of the effects on single-particle dynamics of correlated motion of many particles.

C. Singlet Field Equation

As a first example, we consider the reaction $A + B \rightleftarrows C + B$. The phenomenological description was given in Section II. We now derive the kinetic equation appropriate for this case and restrict our considerations to explicit inclusion of up to doublet phase-space fields. In Section VII.D we consider the effects of the higher order fields. The appropriate fields can be selected using the following considerations: We are interested in the time development of the singlet solute densities δn_α ($\alpha = A, B, C$). The action of the $\mathcal{L}_+$ operator on these densities leads to a coupling to the doublet fields $\delta n_{\alpha S}$ ($\alpha = A, B, C$), δn_{AB} and δn_{CB} (cf. Appendix B) for this dilute reacting system. A straightforward calculation of the matrix elements in (7.19) then yields the desired kinetic equation. We give these details in Appendix C and simply present and discuss the resulting kinetic equation here.

To present the results in the clearest possible fashion, we give only the kinetic equation for the irreversible reaction

$$A + B \overset{k_f}{\rightarrow} C + B$$

Our description of reactions in terms of fluctuations about equilibrium is consistent with this irreversible case under some conditions. We consider a small fluctuation from complete equilibrium. In the subsequent decay of this fluctuation back to equilibrium we neglect reverse rate processes. By the regression hypothesis, the equations governing the decay of these fluctuations should have the same form as the macroscopic equations, with the same neglect of reverse rate processes. This will be a useful description of the chemical relaxation, provided the equilibrium constant is very large. (Modifications of the results when the reverse reaction is included are given in Ref. 51.)

The equations in this form are especially suitable for comparison with diffusion equation approaches, which are most often applied to the irreversible reaction case. The reaction scheme we have selected is also convenient because B clearly plays the role of the "sink" in the diffusion equation approaches, and a rather direct comparison with these methods is possible.

The kinetic equation for the A species phase-space correlation function is

$$\{z + \mathbf{v}_1 \cdot \nabla_1 - K_{AA}(1; z)\} C_{AA}(1, 1'; z) = C_{AA}(1, 1') \qquad (7.21a)$$

where the collision operator is given by

$$K_{AA}(1; z) = \Lambda_-^{AS}(1) - \Lambda_{Rf-}^{AB}(1) + R^{AS}(1; z) + R^{AB}(1; z) \qquad (7.21b)$$

We also write (7.21a) as

$$\{z - L_R(1; z)\} C_{AA}(1, 1'; z) = C_{AA}(1, 1') \qquad (7.22)$$

The contributions to the collision operator $K_{AA}(1; z)$ describe the following types of dynamic event: the Λ operators are Enskog collision operators and describe uncorrelated binary collision events; Λ_-^{AS} describes uncorrelated elastic collisions of A with solvent molecules

$$\Lambda_-^{AS}(1) = n_{eq}^s g^{AS}(\sigma_{AS}) \int d\mathbf{x}_2 \, T_-^{AS}(12) \phi_s(v_2) \qquad (7.23a)$$

and Λ_{Rf-}^{AB} describes the reactive collisions of A with B,

$$\Lambda_{Rf-}^{AB}(12) = n_{eq}^B g^{AB}(\sigma_{AB}) \int d\mathbf{x}_2 \, T_{Rf-}^{AB}(12) \phi_B(v_2) \qquad (7.23b)$$

Enskog level theories for reactions have also been constructed by Xystris and Dahler[81] and Bose and Ortoleva.[52]

The R operators are repeated ring operators and account for the correlated collision events discussed above. The $R^{AS}(1; z)$ operator describes sequences of correlated elastic collisions of the solute molecule A with the solvent. This operator appears in the kinetic theory for the self-diffusion coefficient and has been discussed elsewhere.[63, 70] Since it does not enter in the rate coefficient expressions with which we are concerned, its structure is not studied in detail. To put the subsequent discussion into perspective, however, we briefly review its physical content. For this purpose we consider the simplified form in (C.14), then

$$R^{AS}(1; z) = n_{eq}^{s} \int dx_2 \, \overline{T}_{-}^{AS}(12) \{ G_{AS}^{0}(12; z)^{-1} - \overline{T}_{-}^{AS}(12) \}^{-1} T_{-}^{AS}(12) \phi_s(v_2)$$

(7.24)

with $G_{AS}^{0}(12; z)$ the Enskog propagator for a pair of independent A and S molecules

$$G_{AS}^{0}(12; z)^{-1} = \{ z + \mathcal{L}_{0}^{AS}(12) - \Lambda_{-}^{AS}(1) - \Lambda_{-}^{SS}(2) \}$$

where $\mathcal{L}_{0}^{AS}(12) = \mathcal{L}_{0}^{A}(1) + \mathcal{L}_{0}^{S}(2)$.

The end vertices $T^{AS}(12)$ and $\overline{T}^{AS}(12)$ specify the solute-solvent molecule collisions that initiate and terminate the sequence of correlated events. The propagator

$$\{ G_{AS}^{0}(12; z)^{-1} - \overline{T}_{-}^{AS}(12) \}^{-1}$$

describes the motion between these collisions. Expanding this propagator in powers of $\overline{T}_{-}^{AS}$ in (7.24) leads to a series for R^{AS} in terms of a sequence of correlated collisions with independent propagation in between these collisions. The nth term in the series has the form

$$\int_0^t dt_1 \int_0^{t_1} dt_2 \cdots \int_0^{t_{n-1}} dt_n \, \overline{T}_{-}^{AS}(12) G_{A}^{0}(1; t - t_1)$$

$$G_{S}^{0}(2; t - t_1) \overline{T}_{-}^{AS}(12) G_{A}^{0}(1; t_1 - t_2)$$

$$G_{S}^{0}(2; t_1 - t_2) \overline{T}_{-}^{AS}(12) \cdots G_{A}^{0}(1; t_n) G_{S}^{0}(2; t_n) T_{-}^{AS}(12)$$

where

$$G_{A}^{0}(1; t) = \exp \{ -(\mathcal{L}_{0}^{A}(1) - \Lambda_{-}^{AS}(1)) t \}$$

with a similar definition for $G_{S}^{0}(2; t)$.

If we represent each $\overline{T}_{-}^{AS}$ vertex by a dot and each G^0 propagator by a line, the series can be represented diagramatically and is shown in Fig. 7.3. We have written the series in t-space, since the G_{AS}^{0} propagators factor. In the diagrams, t increases from right to left. This sequence of collision events provides the route for the coupling of the test particle motion to the fluid collective fields. It contains a variety of important effects, such as the origin of the asymptotic power law $(t^{-3/2})$ decay of the velocity autocorrelation function and provides the basis for a microscopic analysis of Stokes' law.

The repeated ring operator R^{AB} in (7.21b) is a new operator that appears in the theory on account of the reactive collision events. It has a form analogous to that of R^{AS},

$$R^{AB}(1;z) = \int dx_2\, \mathcal{V}_{A,AB}(12) G_{AB}(12;z) \mathcal{V}_{AB,A}(12) \qquad (7.25)$$

with the vertices defined (here, we present a more detailed form than that given for R^{AS}) by

$$\mathcal{V}_{A,AB}(12) = \bar{T}_{-}^{AB}(12) + \bar{C}_{-}^{AS}(12) \qquad (7.26)$$

and

$$\mathcal{V}_{AB,A}(12) = n_{eq}^{B}\left[T_{-}^{AB}(12) + C_{-}^{AS}(12) \right] g^{AB}(r_{12}) \phi_B(v_2) \qquad (7.27)$$

where

$$\bar{C}_{-}^{AS}(12) = n_{eq}^{s}\int dx_3 \bar{T}_{-}^{AS}(13)\left[g^{AB}(r_{12})^{-1} g^{ABS}(\mathbf{r}_1,\mathbf{r}_2,\mathbf{r}_3) - g^{AS}(r_{13}) \right] \phi_s(v_3) \qquad (7.28)$$

The explicit expression for the $G_{AB}(12;z)$ propagator is

$$G_{AB}(12;z) = \left\{ z + \mathcal{L}_0(12) - \Lambda_{-}^{AS}(1) - \Lambda_{-}^{BS}(2) \right.$$

$$\left. - \bar{T}_{-}^{AB}(12) - \bar{C}_{-}^{AS}(12) - \bar{C}_{-}^{BS}(12) \right\}^{-1}$$

$$\equiv \left\{ G_{AB}^{0}(12;z)^{-1} - \mathcal{V}_{AB}(12) \right\}^{-1} \qquad (7.29)$$

Here, $G^0_{AB}(12; z)$ is the Enskog propagator for the independent motion of the A and B molecules. Its explicit form is given in (C.22).

Thus R^{AB} is also a repeated ring-type operator, and its interpretation is similar to that of R^{AS} described above. However, there are some new features, including two contributions to the vertex:

1. There is a contribution from a direct collision between the solute A and B molecules. Since

$$\bar{T}^{AB}_-(12) = \bar{T}^{AB}_{E-}(12) - T^{AB}_{Rf-}(12) \tag{7.30}$$

   this collision may be either elastic or reactive.

2. The correlated collisions may be between a solute A (or B) molecule and a solvent molecule, while A and B are *statically* correlated. The scattering processes that contribute to the vertices are shown schematically in Fig. 7.4. Once again, expansion of the propagator in (7.29) in powers of $\mathcal{V}_{AB}$ yields a repeated ring series, but now one with a richer structure due to the variety of collisional processes that enter into each vertex. We explore the structure of this operator in more detail in connection with the rate coefficient.

Thus the principal feature distinguishing this kinetic theory of chemical reactions from earlier theories is the inclusion of the correlated collision events contained in the R collision operators. These terms are crucial for a description of the dynamics in the condensed phase.

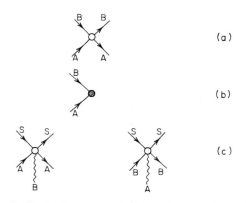

Fig. 7.4. Schematic representation of collision events contributing to $\mathcal{V}_{AB}$. (*a*) Elastic scattering of A and B. (*b*) Forward reactive collision. (*c*) Elastic collision of A(B) with S while A(B) is statically correlated to B(A).

D. Pair Kinetic Equation

We previously noted that many aspects of chemical reactions are often conveniently discussed in terms of the dynamics of a pair of reactive molecules. An example was provided earlier in the discussion of the application of the Smoluchowski equation for the pair probability given in Section III.B. To study these pair approaches from a microscopic point of view, we present a kinetic equation for the solute pair phase-space correlation function.[53] Again, by the regression hypotheses, we expect, for sufficiently long times and small spatial variations, that the equation of motion for this correlation function will have the same form as the macroscopic law. Although the singlet kinetic equation is useful for discussing the macroscopic chemical rate law and rate coefficients, the pair kinetic equation is especially convenient for studying the microscopic basis of the sink Smoluchowski equation and also the rate kernel.

We consider the description of the general bimolecular reaction $A + B \rightleftarrows C + D$. The primary variables of interest are now the reactive pair phase-space densities $\delta n_{AB}(x_1 x_2)$ and $\delta n_{CD}(x_1 x_2)$. These two fields are coupled by the reaction, but because of the diluteness of the solute species they are not coupled to other pair fields by the action of the pseudo-Liouville operator. In the present formulation singlet fields are not included in the description. Thus the pair fields do not have to be orthogonalized to the singlet fields and are simply given by, for example,

$$n_{AB}(x_1 x_2) = \sum_{i,j=1}^{N} \delta(x_1 - X_i)\delta(x_2 - X_j)\Theta_i^A \Theta_j^B \qquad (7.31)$$

and

$$\delta n_{AB}(x_1 x_2) = n_{AB}(x_1 x_2) - \langle n_{AB}(x_1 x_2) \rangle \qquad (7.31a)$$

The reactive pair phase-space correlation functions can be constructed from these fields as

$$C_{\alpha\beta,\gamma\delta}(12,1'2';t) = \langle \delta n_{\alpha\beta}(x_1 x_2;t)\delta n_{\gamma\delta}(x_1' x_2') \rangle$$

The pseudo-Liouville operator does couple these doublet fields to triplet fields such as δn_{ABS} and δn_{CDS} involving the solvent molecules. Thus one of the simplest forms for the pair kinetic equation can be obtained by explicitly including doublet and triplet fields in the generalized Langevin equation. This procedure yields a treatment of the effects of solvent dynamics on the motion of the reactive pair that is much more sophisticated than that given in the singlet kinetic equation discussed in the preceding section. To achieve a similar level of description, triplet phase-space fields must also be incorporated there.[67]

The results given here will also allow for the possibility of soft forces among the *solute* molecules. Hence, if we shut off the hard elastic and reactive terms among these species, our kinetic equation can describe the motion of a pair of particles in an arbitrary potential, thus providing a possible model for atom recombination studies.

Since the derivation of the pair equation exactly parallels that for the singlet kinetic equation, the details are sketched in Appendix D and not given here. It is quite easy to derive a kinetic equation for the general reversible reaction case; the calculations need only be carried out in matrix form.[53] To avoid this more complex notation and to present the results in simple form, however, we again give only the results for the irreversible decay of the AB pair field.

The pair kinetic equation is

$$\{z + \mathcal{L}_0^{AB}(12) - \mathcal{L}_s(12) - \mathbf{v}_{12} \cdot \beta \mathbf{F}_H - K_{AB,AB}(12; z)\} C_{AB,AB}(12, 1'2'; z)$$
$$= C_{AB,AB}(12, 1'2') \qquad (7.32)$$

where $\mathbf{F}_H$ is the hard-sphere mean force,

$$\beta \mathbf{F}_H = \mathbf{\nabla}_{12} \ln g_H(r_{12})$$

with

$$g(r_{12}) = e^{-\beta V_s(r_{12})} g_H(r_{12})$$

The soft-force part of the Liouville operator is $\mathcal{L}_s$ [cf. (6.9)], now in field point space. If only a soft force acts between the A and B solute molecules, then

$$g(r_{12}) = e^{-\beta V_s(r_{12})} y_H(r_{12})$$

where $y_H(r_{12})$, the radial distribution function for a pair of cavities in a hard-sphere fluid, is related to the cavity potential W_c by

$$W_c(r_{12}) = -k_B T \ln y_H(r_{12})$$

The collision operator is given by

$$K_{AB,AB}(12; z) = \Lambda_-^{AS}(1) + \Lambda_-^{BS}(2) + T_-^{AB}(12)$$
$$+ C_-^{AS}(12) + C_-^{BS}(21) + R^{ABS}(12; z) \qquad (7.33)$$

In the kinetic equation, one might have expected the streaming term to contain

$$\sum_{i=1}^{2} \frac{\mathbf{F}_i}{m_i} \cdot \nabla_{v_i}$$

with $\mathbf{F}_i$ the mean force on molecule i. Because the solute molecules undergo hard-sphere elastic collisions with the solvent (and also with each other), however, the contributions to the streaming term from the hard collisions and soft forces have a different structure, as indicated in (7.32). A more explicit discussion of how the mean force enters into the kinetic equation is given in Appendix D.

The new feature in the collision operator is the repeated ring operator R^{ABS}, which describes sequences of correlated collisions of A and B with each other and the solvent molecules.

$$R^{ABS}(12; z) = \bar{\tau}_{-}^{ABS}(12\bar{3})G_{ABS}(12\bar{3}; z)\tau_{-}^{ABS}(12\bar{3})\omega_0^s(\bar{3}) \qquad (7.34)$$

(Some comments on the origin of this term are given in Appendix D.) Here the correlated sequence of events is initiated by a collision between A and S or B and S,

$$\tau_{-}^{ABS}(123) = g^{AS}(\sigma_{AS})T_{-}^{AS}(13) + g^{BS}(\sigma_{BS})T_{-}^{BS}(23) \qquad (7.35)$$

After this initial collision, the molecules propagate via

$$G_{ABS}(123; z) = \left\{ z + \mathcal{L}_e^{ABS}(123) - \mathcal{V}_{ABS}(123) \right\}^{-1}$$

$$\equiv \left\{ G_{ABS}^0(123; z)^{-1} - \mathcal{V}_{ABS}(123) \right\}^{-1} \qquad (7.36)$$

where $G_{ABS}^0(123; z)$ is the propagator for the independent motion of the A, B, and S species

$$G_{ABS}^0(123; z) = \left\{ z + \mathcal{L}_0^{ABS}(123) - \Lambda_{-}^{AS}(1) - \Lambda_{-}^{BS}(2) - \Lambda_{-}^{SS}(3) \right\}^{-1} \qquad (7.37)$$

and the vertex $\mathcal{V}_{ABS}(123)$ couples this motion through soft-force interactions and hard collisions between the A and B molecules and elastic collisions of A with S and B with S,

$$\mathcal{V}_{ABS}(123) = g^{ABS}(\mathbf{r}_1\mathbf{r}_2\mathbf{r}_3)\mathcal{L}_s(12) + g^{AB}(\sigma_{AB})\bar{T}_{-}^{AB}(12)$$

$$+ g^{AS}(\sigma_{AS})\bar{T}_{-}^{AS}(13) + g^{BS}(\sigma_{BS})\bar{T}_{-}^{BS}(23) \qquad (7.38)$$

Finally, the sequence of correlated events is terminated by a collision of A with S or B with S,

$$\bar{\tau}_-^{ABS}(123) = g^{AS}(\sigma_{AS})\bar{T}_-^{AS}(13) + g^{BS}(\sigma_{BS})\bar{T}_-^{BS}(23) \qquad (7.39)$$

The structure of this repeated ring operator is much richer than that of the repeated ring operators discussed earlier. The various classes of collision events can be examined by expanding the G_{ABS} propagator in powers of $\mathcal{V}_{ABS}$. Consider, for instance, the one-ring term

$$\left[\bar{T}_-^{AS}(1\bar{3}) + \bar{T}_-^{BS}(2\bar{3})\right] G_A^0(1;t) G_B^0(2;t) G_S^0(\bar{3};t) \left[T_-^{AS}(1\bar{3}) + T_-^{BS}(2\bar{3})\right]$$

which yields four contributions. The two diagonal contributions have exactly the structure of the first terms of R^{AS} discussed earlier and serve to "renormalize" the independent single-particle motion. The off-diagonal terms lead to a coupling of the motion of the A and B molecules via collisions with the solvent. We shall defer a more detailed discussion of these contributions until the rate coefficient expressions are derived from the kinetic equation.

The specific approximations to the static structural correlations that were made to obtain the result for R^{ABS} are described in detail in Ref. 53. We have presented the results in a form that describes the binary collision events at the Enskog level of approximation.

E. The BGK Equation

The kinetic equations (7.21) and (7.32) provide a microscopic description of the coupled motions of the solute and solvent molecules and the effects of this coupling on the time evolution of the solute molecule distribution functions or correlation functions. As such, they give a much more explicit treatment of the collision dynamics than other more phenomenological kinetic theories. One of the most useful of these phenomenological equations is the Bhatnagar, Gross, and Krook (BGK) equation,[82] and it is interesting to compare the results of the last two sections with the corresponding BGK kinetic equations.

Consider first the kinetic equation for the singlet field in a nonreactive system. It has the general form given in (7.2a), which now reduces (suppressing species labels) to

$$\frac{\partial F(1,t)}{\partial t} = \left\{ -\mathbf{v}_1 \cdot \nabla_1 - \frac{\mathbf{F}_{ex}}{m_1} \cdot \nabla_{v_1} \right\} F(1,t)$$
$$+ \int_0^t dt'\, K(1,\bar{1}';t') F(\bar{1}', t-t') \qquad (7.40)$$

If memory effects in the collision kernel are neglected,

$$K(1,1';t') \simeq 2K(1,1';z=0)\delta(t')$$

then

$$\frac{\partial F(1,t)}{\partial t} = \left\{ -\mathbf{v}_1 \cdot \nabla_1 - \frac{\mathbf{F}_{ex}}{m_1} \cdot \nabla_{v_1} \right\} F(1,t) + K(1,\bar{1}';z=0)F(\bar{1}',t) \quad (7.41)$$

The BGK model corresponding to this kinetic equation may be constructed as follows. The collision term may, in general, be written as a difference of gain and loss terms,

$$\int d\mathbf{x}_1' K(1,1';z=0)F(1';t) = \int d\mathbf{x}_1' W(1',1)F(1',t)$$

$$- \int d\mathbf{x}_1' W(1,1')F(1,t) \quad (7.42)$$

where $W(1,1')d\mathbf{x}_1'$ is the transition probability per unit time for the transition from $\mathbf{x}_1$ to $\mathbf{x}_1'$. A BGK model for the transition probability is

$$W(1',1) = \alpha_c \phi(v_1)\delta(\mathbf{r}_1 - \mathbf{r}_1') \quad (7.43)$$

Thus the transition probability is taken to be local in space and is characterized by a mean collision frequency α_c. It follows that the BGK kinetic equation is

$$\frac{\partial F(1,t)}{\partial t} = \left\{ -\mathbf{v}_1 \cdot \nabla_1 - \frac{\mathbf{F}_{ex}}{m_1} \cdot \nabla_{v_1} + K_{BGK}(1) \right\} F(1,t) \quad (7.44)$$

with

$$K_{BGK}(1) = \alpha_c \left[\phi(v_1) \int d\mathbf{v}_1 - 1 \right] \quad (7.45)$$

This is certainly an idealized model for the true collision dynamics; it assumes that the velocity is randomized at each collision. All characteristics of the solvent and solute molecule properties, including the effects of the uncorrelated and correlated collision events described earlier, are implicitly contained in the collision frequency parameter. Nevertheless, the virtue of this model lies in its simplicity; the full collision operator is rather complex [cf. (7.21)]. Even the Boltzmann-Enskog collision operator is not especially simple. When it is written explicitly in the form of (7.42), $W(1', 1)$ is seen to be a complicated function of the collision parameters and the masses of the solute and solvent molecules.[83]

Applications of this kinetic equation for isomerization dynamics have been carried out by considering the motion of a "particle" in the external force corresponding to the double minimum potential in Fig. 3.2. Since this model treats the free streaming in the potential correctly and specifies a not unreasonable model for the collisions, which provide the energy dissipation, interesting results for the dynamics of the reaction can be obtained.[33, 84] Other more complex collision models, which contain solute and solvent molecule mass effects explicitly, have also been studied.[85] We discuss some of these results in Section XII.

Analogous BGK models can be constructed at the pair kinetic equation level. The simplest model of this type would approximate the collision operator in (7.32) as

$$K(12; z) \simeq K_{BGK}(1) + K_{BGK}(2) \tag{7.46}$$

This assumes that the two particles undergo independent BGK collisions with the solvent; all dynamic correlations are neglected except those that are crudely accounted for in the collision frequency parameters. More elaborate models may be constructed.

VIII. KINETIC THEORY EXPRESSION FOR THE RATE KERNEL

An equation with the form of the macroscopic law in (2.16) can be obtained from the singlet field kinetic equation by projecting out the velocity dependence of the phase-space correlation functions. A comparison of the resulting equation with this macroscopic law can then yield a microscopic correlation function expression for the rate kernel.

We begin with (7.22),

$$\{z - L_R(1; z)\} C_{AA}(1, 1'; z) = C_{AA}(1, 1')$$

and define a projection operator $\mathcal{P}_A$ by

$$\mathcal{P}_A A(\mathbf{v}_1) = \phi_A(v_1) \int d\mathbf{v}_1 A(\mathbf{v}_1) \tag{8.1}$$

where $A(\mathbf{v}_1)$ is any function of the velocity, and let Q_A be the complement of $\mathcal{P}_A$. Since the operator L_R does not act on the primed variables, we may integrate this equation over the primed velocities directly. We then have

$$\mathcal{P}_A \int dv_1' C_{AA}(1,1';z) = \phi_A(v_1) C_{AA}(\mathbf{r}_1 - \mathbf{r}_1';z) \qquad (8.2)$$

where we have used (2.15). An equation for $C_{AA}(\mathbf{r}_1 - \mathbf{r}_1';z)$ then follows directly after the usual projection operator algebra,

$$\left\{ z - \int d\mathbf{v}_1 L_R(1;z)\phi_A(v_1) \right.$$

$$\left. - \int d\mathbf{v}_1 L_R(1;z)Q_A[z - Q_A L_R(1;z)]^{-1}Q_A L_R(1;z)\phi_A(v_1) \right\}$$

$$C_{AA}(\mathbf{r}_1 - \mathbf{r}_1';z) = C_{AA}(\mathbf{r}_1 - \mathbf{r}_1') \qquad (8.3)$$

Using the explicit form of L_R which follows from (7.21), and number conservation in elastic collisions, it follows that

$$- \int d\mathbf{v}_1 L_R(1;z)\phi_A(v_1) = n_{eq}^B k_f^0 - \int d\mathbf{v}_1 R^{AB}(1;z)\phi_A(v_1) \qquad (8.4)$$

In the analysis of the other term in (8.3), we recall that for small solute densities we only need the solute-density-independent rate kernel. Thus, using number conservation again, we may write

$$Q_A L_R(1;z)\phi_A(v_1) = \mathbf{v}_1 \cdot \nabla_1 - Q_A(\Lambda_{Rf-}^{AB}(1) - R^{AB}(1;z))\phi_A(v_1)$$

Since the second term is $\mathcal{O}(n_{eq}^B)$ we have, to lowest order in solute densities with neglect of cross-diffusion–reaction terms as in (5.11),

$$\int d\mathbf{v}_1 L_R Q_A[z - Q_A L]^{-1}Q_A L_R \phi_A(v_1) = \nabla_1 \cdot \mathbf{D}_A(\mathbf{r}_1;z) \cdot \nabla_1 \qquad (8.5)$$

Here $\mathbf{D}_A(\mathbf{r}_1;z)$ is a diffusion operator defined by

$$\mathbf{D}_A(\mathbf{r}_1;z) = \int d\mathbf{v}_1 \mathbf{v}_1[z - Q_A L(1;z)]^{-1}\mathbf{v}_1\phi_A(v_1) \qquad (8.6)$$

where $L(1;z)$ is the $L_R(1;z)$ operator with reactive terms, which are $\mathcal{O}(n_{eq}^B)$, neglected. For small spatial gradients and long times, $\mathbf{D}_A(\mathbf{r}_1;z)$ reduces to the usual self-diffusion coefficient, $D_A \mathbf{1}$. Using these results, we may now write (8.3), to second order in spatial gradients, as

$$\{z - D_A(z)\nabla_1^2 + k_{AA}(z)\}C_{AA}(\mathbf{r}_1 - \mathbf{r}_1'; z) = C_{AA}(\mathbf{r}_1 - \mathbf{r}_1') \qquad (8.7)$$

with $k_{AA}(z)$ given by

$$k_{AA}(z) \equiv n_{eq}^B k_f(z) = n_{eq}^B k_f^0 - V^{-1}\int d\mathbf{x}_1\, R^{AB}(1; z)\phi_A(v_1) \qquad (8.8)$$

This expression should be compared with the more formal result in Section VI. Using the explicit form of R^{AB} in (7.25), the rate kernel may now be written in the form

$$k_f(z) = k_f^0 - V^{-1}\int d\mathbf{r}_1 d\mathbf{r}_2 \langle \overline{T}_{Rf-}^{AB}(12)G_{AB}(12; z)T_{Rf-}^{AB}(12)\rangle_0 g^{AB}(r_{12})$$

$$(8.9)$$

The angular brackets denote an equilibrium velocity average in two-particle space,

$$\langle \cdots \rangle_0 = \int d\mathbf{v}_1 d\mathbf{v}_2 \cdots \phi_A(v_1)\phi_B(v_2) \qquad (8.10)$$

The kinetic theory result for the rate kernel given above provides a much more explicit and tractable description of the dynamic processes that contribute to the rate coefficient in a dense fluid. We analyze these processes in Section X.

IX. CONFIGURATION SPACE EQUATIONS

The pair kinetic theory equation given in Section VII.D can be used to extend the Smoluchowski results outlined earlier. In this section, we present the microscopic derivation of the Smoluchlowski equation from the kinetic theory and also obtain expressions for the space and time nonlocal diffusion and friction tensors, which appear in this theory.

A. Projection onto Pair Configuration Space

The general pair phase-space kinetic equation is given in (7.32). Letting

$$L_R(12; z) = -\mathcal{L}_0^{AB}(12) + \Lambda_-^{AS}(1) + \Lambda_-^{BS}(2) + \overline{T}_-^{AB}(12) + \overline{C}_-^{AS}(12)$$

$$+ \overline{C}_-^{BS}(21) + \mathcal{L}_s(12) + R^{ABS}(12; z) \qquad (9.1)$$

the pair equation can be written compactly as

$$\{z - L_R(12; z)\} C_{AB,AB}(12, 1'2'; z) = C_{AB,AB}(12, 1'2') \tag{9.2}$$

An equation with the form of the sink Smoluchowski equation may now be derived by introducing an operator that projects out the velocities of the solute molecules.[86] This is just the pair version of the derivation presented in the previous section.

We introduce a projection operator $\mathscr{P}_{AB}$ by

$$\mathscr{P}_{AB} A(12) = \phi_A(v_1) \phi_B(v_2) \int dv_1 dv_2 A(12) \tag{9.3}$$

where again $A(12)$ is any function of the phase points of the two solute molecules. The complement of $\mathscr{P}_{AB}$ is Q_{AB}. We note that

$$\mathscr{P}_{AB} \int dv_1' dv_2' C_{AB,AB}(12, 1'2'; z) = \phi_A(v_1) \phi_B(v_2) P_{AB,AB}(\mathbf{r}_1\mathbf{r}_2, \mathbf{r}_1'\mathbf{r}_2'; z) \tag{9.4}$$

where the position space correlation function $P_{AB,AB}(\mathbf{r}_1\mathbf{r}_2, \mathbf{r}_1'\mathbf{r}_2'; t)$ is the correlation between the number of AB pairs initially at $(\mathbf{r}_1'\mathbf{r}_2')$ with the number of AB pairs at $(\mathbf{r}_1\mathbf{r}_2)$ at time t. We expect that $P_{AB,AB}$ will satisfy an equation with the same form as the Smoluchowski equation for the pair probability because for low solute concentrations, the dynamics of the individual pairs are independent. To simplify the notation, we henceforth drop the AB subscripts on $P_{AB,AB}$ and Q_{AB}.

An equation for P follows directly upon integration of (9.2) over primed velocities and application of projection operator techniques. We find

$$\{z - \langle L_R(12; z)\rangle_0 - \langle L_R(12; z)[z - QL_R(12; z)]^{-1} QL_R(12; z)\rangle_0\}$$

$$\times P(\mathbf{r}_1\mathbf{r}_2, \mathbf{r}_1'\mathbf{r}_2'; z) = P(\mathbf{r}_1\mathbf{r}_2, \mathbf{r}_1'\mathbf{r}_2') \tag{9.5}$$

This equation may be reduced further. First, explicitly evaluating $\langle L_R(12; z)\rangle_0$, we find

$$\langle L_R(12; z)\rangle_0 = -\frac{k_f^{eq}}{4\pi\sigma_{AB}^2} \delta(r_{12} - \sigma_{AB}) \equiv -S(r_{12}) \tag{9.6}$$

where k_f^{eq} is the equilibrium forward reactive collision frequency defined by

$$k_f^{eq} = \int d\mathbf{r}_{12} \int d\mathbf{v}_1 \, d\mathbf{v}_2 \, T_{Rf-}^{AB}(12)\phi_A(v_1)\phi_B(v_2) \tag{9.7}$$

Using the simple model for $T_{Rf-}^{AB}(12)$ in (6.18), k_f^{eq} is given by

$$k_f^{eq} = \alpha_R \sigma_{AB}^2 \left(\frac{8\pi k_B T}{\mu_{AB}} \right)^{1/2} \tag{9.8}$$

The second term in the brackets in (9.5) may also be reduced further. Using number conservation in elastic collisions, one may show that

$$\langle A(12) L_R(12; z) \rangle_0 = - \left\langle A(12) \left\{ T_{Rf-}^{AB}(12) + \sum_{j=1}^{2} \mathbf{v}_j \cdot [\nabla_j + \beta(\nabla_j W(r_{12}))] \right\} \right\rangle_0$$

for any $A(12)$.

Here W is the potential of mean force. Also,

$$\langle L_R(12; z) A(12) \rangle_0 = - \left\langle \left\{ T_{Rf-}^{AB}(12) + \sum_{i=1}^{2} \nabla_i \cdot \mathbf{v}_i \right\} A(12) \right\rangle_0 \tag{9.10}$$

Using these results in (9.5), we find

$$\left\{ z + S(r_{12}) - \left\langle \left\{ \delta S(12) + \sum_{i=1}^{2} \nabla_i \cdot \mathbf{v}_i \right\} [z - QL_R(12; z)]^{-1} \right. \right.$$

$$\left. \left. \times \left\{ \delta S(12) + \sum_{j=1}^{2} \mathbf{v}_j \cdot [\nabla_j + (\beta \nabla_j W(r_{12}))] \right\} \right\rangle_0 \right\} P(\mathbf{r}_1\mathbf{r}_2, \mathbf{r}_1'\mathbf{r}_2'; z)$$

$$= P(\mathbf{r}_1\mathbf{r}_2, \mathbf{r}_1'\mathbf{r}_2') \tag{9.11}$$

The quantity $\delta S(12)$ is the deviation of $T_{Rf-}^{AB}(12)$ from its velocity averaged value,

$$\delta S(12) = T_{Rf-}^{AB}(12) - S(r_{12}) \tag{9.12}$$

This equation is considerably more complex than the Smoluchowski equation discussed in Section III.B. Several approximations need to be made before the reduction is complete. We analyze (9.11) with reactive terms dropped, and then return to the case with reactive contributions.

B. Smoluchowski Equation and Diffusion Tensor

We have mentioned several times that, when describing certain types of chemical reaction in liquids, it is often more convenient to consider the motion of the particles in some potential and then introduce species operators that divide up the phase space. As described in detail in Section IV, this is a useful way of treating isomerization and atomic recombination reactions. Our pair description is especially appropriate for the latter case.

If we set the reactive terms in (9.11) to zero, the resulting equation

$$\left\{ z - \sum_{i,j=1}^{2} \nabla_i \cdot \mathbf{D}_{ij}(\mathbf{r}_1\mathbf{r}_2; z) \cdot \left[\nabla_j + (\beta\nabla_j W(r_{12})) \right] \right\} P(\mathbf{r}_1\mathbf{r}_2, \mathbf{r}_1'\mathbf{r}_2'; z)$$

$$= P(\mathbf{r}_1\mathbf{r}_2, \mathbf{r}_1'\mathbf{r}_2') \tag{9.13}$$

describes the motion of the pair of particles in the potential W. This has the form of a Smoluchowski equation with a generalized diffusion tensor $\mathbf{D}_{ij}(\mathbf{r}_1,\mathbf{r}_2; z)$. From (9.11), we see that this operator has the microscopic definition

$$\mathbf{D}_{ij}(\mathbf{r}_1,\mathbf{r}_2; z) = \langle \mathbf{v}_i [z - QL(12; z)]^{-1} \mathbf{v}_j \rangle_0 \tag{9.14}$$

where $L(12; z)$ is equal to $L_R(12; z)$ [cf. (9.1)] with reactive terms dropped. This result for $\mathbf{D}$ has the expected form of a velocity correlation function. The kinetic theory result for L allows the possibility of examining in detail the effects of solvent coupling on the pair diffusive motion.

A derivation of this result for the case when all interactions are soft is given in Ref. 87. In general, the relative and center of mass motions of the pair are coupled. If we neglect such coupling, a Smoluchowski equation for the relative motion of the pair is easily written. In this circumstance

$$P(\mathbf{r}_1\mathbf{r}_2, \mathbf{r}_1'\mathbf{r}_2'; t) = P(\mathbf{r}_{12}, \mathbf{r}_{12}'; t)P(\mathbf{R} - \mathbf{R}'; t) \tag{9.15}$$

where $\mathbf{r}_{12}$ and $\mathbf{R}$ are the relative and center of mass positions, respectively. With this assumption and (9.13), we have, in t-space,

$$\frac{\partial P(\mathbf{r}_{12}, \mathbf{r}_{12}'; t)}{\partial t} = \nabla_{12} \cdot \int_0^t dt' \, \mathbf{D}(\mathbf{r}_{12}; t') \cdot \left[\nabla_{12} + (\nabla_{12}\beta W(r_{12})) \right]$$

$$\times P(\mathbf{r}_{12}, \mathbf{r}_{12}'; t-t') \tag{9.16}$$

with $D(\mathbf{r}_{12}, z) = \int_0^\infty dt\, e^{-zt} \mathbf{D}(\mathbf{r}_{12}; t)$, given by

$$\mathbf{D}(\mathbf{r}_{12}, z) = \langle \mathbf{v}_{12} [z - QL(12; z)]^{-1} \mathbf{v}_{12} \rangle_0 \qquad (9.17)$$

The expression for $\mathbf{D}$ is rather complex because it is an operator in configuration space. This feature can be exposed by writing

$$L(12; z) = L_s(12) + L_c(12; z) \qquad (9.18)$$

where the streaming operator L_s is

$$L_s(12) = -\mathcal{L}_0^{AB}(12) + \mathcal{L}_s(12) = -\sum_{i=1}^{2} \left(\mathbf{v}_i \cdot \nabla_i + \frac{\mathbf{F}_i}{m_i} \cdot \nabla_{v_i} \right) \qquad (9.19)$$

and the collision operator $L_c(12; z)$ is

$$L_c(12; z) = \Lambda_-^{AS}(1) + \Lambda_-^{BS}(2) + \bar{C}_-^{AS}(12) + \bar{C}_-^{BS}(21) + R^{ABS}(12; z) \qquad (9.20)$$

The propagator in (9.17) may then be expanded in powers of L_s. The analysis is similar to that used in the reduction of the Fokker-Planck equation to a generalized Smoluchowski equation.[88, 89] We obtain

$$\mathbf{D}(\mathbf{r}_{12}; z) = \sum_{n=0}^{\infty} \langle \mathbf{v}_{12} G_c(12; z) [QL_s(12) G_c(12; z)]^n \mathbf{v}_{12} \rangle_0$$

$$\equiv \sum_{n=0}^{\infty} \mathbf{D}^{(n)}(\mathbf{r}_{12}; z) \qquad (9.21)$$

where

$$G_c(12; z) = [z - QL_c(12; z)]^{-1} \qquad (9.22)$$

Expressions of this form for $\mathbf{D}$ have also been derived earlier by Skinner and Wolynes[90] for BGK and Fokker-Planck models of chemically reacting systems. In those circumstances, one could use the knowledge of the eigenfunctions of these simple collision operators to reduce the result for $\mathbf{D}$ further. This is not possible for the more complex collision operator $L_c(12; z)$, and one must resort to more approximate methods. We may, for example, evaluate the $\mathbf{D}^{(n)}$ by inserting a complete set of velocity states, which we denote by $H_i((\mu\beta)^{1/2}v_{12})$, where[91]

$$H_i(\mathbf{C}) = (i!)^{-1/2} e^{C^2/2} \left[-\frac{\partial}{\partial\mathbf{C}} \right]^i e^{-C^2/2} \qquad (9.23)$$

Then

$$\mathbf{D}^{(n)} = \sum_{i_1 \cdots i_{2n}} \langle \mathbf{v}_{12} G_c \mathbf{H}_{i_1} \rangle_0 \langle \mathbf{H}_{i_1} QL_s \mathbf{H}_{i_2} \rangle_0 \cdots \langle \mathbf{H}_{i_{2n}} G_c \mathbf{v}_{12} \rangle_0 \qquad (9.24)$$

This expression can be brought into a more tractable form if we make the approximation that only the diagonal matrix elements $G_c^{ii} \equiv \langle \mathbf{H}_i G_c \mathbf{H}_i \rangle_0$ are kept, and use the fact that QL_s couples only nearest-neighbor velocity states, for example,

$$(\beta\mu)^{1/2}\langle H_{1\alpha} QL_s H_{2\beta\nu} \rangle_0 = -2^{-1/2}(\delta_{\alpha\beta}\delta_{\mu\nu} + \delta_{\alpha\nu}\delta_{\mu\beta})\nabla_{12\mu}$$

and

$$(\beta\mu)^{1/2}\langle H_{2\beta\nu} QL_s H_{1\alpha} \rangle_0 = -2^{-1/2}(\delta_{\alpha\beta}\delta_{\mu\nu} + \delta_{\alpha\nu}\delta_{\mu\beta})(\nabla_{12\mu} - \beta F_\mu)$$

In this approximation, we find $\mathbf{D}^{(2i+1)} = 0$ and

$$\mathbf{D}^{(2)}(\mathbf{r}_{12}; z) \simeq 2(\beta\mu)^{-2} G_c^{11}(\mathbf{r}_{12}; z)$$

$$\cdot \{ \nabla_{12} \cdot G_c^{22}(\mathbf{r}_{12}; z) \cdot [\nabla_{12} - \beta\mathbf{F}] \} \cdot G_c^{11}(\mathbf{r}_{12}; z) \qquad (9.25)$$

with similar expressions for the higher order terms. The operator character of $\mathbf{D}$ makes an investigation of nonlocal effects on pair diffusive motion a difficult task, which will require a good deal more study.

We may also examine the content of (9.14) for $\mathbf{D}_{ij}$ in more detail by writing it in terms of a generalized friction coefficient and examining the structure of this quantity.[86] The expression for $\mathbf{D}_{ij}$ may be written in a more compact form by introducing a matrix notation; we let $\mathbf{v}$ be the column vector $\mathbf{v} = (\mathbf{v}_1, \mathbf{v}_2)^T$ and $\mathbf{v}^T$ a row vector. Then,

$$\mathbf{D}(\mathbf{r}_1, \mathbf{r}_2; z) = \langle \mathbf{v}[z - QL(12; z)]^{-1}\mathbf{v}^T \rangle$$

$$\equiv \int d\mathbf{v}_1 d\mathbf{v}_2 \, \mathbf{v}\mathbf{h}(12; z) \qquad (9.26)$$

The second equality defines **h**, which, by construction, satisfies the equation

$$[z - QL(12; z)]\mathbf{h}(12; z) = \mathbf{v}^T \phi_A(v_1) \phi_B(v_2) \qquad (9.27)$$

For simplicity, in what follows, we consider the case of A and B identical. We may then introduce a projection operator $\mathcal{P}_v$ by

$$\mathcal{P}_v = \mathbf{v}^T \phi_A(v_1) \phi_A(v_2) \frac{m_1}{k_B T} \int d\mathbf{v}_1 d\mathbf{v}_2 \mathbf{v} \qquad (9.28)$$

where m_1 is the mass of A. The complement of $\mathcal{P}_v$ is Q_v. The action of $\mathcal{P}_v$ on **h** then yields

$$\mathcal{P}_v \mathbf{h}(12; z) = \mathbf{v}^T \phi_A(v_1) \phi_A(v_2) \frac{m_1}{k_B T} \mathbf{D}(\mathbf{r}_1 \mathbf{r}_2; z) \qquad (9.29)$$

Thus we may apply $\mathcal{P}_v$ to (9.27) to write

$$\mathbf{D}(\mathbf{r}_1, \mathbf{r}_2; z) = \left[\mathbf{z} + \frac{\zeta(\mathbf{r}_1, \mathbf{r}_2; z)}{m_1} \right]^{-1} \frac{k_B T}{m_1} \qquad (9.30)$$

where the generalized friction tensor ζ is defined by

$$\frac{k_B T}{m_1^2} \zeta(\mathbf{r}_1, \mathbf{r}_2; z) = -\langle \mathbf{v} Q L(12; z) \mathbf{v}^T \rangle_0$$

$$- \langle \mathbf{v} Q L(12; z) [z - Q_v Q L(12; z)]^{-1} Q_v Q L(12; z) \mathbf{v}^T \rangle_0$$

$$\equiv \frac{k_B T}{m_1^2} (\zeta^{(1)}(\mathbf{r}_1, \mathbf{r}_2; z) + \zeta^{(2)}(\mathbf{r}_1, \mathbf{r}_2; z)) \qquad (9.31)$$

Although a full analysis of this result would be quite involved and has not yet been carried out, its physical content is easily appreciated. The form of the $L(12; z)$ operator given above allows for the possibility of hard elastic collisions between the solute molecules. We omit these contributions below, although no new difficulties appear when they are included. Our considerations then apply to the case of solute molecules that interact with each other via soft forces and with the solvent via hard-sphere interactions.

Consider the first term in (9.31). We may write

$$
\frac{k_B T}{m_1^2} \zeta^{(1)}(\mathbf{r}_1, \mathbf{r}_2; z) = -\langle \mathbf{v} L_c \mathbf{v}^T \rangle =
$$
$$
- \langle \mathbf{v} \big[\Lambda_-^{AS}(1) + \Lambda_-^{AS}(2) + \overline{C}_-^{AS}(12) + \overline{C}_-^{AS}(21)
$$
$$
+ R^{AAS}(12; z) \big] \mathbf{v}^T \rangle_0 \qquad (9.32)
$$

The uncorrelated collision contributions are easily calculated to give

$$
\frac{m_1^2}{k_B T} \langle v_i \big(\Lambda_-^{AS}(1) + \Lambda_-^{AS}(2) + \overline{C}_-^{AS}(12) + \overline{C}_-^{AS}(21) \big) v_j \rangle_0
$$
$$
= -\zeta_E 1 \delta_{ij} - \zeta_E \delta_{ij} \frac{3}{4\pi\sigma_{AS}^2} \int d\mathbf{r}_3 \left[\frac{g^{AAS}(\mathbf{r}_1\mathbf{r}_2\mathbf{r}_3)}{g^{AS}(r_{13}) g^{AA}(r_{12})} - 1 \right]
$$
$$
\times \delta(r_{i3} - \sigma_{AS}) \hat{\mathbf{r}}_{i3} \hat{\mathbf{r}}_{i3} \qquad (9.33)
$$

where ζ_E is the Enskog value of the single-particle friction coefficient

$$
\zeta_E = \frac{8}{3} (2\pi k_B T \mu)^{1/2} n_{eq}^s \sigma_{AS}^2 g^{AS}(\sigma_{AS})
$$

Although this term arises only from dynamically uncorrelated collisions between the solute molecules and the solvent, we see that static structural correlations couple the motions of the two solute molecules. This contribution to the friction coefficient is not difficult to calculate if expressions for $g^{AAS}(\mathbf{r}_1\mathbf{r}_2\mathbf{r}_3)$ are available. The results display oscillation arising from the static structural correlations at distances greater than 2σ (we assume that solute and solvent diameters are equal.) At distances less than 2σ, where a solvent molecule can no longer intervene, the friction falls. This is a "shadowing" effect insofar as one solute molecule screens the other from collisions with the solvent. At shorter separations ($\sim\sigma$) the friction must diverge because the solute molecules are impenetrable. A detailed discussion of these results can be found in Ref. 92.

The correlated collision term contains several effects, which are exposed by using the definition of R^{AAS} [cf. (7.34)],

$$
\langle v_i R^{AAS}(12; z) v_j \rangle_0 = \Big\langle v_i \Big[\tilde{\overline{T}}_-^{AS}(1\bar{3}) + \tilde{\overline{T}}_-^{AS}(2\bar{3}) \Big]
$$
$$
\times \Big\{ G_{AAS}^0(12\bar{3}; z)^{-1} - g^{AAS}(12\bar{3}) \mathcal{L}_S(12)
$$
$$
+ \tilde{\overline{T}}_-^{AS}(1\bar{3}) + \tilde{\overline{T}}_-^{AS}(2\bar{3}) \Big\}^{-1}
$$
$$
\times \Big[\tilde{T}_-^{AS}(1\bar{3}) + T_-^{AS}(2\bar{3}) \Big] \omega_0(\bar{3}) v_j \Big\rangle_0 \qquad (9.34)
$$

Here, we let

$$\tilde{T}^{\alpha\beta}(ij) = g^{\alpha\beta}(\sigma_{\alpha\beta})T^{\alpha\beta}(ij)$$

Expansion of the propagator about G^0_{AAS} leads to a variety of collision events. The series of terms with the form

$$\left\langle \mathbf{v}_1\left(\underline{\tilde{T}}^{AS}(1\bar{3})G^0_{AAS}(12\bar{3};z)\underline{\tilde{T}}^{AS}(1\bar{3})\cdots G^0_{AAS}(12\bar{3};z)\underline{\tilde{T}}^{AS}(1\bar{3})\right)\omega_0(\bar{3})\mathbf{v}_1 \right\rangle_0$$

give rise to a modification of the single-particle friction due to correlated collision events. These have been discussed earlier and lead to a Stokes's law expression for the single-particle friction when the solute molecules are large compared to the solvent molecules. There are also cross terms that lead to a dynamic coupling between the motions of the solute molecules. The simplest of these is

$$\left\langle \mathbf{v}_1\underline{\tilde{T}}^{AS}(1\bar{3})G^0_{AAS}(12\bar{3};z)\underline{\tilde{T}}^{BS}(2\bar{3})\omega_0(\bar{3})\mathbf{v}_2 \right\rangle_0$$

which corresponds to an event in which the correlation is initiated by a collision between solute molecule 2 and a solvent molecule 3. After independent propagation described by G^0_{AAS}, the solvent molecule 3, or one with which it has collided, collides with the solute molecule 1 to terminate the correlation (see Fig. 9.1). Events of this type lead to the usual Oseen interaction between the solute molecules at large separations[93, 94]. In addition, (9.34) contains the effects of the soft force on these types of events, as well as the generalization of the hydrodynamic Oseen interaction effects to short distance scales.

In the terms discussed above, the soft force between the solute molecule enters only in an indirect fashion. It does not enter at all in (9.33), and only through the propagator in (9.34). The second contribution to the friction tensor $\zeta^{(2)}$ contains this force in a direct and explicit manner. Hence, when a strong direct chemical force operates between the molecules, one might expect this term to play an important role. If we evaluate the action of the

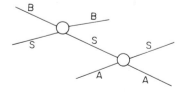

Fig. 9.1.   A sequence of collisions that gives rise to solvent coupling of the motions of the A and B solute molecules.

L operators on both ends of the expression for $\zeta^{(2)}$, we find

$$\zeta_{ij}^{(2)}(\mathbf{r}_1,\mathbf{r}_2;z)=\left\langle\left\{\sum_{k=1}^{2}\boldsymbol{\nabla}_k\cdot\mathbf{H}_{ki}+v_iQ_vL_c\right\}[z-Q_vQL]^{-1}\right.$$
$$\left.\times\left\{\sum_{l=1}^{2}\mathbf{H}_{jl}\cdot(\boldsymbol{\nabla}_l-\beta\mathbf{F}_l)+Q_vL_c v_j\right\}\right\rangle_0 \qquad (9.35)$$

where we defined

$$\mathbf{H}_{ij}=v_iv_j-(m_1\beta)^{-1}\delta_{ij}\mathbf{1} \qquad (9.36)$$

The terms involving $L_c v_j$ are contributions that arise from fluctuations about the average of this term given by $\zeta^{(1)}$. If we neglect such fluctuations, then

$$\zeta_{ij}^{(2)}\simeq\sum_{k,l=1}^{2}\boldsymbol{\nabla}_k\cdot\left\langle\mathbf{H}_{ki}[z-Q_vQL]^{-1}\mathbf{H}_{jl}\right\rangle_0\cdot(\boldsymbol{\nabla}_l-\beta\mathbf{F}_l) \qquad (9.37)$$

As anticipated from the discussion of $\mathbf{D}$ given earlier, the operator character of the two-particle friction assumes an important role in the description of nonlocal effects in the strong force region.

We see that the question of the nature of the nonlocality of the friction tensor is indeed a complex one. Spatial nonlocality can arise from a variety of effects such as static solvent structural correlations, dynamic solvent effects that give rise to Oseen interactions at large distance, and contributions from the direct forces between the molecules.

C. Sink Smoluchowski Equation

In Section III, we described how the Smoluchowski equation could be used in conjunction with boundary conditions or sink terms to describe chemical reactions. We now return to (9.11) and consider its relation to sink Smoluchowski equation.

It is clear that if all reactive contributions to the third term in the brackets of (9.11) are neglected, then

$$\left\{z+S(r_{12})-\sum_{i,j=1}^{2}\boldsymbol{\nabla}_i\cdot\mathbf{D}_{ij}(\mathbf{r}_1,\mathbf{r}_2;z)\cdot[\boldsymbol{\nabla}_j-\beta\mathbf{F}_j]\right\}P(\mathbf{r}_1\mathbf{r}_2,\mathbf{r}_1'\mathbf{r}_2';z)$$

$$=P(\mathbf{r}_1\mathbf{r}_2,\mathbf{r}_1'\mathbf{r}_2') \qquad (9.38)$$

This equation has the form of the sink Smoluchowski equation discussed earlier, apart from the nonlocal diffusion tensor. We now need to examine a bit more closely the circumstances under which such a result is likely to be useful. We expect that the Smoluchowski equation description will be valid, provided velocity relaxation effects can be neglected. To see what such a statement entails, we consider the structure of the neglected terms.

First, there are terms of the form $\langle \delta S(12)[z - QL_R(12; z)]^{-1}\delta S(12)\rangle_0$. From its definition in (9.12), we see that $\delta S(12)$ is the deviation of the reactive operator from its velocity average. The correlation function above characterizes the time evolution (Laplace transformed) of these fluctuations. If the chemical reaction is slow, we expect that perturbations of the velocity distribution induced by the reaction will be small; hence such contributions may be safely neglected in this limit. This argument may be made more formal using limiting procedures analogous to those described in Section V. In principle, one may also use this term to introduce a modification to k_f^{eq} in $S(r_{12})$ due to velocity relaxation effects. This will lead to some effective reactive collision frequency in place of k_f^{eq}.

Second, there are cross terms $\langle \delta S(12)[z - QL_R(12; z)]^{-1}v_j\rangle_0$. These correspond to a coupling between reaction and diffusion, which arises from the perturbation in velocity space induced by the reaction (or the reverse process). To examine this term in more detail, we consider the simple model for T_{Rf-}^{AB} in (6.18), and neglect the reactive terms in the propagator. Then

$$\langle \delta S(12)[z - QL_R(12; z)]^{-1}v_j\rangle_0$$

$$= \alpha_R \delta(r_{12} - \sigma_{AB})\hat{r}_{12}\cdot\langle v_{12}\theta(v_{12}\cdot\hat{r}_{12})[z - QL(12; z)]^{-1}v_j\rangle_0$$

$$\simeq \frac{\alpha_R}{2}\delta(r_{12} - \sigma_{AB})\hat{r}_{12}\cdot(D_{1j} - D_{2j}) \tag{9.39}$$

The contribution of this term to (9.11) is then

$$\frac{\alpha_R}{2}\delta(r_{12} - \sigma_{AB})\hat{r}_{12}\cdot\sum_{j=1}^{2}(D_{1j} - D_{2j})\cdot[\nabla_j - \beta F_j]P(r_1 r_2, r_1' r_2'; z)$$

$$\simeq \frac{\alpha_R}{2}\delta(r_{12} - \sigma_{AB})\hat{r}_{12}\cdot D[\nabla_{12} - \beta F]P(r_{12}, r_{12}'; z) \tag{9.40}$$

where, in the second line, we neglected coupling of the center of mass and relative motion, and introduced the relative diffusion coefficient D by

$$D1 = D_{11} - D_{12} - D_{21} + D_{22} \tag{9.41}$$

For the barrier reactions for which this simple model is suitable, the probability of reaction upon collision α_R is $\mathcal{O}(\exp(-E_f^*))$, with E_f^* barrier height. Slow reactions therefore have small α_R. The order of magnitude of $\delta(r_{12} - \sigma_{AB})\hat{r}_{12} \cdot D[\nabla_{12} - \beta F]P$ can be estimated from the arguments in Section III as $k_f^{eq}(4\pi\sigma_{AB}^2)^{-1}\delta(r_{12} - \sigma_{AB})P$. Thus this cross-diffusion–reaction term is down by a factor of $\exp(-\beta E_f^*)$ from the $S(r_{12})$ term. It will be a small contribution for a slow reaction. A different analysis with similar content has been given by Northrup and Hynes.[21]

In summary then, we expect the usual sink Smoluchowski description to be valid for a slow (on the order of the diffusion rate) reaction. The usual description also entails neglect of the operator character of D_{ij} (high friction limit) and assumes that velocity correlations relax rapidly so that the $z = 0$ limit of D_{ij} can be taken. The coupling between the center of mass and relative motion is also neglected in the usual formulations. These latter conditions reduce (9.38) to (now in t-space)

$$\frac{\partial P(\mathbf{r}_{12}, \mathbf{r}'_{12}; t)}{\partial t} = D\nabla_{12} \cdot [\nabla_{12} - \beta F] P(\mathbf{r}_{12}, \mathbf{r}'_{12}; t)$$

$$- \frac{k_f^{eq}}{4\pi\sigma_{AB}^2} \delta(r_{12} - \sigma_{AB}) P(\mathbf{r}_{12}, \mathbf{r}'_{12}; t) \qquad (9.42)$$

Thus the configuration space correlation function $P(\mathbf{r}_{12}, \mathbf{r}'_{12}; t)$ satisfies the "sink" Smoluchowski equation under the conditions discussed above.

D. Rate Kernel via Pair Theory

Before concluding this section on the implications of the pair kinetic theory for configuration space descriptions, we show that the kinetic equation may also be used to obtain the kinetic theory result for the rate kernel. This can be accomplished by projecting out the position and velocity dependence of the pair phase-space correlation function $C_{AB,AB}(12, 1'2'; t)$ to obtain an equation for

$$\int d1\, d2\, d1'\, d2'\, C_{AB,AB}(12, 1'2'; t) = P_{AB,AB}(t) \equiv P(t) \qquad (9.43)$$

the correlation function for the number of AB pairs at time t with the initial number of AB pairs. We first integrate (9.2) over $(1'2')$, and apply projection operator methods using

$$\mathcal{P}A(12) = V^{-2}g_{AB}(r_{12})\phi_A(v_1)\phi_B(v_2)\int d1\, d2\, A(12) \qquad (9.44)$$

to find

$$\{z + V^{-1}k_f(z)\}P(z) = P \tag{9.45}$$

where $k_f(z)$ is the rate kernel, which, in the large volume limit, is given by

$$k_f(z) = k_f^0 - V^{-1}\int d\mathbf{r}_1 d\mathbf{r}_2 \langle T_{Rf-}^{AB}(12)[z - L_R(12;z)]^{-1} T_{Rf-}^{AB}(12)\rangle_0 g(r_{12}) \tag{9.46}$$

This result is the same as that in (8.9) except that the propagator involving L_R is somewhat more elaborate than $G_{AB}(12;z)$ because the effects of the triple fields have been explicitly included. To avoid proliferation of notation, we also let $G_{AB}(12;z)$ denote this more general propagator.

X. ANALYSIS OF THE RATE KERNEL

The dynamic process that enter into the rate kernal expression [(9.46) or (8.9)] are, of course, those that have been included in the kinetic equation, as discussed briefly in Section VII. We discuss now the specific processes, which are relevant for the rate kernel, in more detail. The kinetic theory expression contains all the collision events that one might anticipate would be important for liquid state reactions. The analysis of the rate kernel in the limit where velocity relaxation effects are neglected bears a strong similarity to the derivation of Stokes' law from kinetic theory, and we also explore this relationship.

A. Dynamic Processes Contributing to the Rate Kernel

The rate kernel can be analyzed in a variety of ways depending on how the propagator $[z - L_R]^{-1}$ is represented. In the present discussion, we consider the simple reactive collision model of Section VI and drop the soft-force terms. We write $L_R(12;z)$ of (9.1) in the form

$$L_R(12;z) = -\mathcal{L}_0^{AB}(12) + \Lambda_-^{AS}(1) + \Lambda_-^{BS}(2) + R^{ABS}(12;z) + \mathcal{V}_{AB}(12) \tag{10.1}$$

where $\mathcal{V}_{AB}(12)$ is the vertex defined in (7.29).

We may then write the propagator in a form analogous to (7.29),

$$[z - L_R(12;z)]^{-1} \equiv G_{AB}(12;z) = \{\tilde{G}_{AB}^0(12;z)^{-1} - \mathcal{V}_{AB}(12)\}^{-1} \tag{10.2}$$

where now

$$\tilde{G}_{AB}^0(12;z)^{-1} = z + \mathcal{L}_0^{AB}(12) - \Lambda_-^{AS}(1) - \Lambda_-^{BS}(2) - R^{ABS}(12;z) \tag{10.3}$$

This propagator differs from G_{AB}^0 introduced in (7.29) in that $R^{ABS}(12; z)$ is now present; hence the correlated collision terms described in Section VII.D are included. In the general case, $R^{ABS}(12; z)$ contains reactive terms via the $\tilde{T}_-^{AB}(12)$ operator in the $\mathcal{V}_{ABS}(123)$ vertex [cf. (7.38)]. For the purposes of the present analysis, we drop these terms so that $\tilde{G}_{AB}^0(12; z)$ describes the correlated motion of an AB pair of nonreactive solute molecules. Given this type of representation of the dynamics, we may then expand (9.46), the expression for the rate kernel, in a series in $\mathcal{V}_{AB}(12)$,

$$k_f(z) = k_f^0 - V^{-1} \int d\mathbf{r}_1 d\mathbf{r}_2 \langle T_{Rf-}^{AB}(12)$$

$$\times \left\{ \tilde{G}_{AB}^0(12; z) + \tilde{G}_{AB}^0(12; z) \mathcal{V}_{AB}(12) \tilde{G}_{AB}^0(12; z) + \cdots \right\}$$

$$\times T_{Rf-}^{AB}(12) \rangle_0 g(r_{12})$$

$$= k_f^0 + \sum_{i=1}^{\infty} \Delta k_f^{(i)}(z) \tag{10.4}$$

The terms in this expansion correspond to a fairly sophisticated representation of the dynamics: Each vertex $\mathcal{V}_{AB}$ describes the collision events schematically depicted in Fig. 7.4, while the propagation between these collisions contains effects of the correlated motion of the two solute molecules in the solvent. To see this more clearly, consider $\Delta k_f^{(1)}(z)$,

$$\Delta k_f^{(1)}(z) = -V^{-1} \int d\mathbf{r}_1 d\mathbf{r}_2 \langle T_{Rf-}^{AB}(12) \tilde{G}_{AB}^0(12; z) T_{Rf-}^{AB}(12) \rangle_0 g(r_{12})$$

$$\tag{10.5}$$

We discussed the structure of the repeated ring operator R^{ABS} in Section VII.D and pointed out that it contains a variety of dynamic events such as a series of correlated collisions identical to those that appear in the singlet theory via the operators R^{AS} and R^{BS}. These operators represent the correlated collisions of a *single* solute molecule with the solvent and serve to "renormalize" the single-particle motion. Other events in R^{ABS} represent the coupling of the motion of the two solute molecules. In view of this, it is convenient to introduce the propagator for *independent* motion of the pair

$$G_{AB}^I(12; z) = \left\{ z + \mathcal{L}_0^{AB}(12) - \Lambda_-^{AS}(1) - R^{AS}(1; z) - \Lambda_-^{BS}(2) - R^{BS}(2; z) \right\}^{-1}$$

$$\equiv \left\{ z - L_I(12; z) \right\}^{-1} \tag{10.6}$$

In terms of this propagator, we may write $\tilde{G}^0_{AB}$ as

$$\tilde{G}^0_{AB}(12;z) = \left\{ G^I_{AB}(12;z)^{-1} - R_c(12;z) \right\}^{-1} \qquad (10.7)$$

Here,

$$R_c(12;z) = R^{ABS}(12;z) - R^{AS}(1;z) - R^{BS}(2;z) \qquad (10.8)$$

and thus represents all the collision events that correlate the motions of the pair (one such event was shown in Fig. 9.1). The expression for $\Delta k_f^{(1)}(z)$ may then be written as a series of terms involving the independent propagation of the pair,

$$\Delta k_f^{(1)}(z) = -V^{-1} \int d\mathbf{r}_1\, d\mathbf{r}_2 \langle T^{AB}_{Rf-}(12)$$

$$\times \left\{ G^I_{AB}(12;z) + G^I_{AB}(12;z) R_c(12;z) G^I_{AB}(12;z) + \cdots \right\}$$

$$\times T^{AB}_{Rf-}(12) \rangle_0\, g(r_{12})$$

$$= \sum_{j=1}^{\infty} \Delta k_f^{(1,j)}(z) \qquad (10.9)$$

To better appreciate the types of collision event that are involved, we examine the first two terms. The first term $\Delta k_f^{(1,1)}(z)$ is simple,

$$\Delta k_f^{(1,1)}(z) = -[Vg(\sigma)]^{-1} \int d\mathbf{r}_1\, d\mathbf{r}_2$$

$$\times \int_0^\infty dt\, e^{-zt} \langle \tilde{T}^{AB}_{Rf-}(12) G_A(1;t) G_B(2;t) \tilde{T}^{AB}_{Rf-}(12) \rangle_0 \qquad (10.10)$$

This represents, reading from right to left, a reactive collision followed by independent propagation of A and B. A reactive event terminates the correlation. The next term, $\Delta k_f^{(1,2)}$, explicitly displays the solvent coupling of the diffusive motion. With a one-ring approximation for R_c, we may write

$$\Delta k_f^{(1,2)}(z) \simeq -[Vg(\sigma)]^{-1} \int d\mathbf{r}_1\, d\mathbf{r}_2 \int_0^\infty dt\, e^{-zt} \int_0^t dt_1$$

$$\times \int_0^{t_1} dt_2 \left\langle \tilde{T}^{AB}_{Rf-}(12) \left\{ G_A(1;t-t_1) G_B(2;t-t_2) \tilde{T}^{AS}_{-}(1\bar{3}) \right. \right.$$

$$\times G_s(\bar{3};t_1-t_2) \tilde{T}^{BS}_{-}(2\bar{3}) G_A(1;t_1) G_B(2;t_2)$$

$$+ G_A(1;t-t_2) G_B(2;t-t_1) \tilde{T}^{BS}_{-}(2\bar{3}) G_s(\bar{3};t_1-t_2)$$

$$\left. \left. \times \tilde{T}^{AS}_{-}(1\bar{3}) G_A(1;t_2) G_B(2;t_1) \right\} \tilde{T}^{AB}_{Rf-}(12) \right\rangle_0 \qquad (10.11)$$

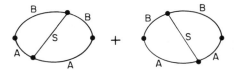

Fig. 10.1. Diagrammatic representation of a contribution to the rate kernel arising from the solvent coupling of the solute dynamics.

In the first term above, a reactive collision again initiates the correlation. The A and B molecules then propagate independently of each other, until at time t_2 the B molecule collides with a solvent molecule. After this collision the B and S molecules continue to undergo independent propagation. At time t_1, molecule A, which was originally correlated to B, undergoes a collision with the solvent molecule S with which B had collided, or any other solvent molecule that is collisionally correlated to S. Molecule A then continues its independent propagation until it collides reactively with B, terminating the correlation.

This sequence of collision events clearly shows how collisions of the solute molecules with the solvent can lead to a coupling between the motions of the solute molecules. This notion of solvent coupling of solute motion in the liquid is reminiscent of the hydrodynamic interaction effect on the friction coefficient of a pair of molecules, briefly discussed in the preceding section. In fact, the terms explicitly written in (10.11) simply represent, from a microscopic collisional point of view, the effects of such "hydrodynamic" interactions on the rate kernel.

These events are represented diagrammatically in Fig. 10.1, where each "vertex" is represented by a dot and assigned a time, and the independent propagators are designated by lines and connect times indicated by their arguments. Time increases from right to left in each diagram, and an integration over positions, velocities, and intermediate times is implied. These diagrams are very similar to the vertex corrections, which arise in mode-coupling theories of critical phenomena[41, 95] and are much more complex than the simple ring operators introduced earlier. In this language, another way of representing the series is to introduce a "renormalized" vertex (denoted by a heavy dot), which has the following diagrammatic representation:

$$\bullet = \cdot + \triangleleft + \triangleleft + \cdots \qquad (10.12)$$

and then rewrite the series in terms of this new "vertex." The two displayed terms may then be written as follows:

$$\Delta k_f^{(1,1)}(z) + \Delta k_f^{(1,2)}(z) \simeq \bigcirc \qquad (10.13)$$

Although the analysis in terms of the propagators for independent motion G_{AB}^I is convenient for displaying the content of the kinetic theory expression for the rate kernel, calculations based on (10.4), which contains the propagator for the correlated motion of the AB pair, are probably more convenient to carry out. In kinetic theory, such rate kernel expressions are usually evaluated by projections onto basis functions in velocity space. (We carry out such a calculation in Section X.B). Hence the problem reduces to calculation of matrix elements of $\tilde{G}_{AB}^0$ (coupled AB motion in a nonreactive system) and subsequent summation of the series. This emphasizes the point that a knowledge of the correlated motion of a pair of molecules for short distance and time scales is crucial for an understanding of the dynamic processes that contribute to the rate kernel.

We have based our discussion of the rate kernel on the simple "hard-sphere" reaction model. This model, which is applicable to high-barrier or sterically constrained reactions, when combined with the fact that the solute-solvent and solvent-solvent collisions are described by hard-sphere interaction, leads to an especially simple picture for the dynamics, since the collision events are well defined and localized even in the dense liquid.

The analysis of the rate kernel may have to be modified if other reaction types are considered. For example, for barrierless reactions, such as atom recombination, the reactive events are best described by partitioning the phase space into reactant and product regions, and then following the dynamics on a potential surface that connects these regions. When the solute-solute interaction is strongly attractive, as it is for these radical reactions, a description in terms of "encounters"[8, 96] is perhaps useful. Noyes defines an encounter as an approach to a distance corresponding to being in the strong force caging region. In this picture, the reaction is described in terms of roughly diffusive motion leading to "encounters," and the complex, definitely nondiffusive motion within the solvent cage.[8, 96, 97] Thus, although the kinetic theory contains the dynamic events that are relevant for reactions in the condensed phase, the most convenient mode of representation depends on the type of reaction under consideration.

This discussion was intended to provide some insight into the dynamic events that are incorporated into the kinetic theory expression for the rate kernel. These include all the events that are normally associated with qualitative ideas concerning caging effects on reaction dynamics. We next indicate how such kinetic theory results might be analyzed further, and how they are related to the diffusion equation results discussed earlier.

B. Projection onto Diffusion Modes

The foregoing discussion of the dynamic processes contributing to the rate kernel was given in terms of phase-space propagators for the A and B motion in the solvent. These propagators describe short- and long-time behaviors on all relevant distance scales. One might expect that if these propagators were projected onto the hydrodynamic diffusion modes, a description similar to that of the configuration space theories might emerge.[66] Here we show that this is indeed the case.

In Section X.A we described the structure of the propagator $[z - L_R(12; z)]^{-1}$ and pointed out that it could be written in terms of the propagator for the correlated motion of the pair $\tilde{G}_{AB}^0$ and a coupling term $\mathcal{V}_{AB}$. The simplest versions of the configuration space diffusion and Smoluchowski theories do not take into account such correlated motion. To make connection with these simpler theories, we therefore write

$$[z - L_R(12; z)]^{-1} \simeq \{ G_{AB}^I(12; z)^{-1} - \mathcal{V}_{AB}(12) \}^{-1}$$

and analyze the approximate rate kernel expression [cf. (9.46)],

$$k_f(z) = k_f^0 - g(\sigma_{AB})V^{-1} \int d\mathbf{r}_1 d\mathbf{r}_2 \langle T_{Rf-}^{AB}(12) \{ G_{AB}^I(12; z)^{-1} - \mathcal{V}_{AB}(12) \}^{-1}$$

$$\times T_{Rf-}^{AB}(12) \rangle_0$$

$$\equiv k_f^0 + \Delta k_f(z) \tag{10.14}$$

The analysis can be carried out more easily if an abstract notation is first introduced.[51] We write an arbitrary operator $A(12)$ in the form

$$\langle \mathbf{v}_1' \mathbf{v}_2' | A(\mathbf{r}_1 \mathbf{r}_2) | \mathbf{v}_1 \mathbf{v}_2 \rangle = A(12)\delta(\mathbf{v}_1 - \mathbf{v}_1')\delta(\mathbf{v}_2 - \mathbf{v}_2') \tag{10.15}$$

We also introduce abstract basis functions, $|I\rangle$, whose v-space matrix elements are related to Hermite polynomials $\mathbf{H}_I(\mathbf{v})$

$$\langle \mathbf{v}_1 | I \rangle = \mathbf{H}_I(\mathbf{v}_1)\phi(v_1)$$

$$\langle I | \mathbf{v}_1 \rangle = \mathbf{H}_I(v_1) \tag{10.16}$$

We also let $|I\rangle |J\rangle = |IJ\rangle$. In this notation, the relaxing part of the rate kernel may be written as

$$\Delta k_f(z) = -g(\sigma_{AB})V^{-1} \int d\mathbf{r}_1 d\mathbf{r}_2 \langle 00| T_{Rf-}^{AB} \{ G_{AB}^I(z)^{-1} - \mathcal{V}_{AB} \}^{-1} T_{Rf-}^{AB} |00\rangle$$

$$\equiv -g(\sigma_{AB})V^{-1} \int d\mathbf{r}_1 d\mathbf{r}_2 \langle 00| T_{Rf-}^{AB}(\mathbf{r}_1\mathbf{r}_2)\Phi(\mathbf{r}_1\mathbf{r}_2; z)|00\rangle \tag{10.17}$$

(We sometimes drop the $\mathbf{r}$ arguments when confusion is unlikely to arise.)

We now calculate Δk_f by projecting onto the eigenstates of the independent diffusive motion of the AB pair. These eigenfunctions are constructed so that

$$\langle D|G^I_{AB}(\mathbf{r}_1\mathbf{r}_2; z)|D\rangle = (z - D\nabla^2_{12})^{-1} \equiv \mathfrak{g}_0(\mathbf{r}_1 - \mathbf{r}_2; z) \qquad (10.18)$$

or less formally,

$$\mathfrak{g}_0(\mathbf{r}; z) = \frac{e^{-\alpha r}}{4\pi Dr}; \qquad \alpha = \left(\frac{z}{D}\right)^{1/2}$$

The diffusion eigenfunction $|D\rangle$ is given, to first order in gradients, as

$$|D\rangle = |00\rangle - D_A(m_A\beta)^{1/2}\nabla_1 \cdot |10\rangle - D_B(m_B\beta)^{1/2}\nabla_2 \cdot |01\rangle \quad (10.19)$$

The details of the construction of this eigenfunction are given in Appendix E. Using the projector $|D\rangle\langle D|$, we write Δk_f as

$$\Delta k_f(z) \simeq -g(\sigma)V^{-1}\int d\mathbf{r}_1 d\mathbf{r}_2 \langle 00|T^{AB}_{Rf-}(\mathbf{r}_1\mathbf{r}_2)|D\rangle\langle D|\Phi(\mathbf{r}_1\mathbf{r}_2; z)|00\rangle$$

$$(10.20)$$

The matrix element involving T^{AB}_{Rf-} is easily calculated to give

$$\langle 00|T^{AB}_{Rf-}|D\rangle = k^{eq}_f\left(4\pi\sigma^2_{AB}\right)^{-1}\delta(r_{12} - \sigma_{AB})$$

$$+ \frac{\alpha_R}{2}D\delta(r_{12} - \sigma_{AB})\hat{\mathbf{r}}_{12} \cdot \nabla_{12} \qquad (10.21)$$

The second term has the same structure as the term neglected in the derivation of the sink Smoluchowski equation in Section IX.C. Therefore, we make the same approximation here to make connection with the configuration space theories. In this approximation, Δk_f becomes

$$\Delta k_f(z) = -k^0_f\overline{\Phi_D(\mathbf{r}; z)}^s \qquad (10.22)$$

where $\Phi_D = \langle D|\Phi|00\rangle$, $\overline{\cdots}^s$ denotes an average over the surface $r = \sigma_{AB}$, and recall $k^0_f = k^{eq}_f g(\sigma_{AB})$. We now need an expression for Φ_D. From its definition in (10.17), Φ is seen to satisfy the equation

$$\left\{G^I_{AB}(z)^{-1} - \mathcal{V}_{AB}\right\}\Phi|00\rangle = T^{AB}_{Rf-}|00\rangle \qquad (10.23)$$

Projecting onto $|D\rangle$ we find

$$\{z - D\nabla^2 - V_{AB}(\mathbf{r})\}\Phi_D(\mathbf{r}; z) = k_f^{eq}(4\pi\sigma_{AB})^{-1}\delta(r - \sigma_{AB}) \quad (10.24)$$

where

$$V_{AB}(\mathbf{r}) = \langle D|\mathcal{V}_{AB}|D\rangle \simeq -k_f^{eq}(4\pi\sigma_{AB}^2)^{-1}\delta(r - \sigma_{AB}) - D\nabla\cdot\beta\mathbf{F} \quad (10.25)$$

Thus

$$\left\{z - D\nabla\cdot[\nabla - \beta\mathbf{F}] + \frac{k_f^{eq}}{4\pi\sigma_{AB}^2}\delta(r - \sigma_{AB})\right\}\Phi_D(\mathbf{r}; z) = \frac{k_f^{eq}}{4\pi\sigma_{AB}^2}\delta(r - \sigma_{AB})$$

$$(10.26)$$

The operator on the left-hand side of (10.26) is exactly the operator that appears in the sink Smoluchowski equation (3.12). The problem is now identical to that considered in Section III and Appendix A, and the same results as the Smoluchowski equation are obtained. As an illustration, we give some details of the solution for the case $g(r) = \theta(r - \sigma_{AB})$ in Appendix E. There we show that

$$\overline{\Phi_D(\mathbf{r}; z)}^s = k_f^0\left(k_f^0 + k_D(z)\right)^{-1} \quad (10.27)$$

and thus $k_f(z)^{-1} = k_f^{0^{-1}} + k_D(z)^{-1}$ as expected.

This calculation is quite similar to that for the derivation of Stokes' law from kinetic theory,[78] where one has an equation for the distribution function similar to (10.23) for Φ. To obtain Stokes' law, one must project the kinetic equation onto the hydrodynamic eigenfunctions, and it is essential to retain terms to first order in the gradients if the proper numerical factor ($\zeta = 4\pi\eta R$ for specular reflection and $\zeta = 6\pi\eta R$ for diffuse reflection) is to be obtained. In our calculation, it is also essential to retain the $\mathcal{O}(\nabla)$ terms in the $|D\rangle$ eigenfunction. If these $\mathcal{O}(\nabla)$ terms are dropped, the result for the rate coefficient $k_f(z = 0)$ still has the form of (3.7), $k_f^{-1} = k_f^{0^{-1}} + k_D^{-1}$, but the $z \neq 0$ result does not agree with (3.6) and (3.8). The gradient terms are essential if one is to obtain the simple z-dependence given by $k_D(z) = k_D(1 + \alpha\sigma_{AB})$. How this comes about is clearly demonstrated by the calculation in Appendix E.

The calculations presented here have simply served to show that a limiting form of the kinetic theory expression for the rate kernel can yield the results of configuration space approaches. However, the real promise of the kinetic theory method lies in the fact that it is not restricted to a description in terms of diffusive propagators, and the consequent motion on these space and time scales.

C. Contributions from Nonhydrodynamic States

We very briefly consider the effects of the terms that were neglected in the course of projecting onto the diffusion mode. Once again, the analysis closely parallels the derivation of Stokes' law. To explicitly consider the effects of nonhydrodynamic (nondiffusion mode) states on the rate kernel, we use an analysis similar to that of van Beijeren and Dorfman[78] and introduce a projection operator $\mathscr{P}_D$

$$\mathscr{P}_D = |D\rangle\langle D|, \tag{10.28}$$

and its complement Q_D. We may then write the kinetic theory expression for the rate kernel, (10.17), as

$$\Delta k_f(z) = -g(\sigma_{AB})V^{-1}\int d\mathbf{r}_1 d\mathbf{r}_2 \{\langle 00|T_{Rf-}^{AB}(\mathbf{r}_1\mathbf{r}_2)|D\rangle\Phi_D(\mathbf{r}_1\mathbf{r}_2;z)$$
$$+ \langle 00|T_{Rf-}^{AB}(\mathbf{r}_1\mathbf{r}_2)Q_D\Phi(\mathbf{r}_1\mathbf{r}_2;z)|00\rangle\} \tag{10.29}$$

Given the integral equation (10.23), we now apply projection operator techniques to obtain formal solutions for Φ_D and $Q_D\Phi$. The equation for $\mathscr{P}_D\Phi|00\rangle = |D\rangle\Phi_D$ is

$$\mathscr{P}_D\{G_{AB}^I(z)^{-1} - \mathcal{V}_{AB}\}\mathscr{P}_D\Phi|00\rangle - \mathscr{P}_D\mathcal{V}_{AB}Q_D\Phi|00\rangle = \mathscr{P}_D T_{Rf-}^{AB}|00\rangle \tag{10.30}$$

while $Q_D\Phi|00\rangle$ can be obtained from

$$Q_D\{G_{AB}^I(z)^{-1} - \mathcal{V}_{AB}\}Q_D\Phi|00\rangle - Q_D\mathcal{V}_{AB}\mathscr{P}_D\Phi|00\rangle = Q_D T_{Rf-}^{AB}|00\rangle \tag{10.31}$$

We have

$$Q_D\Phi|00\rangle = \{Q_D G_{AB}^I(z)^{-1}Q_D - Q_D\mathcal{V}_{AB}Q_D\}^{-1}$$
$$\times [Q_D\mathcal{V}_{AB}\mathscr{P}_D\Phi|00\rangle + Q_D T_{Rf-}^{AB}|00\rangle] \tag{10.32}$$

Inserting this result into (10.30), we obtain

$$\{z - D\nabla^2 - V_{AB}^R(\mathbf{r})\}\Phi_D(\mathbf{r};z) = \mathcal{S}(\mathbf{r}) \tag{10.33}$$

where $V_{AB}^R(\mathbf{r})$ is the diffusion mode matrix element of $\mathcal{V}_{AB}^R$,

$$V_{AB}^R(\mathbf{r}) = \langle D|\mathcal{V}_{AB}^R|D\rangle \tag{10.34}$$

and $\mathcal{V}_{AB}^R$ contains the effects of the nonhydrodynamic states

$$\mathcal{V}_{AB}^R = \mathcal{V}_{AB} + \mathcal{V}_{AB}Q_D G_{AB}^I Q_D \mathcal{V}_{AB}$$
$$+ \mathcal{V}_{AB}Q_D G_{AB}^I Q_D \mathcal{V}_{AB}Q_D G_{AB}^I Q_D \mathcal{V}_{AB} + \cdots \qquad (10.35)$$

The quantity $\mathcal{S}(\mathbf{r})$ may also be written in terms of a modified collision operator $\mathcal{T}^R$,

$$\mathcal{T}^R = T_{Rf-}^{AB} + T_{Rf-}^{AB} Q_D G_{AB}^I Q_D \mathcal{V}_{AB} + T_{Rf-}^{AB} Q_D G_{AB}^I Q_D \mathcal{V}_{AB} Q_D G_{AB}^I Q_D \mathcal{V}_{AB} + \cdots \qquad (10.36)$$

as

$$\mathcal{S}(\mathbf{r}) = \langle D | \mathcal{T}^R | 00 \rangle \qquad (10.37)$$

The rate kernel expression then takes the form

$$\Delta k_f(z) = -g(\sigma_{AB})V^{-1} \int d\mathbf{r}_1 d\mathbf{r}_2 \{ \langle 00| \mathcal{T}^R(\mathbf{r}_1\mathbf{r}_2)| D \rangle \Phi_D(\mathbf{r}_1\mathbf{r}_2; z)$$
$$+ \langle 00| \mathcal{T}^R(\mathbf{r}_1\mathbf{r}_2) - T_R^{AB}(\mathbf{r}_1\mathbf{r}_2)|00 \rangle \}$$

$$(10.38)$$

and is thus expressed in terms of the modified vertex $\mathcal{T}^R$ and Φ_D, which is now obtained by solving (10.33). This equation, which incorporates the effects of nonhydrodynamic states, is structurally similar to (10.24). All nonhydrodynamic effects are contained in the modified vertex. The diffusion mode propagator $\langle D | G_{AB}^I | D \rangle$ describes processes where the two solute molecules make long diffusive excursions into the fluid between the correlated recollision events. It is this sequence of events that the integral equation method of Section X.B takes into account. In contrast to $\langle D | G_{AB}^I | D \rangle$, $Q_D G_{AB}^I Q_D$ describes processes where the particles travel only short distances in the fluid. Hence the modified vertex describes the infinite sequence of such short excursions into the fluid, which occur between the recollision events. Using the language of fluid mechanics, these processes may be described as occurring in the microscopic boundary layer in the vicinity of the solute molecules' surfaces (cf. Fig. 3.1). The calculations in Section X.B, which only take into account diffusion mode propagation, can be formally viewed as an approximation that replaces $\mathcal{V}_{AB}^R$ by $\mathcal{V}_{AB}$, and therefore completely neglects the detailed structure of the boundary layer. In the simple diffusion equation approach, the boundary layer is roughly accounted for by the effective rate coefficient k_i in the radiation boundary condition (cf. Section III.A). The structure of this boundary layer is, of course, crucial for the description of the short-time behavior of the rate kernel.

This reformulation in terms of diffusive propagation and microscopic dynamics in the boundary layer is reminiscent of Noyes's encounter formulation that we briefly described earlier. Now each diffusive "encounter" is interrupted by sequences of collisions and very short excursions into the fluid. The analysis of nonhydrodynamic effects on the rate kernel can, therefore, be discussed naturally in terms of the encounter formalism.

There are other ways of analyzing nonhydrodynamic contributions. Projections onto finite sets of velocity states, in combination with kinetic modeling techniques,[98] have proved useful in the analysis of the small molecule velocity autocorrelation function. These techniques can also be used to calculate the rate kernel.[99]

Before closing this section, we should remark that although this analysis of velocity relaxation effects has focused on a simple collision model, we expect that the detailed structure of the rate kernel for short times will depend on the precise form of the chemical interactions in the system under consideration. It is clear, however, that a number of fundamental questions need to be answered before more specific calculations can be undertaken form the kinetic theory point of view.

XI. INITIAL CONDITION EFFECTS

Often it is not the rate kernel itself that is of interest, but rather the decay of the system from some initial nonequilibrium state. A case in point is the study of atom recombination following a photodissociation event. The initial state is presumed to correspond to a specific phase-space configuration of the pair of atomic radicals in an equilibrium solvent. The decay of this initial state is then monitored in the experiment. This is discussed in more detail in Section XII; here we simply show how the kinetic theory can be formulated to accommodate this situation.

We consider the pair formulation because it is most suitable for this application. The kinetic equation for the doublet distribution function

$$F_{AB}(12, t) = \overline{n_{AB}(x_1 x_2, t)}$$

has the general structure

$$
\begin{aligned}
\frac{\partial F(12, t)}{\partial t} &= \{ -\mathcal{L}_0(12) + \mathcal{L}_s(12) + v_{12} \cdot \beta F_H \} F(12, t) \\
&\quad + \int dt' \, K(12; t') F(12, t - t') \\
&\equiv \int_0^t dt' \, L_R(12; t') F(12, t - t')
\end{aligned}
\tag{11.1}
$$

We have suppressed the species labels. If there are only soft forces operating, $\mathbf{v}_{12} \cdot \beta \mathbf{F}_H$ may be dropped, and the potential in $\mathcal{L}_s$ replaced by the potential of mean force.

An equation for $P(t)$, the number of unreacted AB pairs at time t, can be constructed using the projection operator $\mathcal{P}$ in (9.44), since

$$\int d\mathbf{x}_1 d\mathbf{x}_2 \, F(12, t) = P(t) \tag{11.2}$$

(We use the same symbol as in Section IX.D, but this should not lead to confusion.) Now, however, since $F(12) \equiv F(12, t = 0)$ is arbitrary rather than an equilibrium distribution function, it is no longer true that $QF(12) = 0$. Thus using projection operator methods on the Laplace transform of (11.1), we find

$$\{z + V^{-1}k_f(z)\}P(z) = P - \int d\mathbf{x}_1 d\mathbf{x}_2 \, T_{Rf-}^{AB}(12)[z - QL_R(12; z)]^{-1}QF(12) \tag{11.3}$$

This equation may be compared with (9.45), it differs because the term depending on the nonequilibrium initial condition for $F(12)$ is present. In the large-volume limit we have

$$zP(z) - P = -\int d\mathbf{x}_1 d\mathbf{x}_2 \, T_{Rf-}^{AB}(12)[z - L_R(12; z)]^{-1}F(12)$$

$$\equiv -I(z) \tag{11.4}$$

The initial condition term in turn differs from the relaxing part of the memory kernel in (9.46) in that the initial condition on F replaces $T_{Rf-}^{AB}(12)\phi_A(v_1)\phi_B(v_2)g(r_{12})$. Thus the analysis of the memory kernel discussed in Section X also applies to this initial condition term, except that now the correlation is initiated by the nonequilibrium value of F.[100]

The same manipulations can be carried out using the Smoluchowski equation in place of the kinetic equation. Now the projection operator in (A.4) is used to effect the reduction, and a result with the form of (11.4) is obtained with $I(z)$, given[15, 101] by

$$I_s(z) = \frac{k_f^{eq}}{4\pi\sigma_{AB}^2} \int d\mathbf{r}_{12} \, \delta(r_{12} - \sigma_{AB})[z - \mathcal{L}_{SM}]^{-1}P(\mathbf{r}_{12}) \tag{11.5}$$

This equation can be solved numerically;[101] for simple diffusion equation dynamics analytical results can be obtained[15].

If the initial condition in the kinetic theory formulation [(11.4)] is assumed to have the form

$$F(12) = \phi_A(v_1)\phi_B(v_2)P(\mathbf{r}_{12})V^{-1} \tag{11.6}$$

that is, an initial deviation in relative configuration space only, and $I(z)$ is projected onto the diffusion eigenfunction, one might expect to obtain a result similar to $I_s(z)$. Since the outline of the calculation that follows, closely parallels that given in Section X.B, we can be brief. In terms of the abstract notation of Section X.B, we may write $I(z)$ as

$$I(z) = V^{-1}\int d\mathbf{r}_1\, d\mathbf{r}_2 \langle 00|T_{Rf-}^{AB}\left\{G_{AB}^I(z)^{-1} - \mathcal{V}_{AB}\right\}^{-1}|00\rangle P(\mathbf{r}_{12})$$

$$\equiv V^{-1}\int d\mathbf{r}_1\, d\mathbf{r}_2 \langle 00|T_{Rf-}^{AB}(\mathbf{r}_1\mathbf{r}_2)\Phi'(\mathbf{r}_1\mathbf{r}_2;z)|00\rangle \tag{11.7}$$

If we ignore nonhydrodynamic contributions and project onto the diffusion eigenfunction [(10.19)], we may write

$$I(z) = k_f^{eq}\overline{\Phi'_D(\mathbf{r};z)}^s \tag{11.8}$$

where now $\Phi'_D = \langle D|\Phi'|00\rangle$. We have again neglected the second term in (10.21). The integral equation for Φ' is

$$\left\{G_{AB}^I(z)^{-1} - \mathcal{V}_{AB}\right\}\Phi'|00\rangle = |00\rangle P(\mathbf{r}_{12}) \tag{11.9}$$

Projecting onto $|D\rangle$ and following (10.24) to (10.26), we find

$$\left\{z - D\boldsymbol{\nabla}\cdot[\boldsymbol{\nabla} - \beta\mathbf{F}] + \frac{k_f^{eq}}{4\pi\sigma_{AB}^2}\delta(r - \sigma_{AB})\right\}\Phi'_D(\mathbf{r};z) = P(\mathbf{r}) \tag{11.10}$$

or, formally, using the definition of $\mathcal{L}_{SM}$ in (A.3),

$$\Phi'_D(\mathbf{r};z) = [z - \mathcal{L}_{SM}]^{-1}P(\mathbf{r}) \tag{11.11}$$

Insertion of (11.11) into (11.8) yields the Smoluchowski result, (11.5).

As an illustration of the structure of $I(z)$ [or $I(t)$], we examine the case where simple diffusion equation dynamics is used in place of $\mathcal{L}_{SM}$, that is,

$$\mathcal{L}_{SM} \rightarrow D\nabla^2 - \frac{k_f^{eq}}{4\pi\sigma_{AB}^2}\delta(r - \sigma_{AB}) \tag{11.12}$$

For this simple case, the Laplace inversion may be performed analytically to yield[100]

$$\frac{dP(\tau)}{d\tau} = -I(\tau) = -\frac{\lambda}{\kappa}\left[\frac{1}{\sqrt{\pi\tau}}\exp\left\{-\frac{(\kappa-1)^2}{4\tau}\right\} - (1+\lambda)\right.$$

$$\left. \times \exp\{(1+\lambda)(\kappa-1)\}\exp\{(1+\lambda)^2\tau\}\text{erfc}\left\{(1+\lambda)\sqrt{\tau}+\frac{(\kappa-1)}{2\sqrt{\tau}}\right\}\right] \tag{11.15}$$

Here we defined $\kappa = r_0/\sigma_{AB}$, $\lambda = k_f^0/k_D$, and $\tau = \sigma_{AB}^2 t/D$. We may also calculate $P(\tau)$ directly and obtain

$$P(\tau) = 1 - \frac{\lambda}{\kappa(1+\lambda)}\left\{\text{erfc}\left(\frac{\kappa-1}{2\sqrt{\tau}}\right) - \exp[(1+\lambda)(\kappa-1)]\right.$$

$$\left. \times \exp[(1+\lambda)^2\tau]\text{erfc}\left[(1+\lambda)\sqrt{\tau}+\frac{\kappa-1}{2\sqrt{\tau}}\right]\right\} \tag{11.16}$$

These results illustrate the gross features we expect to obtain. The dependence of $P(\tau)$ on the initial separation of the pairs is shown in Fig. 11.1, which also shows the slow $\mathcal{O}(\tau^{-1/2})$ decay of $P(\tau)$

$$P(\tau) \underset{\tau \text{ large}}{\rightarrow} 1 - \frac{\lambda}{\kappa(1+\lambda)}\left\{1 - \frac{1}{\sqrt{\pi\tau}}\left[(\kappa-1)+\frac{1}{1+\lambda}\exp[(1+\lambda)(\kappa-1)]\right]\right\} \tag{11.17}$$

The presence of forces and spatial dependence of D will, of course, modify these results[101] (cf. Section XII). In addition, according to the arguments presented throughout this chapter, we expect these results to lose all validity in the short-time region. Here the nonhydrodynamic states, discussed in Section X.C, will play a crucial role.

If only soft forces act between the pair of particles, a calculation similar to that in Section V can be carried out. One now projects onto the dynamic variable characterizing the bound (or unbound) species.

This section was designed to make it clear that, with minor modifications, the techniques and discussion given previously can be applied to the initial condition problem.

XII. APPLICATIONS

We have thus far been concerned primarily with general aspects of the microscopic theory of condensed-phase chemical reactions. Our use of

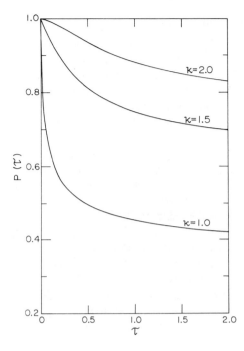

Fig. 11.1. The unreacted pair probability $P(\tau)$ versus τ for several values of the initial separation of the pairs. Simple diffusion equation dynamics is assumed and $\lambda = k_f^0/k_D = 2.0$.

specific models was dictated by a desire to present the results in "simple" form. It is not difficult to give an even more general presentation that replaces the hard-sphere solute-solvent and solvent-solvent interactions by soft forces and makes less use of the simple impulsive reaction model. Our emphasis on general features precluded a discussion of actual calculations. Hence much of the previous discussion was rather formal. There is an even more compelling reason for the absence of reference to specific calculations, namely, very few have been carried out. Only recently have techniques been developed that enable us to deal with chemical reactions in liquids on a microscopic level.

Here, we consider several specific problems, to illustrate some of the complexities encountered when treating actual systems, and also to point out how the general ideas presented earlier apply to these cases. The examples we pick either correspond to reactions that have been discussed in the course of formulating the kinetic theory, or else show how the theory can be implemented for the treatment of actual systems. We do not attempt to comment on the wide variety of reactions studied in the condensed phase, for which a microscopic theory would be desirable. Discussions of other systems can be found in recent reviews.[102]

A. Atom Recombination

The condensed-phase atomic recombination process has been extensively studied from both experimental and theoretical points of view. This apparently simple reaction is actually rather complex, and our knowledge of the process is still very far from being complete. We have already referred to this type of reaction several times to illustrate certain features of the kinetic theory formulation. We now give a more detailed and coherent discussion.[103]

Consider the recombination process following a photodissociation event in a dense inert solvent. To make the discussion concrete, we examine iodine atom recombination. A schematic representation of the processes involved is shown in Fig. 12.1. The excitation to the bound excited electronic state is followed by a predissociation to a repulsive state, which leads to ground-state atoms. Curve crossing to the ground-state potential energy surface can occur at a variety of internuclear separations (indicated by the downward arrows in Fig. 12.1). The atoms produced in this way may either recombine (geminate recombination), or escape (eventually to recombine with other I atoms). We focus solely on this geminate recombination process. The actual situation is considerably more complex than that shown in Fig. 12.1 because many more electronic energy states may participate in the process.[104]

Since we are concerned with recombination in the liquid phase, the appropriate potential for the description of atom-atom interactions is the potential of mean force $W(r) = -k_B T \ln g(r)$, schematically displayed in Fig. 4.1. If we adopt the model described in Section VI, in which the solvent-solvent and solute-solvent forces are approximated by hard-sphere interactions, then

$$W(r) = V(r) + W_c(r) \tag{12.1}$$

where

$$W_c(r) = -k_B T \ln y_{HS}(r) \tag{12.2}$$

where y_{HS} is the radial distribution function for a pair of cavities in a hard-sphere fluid (see Section VII) and $V(r)$ is the direct I–I interaction, which we can approximate by a Morse potential

$$V(r) = D_e \left[\exp(-2b(r-r_e)) - 2\exp(-b(r-r_e)) \right] \tag{12.3}$$

For I_2, the equilibrium separation $r_e = 2.6668$ Å, $b = 1.8674$ Å^{-1}, and the dissociation energy $D_e/k_B T = 60.1487$ at $T = 300°$K. We display this mean

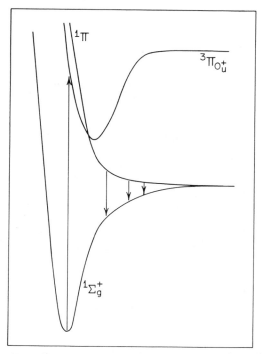

Fig. 12.1.   Representation of some of the principal features of the iodine recombination process following photodissociation. The upward arrow indicates excitation to a bound electronic state. After predissociation to a repulsive excited state, transitions to the ground state may occur at a variety of internuclear separations. Other potential energy surfaces not explicitly shown may also participate in the recombination process.

potential for several iodine atom-to-solvent molecule ratios in Fig. 12.2, and compare it to the direct I–I Morse interaction. The effects of solvent structure are modest, but they do, in fact, influence the reaction dynamics. A similar type of mean potential model has been used in the study of isomerization by Chandler and Pratt.[105] Other methods for constructing mean potentials for reaction dynamics studies have also been devised.[106]

In view of these solvent structure effects, it is convenient to classify the recombination events into primary and secondary processes.[107] *Primary* recombination processes are those in which the recombination takes place before the atoms separate to a distance roughly equal to the first maximum in the mean potential (i.e., recombination in the solvent "cage"). *Secondary* recombination involves the recombination of solvent–separated-atom pairs.

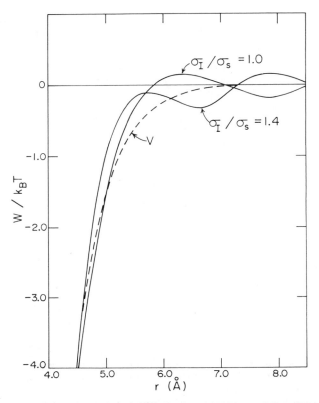

Fig. 12.2. Potential of mean force W for two different solute-solvent diameters. Rough estimates of the effective elastic collision diameters of iodine in inert solvents can be obtained from the known Lennard-Jones σ parameters. The mean force potential is also compared with the Morse potential, V.

Part of the stimulus for research in this area comes from the possibility of probing the dynamics of such processes on short time scales by using picosecond lasers.[108] The standard pulse-and-probe experiments will measure the entire time profile of the recombination and photodissociation processes. An interpretation of such results therefore requires a consideration of the dynamics on several potential energy surfaces for both the primary and secondary recombination processes. The very short time behavior is often obscured by experimental problems (laser rise times etc.), but the secondary recombination process is more easily studied.

Theoretical treatments often focus on either the primary or secondary processes. The traditional approaches that make use of a simple diffusion equation for the pair dynamics are necessarily restricted to a description of the secondary recombination process, since this type of equation loses its validity in the solvent "cage," where the atom pair is not separated by solvent molecules and strong forces operate. Even a Smoluchowski equation cannot describe the dynamics in the solvent cage. On the other hand, recent molecular dynamics simulations have been concerned with a description of the primary recombination process,[109-111] mainly for technical reasons. The simulations are time-consuming, and it is difficult to follow the trajectories for the long times (several hundreds of picoseconds) that characterize the secondary recombination process. It is also difficult to follow the dynamics of the large number of solvent molecules necessary to have solvent-separated pair conditions. In addition, these molecular dynamics simulations have been largely restricted to a treatment of the dynamics on the ground-state potential energy surface.

Before a theory for such reactions can be constructed, it is necessary to look at the nature of the interactions a bit more closely. It is instructive to consider the gas-phase recombination mechanism first. It is usually assumed that recombination occurs by a combination of two mechanisms.[112] The radical complex mechanism (RCM), which is important at room temperature, assumes that recombination takes place by the reactions

$$I + S \rightleftharpoons IS$$
$$IS + I \rightarrow I_2 + S \tag{12.4}$$

The existence of the IS bound pair is clearly dependent on the attractive part of the iodine-solvent interaction. At higher temperatures, the energy transfer mechanism (ET) becomes increasingly important. This mechanism involves stabilization of an unbound quasi-dimer I_2^* by collisions with S,

$$I + I \rightleftharpoons I_2^*$$
$$I_2^* + S \rightarrow I_2 + S \tag{12.5}$$

These reaction schemes represent an attempt to describe the termolecular reaction in terms of binary collision events. Clearly, the model we proposed earlier for the interactions is not appropriate for the gas-phase reaction. In a dense liquid the situation is different. Solvent molecules are always present by virtue of their high density, and thus there is no need to invoke the presence of weak I–S attractive forces (as in the RCM) to provide a route for energy dissipation. Solvent configurations capable of dissipating the energy of the recombining atoms will exist regardless of whether I–S attractive forces are present. In this light, the proposed model for the interactions appears reasonable.

No calculations yet have tackled this problem in all its complexity. Typically, it is assumed that the photodissociation produces some initial distribution of pairs, and then the subsequent time evolution of the unreacted pair probability is calculated. Even this more modest program has been carried out only at the diffusion[103, 113] and Langevin equation[103] levels. We briefly comment on these results, since they indicate the magnitude of the solvent and velocity relaxation effects.

The quantity of interest is the unreacted pair probability, and the strategy of the calculation was outlined in Section XI. To explicitly carry out the calculation with a Langevin equation description of the dynamics, we use the following procedure. The Langevin equation (3.25) is now taken to describe the motion of the relative velocity of the pair,

$$\mu \frac{d\mathbf{v}}{dt}(t) = -\zeta \mathbf{v}(t) + \mathbf{F}(\mathbf{r}(t)) + \mathbf{f}(t) \tag{12.6}$$

where ζ is the relative friction coefficient, just half of the independent atom friction coefficient in this simple r-independent friction model. The independent pairs are assumed to have initial separation r_0 and an equilibrium distribution of velocities. This is equivalent to taking

$$F(12) = \phi_I(v_1)\phi_I(v_2)\left(4\pi r_0^2 V\right)^{-1}\delta(r_0 - r_{12}) \tag{12.7}$$

The Langevin equation is then integrated[103, 114] for each member of the ensemble of pairs until reaction (formation of stable I_2) or escape occurs. The reaction distance, the relative separation at which irreversible formation of stable I_2 is almost certain to occur, must be determined empirically. For the results in Ref. 103, 4 Å was found to be appropriate for this system. The probability of reaction at time t, $P_R(t)$, is just the number of pairs that have reacted up to time t divided by the total number of pairs in the ensemble. The unreacted pair probability is then $P(t) = 1 - P_R(t)$.

Results that illustrate some features of the "cage" effect are given in Fig. 12.3: $P(t)$ is shown for two mean potentials and compared with results using only the direct Morse interaction potential. All calculations were carried out starting at $r_0 = 5.5$ Å. For the small solvent case, this corresponds to a separation near the first maximum in the mean potential, and well within the solvent "cage" for the equal size solvent case. We expect the reaction probability to be enhanced when the initial separation is small (inside the "cage"), and this is indeed the case. For the small solvent case, the reaction probability is decreased. Although this initial separation is just to the left of the barrier, some molecules move to the right and then experience difficulty in recombining because of the small solvent barrier. The effects are not large, approximately 10 to 30%.

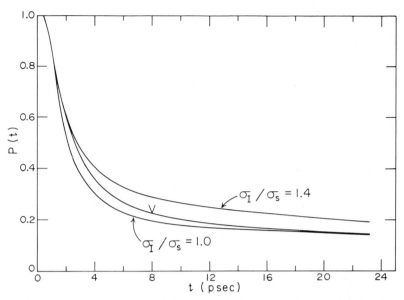

Fig. 12.3. To illustrate the effects of the solvent on the unreacted pair probability, $P(t)$ is plotted versus t. The initial separation is $r_0 = 5.5$ Å and $\mu/\zeta = 0.05$ psec.

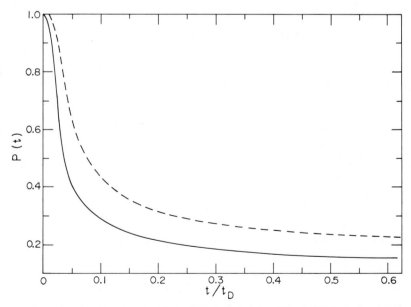

Fig. 12.4. Comparison of Langevin (solid curve) and Smoluchowski (dashed curve) equation solutions for $P(t)$. The parameters for the calculations are: $r_0 = 5.5$ Å, $\zeta/\mu = 0.25$ psec, $t_D = 16$ psec, and $\sigma_I/\sigma_S = 1.0$.

We may also use such calculations to partially assess the validity of configuration space approaches. We expect that if the friction is high, the Langevin equation and Smoluchowski equation methods will give equivalent results. The Smoluchowski equation can be applied to this problem, as outlined previously. The "reaction" at $r = 4$ Å can be incorporated by using a radiation boundary condition at this separation, and the escape can be modeled by a complete absorption boundary condition at some large distance (12 Å in these calculations). For high friction, the good agreement between the Langevin and Smoluchowski equation predictions is indeed found. If the friction is reduced, however, the Smoluchowski equation breaks down rather badly (see Fig. 12.4).

Kinetic theory can be used to extend this simple description in several ways. First, even if the calculation is carried out at the Langevin equation level, the kinetic theory results for the friction coefficient can be used to go beyond the simple (and possibly inadequate) approximation of a constant friction. Second, a direct solution of the kinetic equation rather than the Langevin equation can be carried out. Both types of calculation should increase our understanding of the microscopic dynamics of these processes.

B. Isomerization

We would be remiss if we did not comment on isomerization dynamics in liquids, since they form a very important and widely studied class of reactions. There have been many theoretical models for isomerization reactions in liquids. In Section III.C we briefly outlined Kramers' approach to this problem. Since that time much more extensive studies of such barrier-crossing problems have been carried out. These studies have been concerned mainly with obtaining expressions for the rates in the transition region between the low- and high-friction cases, or with effects arising from the nonlocality of the diffusion coefficient in the context of Smoluchowski equation descriptions, applications to polymeric systems, and so on.[115]

A fully microscopic treatment of this problem is a very difficult task. It is usually the motion of some internal coordinate of a complex molecule that is important for the description of the isomerization reaction (cf. Sections III and IV). A microscopic theory at the same level as that for the bimolecular processes described in the previous sections would entail a full description (or model) of the internal structure of the molecule and its interactions with the surrounding solvent. The collision dynamics for such a process are necessarily complex, but a theory at this detailed level is not out of the question for some models of small molecule isomerization reactions. However, it is probably premature to embark on such a program, since the implications of the kinetic theory for the reactions for which it is more easily formulated have not yet been fully explored.

Simpler BGK kinetic theory models have, however, been applied to the study of isomerization dynamics.[33, 84, 85] The solutions to the kinetic equation have been carried out either by expansions in eigenfunctions of the BGK collision operator[33, 85] (these are similar in spirit to the discussion in Section IX.B) or by stochastic simulation of the kinetic equation.[84] The stochastic trajectory simulation of the BGK kinetic equation involves the calculation of the trajectories of an ensemble of particles as in the Brownian dynamics method described earlier.

The time development is computed differently. The positions and velocities are taken to evolve in time according to Hamilton's equations of motion. However, at random time intervals, which are sampled from an exponential distribution with decay constant equal to the collision frequency α_c (cf. Section VII.E), the particles undergo collisions that randomize the velocity. (The new velocity is selected from a Boltzmann distribution of velocities.) This prescription is equivalent to the solution of the BGK kinetic equation (7.44).[84] This provides a more reasonable description of the dynamics when the friction (collision frequency) is small.

Model kinetic equation approaches of these types should probably be more thoroughly investigated for such complex systems before more elaborate kinetic theories are constructed. Ultimately, however, difficult problems such as the nature of the friction coefficient or collision frequency associated with an internal coordinate must be solved. What, for instance, is the form of its space and time nonlocality? The solution of this problem will involve a more complex calculation than that outlined in Section IX.B for the two-particle friction tensor.

XIII. CONCLUSIONS

We have presented a framework in which condensed-phase chemical reactions can be described from a microscopic point of view. The problem is difficult for obvious reasons; one is faced with the full complexity of a reactive event, a problem that has taxed gas-phase reaction kineticists for many years, and the necessity to describe the correlated motion of several particles in a dense fluid on a microscopic level, a problem in nonequilibrium statistical mechanics that has not yet been completely solved. Admittedly, there may be some simplifications. For example, in a dense fluid some of the finer details of the reactive event may not be as crucial to the outcome as they are in the gas phase because of the energy dissipation by the solvent. We have in fact used this feature to construct some of the reaction models that were presented. However, this simplifying feature is more than compensated for by the necessity to describe the precise nature of this energy dissipation by the solvent for very small separations between the potentially reactive solute molecules.

In spite of the inherent complexity of the problem, it is encouraging to see that it is possible to incorporate many of these relevant features into a kinetic theory that treats the solute and solvent molecules on the same footing. The existing more phenomenological diffusion equation theories follow in a natural way as limiting cases of the kinetic theory. Although all problems that arise in the course of attempting to construct a theory of reactions in liquids are not solved, the kinetic theory, at the least, allows one to state these problems precisely. For example, the discussion of the form of the two-particle friction, which is crucial for determining the energy dissipation as the molecules approach each other, is a well-posed calculational problem that in certain limiting situations has been solved (Section IX). In a similar vein, the problem of the nature of the breakdown of the diffusion equation description of the rate kernel on short distance and time scales can also be given a precise mathematical formulation (Section X). Further progress in resolving some of these questions for model reacting systems should not be too difficult to achieve.

There is still a gap between our models of liquid-state reactions and the often bewildering complexity of real chemical systems. Progress in shortening the gap will probably come only from the application of a variety of methods to this problem. The full promise of picosecond spectroscopy techniques for studying the details of the dynamics of reactive events in liquids has yet to be realized. How deeply can these methods probe the dynamics? Computer simulations, another source of "experimental" information in reacting systems, are only beginning to be exploited.[109-111] The description by direct computer simulation of *both* primary and secondary recombination dynamics, for example, would yield a wealth of information that could be used to test theories.

Although the development of a fully microscopic theory of chemical reactions in liquids will remain a challenge for some time, future progress is expected. We have at our disposal a variety of theoretical and experimental tools that should allow us to unravel this complex problem.

APPENDIX A

Smoluchowski Equation Result for the Rate Kernel

The rate kernel expression in (3.13) can be obtained by using projection operator methods to derive an equation for the unreacted pair probability $P(t)$,

$$P(t) = \int_V d\mathbf{r}\, P(\mathbf{r}, t) \tag{A.1}$$

from the Smoluchowski equation. We write the Fourier transform of the sink Smoluchowski equation (3.12) compactly as

$$(-i\omega - \mathcal{L}_{SM})P(r;\omega) = 0 \qquad (A.2)$$

where

$$\mathcal{L}_{SM} = \nabla \cdot D \cdot [\nabla - \beta F] - \frac{k_i}{4\pi\sigma^2}\delta(r-\sigma) \qquad (A.3)$$

and apply projection operator methods using the projection operator

$$\mathcal{P} = V^{-1}g(r)\int_V d\mathbf{r} \qquad (A.4)$$

By making use of the operator identity

$$(-i\omega - \mathcal{L}_{SM})^{-1} = (-i\omega - Q\mathcal{L}_{SM})^{-1}$$
$$+ (-i\omega - \mathcal{L}_{SM})^{-1}\mathcal{P}\mathcal{L}_{SM}(-i\omega - Q\mathcal{L}_{SM})^{-1}$$

with $Q = 1 - \mathcal{P}$, we may write (A.2) as

$$\{z + V^{-1}k_f(\omega)\}P(\omega) = 0 \qquad (A.5)$$

Here,

$$k_f(\omega) = k_f^0 - (k_f^0)^2[g(\sigma)]^{-1}\overline{\mathcal{G}_R(\mathbf{r}|\mathbf{r}';\omega)}^{S,S'} \qquad (A.6)$$

where formally

$$\mathcal{G}_R(\mathbf{r};\omega) = (-i\omega - \mathcal{L}_{SM})^{-1} \qquad (A.7)$$

and the surface average is defined by

$$\overline{A(\mathbf{r})}^S = (4\pi\sigma^2)^{-1}\int d\mathbf{r}\,\delta(r-\sigma)A(\mathbf{r}) \qquad (A.8)$$

Using the definitions of $\mathcal{G}$ and $\mathcal{G}_R$ in (3.16) and (A.7), respectively, we may write the formal integral equation,

$$\mathcal{G}_R(\mathbf{r}|\mathbf{r}';\omega) = \mathcal{G}(\mathbf{r}|\mathbf{r}';\omega) - \frac{k_f^0}{4\pi\sigma^2 g(\sigma)}\int d\mathbf{r}''\,\mathcal{G}(\mathbf{r}|\mathbf{r}'';\omega)\delta(r''-\sigma)\mathcal{G}_R(\mathbf{r}''|\mathbf{r}';\omega)$$

$$(A.9)$$

Taking the double surface average of this equation immediately leads to the result

$$\overline{\mathcal{G}_R(\mathbf{r}|\mathbf{r}';\omega)}^{S,\,S'} = \left[1 + \frac{k_f^0}{g(\sigma)} \overline{\mathcal{G}(\mathbf{r}|\mathbf{r}';\omega)}^{S,\,S'}\right]^{-1} \overline{\mathcal{G}(\mathbf{r}|\mathbf{r}';\omega)}^{S,\,S'} \qquad (A.10)$$

Substitution into (A.6) and some rearrangement then yields the desired result, (3.13).

APPENDIX B

Field Point Formulation and General Properties

We first show how the pseudo-Liouville operator in phase space can be written in terms of an operator in field point space when it acts on the phase space densities. Consider

$$\mathcal{L}_+ n_A(1) = (\mathcal{L}_0 + \mathcal{L}'_{e+} + \mathcal{L}'_{R+}) n_A(1) \qquad (B.1)$$

where we have written the collisional part of $\mathcal{L}_+$ as a sum of elastic and reactive parts. It is clear that

$$\mathcal{L}_0 n_A(1) = -\mathbf{v}_1 \cdot \nabla_1 n_A(1) \qquad (B.2)$$

The elastic collision term may be analyzed as follows, using the fact that the solute molecules are dilute. Let

$$\mathcal{L}'_{e+} n_A(1) = \sum_{i,\,j} \left[T_+^{AS}(ij) + T_{E+}^{AB}(ij) \right] \delta(\mathbf{x}_1 - \mathbf{X}_i) \Theta_i^A$$

$$= \int d\mathbf{x}_2 \left[\sum_{i,\,j} T_+^{AS}(ij)\delta(\mathbf{x}_1 - \mathbf{X}_i)\delta(\mathbf{x}_2 - \mathbf{X}_j)\Theta_i^A \Theta_j^S \right.$$

$$\left. + \sum_{i,\,j} T_{E+}^{AB}(ij)\delta(\mathbf{x}_1 - \mathbf{X}_i)\delta(\mathbf{x}_2 - \mathbf{X}_j)\Theta_i^A \Theta_j^B \right] \qquad (B.3)$$

Consider the calculation of the first term on the right-hand side,

$$\sum_{i,\,j} T_+^{AS}(ij)\delta(\mathbf{x}_1 - \mathbf{X}_i)\Theta_i^A$$

$$= \int d\mathbf{x}_2 \sum_{i,\,j} \Theta(-\mathbf{V}_{ij}\cdot\hat{\mathbf{R}}_{ij})|\mathbf{V}_{ij}\cdot\hat{\mathbf{R}}_{ij}|\delta(R_{ij} - \sigma_{AS})\Theta_i^A \Theta_j^S$$

$$\times \left[\delta(\mathbf{v}_1 - \mathbf{V}_i^*)\delta(\mathbf{v}_2 - \mathbf{V}_j^*) - \delta(\mathbf{v}_1 - \mathbf{V}_i)\delta(\mathbf{v}_2 - \mathbf{V}_j) \right] \delta(\mathbf{r}_1 - \mathbf{R}_i)\delta(\mathbf{r}_2 - \mathbf{R}_j)$$

$$\qquad (B.4)$$

Using the hard-sphere collision dynamics [(6.7)] and the properties of the delta functions, one may show that

$$\mathbf{V}_{ij}\cdot\hat{\mathbf{R}}_{ij} = -\mathbf{v}_{12}\cdot\hat{\mathbf{R}}_{ij} \tag{B.5}$$

and

$$b_{ij}\delta(\mathbf{v}_1 - \mathbf{V}_i)\delta(\mathbf{v}_2 - \mathbf{V}_j) \equiv \delta(\mathbf{v}_1 - \mathbf{V}_i^*)\delta(\mathbf{v}_2 - \mathbf{V}_j^*)$$

$$= \hat{b}_{12}\delta(\mathbf{v}_1 - \mathbf{V}_i)\delta(\mathbf{v}_2 - \mathbf{V}_j) \tag{B.6}$$

where $\hat{b}_{12}$ is an operator like b_{ij} except it acts on the field point velocities. Equation B.4 may now be written as

$$\sum_{i,j} T_+^{AS}(ij)\delta(\mathbf{x}_1 - \mathbf{X}_i)\Theta_i^A$$

$$= \int d\mathbf{x}_2 |\mathbf{v}_{12}\cdot\hat{\mathbf{r}}_{12}|\delta(r_{12} - \sigma_{AS})\left[\theta(\mathbf{v}_{12}\cdot\hat{\mathbf{r}}_{12})\hat{b}_{12} - \theta(-\mathbf{v}_{12}\cdot\hat{\mathbf{r}}_{12})\right]n_{AS}(12)$$

$$\tag{B.7}$$

$$\equiv \bar{T}_-^{AS}(1\bar{2})n_{AS}(1\bar{2}) \tag{B.8}$$

Since $|\mathbf{v}_{12}\cdot\hat{\mathbf{r}}_{12}|\theta(-\mathbf{v}_{12}\cdot\hat{\mathbf{r}}_{12}) = -\mathbf{v}_{12}\cdot\hat{\mathbf{r}}_{12} + (\mathbf{v}_{12}\cdot\hat{\mathbf{r}}_{12})\theta(\mathbf{v}_{12}\cdot\hat{\mathbf{r}}_{12})$, we may also write

$$\bar{T}_-^{AS}(12) = T_-^{AS}(12) + (\mathbf{v}_{12}\cdot\hat{\mathbf{r}}_{12})\delta(r_{12} - \sigma_{AS}) \tag{B.9}$$

where $T_-^{AS}(12)$ is an operator with the same form as (6.5), but now in field point space. A similar calculation can be done for the AB elastic collision part. Thus,

$$\mathcal{L}'_{e+}n_A(1) = \bar{T}_-^{AS}(1\bar{2})n_{AS}(1\bar{2}) + \bar{T}_{E-}^{AB}(1\bar{2})n_{AB}(1\bar{2}) \tag{B.10}$$

The reactive contribution may be calculated in a similar way. Consider the reactive operator defined in (6.13),

$$\mathcal{L}'_{R+}n_A(1) = \sum_{i,j} T_{R+}^{AB}(ij)n_A(1) = \int d\mathbf{x}_2 \sum_{i,j} T_{R+}^{AB}(ij)\delta(\mathbf{x}_1 - \mathbf{X}_i)\delta(\mathbf{x}_2 - \mathbf{X}_j)$$

$$= \int d\mathbf{x}_2 \sum_{i,j} \alpha_R |\mathbf{V}_{ij}\cdot\hat{\mathbf{R}}_{ij}|\theta(-\mathbf{V}_{ij}\cdot\hat{\mathbf{R}}_{ij})\delta(R_{ij} - \sigma_{AB})\mathcal{P}_i^{AC}\mathcal{P}_j^{BD}$$

$$\times \delta(\mathbf{v}_1 - \mathbf{V}_i^*)\delta(\mathbf{r}_1 - \mathbf{R}_i)\delta(\mathbf{v}_2 - \mathbf{V}_j^*)\delta(\mathbf{r}_2 - \mathbf{R}_j)\Theta_i^A\Theta_j^B$$

$$- \int d\mathbf{x}_2 \sum_{i,j} \alpha_R |\mathbf{V}_{ij}\cdot\hat{\mathbf{R}}_{ij}|\theta(-\mathbf{V}_{ij}\cdot\hat{\mathbf{R}}_{ij})\delta(R_{ij} - \sigma_{AB})\delta(\mathbf{v}_1 - \mathbf{V}_i^*)$$

$$\times \delta(\mathbf{r}_1 - \mathbf{R}_i)\delta(\mathbf{v}_2 - \mathbf{V}_j^*)\delta(\mathbf{r}_2 - \mathbf{R}_j)\Theta_i^A\Theta_j^B \tag{B.11}$$

Using techniques similar to those in the analysis of the elastic operator, we find

$$
\begin{aligned}
\mathcal{L}'_{R+} n_A(1) &= \int dx_2\, \alpha_R |\mathbf{v}_{12} \cdot \hat{\mathbf{r}}_{12}| \theta(\mathbf{v}_{12} \cdot \hat{\mathbf{r}}_{12}) \\
&\quad \times \delta(r_{12} - \sigma_{AB}) \hat{b}_{12} \big[\mathcal{P}_1^{AC} \mathcal{P}_2^{BD} - 1 \big] \Theta_1^A \Theta_2^B n_{AB}(\mathbf{x}_1 \mathbf{x}_2) \\
&\equiv T_{R-}^{AB}(1\bar{2}) n_{AB}(1\bar{2})
\end{aligned}
\tag{B.12}
$$

Putting these results together, we have

$$
\begin{aligned}
\mathcal{L}_+ n_A(1) &= -\mathcal{L}_0^A(1) n_A(1) + \overline{T}_-^{AS}(1\bar{2}) n_{AS}(1\bar{2}) \\
&\quad + \big[\overline{T}_{E-}^{AB}(1\bar{2}) + T_{R-}^{AB}(1\bar{2}) \big] n_{AB}(1\bar{2})
\end{aligned}
\tag{B.13}
$$

Thus we see that the action of the $\mathcal{L}_+$ operator on $n_A(1)$ couples it to doublet phase-space fields.

Calculations involving higher order fields may be done in the same way. For example,

$$
\begin{aligned}
\mathcal{L}_+ n_{AB}(12) &= \big[-\mathcal{L}_0^{AB}(12) + \overline{T}_{E-}^{AB}(12) + T_{R-}^{AB}(12) \big] n_{AB}(12) \\
&\quad + \big[\overline{T}_-^{AS}(1\bar{3}) + \overline{T}_-^{BS}(2\bar{3}) \big] n_{ABS}(12\bar{3})
\end{aligned}
\tag{B.14}
$$

As in (6.12), we henceforth denote

$$
\overline{T}_{\pm}^{AB}(12) = \overline{T}_{E\pm}^{AB}(12) + T_{R\pm}^{AB}(12)
\tag{B.15}
$$

The general forms of (B.13) and (B.14) are easily deduced,

$$
\begin{aligned}
\mathcal{L}_+ n(12\cdots l) &= \big[-\mathcal{L}_0(12\cdots l) + \overline{T}_-(12\cdots l) \big] n(12\cdots l) \\
&\quad + \sum_{i=1}^{l} \overline{T}_-(i\overline{l+1}) n(12\cdots \overline{l+1}),
\end{aligned}
\tag{B.16}
$$

where we have defined $\overline{T}_-(12\cdots l)$ as the sum of all $\binom{l}{2}$ binary collision operators that can be constructed from the n field points. Species labels have been suppressed for simplicity.

It is now straightforward to show that if phase-space fields up to the lth order are included in the generalized Langevin equation description, only the lth-order random force will be nonzero. Consider a column vector made up of the l phase-space fields,

$$\mathbf{n}^{(l)} = (n(1), n(12), \ldots, n(12 \cdots l))^T \tag{B.17}$$

Let $\mathbf{C}$ be the overlap matrix of these fields. If $\mathscr{P}$ projects onto $\mathbf{n}^{(l)}$, that is, if

$$\mathscr{P}\mathbf{A} = \langle \mathbf{A}(\mathbf{n}^{(l)})^T \rangle \mathbf{C}^{-1}\mathbf{n}^{(l)} \tag{B.18}$$

the random "force" corresponding to $\mathbf{n}^{(l)}, \mathbf{f}_{+}^{(l)}$, is

$$\mathbf{f}_{+}^{(l)} = (1 - \mathscr{P})\mathcal{L}_{+}\mathbf{n}^{(l)} \tag{B.19}$$

We may write

$$\mathcal{L}_{+}\mathbf{n}^{(l)} = \mathbf{M}\mathbf{n}^{(l)} + \mathbf{N}n(12 \cdots l + 1) \tag{B.20}$$

where $\mathbf{M}$ and $\mathbf{N}$ are defined by comparison with (B.16). The essential point is that $\mathbf{N}$ is a column vector with zeros in all positions but the lth. Insertion of (B.20) into (B.19) leads to

$$\mathbf{f}_{+}^{(l)} = \mathbf{N}\left(n(12 \cdots l + 1) - \langle n(12 \cdots l + 1)(\mathbf{n}^{(l)})^T \rangle \mathbf{C}^{-1}\mathbf{n}^{(l)} \right) \tag{B.21}$$

Because of the form of $\mathbf{N}$, we have

$$f_{+}(1) = \cdots = f_{+}(12 \cdots l - 1) = 0 \tag{B.22}$$

APPENDIX C

Derivation of the Singlet Kinetic Equation

The kinetic equation description of the reaction $A + B \rightleftarrows C + B$ at the singlet-doublet level involves the set of fields $\{\delta n_{\alpha}(\alpha = A, B, C), \delta n_{AB}, \delta n_{\alpha S}(\alpha = A, B, C)\}$. The doublet fields are generated by the action of $\mathcal{L}_{+}$ on δn_{α} (Appendix B). If we restrict our calculations to forward reaction terms, we need to calculate only matrix elements involving the reduced set of fields $\{\delta n_A, \delta n_B, \delta n_{AB}, \delta n_{AS}, \delta n_{BS}\}$. The singlet A, B, and C fields are orthogonal if the solute density is small. The doublet fields δn_{AB} and δn_{CB} are orthogonal to each other but must be orthogonalized to the singlet fields; the doublet fields $\delta n_{\alpha S}$ must be orthogonalized to the solute doublet fields and singlet fields. Thus

$$\delta n_{AB}(12) = n_{AB}(12) - \langle n_{AB}(12) \rangle - \sum_{\alpha = A}^{C} \langle n_{AB}(12)\delta n_{\alpha}(\bar{1}') \rangle \left[\omega_0^{\alpha}(\bar{1}') \right]^{-1} \delta n_{\alpha}(\bar{1}')$$

$$\tag{C.1}$$

where $\omega_0^{\alpha\beta\gamma\cdots}(123\cdots) = n_{eq}^{\alpha}\phi_{\alpha}(v_1)n_{eq}^{\beta}\phi_{\beta}(v_2)n_{eq}^{\gamma}\phi_{\gamma}(v_3)\cdots g^{\alpha\beta\gamma\cdots}(\mathbf{r}_1\mathbf{r}_2\mathbf{r}_3\cdots)$, and

$$\delta n_{\alpha S}(12) = n_{\alpha S}(12) - \langle n_{\alpha S}(12)\rangle$$
$$- \sum_{\mu\nu=AB}^{CB} \langle n_{\alpha S}(12)\delta n_{\mu\nu}(\bar{1}'\bar{2}')\rangle\left[\omega_0^{\mu\nu}(\bar{1}'\bar{2}')\right]^{-1}\delta n_{\mu\nu}(\bar{1}'\bar{2}')$$
$$- \sum_{\beta=A}^{C} \langle n_{\alpha S}(12)\delta n_{\beta}(\bar{1}')\rangle\left[\omega_0^{\beta}(\bar{1}')\right]^{-1}\delta n_{\beta}(\bar{1}') \qquad (C.2)$$

and the averages are to be calculated to lowest order in the solute densities. The calculation of a few typical matrix elements is outlined below. First consider the static memory function ϕ_{AA}^{s}. This may be written as

$$\phi_{AA}^{s}(1,1') = -\langle\left[\mathcal{L}_+\delta n_A(1)\right]\delta n_A(1')\rangle\left[\omega_0^A(1')\right]^{-1}$$
$$= \mathcal{L}_0^A(1)\delta(1-1') - \bar{T}_-^{AB}(1\bar{2})\langle n_{AB}(1\bar{2})\delta n_A(1')\rangle\left[\omega_0^A(1')\right]^{-1}$$
$$- \bar{T}_-^{AS}(1\bar{2})\langle n^{AS}(1\bar{2})\delta n_A(1')\rangle\left[\omega_0^A(1')\right]^{-1}$$
$$\simeq \left\{\mathcal{L}_0^A(1) + \bar{T}_{Rf-}^{AB}(1\bar{2})n_{eq}^B\phi_B(v_{\bar{2}})g^{AB}(r_{1\bar{2}})\right.$$
$$\left. - \bar{T}_-^{AS}(1\bar{2})n_{eq}^S\phi_S(v_{\bar{2}})g^{AS}(r_{1\bar{2}})\right\}\delta(1-1')$$
$$\equiv \left\{\mathcal{L}_0^A(1) + \Lambda_{Rf-}^{AB}(1) - \Lambda_-^{AS}(1)\right\}\delta(1-1') \qquad (C.3)$$

where the second line follows from the results in Appendix B, and the last line defines the Enskog collision operators Λ_-^{AS} and Λ_{Rf-}^{AB},

$$\Lambda_-^{AS}(1) = g^{AS}(\sigma_{AS})n_{eq}^S\int dx_2\,\bar{T}_-^{AS}(12)\phi_S(v_2)$$
$$= g^{AS}(\sigma_{AS})n_{eq}^S\int dx_2\,T_-^{AS}(12)\phi_S(v_2) \qquad (C.4)$$

and

$$\Lambda_{Rf-}^{AB}(1) = g^{AB}(\sigma_{AB})n_{eq}^B\int dx_2\,T_{Rf-}^{AB}(12)\phi_B(v_2) \qquad (C.5)$$

The static memory function that couples the δn_A and δn_{AB} fields is also easily evaluated using the same method. We have

$$
\begin{aligned}
\phi^{S}_{A,AB}(1,1'2') &= -\left\langle\left[\mathcal{L}_{+}\delta n_{A}(1)\right]\delta n_{AB}(1'2')\right\rangle\left[\omega^{AB}_{0}(1'2')\right]^{-1} \\
&= -\overline{T}^{AB}_{-}(1\bar{2})\delta(1-1')\delta(\bar{2}-2') \\
&\quad -\overline{T}^{AS}_{-}(1\bar{3})\left\langle n_{AS}(1\bar{3})\delta n_{AB}(1'2')\right\rangle\left[\omega^{AB}_{0}(1'2')\right]^{-1} \\
&= \left\{-\overline{T}^{AB}_{-}(1\bar{2})\delta(\bar{2}-2')-\overline{T}^{AS}_{-}(1\bar{3})\left[\omega^{ABS}_{0}(12'\bar{3})\left[\omega^{AB}_{0}(12')\right]^{-1}\right.\right. \\
&\quad \left.\left.-\omega^{AS}_{0}(1\bar{3})\left[\omega^{A}_{0}(1)\right]^{-1}\right]\right\}\delta(1-1') \\
&= -\left\{\overline{T}^{AB}_{-}(1\bar{2})\delta(\bar{2}-2')+\overline{C}^{AS}_{-}(12')\right\}\delta(1-1') \\
&\equiv \mathcal{V}_{A,AB}(12')\delta(1-1')
\end{aligned}
\tag{C.6}
$$

where

$$
\overline{C}^{AS}_{-}(12) = \overline{T}^{AS}_{-}(1\bar{3})\left\{\omega^{ABS}_{0}(12\bar{3})\left[\omega^{AB}_{0}(12)\right]^{-1}-\omega^{AS}_{0}(1\bar{3})\left[\omega^{A}_{0}(1)\right]^{-1}\right\}
\tag{C.7}
$$

and the last line defines $\mathcal{V}_{A,AB}$. The coupling matrix element $\phi^{s}_{A,AS}$ is given by

$$
\begin{aligned}
\phi^{s}_{A,AS}(1,1'2') &= -\left\langle\left[\mathcal{L}_{+}\delta n_{A}(1)\right]\delta n_{AS}(\bar{1}''\bar{2}'')\right\rangle C^{-1}_{AS,AS}(\bar{1}''\bar{2}'',1'2') \\
&= -\overline{T}^{AS}_{-}(1\bar{2})\delta(1-1')\delta(\bar{2}-2') \\
&= \mathcal{V}_{A,AS}(12')\delta(1-1')
\end{aligned}
\tag{C.8}
$$

Using these results, the kinetic equation for the A singlet phase-space correlation function takes the form

$$
\begin{aligned}
\left\{z+\mathcal{L}^{A}_{0}(1)+\Lambda^{AB}_{Rf-}(1)-\Lambda^{AS}_{-}(1)\right\}C_{AA}(1,1';z) \\
= C_{AA}(1,1') \\
+\mathcal{V}_{A,AB}(1\bar{2})C_{AB,A}(1\bar{2},1';z)+\mathcal{V}_{A,AS}(1\bar{2})C_{AS,A}(1\bar{2},1';z)
\end{aligned}
\tag{C.9}
$$

We have made an addition approximation in writing (C.9); the coupling between the singlet A and singlet B fields has been dropped. This coupling is due to elastic collisions between A and B molecules, and it gives rise to cross-diffusion coefficients, which can be neglected in a dilute system. Most elastic collisions are with solvent molecules.

To complete the kinetic equation description, equations for the correlation functions involving doublet fields must be constructed. Once again, we calculate some of the matrix elements as an illustration. Consider first the coupling matrix element $\phi_{AB,A}^s(12,1')$

$$
\begin{aligned}
\phi_{AB,A}^s(12,1') &= -\left\langle\left[\mathcal{L}_+\delta n_{AB}(12)\right]\delta n_A(1')\right\rangle\left[\omega_0^A(1')\right]^{-1}\\
&= \left\langle\delta n_{AB}(12)\mathcal{L}_-\delta n_A(1')\right\rangle\left[\omega_0^A(1')\right]^{-1}\\
&= -\bar{T}_+^{AB}(1'\bar{2})\omega_0^{AB}(12)\left[\omega_0^A(1)\right]^{-1}\delta(1-1')\delta(2-\bar{2}')\\
&\quad -\bar{T}_+^{AS}(1'\bar{2})\left\{\omega_0^{ABS}(12\bar{2}')-\omega_0^{AS}(1\bar{2}')\frac{\omega_0^{AB}(12)}{\omega_0^A(1)}\right\}\\
&\quad\left[\omega_0^A(1)\right]^{-1}\delta(1-1')\\
&= -\left\{T_-^{AB}(12)+C_-^{AS}(12)\right\}\omega_0^{AB}(12)\left[\omega_0^A(1)\right]^{-1}\delta(1-1')\\
&\equiv \mathcal{V}_{AB,A}(12)\delta(1-1') \qquad\qquad\qquad\qquad (\text{C}.10)
\end{aligned}
$$

Here, $C_-^{AS}(12)$ has the same form as $\bar{C}_-^{AS}$ with $\bar{T}_-^{AS}$ replaced by T_-^{AS}. We next consider $\phi_{AB,AB}^s(12,1'2')$, which can be written as

$$
\begin{aligned}
\phi_{AB,AB}^s(12,1'2') &= -\left\langle\left[\mathcal{L}_+\delta n_{AB}(12)\right]\delta n_{AB}(1'2')\right\rangle\left[\omega_0^{AB}(1'2')\right]^{-1}\\
&= \left\{\mathcal{L}_0^{AB}(12)-\bar{T}_-^{AB}(12)\right\}\delta(1-1')\delta(2-2')\\
&\quad -\left[\bar{T}_-^{AS}(1\bar{3})+\bar{T}_-^{BS}(2\bar{3})\right]\omega_0^{ABS}(12\bar{3})\left[\omega_0^{AB}(12)\right]^{-1}\\
&\quad \delta(1-1')\delta(2-2')\\
&\equiv \left\{\mathcal{L}_0^{AB}(12)-\Lambda_-^{AS}(1)-\Lambda^{BS}(2)-\bar{T}_-^{AB}(12)-\bar{C}_-^{AS}(12)\right.\\
&\quad \left. -\bar{C}_-^{BS}(21)\right\}\delta(1-1')\delta(2-2') \qquad\qquad (\text{C}.11)
\end{aligned}
$$

With these results, the kinetic equation for $C_{AB,A}(12,1';z)$ can be written as

$$
\begin{aligned}
&\left\{z+\phi_{AB,AB}^s(12,\overline{12})+\phi_{AB,AB}^d(12,\overline{12};z)\right\}C_{AB,A}(\overline{12},1';z)\\
&= \mathcal{V}_{AB,A}(12)C_{AA}(1,1';z) \qquad\qquad\qquad\qquad (\text{C}.12)
\end{aligned}
$$

In general, there is a coupling between the AB and AS fields due to elastic collisions between A and S in the presence of B. This coupling, which is proportional to n_{eq}^B, has been dropped. We have also not explicitly calculated the dynamic memory function $\phi_{AB,AB}^d$. We examine its structure in Section VII.D, where coupling to triplet phase space fields is considered.

The kinetic equation for $C_{AS,A}(12,1';z)$ can be found using the same methods; the structure is somewhat more complex owing to the solvent static structure correlations. For instance, by a straightforward calculation, we find

$$\phi_{AS,AS}^s(12,1'2') = \left\{ \mathcal{L}_0^{AS}(12) - \bar{T}_-^{AS}(12) \right\} \delta(1-1')\delta(2-2')$$

$$- \bar{T}_-^{AS}(1\bar{3}) \left[\langle n_{ASS}(12\bar{3})\delta n_{AS}(\bar{1}\bar{2}) \rangle C_{AS,AS}^{-1}(\bar{1}\bar{2},1'2') \right.$$

$$\left. - \frac{\omega_0^{AS}(12)}{\omega_0^A(1)} \delta(1-1')\delta(\bar{3}-2') \right]$$

$$- \bar{T}_-^{SS}(2\bar{3}) \langle n_{ASS}(12\bar{3})\delta n_{AS}(\bar{1}\bar{2}) \rangle C_{AS,AS}^{-1}(\bar{1}\bar{2},1'2')$$

$$(C.13)$$

where all correlation functions are to be evaluated in the low *solute* density limit. This operator looks complex. In the low solvent density limit it takes a more familiar form,[50]

$$\phi_{AS,AS}^s(12,1'2') = \left\{ \mathcal{L}_0^{AS}(12) - \Lambda_-^{AS}(1) - \Lambda_-^{SS}(2) - \bar{T}_-^{AS}(12) \right\} \delta(1-1')\delta(2-2')$$

$$(C.14)$$

where $\Lambda_-^{SS}(2)$ is a solvent-solvent collision operator

$$\Lambda_-^{SS}(2) = n_{eq}^S \int d\mathbf{x}_3 T_-^{SS}(23)\phi_S(v_3)(1+P_{23}) \qquad (C.15)$$

and P_{23} changes 2 to 3. Other more useful approximations to the solvent static structure are given in Section VII.D. These results do, however, illustrate the general structure and physical content of the result. The vertex coupling δn_{AS} to δn_A is

$$\phi_{AS,A}^s(12,1') = - T_-^{AS}(1\bar{2})C_{AS,AS}(1\bar{2},1'2)\left[\omega_0^A(1') \right]^{-1}$$

$$= \mathcal{V}_{AS,A}(12)\delta(1-1') \qquad (C.16)$$

Using these results, the approximate kinetic equation for $C_{AS,A}(12, 1'; z)$ can be written as

$$\{z + \phi^s_{AS,AS}(12, \overline{12}) + \phi^d_{AS,AS}(12, \overline{12})\} C_{AS,A}(\overline{12}, 1'; z)$$
$$= \mathcal{V}_{AS,A}(12) C_{AA}(1, 1'; z) \tag{C.17}$$

If we let

$$G_{AB}(12, 1'2'; z)^{-1} = z + \phi^s_{AB,AB}(12, 1'2') + \phi^d_{AB,AB}(12, 1'2'; z) \tag{C.18}$$

with a similar definition for G_{AS}^{-1}, and insert the formal solutions of (C.12) and (C.17) into (C.9), we find

$$\{z + \mathcal{L}_0^A(1) + \Lambda^{AB}_{Rf-}(1) - \Lambda^{AS}_-(1) - R^{AB}(1; z) - R^{AS}(1; z)\} C_{A,A}(1, 1'; z)$$
$$= C_{A,A}(1, 1') \tag{C.19}$$

where

$$R^{AB}(1; z) = \mathcal{V}_{A,AB}(1\overline{2}) G_{AB}(1\overline{2}, \overline{1'2'}; z) \mathcal{V}_{AB,A}(\overline{1'2'}) \tag{C.20}$$

and $R^{AS}(1; z)$ has the same form.

The simplest approximation to the repeated ring operator R^{AB} is obtained by dropping the dynamic memory function $\phi^d_{AB,AB}$. Using (C.11), R^{AB} in this approximation can be written in the operator form as

$$R^{AB}(1; z) = \mathcal{V}_{A,AB}(1\overline{2}) \{G^0_{AB}(1\overline{2}; z)^{-1} - \mathcal{V}_{AB}(1\overline{2})\}^{-1} \mathcal{V}_{AB,A}(1\overline{2}) \tag{C.21}$$

where G^0_{AB} is the propagator for the *independent* motion of A and B in the solvent,

$$G^0_{AB}(12; z)^{-1} = z + \mathcal{L}_0^{AB}(12) - \Lambda^{AS}_-(1) - \Lambda^{BS}_-(2) \tag{C.22}$$

and

$$\mathcal{V}_{AB}(12) = \overline{T}^{AB}_-(12) + \overline{C}^{AS}_-(12) + \overline{C}^{BS}_-(12) \tag{C.23}$$

This approximate form for R^{AB} treats the intermediate propagation in a rather primitive fashion, but it is sufficiently complex to illustrate many features of the condensed-phase reaction dynamics. A more complex expression is presented in the context of the pair theory.

APPENDIX D

Pair Kinetic Equation

Since the derivation of the pair kinetic equation is similar to that given in Appendix C for the singlet kinetic equation, we only outline the calculation. We again restrict the calculation to the irreversible reaction; the details of the full reversible reaction case are given in Ref. 53.

From (B.14), we see that when $\mathcal{L}_+$ acts on the doublet field $n_{AB}(12)$, a contribution proportional to the triplet field $n_{ABS}(123)$ is obtained. Thus according to the general formulation set out in Section VII.B, we consider a description based on the two fields $\{\delta n_{AB}, \delta n_{ABS}\}$. Since the only non-zero damping matrix occurs in the triplet field equation, we may immediately write the following two coupled equations for the phase-space correlation functions. For the doublet field, we have

$$
z C_{AB,AB}(12,1'2';z) + \phi_{AB,AB}^s(12,\overline{12}) C_{AB,AB}(\overline{12},1'2';z)
$$
$$
= -\phi_{AB,ABS}^s(12,\overline{123}) C_{ABS,AB}(\overline{123},1'2';z) + C_{AB,AB}(12,1'2')
$$
(D.1)

and the correlation function involving the triplet field is

$$
z C_{ABS,AB}(123,1'2';z) + \left[\phi_{ABS,ABS}^s(123,\overline{123}) + \phi_{ABS,ABS}^d(123,\overline{123};z) \right]
$$
$$
C_{ABS,AB}(\overline{123},1'2';z) = -\phi_{ABS,AB}^s(123,\overline{12}) C_{AB,AB}(\overline{12};1'2';z)
$$
(D.2)

The static memory functions are easily calculated. After a straightforward calculation using the results in Appendix B, we have (see also Ref. 67)

$$
\phi_{AB,AB}^s(12,1'2') = -\left[-\mathcal{L}_0^{AB}(12) + \overline{T}_-^{AB}(12) \right] \delta(1-1')\delta(2-2')
$$
$$
-\left[\overline{T}_-^{AS}(13) + \overline{T}_-^{BS}(23) \right] \langle n_{ABS}(12\overline{3})\delta n_{AB}(\overline{1'2'}) \rangle
$$
$$
\times C_{AB,AB}^{-1}(\overline{1'2'},1'2')
$$
(D.3)

$$
\phi_{AB,ABS}^s(12,1'2'3') = -\left[\overline{T}_-^{AS}(13') + \overline{T}_-^{BS}(23') \right] \delta(1-1')\delta(2-2')
$$
(D.4)

$$
\phi_{ABS,AB}^s(123,1'2') = \left[T_-^{AS}(13) + T_-^{BS}(23) \right] C_{ABS,ABS}(123,\overline{123})
$$
$$
\times C_{AB,AB}^{-1}(\overline{12},1'2')
$$
(D.5)

and

$$\phi^s_{ABS,ABS}(123,1'2'3') = -\left[-\mathcal{L}_0^{ABS}(123) + \overline{T}^{AB}_-(12) + \overline{T}^{AS}_-(13) + \overline{T}^{BS}_-(23) \right]$$

$$\times \delta(1-1')\delta(2-2')\delta(3-3')$$

$$-\left[\overline{T}^{AS}_-(1\overline{4}) + \overline{T}^{BS}_-(2\overline{4}) + \overline{T}^{SS}_-(3\overline{4}) \right]$$

$$\times \langle n_{ABSS}(123\overline{4})\delta n_{ABS}(\overline{1}'\overline{2}'\overline{3}') \rangle C^{-1}_{ABS,ABS}(\overline{1}'\overline{2}'\overline{3}',1'2'3')$$

$$\tag{D.6}$$

The pair kinetic equation in Section VII.D follows directly from these results if the dynamic memory function $\phi^d_{ABS,ABS}$ is neglected, and the static structural correlations in (D.3) to (D.6) are approximated so that all binary collisions are calculated in the Enskog approximation. [This is the singly independent disconnected (SID) approximation, which is discussed in detail in Ref. 53.] We have also used the static hierarchy to obtain the final form involving the mean force, given in (7.32). This latter reduction involving the static hierarchy is carried out below in the context of a comparison of the singlet and doublet formulations.

To examine the relation between the pair kinetic equation (7.32) and the corresponding propagator for the doublet field that enters into the singlet field equation derived in Appendix C, consider (C.12). The static memory kernel $\phi^s_{AB,AB}$ defined in (C.11) may be written in a form closely related to that in (7.32) by using the static hierarchy. For a hard-sphere system, the static hierarchy takes the form[116]

$$\nabla_1 g^{AB}_H(r_{12}) = \hat{\mathbf{r}}_{12}\delta(r_{12}-\sigma_{AB})g^{AB}_H(\sigma_{AB})$$

$$+ n^s_{eq}\int d\mathbf{r}_3 \hat{\mathbf{r}}_{13}\delta(r_{13}-\sigma_{AS})g^{ABS}_H(\mathbf{r}_1\mathbf{r}_2\mathbf{r}_3) \tag{D.7}$$

Next, using the fact that

$$\overline{T}_-(12) = T_-(12) + \mathbf{v}_{12}\cdot\hat{\mathbf{r}}_{12}\delta(r_{12}-\sigma)$$

and (D.7), we may write

$$\overline{T}^{AB}_-(12) + \overline{C}^{AS}_-(12) + \overline{C}^{BS}_-(21) = \mathbf{v}_{12}\cdot\beta\mathbf{F}_H + T^{AB}_-(12) + C^{AS}_-(12) + C^{BS}_-(21)$$

$$\tag{D.8}$$

This is the origin of the mean force term in (7.32). Using (D.8), we have

$$\left\{z + \phi^s_{AB,AB}(12, \overline{12}) + \phi^d_{AB,AB}(12, \overline{12}; z)\right\} C_{AB,A}(\overline{12}, 1'; z)$$
$$= \left\{z + \mathcal{L}^{AB}_0(12) - \mathbf{v}_{12} \cdot \beta \mathbf{F}_H - \Lambda^{AS}_-(1) - \Lambda^{BS}_-(2) - T^{AB}_-(12)\right.$$
$$\left. - C^{AS}_-(12) - C^{BS}_-(21)\right\} C_{AB,A}(12, 1'; z)$$
$$+ \phi^d_{AB,AB}(12, \overline{12}; z) C_{AB,A}(\overline{12}, 1'; z) \qquad (D.9)$$

The operator on this correlation function, involving the doublet field $\delta n_{AB}(12)$, may now be compared directly with the operator in the pair kinetic equation (7.32). There, of course, the possibility of soft forces between the solute species was also taken into account. The ring operator in (7.33) and (7.34) takes the place of $\phi^d_{AB,AB}$ above. In the singlet kinetic equation that we used in Section VII.C, we ignored $\phi^d_{AB,AB}$. We now see that inclusion of the triplet phase space field δn_{ABS} leads to an explicit form for this dynamic memory function.

Another way of stating the foregoing results is to consider the pair field $\delta n_{AB}(\mathbf{x}_1 \mathbf{x}_2)$. The equation of motion for the $C_{AB,AB}(12, 1'2'; z)$ autocorrelation function can be immediately written as

$$\left\{z + \phi^s_{AB,AB}(12, \overline{12}) + \phi^d_{AB,AB}(12, \overline{12}; z)\right\} C_{AB,AB}(\overline{12}, 1'2'; z) = C_{AB,AB}(12, 1'2')$$
$$(D.10)$$

This follows directly from the projection operator algebra, where the projections are carried out onto the δn_{AB} field only. The expression for $\phi^s_{AB,AB}$ is the same as that in (D.9), and $\phi^d_{AB,AB}$ is an autocorrelation function involving the random "force" corresponding to δn_{AB}. A more useful form for $\phi^d_{AB,AB}$ can be obtained by explicitly including the triplet field as shown in the first part of this appendix.

APPENDIX E

Diffusion Eigenfunction and Integral Equation Solution

Construction of Diffusion Eigenfunction

The eigenfunction $|D\rangle$ is defined as the solution to the equation (cf. (10.6)),

$$-L_I(\mathbf{r}_1 \mathbf{r}_2)|D\rangle = \lambda |D\rangle \qquad (E.1)$$

We consider L_I in the $z=0$ limit here. The eigenvalue problem is solved by expanding $|D\rangle$ in two-particle velocity states. To linear order in the momentum, these states are $|00\rangle$, $|10\rangle$, and $|01\rangle$ which we denote by $|\mathbf{I}\rangle$ ($\mathbf{I}=1,2,3$). Thus

$$|D\rangle = \sum_{I=1}^{3} |\mathbf{I}\rangle \cdot \mathbf{a}_I \tag{E.2}$$

and

$$\sum_I \left[\langle \mathbf{J}| - L_I(\mathbf{r}_1\mathbf{r}_2)|\mathbf{I}\rangle - \lambda\delta_{IJ}\right] \cdot \mathbf{a}_I = 0 \tag{E.3}$$

The eigenvalues are given by the roots of the secular equation

$$\begin{vmatrix} -\lambda & (m_A\beta)^{-1/2}\nabla_1 & (m_B\beta)^{-1/2}\nabla_2 \\ (m_A\beta)^{-1/2}\nabla_1 & (m_A\beta D_A)^{-1}-\lambda & 0 \\ (m_B\beta)^{-1/2}\nabla_2 & 0 & (m_B\beta D_B)^{-1}-\lambda \end{vmatrix} = 0 \tag{E.4}$$

The diffusion eigenfunction is the root proportional to ∇^2,

$$\lambda = = -(D_A+D_B)\nabla^2 = -D\nabla^2 \tag{E.5}$$

The $\mathbf{a}_I$ coefficients then follow directly from the solution to the homogeneous equation (E.3) with $\lambda = -D\nabla^2$. We find

$$|D\rangle = |00\rangle - D_A(m_A\beta)^{1/2}\nabla_1 \cdot |10\rangle - D_B(m_B\beta)^{1/2}\nabla_2 \cdot |01\rangle \tag{E.6}$$

Since $G_{AB}^I = [z-L_I]^{-1}$, we have the desired result

$$\langle D|G_{AB}^I|D\rangle = (z-D\nabla^2)^{-1} \tag{E.7}$$

Solution of the Φ_D Integral Equation

If the details of the mean force are neglected, and only the particle exclusion effect is taken into account, we may write $g(r)=\theta(r-\sigma)$. In this case, the equation (10.26) for Φ_D simplifies to

$$\left\{ z-D\nabla^2 + \frac{k_f^0}{4\pi\sigma_{AB}^2}\delta(r-\sigma_{AB}) - D\nabla\cdot\hat{r}\delta(r-\sigma_{AB}) \right\}\Phi_D(\mathbf{r};z)$$

$$\equiv \left\{ z-D\nabla^2 - V_{AB}(\mathbf{r}) \right\}\Phi_D(\mathbf{r};z) = \frac{k_f^0}{4\pi\sigma_{AB}^2}\delta(r-\sigma_{AB}) \tag{E.8}$$

which may be written as the integral equation,

$$\Phi_D(\mathbf{r}; z) = \Phi_D^0(\mathbf{r}; z) + \int d\mathbf{r}' \, \mathcal{G}_0(\mathbf{r} - \mathbf{r}'; z) V_{AB}(\mathbf{r}') \Phi_D(\mathbf{r}'; z) \tag{E.9}$$

where

$$\Phi_D^0(\mathbf{r}; z) = \frac{k_f^0}{4\pi\sigma_{AB}^2} \int d\mathbf{r}' \, \mathcal{G}_0(\mathbf{r} - \mathbf{r}'; z) \delta(r' - \sigma_{AB}) \tag{E.10}$$

According to (10.22), we only need $\overline{\Phi_D}^S$ to calculate the rate kernel. The delta functions in the collision operators are to be interpreted in the limit as the sphere is approached from the outside,[78] so we average (E.10) over a surface at $r = \sigma_{AB}^+$ to obtain

$$\overline{\Phi_D(\mathbf{r}; z)}^{S+} = \frac{k_f^0}{k_D} e^{-\alpha\sigma_{AB}} \frac{\sinh \alpha\sigma_{AB}}{\alpha\sigma_{AB}} \left(1 - \overline{\Phi_D(\mathbf{r}; z)}^{S+}\right)$$

$$+ D \int d\mathbf{r}' \overline{\mathcal{G}_0(\mathbf{r} - \mathbf{r}'; z)}^{S+} \nabla' \cdot \hat{r}' \delta(r' - \sigma_{AB}) \Phi_D(\mathbf{r}'; z) \tag{E.11}$$

Using the fact[117] that

$$\overline{\mathcal{G}_0(\mathbf{r} - \mathbf{r}'; z)}^S = \begin{cases} \mathcal{G}_0(\sigma_{AB}; z)(\alpha r)^{-1} \sinh \alpha r & \text{for} \quad r \leq \sigma_{AB} \\ \mathcal{G}_0(\mathbf{r}; z)(\alpha\sigma_{AB})^{-1} \sinh \alpha\sigma_{AB}' & \text{for} \quad r \geq \sigma_{AB} \end{cases} \tag{E.12}$$

in (E.11), and integrating by parts, we find

$$\overline{\Phi_D(\mathbf{r}; z)}^S = k_f^0 \left(k_f^0 + k_D(z)\right)^{-1} \tag{E.13}$$

Acknowledgments

I have had the pleasure of working on this problem with many colleagues. Progress would have been very slow indeed without their collaboration.

This research was supported in part by grants from the Natural Science and Engineering Research Council of Canada and the Petroleum Research Fund.

References

1. M. von Smoluchowski, *Ann. Phys.*, **48**, 1003 (1915); *Phys. Z.*, **17**, 557 (1916); *Z. Phys. Chem.*, **92**, 129 (1917).
2. N. S. Snider, *J. Chem. Phys.*, **42**, 548 (1965).
3. B. Widom, *Science*, **148**, 1555 (1965); *J. Chem. Phys.*, **55**, 44 (1971).
4. N. S. Snider and J. Ross, *J. Chem. Phys.*, **44**, 1987 (1966).
5. R. Kubo, in S. F. Edwards, Ed., *Many Body Problems*, Benjamin, Reading, MA, 1969, p. 235.
6. R. Zwanzig, *J. Chem. Phys.*, **40**, 2527 (1964).
7. B. J. Berne and R. Pecora, *Dynamic Light Scattering*, Interscience, New York, 1976.
8. R. M. Noyes, *Prog. Reac. Kinet.*, **1**, 128 (1961).
9. S. R. de Groot and P. Mazur, *Non-Equilibrium Thermodynamics*, North-Holland, Amsterdam, 1962.
10. D. Forster, *Hydrodynamic Fluctuations, Broken Symmetry, and Correlation Functions*, Benjamin, Reading, MA, 1975.
11. G. Wilemski and M. Fixman, *J. Chem. Phys.*, **58**, 4009 (1973).
12. H. L. Frisch and F. C. Collins, *J. Chem. Phys.*, **20**, 1797 (1952).
13. See, for instance, A. E. Nielsen, *Kinetics of Precipitation*, Pergamon Press, New York, 1964.
14. F. C. Collins and G. E. Kimball, *J. Colloid Sci.*, **4**, 425 (1949); H. S. Carslaw and J. C. Jaeger, *Conduction of Heat in Solids*, Clarendon Press, Oxford, 1947.
15. S. H. Northrup and J. T. Hynes, *Chem. Phys. Lett.*, **54**, 244 (1978).
16. L. Monchick, J. L. Magee, and A. H. Samuel, *J. Chem. Phys.*, **26**, 935 (1957).
17. T. R. Waite, *Phys. Rev.*, **107**, 463 (1957).
18. B. U. Felderhof and J. M. Deutch, *J. Chem. Phys.*, **64**, 455 (1976).
19. J. R. Lebenhaft and R. Kapral, *J. Stat. Phys.*, **20**, 25 (1979).
20. L. Monchick, *J. Chem. Phys.*, **24**, 381 (1956).
21. S. H. Northrup and J. T. Hynes, *J. Chem. Phys.*, **68**, 3203 (1978).
22. P. Debye, *Trans. Electrochem. Soc.*, **82**, 265 (1942).
23. B. V. Derjaguin and V. M. Muller, *Dokl. Akad. Nauk SSSR*, **176**, 738 (1967).
24. C. A. Emeis and P. L. Fehder, *J. Am. Chem. Soc.*, **92**, 2246 (1970).
25. S. Chandrasekar, *Rev. Mod. Phys.*, **15**, 1 (1943).
26. H. A. Kramers, *Physica*, **7**, 284 (1940).
27. S. Glasstone, K. J. Laidler, and H. Eyring, *The Theory of Rate Processes*, McGraw-Hill, New York, 1941.
28. S. Golden, *Quantum Statistical Foundations of Chemical Kinetics*, Clarendon Press, Oxford, 1969.
29. H. Aroeste, *Adv. Chem. Phys.*, **6**, 1 (1964).
30. L. P. Kadanoff and G. Baym, *Quantum Statistical Mechanics*, Benjamin, Reading, MA, 1962.
31. H. D. Kutz, I. Oppenheim, and A. Ben-Reuven, *J. Chem. Phys.*, **61**, 3313 (1974).
32. D. Chandler, *J. Chem. Phys.*, **68**, 2959 (1978).
33. J. L. Skinner and P. G. Wolynes, *J. Chem. Phys.*, **69**, 2143 (1978).
34. J. C. Keck, *Adv. Chem. Phys.*, **13**, 85 (1967).
35. F. H. Stillinger, in H. Eyring and D. Henderson, Eds., *Theoretical Chemistry Advances and Perspectives*, Vol. 3, Academic Press, New York, 1978.
36. H. Mori, *Prog. Theor. Phys.* (*Kyoto*), **33**, 423 (1965).
37. R. Zwanzig, *J. Chem. Phys.*, **33**, 1338 (1960); *Lect. Theo. Phys.*, **3**, 106 (1961).
38. R. Kapral, *J. Chem. Phys.*, **56**, 1842 (1972); F. Garisto and R. Kapral, *J. Chem. Phys.*, **58**, 3129 (1973).

39. S. A. Rice and P. Gray, *Statistical Mechanics of Simple Liquids*, Wiley, New York, 1965.
40. R. Balescu, *Equilibrium and Nonequilibrium Statistical Mechanics*, Wiley, New York, 1975.
41. K. Kawasaki, *Anal. Phys. (N.Y.)*, **61**, 1 (1970).
42. R. Kapral, S. Hudson, and J. Ross, *J. Chem. Phys.*, **53**, 4387 (1970).
43. T. Yamamoto, *J. Chem. Phys.*, **33**, 281 (1960).
44. J. Ross and P. Mazur, *J. Chem. Phys.*, **35**, 19 (1961).
45. J. C. Light, J. Ross, and K. E. Shuler, in A. R. Hochstim, Ed., *Kinetic Processes in Gases and Plasmas*, Academic Press New York, 1969, p. 281.
46. H. C. Anderson, *Annu. Rev. Phys. Chem.*, **26**, 145 (1975).
47. J. Kuschick and B. J. Berne, *J. Chem. Phys.*, **59**, 3732 (1973).
48. M. H. Ernst, J. R. Dorfman, W. R. Hoegy, and J. M. J. van Leeuwen, *Physica*, **45**, 127 (1969).
49. P. Résibois and M. De Leener, *Classical Kinetic Theory of Fluids*, Wiley, New York, 1977.
50. R. Kapral, *J. Chem. Phys.*, **68**, 1903 (1978).
51. M. Pagitsas and R. Kapral, *J. Chem. Phys.*, **69**, 2811 (1978).
52. S. Bose and P. Ortoleva, *J. Chem. Phys.*, **70**, 3041 (1979).
53. R. I. Cukier, R. Kapral, J. R. Mehaffey, and K. J. Shin, *J. Chem. Phys.*,**72**, 1830 (1980).
54. P. Pechukas, in W. H. Miller, Ed., *Dynamics of Molecular Collisions*, Part B, Plenum Press, New York, 1976.
55. W. H. Miller, *Acc. Chem. Res.*, **9**, 306 (1976).
56. M. H. Ernst and J. R. Dorfman, *Physica (Utrecht)*, **45**, 127 (1969).
57. P. Résibois and J. L. Lebowitz, *J. Stat. Phys.*, **12**, 483 (1975).
58. P. Résibois, *J. Stat. Phys.*, **13**, 393 (1975).
59. E. P. Gross, *Anal. Phys. (N.Y.)*, **69**, 42 (1972).
60. G. F. Mazenko, *Phys. Rev. A*, 7, 209, 222 (1973).
61. G. F. Mazenko, *Phys. Rev. A*, **9**, 360 (1974).
62. G. F. Mazenko and S. Yip, in B. J. Berne, Ed., *Modern Theoretical Chemistry, Statistical Mechanics*, Vol. 6, Plenum Press, New York, 1977, p. 181.
63. S. Yip, *Annu. Rev. Phys. Chem.*, **30**, 547 (1979).
64. A. Z. Akcasu and J. J. Duderstadt, *Phys. Rev.*, **188**, 479 (1969).
65. C. Boley, *Phys. Rev. A*, **11**, 328 (1975).
66. R. Kapral and K. J. Shin, *J. Chem. Phys.*, **70**, 5623 (1979).
67. M. Pagitsas, Thesis, University of Toronto (1981).
68. J. Franck and E. Rabinowitch, *Trans. Faraday Soc.*, **30**, 120 (1934).
69. B. J. Alder and T. E. Wainwright, *Phys. Rev. Lett.*, **18**, 968 (1967); B. J. Alder, D. M. Gass, and T. E. Wainwright, *J. Chem. Phys.*, **53**, 3813 (1970).
70. Y. Pomeau and P. Résibois, *Phys. Rep.*, **2**, 63 (1975).
71. K. Kawasaki and I. Oppenheim, *Phys. Rev.* **139**, 1763 (1965).
72. J. R. Dorfman and E. G. D. Cohen, *Phys. Rev. Lett.*, **25**, 1257 (1972); *Phys. Rev. A*, **6**, 776 (1972); **12**, 292 (1975).
73. R. Kapral and M. Weinberg, *Phys. Rev. A*, **8**, 1008 (1973); **9**, 1676 (1974).
74. T. Keyes and I. Oppenheim, *Phys. Rev. A*, 7, 1384 (1973); **8**, 973 (1973).
75. J. R. Mehaffey and R. I. Cukier, *Phys. Rev. A*, **17**, 1181 (1978).
76. R. I. Cukier and J. R. Mehaffey, *Phys. Rev. A*, **18**, 1202 (1978).
77. J. T. Hynes, R. Kapral, and M. Weinberg, *J. Chem. Phys.*, **70**, 1456 (1979).
78. J. R. Dorfman, H. van Beijeren, and C. F. McClure, *Arch. Mech. Stos.*, **28**, 333 (1976); H. van Beijeren and J. R. Dorfman, *J. Stat. Phys*, **23**, 335 (1980).
79. The $t^{-3/2}$ power law decay for the rate kernel implies that the rate coefficient $k_q(t)$ in

(2.14) decays as $t^{-1/2}$ for long times. This decay has been confirmed experimentally. See, for example, T. L. Nemzak and W. R. Ware, *J. Chem. Phys.*, **62**, 477 (1975).

80. M. Jhon and J. S. Dahler, *J. Stat. Phys.*, **20**, 3 (1979).

81. N. Xystris and J. S. Dahler, *J. Chem. Phys.*, **68**, 354 (1978).

82. P. L. Bhatnagar, E. P. Gross, and M. Krook, *Phys. Rev.*, **94**, 511 (1954); D. Bohm and E. P. Gross, *Phys. Rev.*, **75**, 1864 (1949).

83. L. Monchick and E. A. Mason, *Phys. Fluids*, **10**, 1377 (1967); R. E. Kapral and J. Ross, *J. Chem. Phys.*, **52**, 1238 (1970).

84. J. A. Montgomery, D. Chandler, and B. J. Berne, *J. Chem. Phys.*, **70**, 4056 (1979).

85. J. L. Skinner and P. G. Wolynes, *J. Chem. Phys.*, **72**, 4913 (1980).

86. R. I. Cukier, R. Kapral, J. R. Mehaffey, and K. J. Shin, *J. Chem. Phys.*, **72**, 1844 (1980).

87. R. I. Cukier, J. R. Mehaffey, and R. Kapral, *J. Chem. Phys.*, **69**, 4962 (1978).

88. G. T. Evans, *J. Chem. Phys.*, **65**, 3030 (1976).

89. G. Wilemski, *J. Stat. Phys.*, **14**, 153 (1976).

90. J. L. Skinner and P. G. Wolynes, *Physica*, **96A**, 561 (1979).

91. H. Akama and A. Siegel, *Physica*, **31**, 1493 (1965).

92. R. I. Cukier, R. Kapral, and J. R. Mehaffey, *J. Chem. Phys.*, **73**, 5254 (1980).

93. C. W. Oseen, *Neuere Methoden und Ergebnisse in der Hydrodynamik*, Akademische Verlagsgesellschaft, Leipzig, 1927; J. Happel and H. Brenner, *Low Reynolds Number Hydrodynamics*, Prentice-Hall, Englewood Cliffs, NJ, 1965.

94. J. M. Deutch and I. Oppenheim, *J. Chem. Phys.*, **54**, 3547 (1971); T. J. Murphy and J. L. Aguirre, *J. Chem. Phys.*, **57**, 2908 (1972).

95. S. M. Lo and K. Kawasaki, *Phys. Rev. A*, **5**, 421 (1972).

96. R. M. Noyes, *J. Chem. Phys.*, **72**, 1349 (1954).

97. S. H. Northrup and J. T. Hynes, *J. Chem. Phys.*, **71**, 871 (1979).

98. P. M. Furtado, G. F. Mazenko, and S. Yip, *Phys. Rev. A*, **14**, 869 (1976); J. R. Mehaffey, R. C. Desai, and R. Kapral, *J. Chem. Phys.*, **66**, 1665 (1977).

99. M. Schell and R. Kapral, to be published.

100. K. J. Shin and R. Kapral, *J. Chem. Phys.*, **69**, 3685 (1978).

101. S. H. Northrup and J. T. Hynes, *J. Chem. Phys.*, **71**, 884 (1979).

102. H. Eyring, D. Henderson, and W. Jost, Eds., *Physical Chemistry, An Advanced Treatise*, Vol. VII, Academic Press, New York, 1975; J. H. Freed and J. B. Pedersen, *Adv. Magn. Res.*, **8**, 1 (1976); A. H. Alwattar, M. D. Lumb, and J. D. Birks, in J. D. Birks, Ed., *Organic Molecular Photophysics*, Wiley, New York, 1973, p. 403.

103. J. T. Hynes, R. Kapral, and G. M. Torrie, *J. Chem. Phys.*, **72**, 177 (1980).

104. J. Tellinghuisen, *J. Chem. Phys.*, **58**, 2821 (1973).

105. D. Chandler and L. Pratt, *J. Chem. Phys.*, **65**, 2925 (1976).

106. B. C. Freasier, D. L. Jolly, N. D. Hamer, and S. Nordholm, *Chem. Phys.*, **38**, 293 (1979).

107. R. M. Noyes, *J. Am. Chem. Soc.*, **77**, 2042 (1955).

108. T. J. Chuang, G. W. Hoffman, and K. B. Eisenthal, *Chem. Phys. Lett.*, **25**, 201 (1974); K. B. Eisenthal, in S. L. Shapiro, Ed. *Topics in Applied Physics*, Vol. 18, Springer, Berlin, 1977, p. 275.

109. D. L. Bunker and B. S. Jacobson, *J. Am. Chem. Soc.*, **94**, 1843 (1972).

110. A. J. Stace and J. N. Murrell, *Mol. Phys.*, **33**, 1 (1977); J. N. Murrell, A. J. Stace, and R. Dammel, *J. Chem. Soc., Faraday Trans. 2*, 1532 (1978); A. J. Stace and J. N. Murrell, *Int. J. Chem. Kinet.*, **10**, 197 (1978).

111. D. L. Jolly, B. C. Freasier, and S. Nordholm, *Chem. Phys.*, **25**, 361 (1977).

112. H. S. Johnston, *Gas Phase Reaction Rate Theory*, Ronald Press, New York, 1966, Ch. 14.

113. G. T. Evans and M. Fixman, *J. Phys. Chem.*, **80**, 1544 (1976).

114. P. Turg, F. Lantelme, and H. L. Friedman, *J. Chem. Phys.*, **66**, 3039 (1976); J. H. Weiner and R. E. Forman, *Phys. Rev. B*, **10**, 315 (1975); S. A. Adelman and B. J. Garrison, *J. Chem. Phys.*, **65**, 3751 (1975); M. Shugard, J. C. Tully, and A. Nitzan, *J. Chem. Phys.*, **66**, 2534 (1977).
115. See, for example, P. B. Vischer, *Phys. Rev. B*, **14**, 347 (1976); N. Blomberg, *Physica (Utrecht)*, **A86**, 49 (1977); N. G. van Kampen, *J. Stat. Phys.*, **17**, 71 (1977); S. H. Northrup and J. T. Hynes, *J. Chem. Phys.*, **69**, 5246 (1978); E. Helfand, *J. Chem. Phys.*, **69**, 1010 (1978).
116. H. H. U. Konijnendijk and J. M. J. van Leeuween, *Physica (Utrecht)*, **64**, 342 (1973).
117. P. Mazur and D. Bedeaux, *Physica (Utrecht)*, **76**, 235 (1974).

DIELECTRIC CONSTANTS OF FLUID MODELS: STATISTICAL MECHANICAL THEORY AND ITS QUANTITATIVE IMPLEMENTATION

G. STELL

*Departments of Mechanical Engineering and Chemistry
State University of New York
Stony Brook, New York*

G. N. PATEY

*Department of Chemistry
University of British Columbia
Vancouver, British Columbia, Canada*

AND

J. S. HØYE

*Institutt for Teoretisk Fysikk
Universitetet i Trondheim
Trondheim, Norway*

CONTENTS

I. INTRODUCTION

The statistical mechanical theory of the dielectric constant ε is in the midst of a period of rapid development. Our purpose here is to summarize the aspects of this progress with which we have been most directly concerned over the past few years. To keep this monograph manageable in size and scope, however, we have imposed strong limitations on our subject.

First of all we restrict ourselves to Hamiltonian models that can be treated entirely within the framework of statistical mechanical techniques, without recourse to phenomenological or heuristic input. Thus we do not dwell on the enormous amount of work that has been done in the context of semimacroscopic "continuum" approximations, except where we are able to make statistical mechanical contact in certain limits or approximations with such approaches, as in Section II.D. Unfortunately this restriction also means that we are still limited in our quantitative treatment to a small set of models of artificial simplicity.

Our second self-imposed restriction consists of taking the phrase "dielectric constant" in its narrowest sense, and confining ourselves to the relation between polarization and electric field in the strict zero-frequency, zero-field limit. Thus, for example, we do not consider index of refraction, dielectric saturation, or dielectric relaxation.

Our third restriction is to a purely classical (nonquantal, nonrelativistic) description of our systems.

This is not a review chapter in the usual sense. An appreciable amount of the material describes heretofore unpublished work of the authors. In particular, Sections II.E and IV.E include previously unpublished work of Høye and Stell, and Section IV.B includes new quantitative results of Patey and Stell. On the other hand, we have made no attempt to survey the open literature concerning our subject. (A splendid review of material that overlaps substantially with our own has recently been given by Wertheim.[1]) Our specific objectives have been as follows:

1. To present a summary of the most important exact expressions giving ε in terms of microscopic correlation functions.
2. To show how certain new formalisms naturally lead to new classes of approximation schemes.
3. To provide a do-it-yourself manual for the quantitative implementation of these schemes, along with representative numerical results that have come out of them.
4. To summarize some techniques and results of computer simulations that bear on the evaluation of ε.

In addition, we have endeavored to organize the chapter so that the sections that pertain to quantitative results are as self-contained as possible, to ensure that they will be intelligible and useful without obliging the reader to go through the quite abstract development of the theory that yields the final recipes. Finally, we have tried throughout to use notation and terminology that facilitates close contact with the primary literature pertaining to each of the items above.

A serious notational dilemma arose because there is no standard notation. In particular, each of the several key series of papers on which we lean very heavily, employs its own notation. Thus the goal of maintaining close contact with the primary literature is essentially inconsistent with the strict use of a single standard notation throughout. It is compatible with the goal of keeping the material self-contained, though, and we have adopted the following strategy.

Section II, which exploits the formalism of Høye and Stell as a unifying approach, uses (with minor variations in the interest of overall notational consistency) the notation of Høye and Stell, with keys provided to the notation of other workers (see Table I) where such keys illuminate conceptual correspondence. Section III focuses on implementation of theory, largely from the standpoint of the approach taken by Patey, Levesque, and Weis. It uses the notation of those workers. Section III is self-contained in the sense that all exact results used therein are first fully defined.

The differences in notation between Sections II and III (and the primary references on which they are based) are not great, and for the most part they are not gratuitous, but follow from differences in substance and purpose. For example, all the expressions giving ε in terms of two-point correlations discussed in Section II can be expressed in terms of the Δ and D components of correlation introduced there. To quantitatively evaluate these expressions, however, many other components of correlation must be taken into account (in terms of their effect on the Δ and D components), and more complete rotational-invariant subscript notation is introduced in Section III to accommodate such evaluation.

The difference in notation also extends to the generalized Fourier transforms of the rotational invariants. A few differences in the Høye-Stell and Patey-Levesque-Weis notation serve no such organic need but simply reflect common notational variations in the literature. (Høye and Stell often use m for permanent dipole magnitude and $\hat{s}$ for the associated unit vector; Patey et al. use μ and $\hat{\mu}$.) In Section IV, where polarizability is introduced and both theory and its implementation discussed, we have used **m** for total moment, **p** for induced moment, and μ for permanent moment. Where we discuss Wertheim's results in detail, we follow his notation as closely as possible (again subject to minor variations in the interest of overall notational consistency).

There are various ways of defining the static dielectric constant ε of a fluid. Perhaps the most fundamental is through the equation

$$(\varepsilon - 1)\mathbf{E} = 4\pi\mathbf{P} \tag{1.1}$$

where $\mathbf{E}$ is the macroscopic (Maxwell) field and $\mathbf{P}$ the polarization (mean dipole moment per unit volume). Direct use of this expression was made by Debye[2] in 1912 to compute the response of a dilute gas of particles to the presence of an applied external field $\mathbf{E}_0$, using a transcription of an earlier computation by Langevin of the susceptibility of a paramagnetic system in a magnetic field. Although this method remains a standard textbook introduction to the microscopic theory of ε, subsequent developments in that theory have for the most part been mainly concerned with approaches that relate ε to the properties of a system of particles in the absence of an applied field. These approaches all hinge on the fact that each particle can be regarded as the source of an external field acting on the system consisting of all the other particles. The dielectric response of the other particles to this source can thus be used as a measure of ε even in the absence of terms external to the whole system.

Only recently has the theory of ε for systems in the presence of applied fields reached a level at which one can compute ε with satisfactory accuracy for nontrivial Hamiltonian models at liquid-state densities by direct extension of the original Debye-Langevin method. We touch on this extension in Section VI, but for the most part we treat only systems in the absence of external fields.

II. POLAR–NONPOLARIZABLE FLUIDS: THEORY IN THE CONTEXT OF THE HØYE-STELL FORMULATION

We start with a model of polar molecules in which the effects of polarizability are neglected. More precisely, we assume that in the absence of external fields, the potential energy associated with N particles is a sum of pair potentials $\phi(ij)$, each of which depends on the positions $\mathbf{r}_i$ and $\mathbf{r}_j$ and orientations Ω_i and Ω_j of particles i and j. Thus the particles are regarded as rigid, with no internal coordinates, and we assume for simplicity that they are all identical. Extensions of the results of Section II to mixtures are for the most part straightforward, as discussed by Høye and Stell[3] and in references they cite. Pertinent references to the mean spherical approximation generalized to mixtures are also given at an appropriate point in this chapter.

A. Exact Expressions for ε for Rigid Polar Molecules in Terms of the Pair Correlation Function

As mentioned above, in the absence of an external field, ε can be expressed in terms of the response of each particle to the field set up by the others. In the model under consideration that field is a sum of pair terms, so the key statistical mechanical quantity involved in the expression of ε is the two-body correlation function. Its systematic use unavoidably entails a heavy dose of terminology and notation, which we now introduce in the language of Refs. 4 and 5.

For an infinite fluid of particles in the absence of an external field the probability density $\rho(i)$ associated with a particle at position $\mathbf{r}_i$ with orientation Ω_i is just $\rho\Omega^{-1}$, where ρ is the number density and $\Omega = \int d\Omega_i$. [More generally we would have $\rho = \int\rho(i)d\Omega_i$ if the fluid were orientationally ordered and if a distinguished orientation were fixed, e.g., by an infinitesimal external field.]

We let the joint probability density associated with a particle at $\mathbf{r}_1$, Ω_1 and a particle (possibly the same one) at $\mathbf{r}_2$, Ω_2 be $\rho_\delta(12)$. This is simply related to the joint probability density $\rho(12)$ for distinct particles by an additive δ-function term:

$$\rho_\delta(12) = \rho(12) + \delta(12)\rho(1) \tag{2.1}$$

where $\delta(12)$ is a delta function in both position and orientation: $\delta(12) = \delta(\mathbf{r}_1, \mathbf{r}_2)\delta(\Omega_1, \Omega_2)$. Letting $F_\delta(12)$ and $F(12)$ be the cluster or Ursell functions associated with $\rho_\delta(12)$ and $\rho(12)$, respectively,

$$F_\delta(12) = \rho_\delta(12) - \rho(1)\rho(2), \qquad F(12) = \rho(12) - \rho(1)\rho(2) \tag{2.2}$$

we follow Lebowitz, Stell, and Baer[6] in introducing a "self-energy" function $\Sigma(12)$, and the related $W(12)$,

$$\Sigma(12) = W(12) + \rho(1)\delta(12) \tag{2.3}$$

These functions are defined in terms of a decomposition of the pair potential $\phi(12)$ into two parts, a reference term $q(12)$, and a perturbing term $w(12)$,

$$\phi(12) = q(12) + w(12) \tag{2.4}$$

The formal relations developed in Ref. 6 are independent of how the decomposition is made. For convenience we introduce the function

$$v(12) = -\beta w(12) \tag{2.5}$$

since $w(12)$ always appears with the factor $-\beta = -(kT)^{-1}$. We can most conveniently define $\Sigma(12)$ in Fourier space. We denote the d-dimensional transform by a tilde. Thus, letting $\mathbf{r} = \mathbf{r}_1 - \mathbf{r}_2$, we have, for example,

$$\tilde{\Sigma}(12) = \int e^{i\mathbf{k}\mathbf{r}} \Sigma(12) \, d\mathbf{r} \qquad (2.6a)$$

and

$$\tilde{v}(12) = \int e^{i\mathbf{k}\mathbf{r}} v(12) \, d\mathbf{r} \qquad (2.6b)$$

We use the notation $\tilde{F}_\delta(\Omega_1 \Omega_2)$, $\tilde{\Sigma}(\Omega_1 \Omega_2)$, and so on, to mean the transforms $\tilde{F}_\delta(12)$, $\tilde{\Sigma}(12)$, and so on, evaluated at $k = 0$. Our definition of $\tilde{\Sigma}(12)$ is given by

$$\tilde{F}_\delta(12) = \tilde{\Sigma}(12) + \int \tilde{\Sigma}(13) \tilde{v}(34) \tilde{F}_\delta(42) \, d\Omega_3 \, d\Omega_4 \qquad (2.7)$$

Our notation here essentially follows Refs. 4 and 5, which in turn closely followed that of Lebowitz, Stell, and Baer[6] and of Stell, Lebowitz, Baer, and Theumann.[7] There are minor differences from paper to paper, however. Our Σ and W here are the $\Omega^{-2}\Sigma$ and $\Omega^{-2}W$ of Ref. 4. In Ref. 6 n^2W is used to denote what we call W here, whereas in Ref. 7, $\hat{W}$ is used to denote our Σ. The $v(12)$ here is the Φ of Refs. 6 and 7. Moreover, the "modified" two-particle functions we denote here with the subscript δ are denoted in the papers above with a caret, and their Fourier transforms carry a bar. (We reserve the caret and bar here to denote certain spherically symmetric functions and Hankel transforms that play a fundamental role in the mathematics of polar fluids.) Finally, in Refs. 4 and 5, $\rho(1)$, $\rho(12)$, and $F(12)$ were written as $\rho_1(1)$, $\rho_2(12)$, and $F_2(12)$, respectively. The subscripts are redundant when one exhibits the arguments, so we drop them here.

It is worth comparing (2.7) with a similar relation between $\tilde{F}_2(12)$ and the transform $\tilde{c}(12)$ of the direct correlation function $c(12)$ defined by the Ornstein-Zernike (OZ) equation, which can be written as

$$\tilde{F}_\delta(12) = \frac{\rho \delta(\Omega_1 \Omega_2)}{\Omega} + \rho \int \frac{\tilde{c}(13) \tilde{F}_\delta(32) \, d\Omega_3}{\Omega} \qquad (2.8)$$

From (2.7) and (2.8) it follows that

$$\tilde{\Sigma}(12) = \frac{\rho}{\Omega} \delta(\Omega_1 \Omega_2) + \frac{\rho}{\Omega} \int \left[\tilde{c}(13) - \tilde{v}(13) \right] \tilde{\Sigma}(23) \, d\Omega_3 \qquad (2.9a)$$

or

$$\tilde{W}(12) = \left(\frac{\rho}{\Omega}\right)^2 [\tilde{c}(12) - \tilde{v}(12)] + \frac{\rho}{\Omega} \int [\tilde{c}(13) - \tilde{v}(13)] \tilde{W}(23) \, d\Omega_3$$

(2.9b)

From (2.8) and (2.9) we see that $\Sigma(12)$ stands to the function $c(12) - v(12)$ precisely as the full $F_\delta(12)$ stands to $c(12)$ itself. Similarly $W(12)$ stands to $c(12) - v(12)$ as $F(12)$ stands to $c(12)$ [where $F_\delta(12) = F(12) + \delta(12)\rho(1)$ from (2.2)]. Thus, although $F_\delta(12)$ and $c(12)$ are clearly independent of our choice of $q(12)$ and $w(12)$, the $W(12)$ and $\Sigma(12)$ are not; they are functionals of $w(12)$.

The simplest choice of $w(12)$ is that of an ideal dipole with a sharp cutoff,

$$w(12) = -m^2 D(12) r^{-3} \quad \text{for} \quad r > d$$
$$w(12) = 0 \quad \text{for} \quad r < d$$

(2.10)

where d is the cutoff diameter, m is the strength of the dipole moment, and

$$D(12) = 3(\hat{s}_1 \cdot \hat{r})(\hat{s}_2 \cdot \hat{r}) - (\hat{s}_1 \cdot \hat{s}_2)$$

(2.11)

Here the $\hat{s}_1$ and $\hat{s}_2$ are unit vectors giving the orientations of the dipoles and $\hat{r}$ is the unit vector in the r_{12} direction. With this choice of $w(12)$, an expression we derived for ε in Ref. 4 can be summarized in compact form in terms of Σ. We have

$$\frac{\varepsilon - 1}{\varepsilon + 2} = z$$

(2.12a)

where

$$z = \frac{4\pi}{9} \beta \rho \mu^I \cdot \mu$$

(2.12b)

Here $\mu = \mu(i)$ is the dipole vector $m\hat{s}_i$ and

$$\mu^I = \int \Sigma(12) \left[\frac{\mu(2)}{\rho(2)}\right] d(2)$$

(2.12c)

where $d(2) = d r_2 \, d\Omega_2$. Thus, since $\rho(i) = \rho/\Omega$, we can write

$$\mu \cdot \mu^I = \frac{\Omega}{\rho} \int \Sigma(12) \mu(1) \cdot \mu(2) \, d(2)$$

(2.12d)

Substituting (2.3) into (2.12), we have the alternative expression

$$\boldsymbol{\mu}(1)\cdot\boldsymbol{\mu}^{J}(1)=\boldsymbol{\mu}(1)\cdot\boldsymbol{\mu}(1)+\frac{\Omega}{\rho}\int W(12)\boldsymbol{\mu}(1)\cdot\boldsymbol{\mu}(2)\,d(2) \qquad (2.12e)$$

Equations 2.12a and 2.12b, with 2.12d or 2.12e, represent a fundamental result for the model under consideration—an exact expression for ε in terms of Σ, or equivalently, W. The Σ and W can be regarded as the "short-ranged parts" of F_δ and F, respectively, from which the long-range effects of $v(12)$ have been subtracted out [upon the subtraction of v from c before forming the OZ equation (2.9)].

The result (2.12) can be reexpressed in several different ways that will prove useful. First of all, only the $\hat{s}_1\cdot\hat{s}_2$ projection of $W(12)$ will contribute to ε in (2.12e), because of the $\boldsymbol{\mu}(1)\cdot\boldsymbol{\mu}(2)$ factor that multiplies $W(12)$ in that expression. If we write the coefficient of that projection of $\tilde{W}(\Omega_1\Omega_2)\Omega^2$ as ρB, then from (2.12b) and (2.12e) we have simply

$$z=y\left(1+\tfrac{1}{3}B\right) \qquad (2.13)$$

where

$$y=\tfrac{4}{9}\pi m^2 \beta\rho \qquad (2.14)$$

which was the expression first given for z in Ref. 4. Equation 2.12 with (2.9) also yields a perfectly general expression for z via (2.12b) in terms of the direct correlation function. In the case in which the reference-system potential shares the cylindrical symmetry of the ideal dipole (including the special case in which it is orientation independent) (2.12) and (2.9) can be further reduced to the expression

$$\frac{\varepsilon-1}{\varepsilon+2}=\frac{y}{1-(\rho/4\pi)I} \qquad (2.15a)$$

where

$$I=\frac{1}{4\pi}\int d\mathbf{r}_{12}\,d\Omega_1\,d\Omega_2\,c(12)\hat{s}_1\cdot\hat{s}_2 \qquad (2.15b)$$

or

$$I=\frac{1}{4\pi}\int d\Omega_1\,d\Omega_2\,c(\Omega_1\Omega_2)\hat{s}_1\cdot\hat{s}_2 \qquad (2.15c)$$

This is an expression originally derived by John Ramshaw[8] under certain approximations that one cannot expect to be exactly satisfied. However, as shown in Ref. 4, these approximations need not be assumed in the case of a cylindrically symmetric reference system to obtain (2.15). In contrast to (2.12), on the other hand, (2.15) will not hold if the reference system has more general symmetry. An Appendix to Ref. 5 gave a general reduction of (2.12) that yields an expression for ε in terms of $c(12)$ for all symmetries. Subsequently Ramshaw himself derived[9] an equivalent general reduced form of (2.12) and (2.9) in somewhat different notation.

In terms of $c_\delta(12) = c(12) - \delta(12)/\rho(1)$, we can write

$$1 - \left(\frac{\rho}{4\pi}\right)I = \frac{\rho}{(4\pi)^2} \int dr_{12}\, d\Omega_1\, d\Omega_2\, c_\delta(12)\hat{\mathbf{s}}_1 \cdot \hat{\mathbf{s}}_2 \qquad (2.15d)$$

Prior to the Høye-Stell work, Nienhuis and Deutch[10] (ND) also derived a very closely related expression that is of a form somewhat different from (2.12). It is in terms of their auxiliary function $G_0^{(2)}$ that they defined graphically. Except for notational differences, their graphical analysis exactly corresponds to that part of the graphical formalism of Lebowitz, Stell, and Baer that involves the LSB function $n^2W(12)$ [our $W(12)$]. For the reader's convenience we have drawn up Table I, which allows one to go from one notation to the other, as well as to the notation we are using here. (In LSB, $i = \mathbf{r}_i$ rather than the full $\mathbf{r}_i$, Ω_i, but the graphical manipulations are completely insensitive to this.)

It can be seen from our Table I that our W and the ND $G_2^{(0)}$ are *identical* objects graphically; yet the ND expression for ε in terms of $G_2^{(0)}$ is clearly different from ours. Theirs is

$$\frac{\varepsilon - 1}{3} = \frac{4\pi}{9}\beta\rho\boldsymbol{\mu} \cdot \boldsymbol{\mu}^{ND} \qquad (2.16a)$$

$$\boldsymbol{\mu} \cdot \boldsymbol{\mu}^{ND} = \boldsymbol{\mu} \cdot \boldsymbol{\mu} + \frac{\Omega}{\rho} \int G_2^{(0)}(2)\boldsymbol{\mu}(1) \cdot \boldsymbol{\mu}(2)\, d(2) \qquad (2.16b)$$

TABLE I

Comparison of Terminology: Lebowitz, Stell, and Baer (LSB),
Nienhuis and Deutch (ND), and Here

LSB	ND	Here
$F_2(12)$	$G_2(12)$	$F(12)$
$n^2(1)W(12)$	$G_2^{(0)}(12)$	$W(12)$
$n^2(1)W(12) + n(1)\delta(12)$	$H_2^{(0)}(12)$	$\Sigma(12)$
$F_2(12) - n^2(1)W(12)$	$G_2^{(1)}(12)$	$F(12) - W(12)$

This seems to be a paradox, since both ND and we use an ideal dipole term as the perturbing potential. The paradox is readily resolved by noting that the ND and Høye-Stell (HS) choices of $w(12)$ are in fact different, with HS choosing a dipole with a sharp cutoff at small r, as in (2.10), and ND choosing the full ideal dipole for all $r \neq 0$ plus a singular term at $r=0$. Their handling of $w(12)$ at the origin can be considered in terms of the choice

$$w(12) = -m^2 D(12) r^{-3} \quad \text{for} \quad r > \gamma^{-1}$$

$$w(12) = +m^2 \Delta(12) \gamma^3 \quad \text{for} \quad r < \gamma^{-1} \tag{2.17}$$

where

$$\Delta(12) = \hat{s}_1 \cdot \hat{s}_2 \tag{2.18}$$

with $D(12)$ given in (2.11). In the limit $\gamma^{-1} \to 0$ (note that this is the *opposite* of the mean field limit $\gamma \to 0$), it is clear that (2.17) is equivalent to the ND method of handling the dipole at $r=0$, since in this limit the Fourier transform $\tilde{w}(12)$ becomes

$$\tilde{w}(12) = \frac{4\pi m^2}{3} \left[D_k(12) + \Delta(12) \right] \tag{2.19a}$$

where $D_k(12)$ is just $D(12)$ with $\hat{r}$ replaced by the unit vector $\hat{k} = k/|k|$:

$$D_k(12) = 3(\hat{s}_1 \cdot \hat{k})(\hat{k} \cdot \hat{s}_2) - \hat{s}_1 \cdot \hat{s}_2 \tag{2.19b}$$

The choice of $w(12)$ given by (2.17) when $\gamma^{-1} \to 0$ can alternatively be expressed as

$$w(12) = -m^2 \hat{s}_1 \cdot \mathbf{T}(\mathbf{r}) \cdot \hat{s}_2 \tag{2.19c}$$

where

$$\mathbf{T}(\mathbf{r}) = \nabla \cdot \nabla (r^{-1}) \tag{2.19d}$$

This representation of the dipole interaction is the one used by ND; it is easily verified that (2.19c) has the same Fourier transform as (2.17) does in the limit $\gamma^{-1} \to 0$. Ramshaw has also considered in detail[11] the relation between the use of (2.10) and (2.17).

To make the connection between (2.12) and (2.16) more precise, as well as to relate (2.12) to other formally exact expressions for ε of still different form, we must set down a few results concerning the general decomposition of an arbitrary two-point function in an expansion in terms of its rotational invariants as well as one fundamental result concerning the structure of $F(12)$ when so decomposed. For our purposes, we need only consider the rotational invariants $D(12)$, given by (2.11), plus $\Delta(12)$, given by (2.18), and unity. For an arbitrary two-point function $a(12)$ we define the projections

$$a_S(r) = \int \frac{a(12)\, d\Omega_1\, d\Omega_2}{\Omega^2}$$

$$a_\Delta(r) = 3 \int \frac{a(12)\Delta(12)\, d\Omega_1\, d\Omega_2}{\Omega^2}$$

$$a_D(r) = \left(\frac{3}{2}\right) \int \frac{a(12)D(12)\, d\Omega_1\, d\Omega_2}{\Omega^2} \tag{2.20}$$

so we can write

$$a(12) = a_S(r) + a_\Delta(r)\Delta(12) + a_D(r)D(12) + \cdots \tag{2.21}$$

Taking the Fourier transform of both sides, we find

$$\tilde{a}(12) = \tilde{a}(k) + \tilde{a}_\Delta(k)\Delta(12) + \bar{a}_D(k)D_k(12) + \cdots \tag{2.22a}$$

with $D_k(12)$ given by (2.19b). The $\tilde{a}_\Delta(k)$ is just the Fourier transform of $a_\Delta(r)$

$$\tilde{a}_\Delta(k) = \int e^{i\mathbf{k}\cdot\mathbf{r}}\tilde{a}_\Delta(r)\, d\mathbf{r} \tag{2.22b}$$

and $\bar{a}_D(k)$ is a Hankel transform

$$\bar{a}_D(k) = -\int j_2(kr)a_D(r)\, d\mathbf{r} \tag{2.22c}$$

Here j_2 is the spherical Bessel function of order 2,

$$j_2(x) = \frac{3\sin x}{x^3} - \frac{3\cos x}{x^2} - \frac{\sin x}{x} \tag{2.23}$$

The $\bar{a}_D(r)$ can be regarded as the Fourier transforms of certain spherically symmetric functions $\hat{a}_D(r)$, which were introduced by Wertheim[12] in his analysis of the mean spherical approximation for dipolar spheres. If $a(r) \to 0$

as $r \to 0$, let

$$\hat{a}(r) = a(r) - 3 \int_r^\infty a(s)s^{-1} ds \qquad (2.24a)$$

This inverts to give

$$a(r) = \hat{a}(r) - 3r^{-3} \int_0^r \hat{a}(s)s^2 ds \qquad (2.24b)$$

As shown by Høye, Lebowitz, and Stell,[13, 14]

$$\bar{a}(k) = \tilde{\hat{a}}(k) \qquad (2.24c)$$

From (2.24b) we have the important conclusion

$$a_D(r) \to -\frac{3\bar{a}_D(0)}{4\pi r^3} \qquad \text{as} \quad r \to \infty \qquad (2.24d)$$

In Section III, we use a more general notation in which $a_S(r)$ is denoted as $a^{000}(r)$, $a_\Delta(r)$ is denoted as $a^{110}(r)$, and $a_D(r)$ as $a^{112}(r)$, while $\tilde{a}^{ijk}(k)$ refers to a generalized Fourier transform that includes the Hankel transform introduced here. In treating simple dipolar models (in which higher ideal multipole terms may or may not be present, but are not explicitly discussed) the more general notation is unnecessary and is not used in Section II.

As Høye and Stell have shown in Refs. 4 and 5, it turns out that all the expressions for ε that we shall be dealing with in our model can be expressed in terms of the Δ-projection and D-projection of two-point correlation functions in Fourier space, evaluated as $k = 0$. We have already seen examples of this in connection with our first key result; (2.13) can be rewritten in the notation we have just introduced as

$$z = y\left[1 + \frac{\Omega^2 \tilde{W}_\Delta(0)}{3\rho}\right] = \frac{y\Omega^2 \tilde{\Sigma}_\Delta(0)}{3\rho} \qquad (2.25a)$$

and (2.15) can be reexpressed as

$$z = y\left[1 - \frac{\rho}{3}\tilde{c}_\Delta(0)\right]^{-1} = -y\left[\frac{\rho}{3}\tilde{c}_{\delta\Delta}(0)\right]^{-1} \qquad (2.25b)$$

There is yet another expression for z, hence for ε through (2.12a), which has turned out to be one that is computationally the most useful and important, along with its generalization to the case of polarizable and charged particles. It was introduced by Wertheim[12] as part of his mean spherical approximation (MSA) for ε, and subsequently reused by him (as an exact expression) in a form appropriate to nonpolar particles with constant polarizability.[15] Independently, Høye, Lebowitz, and Stell[13] and Høye and Stell[16] extended the MSA result for nonpolarizable polar particles to a class of generalized mean spherical results; their extension makes clear the formally exact status of the result in the dipolar case in terms of exact $\hat{c}_\Delta(0)$ and $\hat{c}_D(0)$. One introduces the quantities

$$q_1 = 1 - \frac{\rho\left[\tilde{c}_\Delta(0) + 2\tilde{c}_D(0)\right]}{3}$$

$$q_2 = 1 - \frac{\rho\left[\tilde{c}_\Delta(0) - \tilde{c}_D(0)\right]}{3} \qquad (2.25c)$$

Following the development of Høye, Lebowitz, and Stell, (or equivalently of Ref. 15) one finds

$$3y = q_1 - q_2, \qquad y = \frac{4\pi}{9}\beta\rho m^2 \qquad (2.25d)$$

and

$$z = \frac{q_1 - q_2}{q_1 + 2q_2} \qquad (2.25e)$$

Using the last two equations with (2.12a) immediately yields

$$\varepsilon = \frac{q_1}{q_2} \qquad (2.26)$$

As we shall see in Sections III to V, it is (2.26)—along with (2.25b) to (2.25e) and their appropriate extensions to more general Hamiltonian models—that forms the basis of most currently available analytic calculations of ε based on those models.

We introduce now for convenience the correlation functions $h_\delta(12)$, $h(12)$, $g_\delta(12)$, and $g(12)$, related to the ρ_δ and F_δ of (2.2) and (2.3) as follows:

$$\rho_\delta(12) = \rho(1)\rho(2)g_\delta(12)$$

$$\rho(12) = \rho(1)\rho(2)g(12)$$

$$F_\delta(12) = \rho(1)\rho(2)h_\delta(12)$$

$$F(12) = \rho(1)\rho(2)h(12)$$

In terms of $h(12)$ and $c(12)$, (2.8) takes the more familiar form

$$h(12) = c(12) + \rho \int \frac{c(13)h(23)\, d(3)}{\Omega} \tag{2.27}$$

where

$$h(12) = g(12) - 1 \tag{2.28}$$

Decomposing $\tilde{h}(12)$, it is found in Eq. 50 of Ref. 4 that

$$\lim_{k \to 0} \tilde{h}(12) = AD_k(12) + (B - 2zA)\Delta(12) + (\text{terms not contributing to } \varepsilon)$$
$$\tag{2.29a}$$

So

$$\rho \bar{h}_D(0) = A \tag{2.29b}$$

$$\rho \tilde{h}_\Delta(0) = B - 2zA \tag{2.29c}$$

where

$$A = -3 \frac{z^2}{y(1+2z)(1-z)} \tag{2.29d}$$

and B we have already met in (2.13),

$$B = 3\left(\frac{z}{y} - 1\right) \tag{2.29e}$$

Using (2.12a), (2.29) yields

$$\rho \tilde{h}_\Delta(0) = 3\left[\frac{(\varepsilon-1)(2\varepsilon+1)}{9\varepsilon y} - 1 \right] \tag{2.30a}$$

$$\rho \bar{h}_D(0) = -3\frac{y}{\varepsilon}\left(\frac{\varepsilon-1}{3y} \right)^2 \tag{2.30b}$$

Equation 2.30a can be rewritten in terms of a familiar[17, 18] "g-factor"

$$\frac{(\varepsilon-1)(2\varepsilon+1)}{9\varepsilon} = yg; \qquad g = 1 + \frac{\rho}{3}\tilde{h}_\Delta(0) \tag{2.30c}$$

Onsager's well-known approximation[19] is recovered by setting $\tilde{h}_\Delta(0)=0$,

$$\frac{(\varepsilon-1)(2\varepsilon+1)}{9\varepsilon}=y; \qquad g=1 \tag{2.30d}$$

From (2.24d), with $a_D(r)$ set equal to $h_D(r)$, we find that (2.30b) can be reexpressed in r space as

$$\lim_{r\to\infty} h(12)=\frac{\beta m^2}{\varepsilon}\left(\frac{\varepsilon-1}{3y}\right)^2 D(12)r^{-3} \tag{2.30e}$$

Equation 2.30c was derived decades ago by Kirkwood.[17] Equation 2.30e was first obtained much more recently by Nienhuis and Deutch[10] on the basis of a plausible assumption that the analysis of Høye and Stell outlined here has shown to be fully justified. ND also obtained[20] several important companion equations to (2.30e). One gives the asymptotic form of the correlation function $h_{CD}(12)$ between a single rigid impurity ion (e.g., a charged hard sphere) of charge q and a rigid dipolar impurity particle (or, equivalently, a rigid dipolar molecule of the fluid)

$$\lim_{r\to\infty} h_{CD}(12)=-\frac{\beta}{\varepsilon}\left(\frac{\varepsilon-1}{3y}\right)\phi_{CD}(12) \tag{2.31a}$$

where ϕ_{CD} is the ideal charge-dipole interaction

$$\phi_{CD}(12)=qm\hat{r}\cdot\hat{s}r^{-2} \tag{2.31b}$$

Another gives the ion-ion correlation function $h_{CC}(12)$ for two ions in the same rigid-dipole fluid

$$\lim_{r\to\infty} h_{CC}(12)=-\frac{\beta}{\varepsilon}\phi_{CC}(12) \tag{2.32a}$$

where ϕ_{CC} is the ideal Coulomb interaction

$$\phi_{CC}(12)=q_1q_2r^{-1} \tag{2.32b}$$

(Asymptotic expressions for correlations between quadrupolar impurity and charge impurity and between quadrupolar impurity and dipolar impurity were also obtained by ND.)

These expressions as well as (2.30e) can be written equivalently in terms of the potentials of mean force $W(12)$, where $-\beta W(12)$ is defined as

$\ln[h(12)+1]$. Letting $\phi_{DD}(12)$ be the ideal dipole-dipole potential, we have

$$W_{DD}(12) \underset{r\to\infty}{\to} \frac{1}{\varepsilon}\left(\frac{\varepsilon-1}{3y}\right)^2 \phi_{DD}(12) \tag{2.33a}$$

$$W_{CD}(12) \underset{r\to\infty}{\to} \frac{1}{\varepsilon}\left(\frac{\varepsilon-1}{3y}\right)\phi_{CD}(12) \tag{2.33b}$$

$$W_{CC}(12) \underset{r\to\infty}{\to} \frac{1}{\varepsilon}\phi_{CC}(12) \tag{2.33c}$$

Equation 2.33 answers an interesting question raised by the work of Jepsen and Friedman[21] some time ago. They pointed out that for a macroscopic charged-sphere impurity and a macroscopic dipolar-sphere impurity (both of dielectric constant 1) immersed in a continuum dipolar solvent, one has the expressions

$$W_{DD}(12) \underset{r\to\infty}{\to} \frac{1}{\varepsilon}\left(\frac{3\varepsilon}{2\varepsilon+1}\right)^2 \phi_{DD}(12) \tag{2.34a}$$

$$W_{CD}(12) \underset{r\to\infty}{\to} \frac{1}{\varepsilon}\left(\frac{3\varepsilon}{2\varepsilon+1}\right)\phi_{CD}(12) \tag{2.34b}$$

$$W_{CC}(12) \underset{r\to\infty}{\to} \frac{1}{\varepsilon}\phi_{CC}(12) \tag{2.34c}$$

Jepsen and Friedman[21] found, however, that for microscopic impurities, (2.34a) and (2.34b) — in contrast to (2.34c) — no longer appeared to be satisfied beyond the lowest order in y in the low-density approximation they were considering, which left open the asymptotic form the microscopic results would have. Equation 2.33 reveals that only if the Onsager approximation (2.30d) were satisfied in the molecular solvent would (2.33) and (2.34) be the same. The reason for this will become clear in our discussion of the $\gamma\to0$ limit below, where we show that only in the *Onsager continuum limit*, in which (2.30d) becomes exact, is the dielectric response to each solvent dipole that of a vacuum in a macroscopic sphere surrounding the solvent dipole. Thus only in the Onsager continuum limit are the assumptions satisfied under which one can identify each solvent particle as a macrosphere within which $\varepsilon=1$, and so assure the identity of the full set of ratios in (2.33) to (2.34).

Høye and Stell[22] have generalized (2.33) to the case of a nonzero concentration of ionic particles in a dipolar fluid. They find that in the low λ limit, where λ is a characteristic inverse shielding length $\lambda^2 = 4\pi q^2 \beta \rho_c/\varepsilon$, with ρ_c the ionic concentration, q the ionic charge, and ε the solvent dielectric constant

$$W_{\mathrm{DD}}(12) \underset{r\to\infty}{\longrightarrow} \left(\frac{\varepsilon-1}{3y}\right)\frac{m^2}{\varepsilon}\left[\left(1+\lambda r+\frac{\lambda^2 r^2}{3}\right)\frac{e^{-\lambda r}}{r^3}D(12)+\frac{1}{3}\frac{\lambda^2}{r}e^{-\lambda r}\Delta(12)\right] \tag{2.35a}$$

$$W_{\mathrm{CD}}(12) \underset{r\to\infty}{\longrightarrow} \left(\frac{\varepsilon-1}{3y}\right)\frac{qm}{\varepsilon r^2}(1+\lambda r)e^{-\lambda r}\hat{\mathbf{r}}\cdot\hat{\mathbf{s}} \tag{2.35b}$$

$$W_{\mathrm{CC}}(12) \underset{r\to\infty}{\longrightarrow} \frac{q^2}{\varepsilon r}e^{-\lambda r} \tag{2.35c}$$

We note that (2.35) can be elegantly reexpressed in terms of gradients of the shielded Coulomb interaction, if we use the effective dipole-moment vector $\hat{\mathbf{m}}_i^{\mathrm{eff}}=m\hat{\mathbf{s}}_i(\varepsilon-1)/3y$ introduced by Nienhuis and Deutch.[10, 20] Writing $e(r)=e^{-\lambda r}/\varepsilon r$, we have

$$W_{\mathrm{CC}}=q^2e(r) \tag{2.36a}$$

$$W_{\mathrm{CD}}=q\mathbf{m}_2^{\mathrm{eff}}\cdot\boldsymbol{\nabla}_2 e(r) \tag{2.36b}$$

$$W_{\mathrm{DD}}=\mathbf{m}_2^{\mathrm{eff}}\cdot\boldsymbol{\nabla}_2\left[\mathbf{m}_1^{\mathrm{eff}}\cdot\boldsymbol{\nabla}_1 e(r)\right] \tag{2.36c}$$

All the results we have considered so far are for a system in which the thermodynamic limit has been taken before the correlation functions are assessed. Thus $\tilde{h}_\Delta(0)=\int h_\Delta(r)\,d\mathbf{r}$, for example, refers to a volume integral taken over a volume that is allowed to become infinite *after* one takes the thermodynamic limit. If one has a finite spherical sample in a vacuum, on the other hand, the $h(12)$ for that system will be missing its $D(12)$ term for r larger than the sample diameter. This means that in (2.29a) the A will be zero; but if the sample is sufficiently large, the terms shown will be otherwise unchanged, so that $\rho\tilde{h}_D(0)=0$ and $\rho\tilde{h}_\Delta(0)=B$. Thus (2.12a) with (2.25) can be rewritten as

$$\frac{\varepsilon-1}{\varepsilon+2}=y\left[1+\frac{\rho}{3}\tilde{h}_\Delta(0)\right] \tag{2.37}$$

This is another classic result, implicit in Kirkwood's early work.[17] We note the following points related to (2.37).

1. If one is already in the thermodynamic limit but cuts off the $D(12)$ term of the pair potential $w(12)$ given by (2.10) beyond some radius R, one is again led to (2.37) for reasons similar to those in the finite-sample

case. (Wertheim's recent review[1] provides an especially clear discussion of this in modern terminology.)

2. Let us denote as $\tilde{h}_\Delta^R(k)$ the $\tilde{h}_\Delta(k)$ associated with a finite sample of radius R. Then

$$\lim_{k\to 0} \lim_{R\to\infty} \tilde{h}_\Delta^R(k) \neq \lim_{R\to\infty} \tilde{h}_\Delta^R(0). \qquad (2.38)$$

It is the right-hand side that is to be identified with (2.37) and the left-hand side with (2.30c). The latter equation can be thought of as the equation relevant to a sample of material of macroscopic radius R embedded in an infinite system of the same material (i.e., an infinite system of dielectric constant ε). Moreover, on the macroscale determined by the length unit R, the system external to the sphere can be regarded dielectrically simply as a continuum of dielectric constant ε. Thus (2.37) and (2.30c) are the relevant equations for the same macroscopic spherical sample embedded in continua of dielectric constant 1 and ε, respectively. These results can be generalized to a sample embedded in a continuum of arbitrary dielectric constant ε', as discussed by de Leeuw, Perram, and Smith,[23] who use the generalization to illuminate the status of Ewald summation in systems with periodic boundary conditions. We review their work in Section III.C.

Our catalog of exact representations of ε for the rigid-particle model would be incomplete without the expression for ε in terms of site-site (i.e., atom-atom) correlation functions, defined in terms of $h(12)$ by

$$h_{\alpha\beta}(12) = \int_{r_{\alpha\beta}=\text{const}} \frac{h(12)\, d\Omega_1\, d\Omega_2}{\Omega^2} \qquad (2.39)$$

where $r_{\alpha\beta}$ is the distance between two sites (or atoms) α and β in molecules 1 and 2, respectively. If the dipolar moment of the molecule, m, is prescribed in terms of a set of point charges q_γ at the sites γ, Høye and Stell[24, 25] have shown that one can write

$$\lim_{k\to 0}\left[k^{-2}\rho \sum_{\alpha\beta} q_\alpha q_\beta \tilde{h}_{\alpha\beta}(12) + \frac{m^2}{3} \right] = \left(\frac{\varepsilon-1}{9y\varepsilon} \right) m^2 \qquad (2.40)$$

Here ρ is molecular number density.

B. Core-Parameter Extension of the Expression for ε, and γ Parameterization

In the results summarized above we have not yet introduced a constructive scheme for quantitatively evaluating either the correlation functions already introduced or ε itself. We shall do so by embedding our perturbing potential $w(12)$ in a one-parameter family of potentials parameterized by an inverse range parameter γ such that when $\gamma = d^{-1}$ we recover the $w(12)$ corresponding to the pair potential $\phi(12)$ of actual interest in our model. In the limit $\gamma \to 0$, exact results can be recovered and γ then used as a parameter of smallness in approximating the desired results for $\gamma = d^{-1}$. For a prescribed $\phi(12)$, the choice of $w(12)$ is itself open to considerable latitude, as our comparison between certain Høye-Stell and Nienhuis-Deutch results has already suggested. The formalism of Lebowitz, Stell, and Baer,[6] (and the earlier work of Hemmer[26]) on which the Høye-Stell results rest, is independent of how we choose to break up $\phi(12)$ into $q(12)$ and $w(12)$ within wide limits, as discussed in Ref. 6. It is natural to require

$$\frac{w(12)}{\phi(12)} \to 1 \quad \text{as} \quad r \to \infty \qquad (2.41)$$

which dictates, in a polar system,

$$w(12) \to -m^2 D(12) r^{-3} \quad \text{as} \quad r \to \infty \qquad (2.42)$$

For a potential in which a clearly defined measure of particle diameter d is available [i.e., a hard-core particle in which $\phi(12) = \infty$ for $r < d$, a Stockmayer potential in which the Lennard-Jones σ is the measure of core size, etc.], it is also often convenient (although not necessary) to satisfy (2.41) by requiring

$$w(12) = -m^2 D(12) r^{-3} \quad \text{for} \quad r > d \qquad (2.43)$$

For a $\phi(12)$ with a hard-core diameter d, if $q(12)$ is chosen to include the hard core, so that $q(12) = \infty$ for $r < d$, then clearly $w(12)$ can be set equal to *any* finite function for $r < d$—including functions that are dependent on thermodynamic state—without quantitatively effecting the value of any exact expressions for $h(12)$, $c(12)$, or ε in which $w(12)$ appears. But in the approximate evaluation of such expressions, one must expect that some choices of $w(12)$ for $r < d$ will yield better approximations than others, and this turns out to be the case, as we shall see. Even for $\phi(12)$ that do not have hard cores, it proves useful and natural to exploit our freedom of choice of $w(12)$ for small r. [It is just that without the core, different choices of $w(12)$ now dictate corresponding differences in $q(r)$ for a fixed $\phi(12)$.]

Because only the Δ and D projections of correlation functions appear in our various representations of ε, only the choice of the Δ and D projections of $w(12)$ are of primary dielectric importance in our model.

Let us generalize (2.17) to

$$w(12) = \gamma^3 w_\Delta(\gamma r)\Delta(12) + \gamma^3 w_D(\gamma r)D(12) \qquad (2.44)$$

where the initial value of γ is d^{-1}, but we shall take the limit $\gamma \to 0$ in Section II.D. For $x > 1$, $w_\Delta(x)$ must be zero and $w_D(x)$ must be $-m^2 x^{-3}$ if $w(12)$ is to be the ideal dipole potential outside the hard core, but for $x < 1$, $w_\Delta(x)$ and $w_D(x)$ can be arbitrary. Here $x = \gamma r$.

We introduce the parameter Θ given by

$$\frac{4\pi\Theta m^2}{3} = \int \gamma^3 w_\Delta(\gamma r)\, d\mathbf{r}$$

$$= \int w_\Delta(r)\, d\mathbf{r} \qquad (2.45)$$

The simplest choice of $w_\Delta(x)$ is just a step function,

$$w_\Delta(x) = m^2\Theta \qquad \text{for} \quad x < 1$$

$$w_\Delta(x) = 0 \qquad \text{for} \quad x > 1 \qquad (2.46)$$

and the simplest choice of $w_D(x)$ in the context of our work is

$$w_D(x) = -m^2 x^{-3} \qquad \text{for} \quad x > 1$$

$$w_D(x) = 0 \qquad \text{for} \quad x < 1 \qquad (2.47)$$

so that $d^{-3}\tilde{w}_D(r/d)D(12)$ is the cutoff dipole of (2.10). The Fourier transform of $w(12)$ is just

$$\tilde{w}(12) = \overline{w}_D\left(\frac{k}{\gamma}\right)D_k(12) + \tilde{w}_\Delta\left(\frac{k}{\gamma}\right)\Delta(12) \qquad (2.48)$$

The $\tilde{w}_\Delta(k)$ is the transform of $w_\Delta(r)$; if (2.47) is used, we have

$$\overline{w}_D(k) = \frac{4\pi m^2}{3}f(k)$$

$$f(k) = 3\left[\frac{\sin k}{k^3} - \frac{\cos k}{k^2}\right] \qquad (2.49)$$

Note that $f(0) = 1$. If (2.46) is also used then $\tilde{w}_\Delta(k)$ can also be written in terms of $f(k)$ and we have

$$\tilde{w}(12) = \frac{4\pi m^2}{3} f\left(\frac{k}{\gamma}\right)\left[\tilde{D}_k(12) + \Theta\tilde{\Delta}(12)\right] \tag{2.50a}$$

Since $w(12)$ of (2.44) is of the form γ^3 times a function dependent on r only through γr, it lends itself to the γ-ordering techniques. For $\Theta = 0$ all results must reduce to those based on (2.10). For $\Theta = 1$ and $\gamma \to \infty$, $\tilde{w}(12)$ becomes the Nienhuis-Deutch expression for $\tilde{w}$,

$$\tilde{w}(12) = \frac{4\pi m^2}{3}\left[D_k(12) + \Delta(12)\right] \tag{2.50b}$$

The functions $\Sigma(12)$ and $W(12)$ turn out to depend on $w(12)$ in such a way that for fixed γ, $\Sigma(\Omega_1\Omega_2)$ and $W(\Omega_1\Omega_2)$ depend on $w(12)$ only through Θ. From this it follows that the relation for ε itself depends on $w(12)$ only through Θ and we find, generalizing (2.12) to $\Theta \neq 0$,

$$\frac{\varepsilon - 1}{(1 - \Theta)\varepsilon + 2 + \Theta} = \frac{4\pi\beta\rho}{9}\boldsymbol{\mu}_{\text{eff}}(\Theta)\cdot\boldsymbol{\mu} \tag{2.51a}$$

for all γ and Θ where

$$\boldsymbol{\mu}_{\text{eff}}(\Theta) = \frac{\Omega}{\rho}\int\Sigma(12; w)\mu(2)\,d(2)$$

$$= \frac{\Omega}{\rho}\int\tilde{\Sigma}(\Omega_1\Omega_2; \Theta)\mu(2)\,d\Omega_2 \tag{2.51b}$$

For $\gamma = d^{-1}$ and $\Theta = 0$ we recover (2.12) with $\boldsymbol{\mu}_{\text{eff}}(\Theta) = \boldsymbol{\mu}^I$, and for $\gamma \to \infty$, $\Theta = 1$ we recover (2.16) with $\boldsymbol{\mu}_{\text{eff}}(1) = \boldsymbol{\mu}^{\text{ND}}$. For *any* $\gamma \geq d^{-1}$, and $\Theta = 1$ we would in fact find $\boldsymbol{\mu}_{\text{eff}}(1) = \boldsymbol{\mu}^{\text{ND}}$. More generally for all $\gamma \geq d^{-1}$ and fixed Θ, $\boldsymbol{\mu}_{\text{eff}}(\Theta)$ and ε are independent of γ, reflecting the fact that for a fixed reference potential $q(12)$ with hard core of diameter d, the full potential $q(12) + w(12)$ is independent of γ as long as $\gamma \geq d^{-1}$. Thus the actual physics of the problem must be likewise independent of γ where $\gamma d > 1$.

In the remainder of this section we give various other key results that one obtains in generalizing[5] to nonzero Θ.

In the limit $\gamma \to 0$ the pair correlation function again is given by a simple dipole chain

$$\rho\tilde{h}(12) = \sum_{n=1}^{\infty}(-3y)^n(D_k + \Theta\Delta)^n \tag{2.52}$$

This is easily computed by noting

$$3(D_k + \Theta\Delta) = (2+\Theta)J_1 - (1-\Theta)J_2 \tag{2.53}$$

where $J_1 = \Delta + D_k$ and $J_2 = 2\Delta - D_k$, which from (B.18) of Ref. 4 satisfy $J_1 J_1 = J_1$; $J_2 J_2 = J_2$; $J_1 J_2 = 0$. So we get

$$
\begin{aligned}
\rho\tilde{h}(12) &= \sum_{n=1}^{\infty} \left[(-(2+\Theta)y)^n J_1 + ((1-\Theta)y)^n J_2 \right] \\
&= \frac{-(2+\Theta)y}{1+(2+\Theta)y} J_1 + \frac{(1-\Theta)y}{1-(1-\Theta)y} J_2 \\
&= \frac{-3y}{(1+(2+\Theta)y)(1-(1-\Theta)y)} \left[D_k + (\Theta - (1-\Theta)(2+\Theta))y\Delta \right]
\end{aligned}
\tag{2.54}
$$

Equation 2.54 is to be compared to the general expression (2.29a). One finds that (2.54) and (2.29a) are equal if one puts

$$z = \frac{y}{1+\Theta y} \tag{2.55}$$

So from (2.12a) the dielectric constant for this system in the limit $\gamma \to 0$ will be

$$\frac{\varepsilon - 1}{\varepsilon + 2} = z = \frac{y}{1+\Theta y} \tag{2.56a}$$

or

$$\frac{\varepsilon - 1}{(1-\Theta)\varepsilon + (2+\Theta)} = y; \qquad \varepsilon = \frac{1+(2+\Theta)y}{1-(1-\Theta)y} \tag{2.56b}$$

Expression 2.56b for z can also be seen more directly. According to the treatment in Ref. 4, only the D term in (2.50) belongs to the dipole interaction, while the Δ term belongs to the reference system, and contributes to $W(12)$ given by (2.3), (2.13), and (2.14). As $\gamma \to 0$, $W(12)$ obviously will be given by a "Δ chain":

$$\frac{1}{\rho}\tilde{W}(12) = 3\left(\frac{z}{y} - 1\right)\Delta = \sum_{n=1}^{\infty} (-\Theta y)^n (3\Delta)^n = \sum_{n=1}^{\infty} (-\Theta y)^n 3\Delta$$

$$z = y \sum_{n=0}^{\infty} (-\Theta y)^n = \frac{y}{1+\Theta y} \tag{2.57}$$

References 4 and 5 introduced test dipoles that interact with a potential proportional to that given by Eq. (2.44). The limit of large $R = 1/\gamma \to \infty$ was considered such that pure electrostatics can be used.

To summarize these results, we first compute the field set up by our new test dipole in a dielectric medium. It interacts primarily with a field $\mathbf{K} = -\nabla \psi_K$ where

$$\psi_K = \begin{cases} \dfrac{m_2 \cos \theta}{r^2} & \text{for } r > R \\[2ex] \dfrac{\Theta m_2 r \cos \theta}{R^3} & \text{for } r < R \end{cases} \tag{2.58a}$$

The electric field $\mathbf{F} = -\nabla \psi$ created is given by

$$\psi = \begin{cases} \dfrac{a \cos \theta}{r^2} & \text{for } r > R \\[2ex] b r \cos \theta & \text{for } r < R \end{cases} \tag{2.58b}$$

where a and b are to be determined from the conditions at $r = R$ where the general conditions $\nabla \times \mathbf{F} = 0$ and $\nabla \mathbf{D} = 0$ have to be fulfilled, with electric displacement $\mathbf{D} = \mathbf{F} + 4\pi \mathbf{P}$, and polarization $4\pi \mathbf{P} = (\varepsilon - 1)(\mathbf{F} + \mathbf{K})$. With this $\mathbf{K}$ we then must determine a and b from

$$2\varepsilon a + 2(\varepsilon - 1)m_2 = -\varepsilon b R^3 - (\varepsilon - 1)\Theta m_2$$
$$a = b R^3 \tag{2.59}$$

Equation 2.59 leads to

$$a = b R^3 = -\frac{(2 + \Theta)(\varepsilon - 1)}{3\varepsilon} m_2$$

or
$$\mathbf{F} = -\frac{(2 + \Theta)(\varepsilon - 1)}{3\varepsilon} \mathbf{K} \qquad \text{for } r > R \tag{2.60}$$

So the total electric field set up by the dipole is for $r > R$:

$$\mathbf{E} = \mathbf{F} + \mathbf{K} = \frac{(1 - \Theta)\varepsilon + 2 + \Theta}{3\varepsilon} \mathbf{K} \tag{2.61}$$

C. Three Approximations on the Two-Particle Level

Before further considering the $\gamma \to 0$ limit, we have some general remarks concerning three approximations that are defined by expressions that prove to be exact in the $\gamma \to 0$ limit when different choices for $w(12; \gamma)$ are made. Let us assume that $c(12)$ has the form

$$c(12) = c_S(r) + c_\Delta(r)\Delta(12) + c_D(r)D(12) \tag{2.62}$$

where we recall that for an arbitrary function $a(12)$,

$$a_S(r) = \int \frac{a(12)\,d\Omega_1\,d\Omega_2}{\Omega^2} \tag{2.63}$$

Now if (2.62) holds, (2.27) implies that $h(12)$ has the form

$$h(12) = h_S(r) + h_\Delta(r)\Delta(12) + c_D(r)D(12)$$

and also that

$$c_S(r) = \int \frac{c(12)\,d\Omega_2}{\Omega} \tag{2.64}$$

From condition (2.64) it in turn immediately follows that the relation between h_S and c_S via (2.27) decouples from the relation among h_Δ, h_D, c_Δ, and c_D imposed by that equation. In particular, this means that for any length d, if h_S is prescribed for $r < d$ and c_S for $r > d$, then (2.27) determines h_S for $r > d$ and c_S for $r < d$. Moreover, if h_Δ and h_D are prescribed for $r < l$ and c_Δ and c_D are prescribed for $r > l$, where l is any length, then h_Δ and h_D are determined for $r > l$ and c_Δ and c_D are determined for $r < l$. More specifically, for particles with hard core of diameter d, it follows from the definition of $h(12)$ as a correlation function that $h(12)$ must satisfy the core condition

$$h_S(r) = -1 \quad \text{for} \quad r < d \tag{2.65a}$$

$$h_\Delta(r) = h_D(r) = 0 \quad \text{for} \quad r < d \tag{2.65b}$$

so that if $c(12)$ is prescribed for $r > d$, then h_s is determined for $r > d$ from (2.65a) and (2.27) alone, and h_Δ and h_D can be determined for $r > d$ from (2.65b) and (2.27) alone.

These remarks follow immediately from a straightforward generalization of the techniques of Wertheim,[12] who solved (2.27) subject to (2.65) and the conditions

$$c_S(r) = 0 \quad \text{for} \quad r > d \tag{2.66a}$$

$$c_\Delta(r) = 0, \quad c_D(r) = \beta m^2 r^{-3} \quad \text{for} \quad r > d \tag{2.66b}$$

From what we have just said, one can replace the approximation (2.66a) with any other, or can simply assume that

$$c_S(r) \text{ is exactly given} \qquad (2.67)$$

without altering the determination of h_Λ and h_D via (2.27).

In the mean field limit, $\gamma \to 0$, the relation

$$c(12) = c_S(r) - \beta w(12) \qquad (2.68)$$

becomes exact for a pair potential consisting of a spherically symmetric reference term plus an ideal dipole term $w(12)$. Thus

$$c_\Lambda(r) = 0, \qquad c_D(r) = \beta m^2 r^{-3} \qquad (2.69)$$

is the appropriate mean field result for such a potential. For a hard-core potential of diameter d, however, our freedom to choose $w(12)$ for $r < d$ reflects itself in (2.68) as freedom to choose c_Λ and c_D for $r < d$. The choice (2.10) yields

$$c_\Lambda = 0, \qquad c_D = 0 \qquad \text{for} \quad r < d \qquad (2.70)$$

and the choice (2.44) gives more generally

$$c_\Lambda(r) = -\beta \gamma^3 w_\Lambda(\gamma r) \qquad (2.71a)$$
$$c_D(r) = -\beta \gamma^3 w_D(\gamma r) \qquad (2.71b)$$

where $w_\Lambda(x)$ and $w_D(x)$ can remain undetermined for $x < \gamma d$ and still satisfy the requirement implied by (2.43) that when $\gamma d = 1$,

$$w_\Lambda(x) = 0 \qquad \text{for} \quad x > 1$$
$$w_D(x) = -m^2 x^{-3} \qquad \text{for} \quad x > 1$$

We now call attention to three closely related approximations. All three use the mean field result for c_Λ and c_D outside the core, given by (2.66b). The first also uses (2.70). With the use of (2.27) both h_Λ and h_D can be determined for all r. Both approximations violate the core condition (2.65b). The resulting ε is easy to compute via (2.15) and yields the Clausius-Mossotti relation. The resulting approximation for $h(12)$ could be called the "two-particle Clausius-Mossotti approximation."

The second approximation uses (2.71) with w_Λ and w_D given by (2.46) and (2.47), where $\Theta = (1 - \Theta)(2 + \Theta)y$ and $x = r/d$ (i.e., $\gamma = d^{-1}$). From (2.54) one sees that this yields $\tilde{h}_\Lambda(0) = 0$ and from (2.30) [or (2.56)] that it gives the Onsager approximation for ε. We shall therefore refer to it as the "two-particle Onsager approximation". [In Ref. 5, the approximation with this designation was characterized in a somewhat different way that is not obviously solvable for $h(12)$. The more natural characterization given here has no such problem.]

Finally, if (2.66b) is used along with the core condition (2.65b), then (2.27) yields the Wertheim approximation for c_Λ and c_D inside the core and h_Λ and h_D outside it. Thus it yields the Wertheim ε. The generalization of the Wertheim approximation to mixtures has been discussed in a number of references.[27]

D. The $\gamma \to 0$ Limits

We have just generalized the Clausius-Mossotti and Onsager approximations for ε to fully defined approximations on the pair-correlation level, using a conceptual framework that makes them directly comparable to the Wertheim approximation. We shall refer to these new approximations as the two-particle Clausius-Mossotti and Onsager approximations, respectively.

We now seek a deeper understanding of the meaning and status of these approximations. We shall gain it by noting that each approximation can be associated with a precisely defined "continuum model" of matter, within which it becomes exact. Each of the continuum models can in turn be defined as the $\gamma \to 0$ limit of a molecular model, in which the pair potential is given by (2.4) and (2.44), and in which (2.68) becomes exact.

It is clear from (2.44) that if one lets $\gamma \to 0$ with a fixed reference system core diameter d, one is approaching the limit of completely penetrable particles of macroscopic diameter $l = \gamma^{-1}$. (They are completely penetrable in this limit because the hard-core reference diameter d shrinks to zero compared to l. As a result, particles are free to fully permeate each other.)

We note that one can view this continuum limit either as an observer scaled to a fixed d as $\gamma \to 0$ [the ρd^3 and $\beta m^2/d^3$ are natural to hold fixed to assure a fixed $y = (4\pi/9)\beta m^2 \rho$] or as an observer on a scale of l looking at molecules with cores of diameter d shrinking in the limit $d\gamma \to 0$. (For the latter observer, the number of particles per unit volume becomes infinite, but the m^2 associated with each one becomes infinitesimal; i.e., the particles have become elements of a dielectric "continuum.")

We saw in Ref. 4 that when $w_\Lambda(r)$ is set equal to zero in the pair potential given by (2.44), we recover in the continuum limit the Clausius-Mossotti equation for ε. We can relate this directly to our discussion here by noting that as $\gamma \to 0$, $c(12)$ assumes the form given by (2.62) and (2.71) because

(2.68) becomes exact. The choice

$$w_\Delta(x) = 0 \qquad \text{for all} \quad x$$
$$w_D(x) \qquad \text{given by} \quad (2.47) \tag{2.72}$$

defines one continuum model in the limit $\gamma \to 0$, which we shall call the *Clausius-Mossotti continuum*. From (2.45) one sees that (2.72) implies $\Theta = 0$ for all γ, and in the limit $\gamma \to 0$ in which (2.71) holds, the resulting ε is seen to be the Clausius-Mossotti result from the equations of Section II.B [e.g., from (2.56b)]. For $\gamma = d^{-1}$, (2.72) with (2.71) gives us our two-particle Clausius-Mossotti approximation, which completely defines h_Δ and h_D via (2.27).

We now define an *Onsager continuum* by again choosing

$$\left. \begin{array}{l} w_\Delta(x) = 0 \\ w_D(x) = -m^2 x^{-3} \end{array} \right\} \text{ for } \quad x > 1 \tag{2.73a}$$

but we take

$$\left. \begin{array}{l} w_\Delta(x) = m^2 \Theta \\ w_D(x) = 0 \end{array} \right\} \text{ for } \quad x < 1 \tag{2.73b}$$

$$\Theta - (1 - \Theta)(2 + \Theta)y = 0$$

Equation 2.27, with (2.62), (2.71), and (2.73), determines $h_\Delta(x)$ and $h_D(x)$ for all x associated with this choice in the $\gamma \to 0$ limit, in which (2.71) becomes exact.

To get more insight into the properties of the Onsager continuum, we consider first (2.30c), recalling that to recover Onsager's result (2.30d) one has to put

$$\tilde{h}_\Delta(0) = 0$$

We now turn to our expression (2.54) for the limiting case $\gamma \to 0$ and conclude that for the Δ term of (2.54) to be zero in this limit, we have the additional relation, which we have incorporated into (2.73).

$$\Theta - (1 - \Theta)(2 + \Theta)y = 0 \tag{2.74}$$

Accordingly we get Onsager's result (2.30d) in the limit $\gamma \to 0$ if Θ is chosen such that (2.74) is fulfilled. This can also be shown directly by noting that (2.74) may be written $1 = (1 - \Theta)[1 + (2 + \Theta)y]$ or $2 = (2 + \Theta)[1 - (1 - \Theta)y]$.

So by use of (2.56b)

$$\varepsilon = \frac{2+\Theta}{2(1-\Theta)} \qquad \text{or} \qquad \Theta = \frac{2(\varepsilon-1)}{2\varepsilon+1} \tag{2.75}$$

When expression (2.75) for Θ is put into (2.74) the result (2.30d) again is recovered.

We note that the two-particle Onsager approximation defined in Section II.C is recovered from (2.62), (2.71), and (2.73) on setting $\gamma d = 1$ rather than letting $\gamma d \to 0$.

Next we investigate the physical interpretation of the Onsager continuum. We start by considering the discussion of the field set up by test dipoles in a dielectric medium given in Section II.B. We compute the polarization of the medium for $r < R$ [with $\mathbf{K}$ given by (2.58a)]. From (2.60) and (2.58) we find ($r < R$)

$$\mathbf{F} = \frac{bR^3}{\Theta m_2} \mathbf{K} = -\frac{(2+\Theta)(\varepsilon-1)}{3\Theta\varepsilon} \mathbf{K} \tag{2.76}$$

$$\mathbf{F} + \mathbf{K} = \frac{-2(1-\Theta)\varepsilon + 2 + \Theta}{3\Theta\varepsilon} \mathbf{K} \tag{2.77}$$

Use of (2.75) then gives

$$\mathbf{F} + \mathbf{K} = 0 \qquad \text{for} \quad r < R \tag{2.78}$$

and accordingly from (10) in Ref. 4, the polarization of the medium for $r < R$ is zero; that is, the medium for $r < R$ has no effect on the field set up by a dipole with Θ chosen such that (2.74) is fulfilled. Therefore for $r < R$ the medium acts as a vacuum. And this is just the physical assumption under which Onsager's result is derived.[19] Onsager considered a fluid consisting of hard spheres with dipoles embedded at the center. He then assumed that the hard spheres could be treated as a vacuum and the surrounding medium as a continuum with the sought dielectric constant ε. One finds by direct calculation that the electric field set up outside such a sphere is reduced by a factor $3/(2\varepsilon+1)$ compared to a surrounding vacuum. One sees that our result also agrees with this if (2.75) for Θ is put into (2.61).

Naturally Onsager's derivation cannot be exact for a molecular fluid, since on a microscopic scale the surrounding medium cannot be treated as a continuum. On the other hand we have recovered the same expression as an exact result in the limit $\gamma \to 0$. This is done by choosing a dipole interaction as given by (2.44), where Θ fulfills (2.74). As we have just shown, this dipole interaction has the same effect as a hard sphere of diameter $R = 1/\gamma$ with respect to the electric field set up in the surrounding medium; but in contrast to the hard sphere, this interaction does *not* remove the medium for $r < R$. Therefore we can increase $R(\rightarrow \infty)$ without lowering the particle density. Thus we get a continuum limit ($\gamma \rightarrow 0$) in which Onsager's result is recovered as an exact result. [Note that from (2.74) Θ will vary with y and accordingly with temperature and density. So Θ cannot be kept fixed to yield Onsager's result.]

We come finally to the *Wertheim continuum*, defined by (2.73a) along with the choice

$$w_\Delta(x) \quad \text{and} \quad w_D(x) \quad \text{for} \quad x < 1$$
$$\text{such that} \quad h_\Delta(x) = h_D(x) = 0 \quad \text{for} \quad x < 1 \tag{2.79a}$$

It follows from Wertheim's own work,[12] and our extension of it given in Ref. 13, that (2.45) yields

$$\Theta = \frac{-3 + q(2\xi) + 2q(-\xi)}{3y} \quad \text{with} \quad y = \frac{1}{3}\left[q(2\xi) - q(-\xi)\right]$$

$$\tag{2.79b}$$

and from (2.56b) ε is given by the Wertheim expression

$$\varepsilon = \frac{q(2\xi)}{q(-\xi)} \tag{2.79c}$$

where

$$q(\xi) = \frac{(1 + 2\xi)^2}{(1 - \xi)^4} \tag{2.79d}$$

and (letting $l = \gamma^{-1}$)

$$\xi = \left(\frac{\pi}{6}\right) K \rho l^3 \tag{2.79e}$$

Here

$$3K = \int_l^\infty h_D(r) r^{-3} dr \tag{2.79f}$$

For any γd one recovers the same ε and Θ. For $\gamma d = 1$, one recovers the Wertheim approximation rather than the Wertheim continuum. [When we speak of Wertheim's approximation, we mean a suitably generalized form in which the reference potential need not be a hard-sphere potential and c_s need not be given by (2.66a).] In the Wertheim continuum, one finds $g(12)$ $= 1$ for all r in a macrosphere (i.e., for all $r < l$) because of (2.65), since $h_s(r) \to 0$ for all $r < l$ (neglecting $r \sim d \ll l$). Thus in the Wertheim continuum there is no correlation between elements of the continuum that are within a macrodiameter $l = \gamma^{-1}$ from one another.

Ramshaw[28] has pointed out that in the $\gamma \to 0$ limit, one can regard the core parameter Θ as a geometric parameter describing the degree of ellipticity of a spheroidal cavity used in defining a local field in a phenomenological mean field treatment. Such a treatment, described by Ramshaw, is the analog of the Debye-Hückel treatment of a simple ionic fluid.

E. Higher Order Approximations Based on γ Ordering

We have shown in the preceding section how certain results that become exact as $\gamma \to 0$ define useful approximations for $\gamma = d^{-1}$ as well. In this section we derive some further approximations based on the use of γ ordering that are even more accurate for $\gamma = d^{-1}$. One of these turns out to be identical to the single superchain (SSC) approximation first suggested by Wertheim[15, 29] and to the "reference" version of the linearized hypernetted-chain (LHNC) approximation developed by Patey.[30] We shall also consider several closely related approximations, including the Verlet-Weis[31] "linear" (LIN) and the Høye-Stell analytic direct correlation (ADC) approximations,[32] which have the advantage of being expressible in simple analytic form.

It will be useful to define the "anisotropic piece" of a function $a(12)$ by subtracting its orientational average over Ω_1 and Ω_2 from it. We have been writing the latter simply as $a_S(r)$ or a_S, but we shall have to deal with functions bearing a variety of other subscripts and superscripts in the latter part of this section. To accommodate these, we shall sometimes use bracket notation,

$$[a(12)]_S \equiv \int \frac{a(12)\, d\Omega_1\, d\Omega_2}{\Omega^2} \qquad (2.80)$$

and

$$[a(12)]_A = a(12) - [a(12)]_S \qquad (2.81)$$

although we use the simpler a_S and a_A wherever it seems more appropriate. The average over a single orientation will also be important here, and we shall use the notation

$$\langle a(12) \rangle_2 \equiv \int \frac{a(12)\, d\Omega_2}{\Omega} \qquad (2.82)$$

We note that if $a(12)$ is such that

$$\langle a(12)\rangle_2 = [a(12)]_S \tag{2.83}$$

then

$$\langle [a(12)]_A\rangle_2 = 0 \tag{2.84}$$

If the pair potential $\phi(12)$ is the sum of a spherically symmetric part $q(r)$ plus a linear combination of ideal multipole (IM) terms,

$$\phi(12) = q(r) + \sum_i A_i \phi_i^{IM}(12) \tag{2.85}$$

then for $a(12) = \phi(12)$, (2.83) and (2.84) hold with

$$[\phi(12)]_S = q(r) \tag{2.86}$$

$$[\phi(12)]_A = \sum_i A_i \phi_i^{IM}(12) \tag{2.87}$$

For clarity of presentation we restrict our attention here to potentials of the form (2.85). As discussed earlier in connection with (2.44), if a potential is given as a sum of a reference term $q(12)$ plus a perturbing term $w(12)$, where $q(12)$ has a hard core of diameter d, then $[\phi(12)]_A$ is only well defined for $r > d$. This means that our perturbing potential $w(12)$ need only be identified with the $\sum_i A_i \phi_i^{IM}$ of (2.85) for $r > d$. In the following discussion we always assume that $w(12)$ is chosen (for $r < d$ as well as $r > d$) such that

$$\langle w(12)\rangle_2 = 0 \tag{2.88}$$

which holds in particular if $w(12)$ is simply taken to be $\sum_i A_i \phi_i^{IM}$ for all r. The $w(12)$ is further assumed to be embedded in a γ-parameterized family of potentials $w(12; \gamma)$ with $w(12; d^{-1}) = w(12)$, where d is a characteristic particle size and $w(12; \gamma)$ is of the form

$$w(12; \gamma) = \sum_i \gamma^3 w_i(\gamma r)\phi_i(\Omega_1, \Omega_2) \tag{2.89}$$

with $w_i(\gamma r)$ orientation independent and ϕ_i dependent only on orientation.

In the Lebowitz, Stell, and Baer scheme[6] the decomposition of $\phi(12)$ into its short-ranged (SR) part, which we here identify with $[\phi(12)]_S$, and long-ranged (LR) part, which we identify as $[\phi(12)]_A$, induces a corresponding decomposition of the functions $F_\delta(12)$, $h(12)$, and $c(12)$:

$$F_\delta(12) = F_\delta^{SR}(12) + F_\delta^{LR}(12) \tag{2.90}$$

$$h(12) = h^{SR}(12) + h^{LR}(12) \tag{2.91}$$

$$c(12) = c^{SR}(12) + c^{LR}(12) \tag{2.92}$$

Here $F_\delta(12) = (\rho/\Omega)^2 h(12) + (\rho/\Omega)\delta(12)$, $F_\delta^{SR}(12) = (\rho/\Omega)^2 h^{SR}(12) + (\rho/\Omega)\delta(12)$, and $F_\delta^{LR}(12) = (\rho/\Omega)^2 h^{LR}(12)$. In the limit $\gamma \to 0$,

$$h^{SR}(12) = h(12)_0 \tag{2.93}$$

$$c^{SR}(12) = c(12)_0 \tag{2.94}$$

where the subscript "0" refers to a reference system quantity describing the system at a given temperature and density when we have set $w(12) = 0$ or, equivalently, when we have taken the limit $\gamma \to 0$. In this limit c^{LR} approaches the function

$$c^{LR}(12) = v(12) \tag{2.95}$$

where

$$v(12) = -\beta w(12; \gamma) \tag{2.96}$$

and $(\rho/\Omega)^2 h^{LR}(12)$ approaches a sum of all distinct terms that can be thought of as chains of alternating $F_{\delta,0}$ and v functions, with $F_{\delta,0}$ at each end:

$$\lim_{\gamma \to 0} F_\delta^{LR}(12) = \lim_{\gamma \to 0} \left(\frac{\rho}{\Omega}\right)^2 h^{LR}(12)$$

$$= \int F_\delta(13)_0 v(34) F_\delta(42)_0 \, d(3) \, d(4)$$

$$+ \int F_\delta(13)_0 v(34) F_\delta(45)_0 v(56) F_\delta(62)_0 \, d(3) \, d(4) \, d(5) \, d(6)$$

$$+ \cdots \tag{2.97}$$

where $F_\delta(12)_0 = (\rho/\Omega)^2 h(12)_0 + (\rho/\Omega)\delta(12)$. When (2.88) is fulfilled, the $F_\delta(ij)_0$ on the right-hand side of (2.97) may be replaced by $(\rho/\Omega)\delta(ij)$, since the integrations over $h(ij)_0$ yield zero, to give

$$\lim_{\gamma \to 0} \left(\frac{\rho}{\Omega}\right)^2 h^{LR}(12) = \left(\frac{\rho}{\Omega}\right)^2 v(12) + \left(\frac{\rho}{\Omega}\right)^3 \int v(13) v(42) \, d(3) \, d(4) + \cdots$$

We denote the chain sum of (2.97) as $(\rho/\Omega)^2 \mathcal{C}(12; F_{\delta,0})$ so that

$$\lim_{\gamma \to 0} h^{LR}(12) = \mathcal{C}(12; F_{\delta,0}) \tag{2.98}$$

When $q(r)$ is a hard-sphere potential of diameter d, use of the OZ equation (2.27) plus (2.94) and (2.95), with $v(12)$ adjusted inside the core region $r < d$ to give $g(12) = 0$ for $r < d$, defines an approximation that we call the lowest-order γ-ordered approximation (LOGA). The further approximation that $c_0(12)$ is equal to zero for $r > d$ then yields the MSA. As one of us has discussed in detail elsewhere, the LOGA also results from using the same v in (2.97) along with (2.93) and (2.90).[33-35]

For the simple potential given by (2.85), one can identify at this level of approximation the long-range parts of the correlations with their anisotropic parts and the short-range parts with their symmetric parts, and further identify the latter with pure reference system functions. Thus we have

$$c^{LR} = c_A$$

$$h^{LR} = h_A \tag{2.99}$$

$$c^{SR} = c_S = c_0$$

$$h^{SR} = h_S = h_0 \tag{2.100}$$

at this level of approximation.

We note that in the $\gamma \to 0$ limit c^{LR} and h^{LR} are approaching functions that are of first order in γ^3 (for fixed γr) while c^{SR} and h^{SR} are approaching functions of zeroth order in γ^3 (for fixed r). Thus for arbitrary γ, (2.95) and (2.98) define approximations that are exact through first order in γ^3 for fixed γr, whereas (2.93) and (2.94) are only exact through zeroth order for fixed r. To next-higher order we find, for potentials that satisfy (2.88),

$$c^{SR} = c_0 + h_0 h^{LR} \tag{2.101a}$$

and

$$h^{SR} = h_0 + h_0 h^{LR} \tag{2.101b}$$

These results are exact through $O(\gamma^3)$ for fixed r. Thus, if we use (2.101) instead of (2.100) with (2.95) and (2.98), we might anticipate an improved approximation. Let us investigate this. Adding c^{LR} to both sides of (2.101a) and h^{LR} to both sides of (2.101b), we find [using (2.95)]

$$c = v + c_0 + h_0 h^{LR} \tag{2.102a}$$

and

$$h = h^{LR} + h_0 + h_0 h^{LR}$$
$$= h_0 + g_0 h^{LR} \tag{2.102b}$$

If we stop here and use the LOGA expression for $h^{LR}(12)$ in (2.102b), the result coincides with the LIN approximation, first suggested by Verlet and Weis on somewhat different grounds.[31] It is a useful approximation, to which we shall return. If we instead use the LOGA expression for $h^{LR}(12)$ in (2.102a), we have a closely related approximation, used by Høye and Stell,[32] which we call the analytic direct correlation (ADC) approximation. Since the LOGA h^{LR} can be thought of as being associated with the particular choice of v in (2.97) being given by

$$v = c^{LOGA} - c_0 \tag{2.103a}$$

the ADC approximation can be written as

$$c = c^{LOGA} + h_0 (h^{LR})^{LOGA} \tag{2.103b}$$

[It is worth pointing out that if one puts the ADC result (2.103b) into the OZ equation (2.27), one will not recover the LIN h. The resulting h will have an LR part coinciding with the LIN h^{LR} through $O(\gamma^3)$ for fixed γr an SR part coinciding with the LIN h^{SR} through $O(\gamma^3)$ for fixed r, but it will have higher order terms in γ^3 that do not coincide with those found in the LIN h.]

Instead of stopping with the LIN and ADC results, we shall exploit (2.102) by eliminating h^{LR} from our equations to get a relation among c_A, h_A, v, and h_0. To do this we begin by noting from (2.95) that

$$[c^{LR}]_S = 0 \qquad \text{or} \qquad [c^{LR}]_A = c^{LR} \tag{2.104a}$$

It follows similarly from (2.97) that for potentials that satisfy (2.88),

$$[h^{LR}]_S = 0 \qquad \text{or} \qquad [h^{LR}]_A = h^{LR}$$
$$[h_0 h^{LR}]_S = 0 \qquad \text{or} \qquad [h_0 h^{LR}]_A = h_0 h^{LR} \tag{2.104b}$$

In fact for such potentials, we have the stronger results

$$\langle c^{LR} \rangle_2 = 0 \tag{2.105a}$$
$$\langle h^{LR} \rangle_2 = 0 \tag{2.105b}$$

for the expressions satisfying (2.95) and (2.97). Operating on (2.102) with $[\quad]_S$ and using (2.104), we find immediately

$$c_S = c_0 \tag{2.106a}$$

$$h_S = h_0 \tag{2.106b}$$

For c_A we find from (2.102a) [using either (2.104b) or (2.106a)] that

$$c_A = v + h_0 h^{\mathrm{LR}} \tag{2.107a}$$

and for h_A we have the corresponding result from (2.102b) [and either (2.104b) or (2.106b)]

$$h_A = h^{\mathrm{LR}} + h_0 h^{\mathrm{LR}}$$

$$h_A = g_0 h^{\mathrm{LR}} \tag{2.107b}$$

If one uses the LOGA h^{LR} and v in (2.107a) and (2.107b), one has c_A in the ADC approximation and h_A in the LIN approximation, respectively. Instead we can subtract (2.107a) from (2.107b) to get

$$h^{\mathrm{LR}} = v + h_A - c_A \tag{2.108}$$

Reinserting this into (2.107a) gives

$$c_A = v + h_0 [v + h_A - c_A] \tag{2.109a}$$

which can be rewritten as

$$c_A = g_0 v + h_0 [h_A - c_A] \tag{2.109b}$$

or

$$c_A = v + \left(\frac{h_0}{g_0} \right) h_A \tag{2.109c}$$

or

$$c_A = v + (1 - g_0^{-1}) h_A \tag{2.109d}$$

Along with (2.106) and the OZ equation (2.27), (2.109) yields a closed set of equations for c_A and h_A. Equation 2.109 is identical to Wertheim's SSC approximation for the Hamiltonian under consideration and is also identical to the "reference" form of Patey's LHNC approximation. [Strictly speaking, the LHNC requires that one use the hypernetted chain (HNC) approximation for h_0 in (2.109), but Patey instead uses the best available h_0 values.] The SSC approximation has also been rederived on the basis of a graphical resummation procedure by Henderson and Gray,[36] who call it the generalized mean field (GMF) approximation.

We note that to our level of approximation, (2.109) can just as well be written, from (2.106), as

$$c_A = v + h_S [v + h_A - c_A] \tag{2.110}$$

From (2.107a), (2.97), and (2.88) we find that

$$\langle c_A \rangle_2 = 0 \tag{2.111}$$

For $c(12)$ satisfying (2.111), we can decouple the OZ equation (2.27) into two separate OZ equations

$$h_S(r_{12}) = c_S(r_{12}) + \rho \int d\mathbf{r}_3 h_S(r_{13}) c_S(r_{32}) \tag{2.112a}$$

$$h_A(12) = c_A(12) + \rho \int \frac{d\mathbf{r}_3 d\Omega_3 h_A(13) c_A(23)}{\Omega} \tag{2.112b}$$

The approximation $h_S = h_0$ is known to lack high precision for some systems, such as liquids of strongly polar particles, but (2.110) with (2.112b) may prove useful in such cases if the h_S used in (2.110) has been accurately assessed by some independent method. Similarly, (2.107a) and (2.107b) used with the LOGA h^{LR} may well represent better approximations to c_A and h_A than (2.102a) and (2.102b) represent for the full c and h. We shall call (2.110) with (2.112b) the renormalized simple superchain (RSSC) approximation.

There is a powerful argument that dictates the use of (2.110) rather than (2.109)—and the replacement of $h_0 h^{LR}$ by $h_S h^{LR}$ in (2.101) through (2.103)—when h_S is appreciably different from h_0 (and incidentally provides an example of one important effect of their difference). The isothermal compressibility depends on h only through h_S and is wholly insensitive to h_A, since

$$\beta \frac{\partial P}{\partial \rho} = \left[1 + \int h_S(r) \, d\mathbf{r} \right]^{-1} = 1 - \int c_S(r) \, d\mathbf{r}$$

Thus for dipolar spheres the very existence of a critical point hinges on the difference between h_0 and h_S (or, equivalently, c_0 and c_S), whereas for reference systems that already have a critical point (such as a Lennard-Jones system), the shift in critical parameters as one adds $w(12)$ is entirely due to the difference between h_0 and h_S (or, equivalently, c_0 and c_S). Now for the latter systems, if (2.109) were used in (2.15) to determine ε, then $\partial\varepsilon/\partial T$ has a specific heat anomaly at the *reference system* critical point rather than its true critical point, but if (2.110) is used, the resulting $\partial\varepsilon/\partial T$ has a specific heat anomaly at the true critical point, as expected on the grounds of thermodynamic scaling theories. The superiority of (2.110) over (2.109) when h_0 and h_S differ appreciably suggests that in such cases, (2.101) through (2.103) can also be improved as approximations through replacement of $h_0 h^{LR}$ by $h_S h^{LR}$, since (2.110) rather than (2.109) follows from our derivation without the use of (2.106) if and only if this replacement is made. The replacement of the remaining c_0 and h_0 in (2.101) by c_S and h_S, respectively, on the other hand, does nothing to the form of (2.110), which is in fact insensitive to the replacement of these quantities by any functions that depend on the arguments 1 and 2 only through r_{12}.

We can improve our approximations still further by going one order higher in our γ-ordered result for c^{LR} and h^{LR}. But to do the bookkeeping involved, we must introduce a bit more notation than has so far been necessary. We shall use subscript 1 to refer to first-order results for LR functions in γ^3 (for fixed γr) and first-order results for SR functions in γ^3 (for fixed r). For the corresponding second-order results we shall use a subscript 2. In this notation, for c^{LR} exact through $O(\gamma^6)$ for fixed γr we have [when condition (2.88) holds]

$$c_{(2)}^{LR} = v + \tfrac{1}{2}\left[h_{(1)}^{LR} \right]^2 \tag{2.113}$$

where $h_{(1)}^{LR}$ is given by (2.97), that is,

$$h_{(1)}^{LR} = \mathcal{C}(12; F_{\delta,0}) \tag{2.114}$$

For h^{LR} exact through $O(\gamma^6)$ for fixed γr, we have

$$h_{(2)}^{LR}(12) = \mathcal{C}\left(12; F_{\delta,(1)}^{SR}\right) + \tfrac{1}{2} h_{\delta 0} * \left(h_{(1)}^{LR} \right)^2 * h_{\delta 0} \tag{2.115}$$

where

$$h_{\delta 0} * \left(h_{(1)}^{LR} \right)^2 * h_{\delta 0} = \Omega^{-2} \int h_\delta(13)_0 \left[h_{(1)}^{LR}(34) \right]^2 h_\delta(42)_0 \, d(3) \, d(4)$$

Here $\mathcal{C}(12; F_{\delta,(1)}^{SR})$ denotes the same chain sum (2.97) that defines $\mathcal{C}(12; F_{\delta,0})$ except with $F_{\delta,0}(ij)$ everywhere replaced by $F_{\delta,(1)}^{SR}(ij) = (\rho/\Omega)^2 h^{SR}(ij)_{(1)} +$

$(\rho/\Omega)\delta(ij)$, with $h^{SR}(ij)_{(1)}$ given by

$$h^{SR}_{(1)} = h_0 + h_0 h^{LR}_{(1)} \tag{2.116}$$

The $c^{SR}_{(1)}$ corresponding to (2.116) is

$$c^{SR}_{(1)} = c_0 + h_0 h^{LR}_{(1)} \tag{2.117}$$

The c given by

$$c = c^{SR}_{(1)} + c^{LR}_{(2)} \tag{2.118}$$

is thus, from (2.113) and (2.117), equal to

$$c = v + \tfrac{1}{2}\left(h^{LR}_{(1)} \right)^2 + c_0 + h_0 h^{LR}_{(1)} \tag{2.119}$$

To find the corresponding c_S we operate on (2.119) with $[\quad]_S$, using

$$\left[h_0 h^{LR}_{(1)} \right]_S = 0 \tag{2.120}$$

to get

$$c_S = c_0 + \tfrac{1}{2}\left[\left(h^{LR}_{(1)} \right)^2 \right]_S \tag{2.121}$$

The corresponding c_A is given by

$$c_A = v + \tfrac{1}{2}\left[\left(h^{LR}_{(1)} \right)^2 \right]_A + \left[h_0 h^{LR}_{(1)} \right]_A \tag{2.122}$$

but since

$$\left[h_0 h^{LR}_{(1)} \right]_A = h_0 h^{LR}_{(1)} \tag{2.123}$$

we have

$$c_A = v + \tfrac{1}{2}\left[\left(h^{LR}_{(1)} \right)^2 \right]_A + h_0 h^{LR}_{(1)} \tag{2.124}$$

The LOGA expressions for $h^{LR}_{(1)}$ and v can be used in (2.119) to (2.124). In (2.119) they yield

$$c = c_{LOGA} + \tfrac{1}{2}\left(h^{LR}_{LOGA} \right)^2 + h_0 h^{LR}_{LOGA} \tag{2.125}$$

For h given by

$$h = h_{(1)}^{SR} + h_{(2)}^{LR} \tag{2.126}$$

we have, from (2.116),

$$h = h_{(2)}^{LR} + h_0 + h_0 h_{(1)}^{LR} \tag{2.127}$$

The corresponding h_S is given by

$$h_S = h_0 + \tfrac{1}{2}\left[h_{\delta,0} * \left(h_{(1)}^{LR} \right)^2 * h_{\delta,0} \right]_S \tag{2.128}$$

because

$$\left[\mathcal{C}\left(12; F_{\delta,(1)}^{SR} \right) \right]_S = 0 \tag{2.129}$$

and the corresponding h_A is thus given by

$$h_A = \mathcal{C}\left(12; F_{\delta,(1)}^{SR} \right) + \tfrac{1}{2}\left[h_{\delta,0} * \left(h_{(1)}^{LR} \right)^2 * h_{\delta,0} \right]_A + h_0 h_{(1)}^{LR} \tag{2.130}$$

These expressions for h, h_S, and h_A are not as readily evaluated with the aid of the LOGA results as the corresponding expressions involving c. However, we can take a somewhat different tack and derive a higher order approximation analogous to (2.110) that does not require explicit use of $h_{(2)}^{LR}$ given by (2.115), just as (2.110) did not require use of (2.97). We shall drop the subscripts 1 and 2; the method will treat h^{LR} in second order wherever it appears linearly and in first order wherever it appears quadratically. We use our first-order results (2.101) to relate c^{SR} and h^{SR} to h^{LR}.

To (2.101) we add the second-order result

$$c^{LR} = v + \tfrac{1}{2}(h^{LR})^2 \tag{2.131}$$

The first-order result that we shall use in assessing h^{LR} comes from (2.107b). It is

$$h^{LR} = \frac{h_A}{g_0} \tag{2.132}$$

Equations 2.101, 2.131, and 2.132, along with the general relation

$$c = c^{SR} + c^{LR}$$

give

$$c = c_0 + v + \frac{1}{2} \left(\frac{h_A}{g_0} \right)^2 + \left(\frac{h_0}{g_0} \right) h_A \qquad (2.133)$$

Here we note, as we did in connection with (2.109) and (2.110), that to our level of approximation (2.133) can just as well be written

$$c = c_0 + v + \frac{1}{2} \left(\frac{h_A}{g_S} \right)^2 + \left(\frac{h_S}{g_S} \right) h_A \qquad (2.134)$$

This yields, on application of $[\quad]_A$ and $[\quad]_S$,

$$c_A = v + \frac{\frac{1}{2} \left[h_A^2 \right]_A}{g_S^2} + \left(\frac{h_S}{g_S} \right) h_A \qquad (2.135a)$$

$$c_S = c_0 + \frac{\frac{1}{2} \left[h_A^2 \right]_S}{g_S^2} \qquad (2.135b)$$

When all components in the rotational-invariant expansions of h_A and h_A^2 are neglected except those already in that of the MSA h_A, (2.135) coincides with the relation for c_A given in the QHNC (quadratic hypernetted chain) approximation[38] developed by Patey. The relation for c_S in the QHNC is a bit different from our equation (2.135b). It reduces to the HNC result rather than the exact result when the anisotropic perturbing term $w(12)$ in the potential goes to zero, just as the LHNC result for c_S does. Because of appreciable inaccuracy of the HNC reference system result, Patey actually works with an improved "reference version" of the QHNC, in which c_S is given by an expression that correctly reduces to the exact c_0 when $w(12)=0$; it is (in our notation)

$$c_S = c_0 + (h_S - h_0) - \ln \left(\frac{g_S}{g_0} \right) + \frac{B(r)}{g_S^2} \qquad (2.136)$$

where $B(r)$ represents the contribution to $\frac{1}{2}[h_A^2]_S$ that comes from retaining only the terms in the rotational-invariant expansion of h_A that already are in the MSA h_A.

F. Other Extensions of the Lowest-Order γ-Ordered Results

In the preceding section a pair of approximations were given—the LIN and ADC approximations—that can be expressed directly in terms of the

LOGA or MSA results for $c(12)$ and $h(12)$. In this section we discuss several other approximation schemes that also directly build on the MSA expressions, but involve features that go beyond the use of γ ordering.

Before proceeding further it is worthwhile pointing out the precise difference in form between the LOGA and MSA results. As noted in Section II.E, the difference arises in the assumption concerning the hard-sphere reference system direct correlation function $c_0(r)$ outside the diameter d. The LOGA is defined by (2.27) with

$$h(12) = -1 \quad \text{for} \quad r < d \tag{2.137}$$

$$c(12) = c_0(r) + v(12) \quad \text{for} \quad r > d \tag{2.138}$$

while in the MSA, the $c_0(r)$ is itself taken to be zero for $r > d$ so that (2.138) is replaced by the simpler

$$c(12) = v(12) \quad \text{for} \quad r > d \tag{2.139}$$

The LOGA is identical to the optimized random phase approximation (ORPA) of Andersen and Chandler.[39] The "optimized" v they consider is defined by using (2.138) for $r < d$ as well as $r > d$.

Following the notation of (2.98), we can write the LOGA $h(12)$ as the sum of two terms

$$h^{\text{LOGA}}(12) = h_0(r) + \mathcal{C}(12) \tag{2.140}$$

For perturbing potentials satisfying (2.88), the MSA $h(12)$ is given in terms of the identical function $\mathcal{C}(12)$, and differs only in that $\mathcal{C}(12)$ is added to the MSA approximation to $h_0(r)$ rather than the exact $h_0(r)$:

$$h^{\text{MSA}}(12) = h_0^{\text{MSA}}(r) + \mathcal{C}(12) \tag{2.141}$$

The $h_0^{\text{MSA}}(r)$ [which is identical to the Percus-Yevick $h_0(r)$] is a sufficiently good approximation for the difference in structure predicted by (2.140) and (2.141) in the fluid region to be negligible at all but the highest liquid densities. Moreover, ε computed from $h(12)$ by means of any one of the several paths we have discussed in earlier sections is totally insensitive to differences in $h_0(r)$; that is, the $\varepsilon^{\text{LOGA}}$ and the ε^{MSA} are identical.

As usually formulated, both the LOGA and the MSA are defined only for systems with hard-sphere reference potentials. They can be immediately extended, however, by using one of several perturbation[40, 41] or variational schemes[42] available to relate soft-core reference potentials $q(r)$ to hard-core potentials with state-dependent core diameters. [Of these the prescription by Weeks, Chandler, and Andersen (WCA)[41] is the one that appears to have the greatest overall utility.] Alternatively, one can simply replace (2.137) by

$$h(12) = h_0(r) \qquad \text{for} \quad r < d \tag{2.142}$$

where the d in (2.142), (2.138), and (2.139) is to be regarded as small-r cutoff distance below which the perturbing $v(12)$ is given by $c(12) - c_0(r)$ rather than the actual $-\beta w(12)$. It turns out that LOGA–MSA approximation for ε is strictly independent of this d. The LOGA and MSA $h(12)$ do depend on d, which can be chosen, for example, according to the WCA prescription[41] for an effective hard-sphere diameter.

One way of developing approximations that go beyond LOGA or MSA is to make use of a cluster expansion of $h(12)$ in terms of ρ, $h_0(r)$, and $\mathcal{C}(12)$ that was derived by one of us[34, 43] and independently by Andersen and Chandler.[44] The expansion can be systematically ordered (nodally ordered) according to the number of arguments appearing in the cluster integrals. Through terms with three arguments (nodal order 3), one has

$$\ln\left[\frac{g(12)}{g_0(r)}\right] = \mathcal{C}(12) + \rho \int d(3)\, \mathcal{S}(13)\mathcal{S}(23)$$

$$+ 2\rho \int d(3)\, \mathcal{S}(13)\mathcal{H}(23) \tag{2.143a}$$

where

$$\mathcal{S}(12) = g_0(r)\left[e^{\mathcal{C}(12)} - 1 - \mathcal{C}(12)\right] + \mathcal{C}(12)h_0(r) \tag{2.143b}$$

$$\mathcal{H}(12) = h_0(r) + \mathcal{C}(12) \tag{2.143c}$$

Verlet and Weis argued[31] that one could expect to develop a satisfactory approximation scheme by retaining in the rotationally-invariant expansion of $h(12)$ only the terms that appear on a certain minimal basis, which consists of the terms that enter the MSA. [For a $w(12)$ of ideal dipole form, these are just the S, Δ, and D terms of (2.62), defined by (2.20)]. The linearization of (2.143) that will ensure the retention of these terms and introduce no others is given by

$$g(12) = g_0(r)\left[1 + \mathcal{C}(12) + \rho \int d(3)\, \mathcal{S}(13)\mathcal{S}(23)\right.$$

$$\left. + 2\rho \int d(3)\, \mathcal{S}(13)\mathcal{H}(23)\right] \tag{2.144a}$$

with

$$\mathcal{S}(12) = \mathcal{C}(12)h_0(r) \qquad\qquad (2.144b)$$

$$\mathcal{H}(12) = h_0(r) + \mathcal{C}(12) \qquad\qquad (2.144c)$$

Retention of only the two-node terms yields the "linear" (LIN) approximation

$$g(12) = g_0(r)[1 + \mathcal{C}(12)] \qquad\qquad (2.145)$$

which Verlet and Weis[31] found to be significantly better for dipolar spheres than the full two-node "exponential" (EXP) approximation[44]

$$g(12) = g_0(r) \exp \mathcal{C}(12) \qquad\qquad (2.146)$$

Although the approximation corresponding to (2.144) on the free-energy level was found by Verlet and Weis to be quite accurate for dipolar spheres, (2.144) itself was found to exaggerate the structure of $h(12)$ at the second-nearest-neighbor distance,[45] as discussed in Section III.

A quite different way of improving the MSA result follows from the observation that the function $c(12) - v(12)$ is a short-ranged and relatively structureless function outside of the repulsive core region of interparticle interaction. Høye, Lebowitz, and Stell[13] noted that this suggests a generalized mean spherical approximation (GMSA) procedure, which exploits the fact that (2.27) can be solved analytically for a certain class of short-ranged $c(12)$. Høye and Stell have further considered this idea for dipolar spheres, developing a self-consistent Ornstein-Zernike approximation (SCOZA), a version of the GMSA in which thermodynamic self-consistency is incorporated into the determination of $c(12)$.[16] In their SCOZA for dipolar spheres one has (setting $d = 1$ without loss of generality)

$$h_S(r) = -1, \qquad h_\Delta(r) = h_D(r) = 0 \qquad \text{for} \quad r < 1 \qquad (2.147a)$$

For $r > 1$, one has

$$c_S(r) = \frac{Se^{-\kappa r}}{r} \qquad\qquad (2.147b)$$

$$c_\Delta(r) = \frac{A_1 e^{-\lambda_1 r}}{r} - \frac{A_2 e^{-\lambda_2 r}}{r} \qquad\qquad (2.147c)$$

$$c_D(r) = \frac{\beta m^2}{r^3} + 2A_1 b_1(r) + A_2 b_2(r) \qquad\qquad (2.147d)$$

with

$$b_i(r) = \lambda_i e^{-\lambda_i r} \left[(\lambda_i r)^{-1} + 3(\lambda_i r)^{-2} + 3(\lambda_i r)^{-3} \right]$$

where the six parameters S, κ, A_1, A_2, λ_1, and λ_2 are determined uniquely as follows. Prescribing a dielectric constant ε, contact values $h_\Delta(1+), h_D$ $(1+)$, and configurational internal energy per particle,

$$u = \frac{\rho}{2} \int \left[h(12) + 1 \right] v(12) \, d\mathbf{r}_{12} \frac{d\Omega_1 d\Omega_2}{\Omega^2}$$

determines A_1, A_2, λ_1, and λ_2. This in turn fully determines, via the OZ equation (2.27), $h_\Delta(r)$ and $h_D(r)$ for all r.

Further prescribing that both the inverse compressibility a_S,

$$a_S = 1 - \rho \int c_S(r) \, d\mathbf{r} \qquad (2.148)$$

and the pressure P via the virial theorem [which involves $h_S(1+)$] must yield thermodynamic results consistent with $u = u(\rho, \beta)$ further fixes S and κ.

This in turn fully determines $h_S(r)$ for all r. Strictly speaking, the SCOZA is silent in regard to the value of any component $h_Q Q(12)$ in $h(12)$ orthogonal to the I, Δ, or D [i.e., such that $\int Q(12)P(12) \, d\Omega_1 d\Omega_2 = 0$ for $P(12) = I$, $\Delta(12)$, or $D(12)$]. However, for convenience, it is simply assumed that all such components are zero until there is some good evidence that they cannot be reasonably neglected, and $h(12)$ is approximated by

$$h(12) = h_S(r) + h_\Delta(r)\Delta(12) + h_D(r)D(12) \qquad (2.149)$$

We turn now to a general diagrammatic representation that applies[46] to nearly all the approximations we have considered in Sections II.E and II.F. We write

$$c_A(12) = m(12) + f(r)\eta_A(12) \qquad (2.150)$$

where

$$m(12) = \left[f(r) + 1 \right] d(12) \qquad (2.151)$$

and

$$\eta_A(12) = h_A(12) - c_A(12) \qquad (2.152)$$

while $f(r)$ and $d(12)$ vary from approximation to approximation. For the MSA

$$f(r) = f_0(r), \qquad d(12) = v(12) \tag{2.153}$$

where $f_0(r)$ is the "Mayer f-function" of the reference potential—that is [exp $-\beta q(r)]-1$, where $q(r)$ is the hard-sphere reference potential. In the SSC –LHNC–GMF approximation,

$$f(r) = h_0(r), \qquad d(12) = v(12) \tag{2.154}$$

For the RSSC approximation of Section II.E, defined by (2.110) and (2.112b),

$$f(r) = h_S(r), \qquad d(12) = v(12) \tag{2.155}$$

For the approximation defined by (2.135a),

$$f(r) = h_S(r), \qquad d(12) = v(12) + \frac{\frac{1}{2}\left[h_A^2 \right]_A}{g_S^2} \tag{2.156}$$

and for the SCOZA,

$$f(r) = f_0(r), \qquad d(12) = v(12) + s(12) \tag{2.157}$$

where $s(12)$ is given by the sum of the right-hand-sides of (2.147c) and (2.147d).

The OZ equation for c_A and h_A can be written as

$$\eta_A(12) = \rho\left[c_A(12) * c_A(12) + c_A(12) * \eta_A(12) \right] \tag{2.158}$$

where the asterisk stands for convolution. Equations 2.150 and 2.158 can be combined to read[46]

$$\eta_A = \rho\left[m * m + m * f\eta_A + f\eta_A * m + m * \eta_A \right.$$
$$\left. + f\eta_A * \eta + f\eta_A * f\eta_A \right] \tag{2.159}$$

It immediately follows upon repeated iteration of (2.159) that we have the following diagrammatic descriptions:

h_A = a line representing $m(12)$ plus all distinct planar graphs such that each is obtainable by the following prescription. Take a convex polygon, and label a pair of contiguous vertices 1 and 2, respectively. Then delete or do not delete the line between 1 and 2 and add some or no nonintersecting straight lines between any vertices. Each line of the polygon is an m-bond except the undeleted line between 1 and 2, which is an f-bond, as are all added lines.   (2.160)

$c_A =$ the line representing $m(12)$ plus the sum of all graphs in (2.160) in which the line between 1 and 2 is retained   (2.161)

In the case of (2.154), these prescriptions are essentially the ones used by Henderson and Gray to characterize their GMF.[36]

III. POLAR–NONPOLARIZABLE FLUIDS: ALTERNATIVE FORMULATIONS, QUANTITATIVE RESULTS, AND COMPUTER SIMULATIONS

Section II was mainly concerned with the derivation and statement of formal results, with emphasis on exact expressions for the dielectric constant and the motivation for several approximate theories from a more or less unified point of view. Section III describes these and other theories in more detail, giving alternative derivations as well as numerical results for model systems. We also discuss the computer simulation of polar fluids and compare the various theoretical approximations. There have been a considerable number of computer studies of simple polar fluids[23, 31, 47–54] using both Monte Carlo (MC) and molecular dynamics (MD) techniques. However, such calculations are problematic and the computer simulation of dipolar systems has been the subject of recent debate.[55] The problems stem from the long-range nature of the dipolar interactions and the great difficulty of obtaining estimates of ε. We review some of this work and briefly discuss the problems involved.

Both the theoretical and computer results are presented in a logical rather than historical order, and we do not attempt an exhaustive review of all the data that have been compiled. Instead we attempt to evaluate as far as is possible the accuracy of the different approximations and focus on the physical insights that can be gained from current theory. The numerical examples presented are chosen from this perspective.

A. Models and Formal Relationships

We consider simple models defined by a pair potential of the form

$$u(12) = u^{000}(r) + u_{DD}(12) + u_{DQ}(12) + u_{QQ}(12) + \cdots \qquad (3.1)$$

where $u^{000}(r)$ is a spherically symmetric short-range potential and $u_{DD}(12)$, $u_{DQ}(12)$, and $u_{QQ}(12)$ represent the dipole-dipole (DD), dipole-quadrupole (DQ), and quadrupole-quadrupole (QQ) terms of the multipole expansion for electrostatic interactions. Restricting ourselves to linear or axially symmetric quadrupoles, the multipolar interactions can be conveniently written as

$$u_{DD}(12) = u^{112}(r)\Phi^{112}(12) \tag{3.2a}$$

$$u_{DQ}(12) = u^{123}(r)\Phi^{123}(12) + u^{213}(r)\Phi^{213}(12) \tag{3.2b}$$

and

$$u_{QQ}(12) = u^{224}(r)\Phi^{224}(12) \tag{3.2c}$$

where

$$u^{112}(r) = \frac{-\mu^2}{r^3} \tag{3.2d}$$

$$u^{123}(r) = -u^{213}(r) = \frac{\mu Q}{2r^4} \tag{3.2e}$$

$$u^{224}(r) = \frac{3Q^2}{4r^5} \tag{3.2f}$$

Here* μ and Q are the dipole and quadrupole moments, respectively, and r is the distance between particles 1 and 2. The rotational invariants $\Phi^{mnl}(12)$, exploited by Blum[56] in connection with OZ theory, can be written in the general form for axially symmetric molecules

$$\Phi^{mnl}(12) = \sum_{\mu\nu\lambda} f^{mnl} \begin{pmatrix} m & n & l \\ \mu & \nu & \lambda \end{pmatrix} D_{0\mu}^m(\Omega_1) D_{0\nu}^n(\Omega_2) D_{0\lambda}^l(\Omega_{12}) \tag{3.3a}$$

where Ω_1 and Ω_2 represent the orientational coordinates of particles 1 and 2, Ω_{12} describes the orientation of the vector $\mathbf{r}_{12} = \mathbf{r}_2 - \mathbf{r}_1$, and the Wigner matrix element $D_{mn}^l(\Omega)$ and the $3-j$ symbol are defined in Edmonds.[57] The f^{mnl} can be any arbitrary nonzero constant, and the functions we consider correspond to choosing[38] $f^{220} = -2\sqrt{5}$ and $f^{224} = 8\sqrt{35/2}$; all other

*Q is the zz component of the quadrupole moment tensor given by $Q = \frac{1}{2}\Sigma_i e_i(3z_i^2 - r_i^2)$, where z_i is the z component of the vector $\mathbf{r}_i$ describing the position of the charge e_i relative to the center of mass.

coefficients are given by

$$f^{mnl} = \frac{l!}{\begin{pmatrix} m & n & l \\ 0 & 0 & 0 \end{pmatrix}} \tag{3.3b}$$

In Appendix B the $\Phi^{mnl}(12)$ referred to in this chapter are written in a more explicit and familiar manner. We remark that in Section II only dipolar interactions were considered and a simpler notation was used. However, when higher multipole moments are included, a more systematic approach is necessary, and rotational invariants[56] are particularly convenient and theoretically useful. In the present notation the $D(12)$ and $\Delta(12)$ functions of Section II are $\Phi^{112}(12)$ and $\Phi^{110}(12)$, respectively.

The spherically symmetric short-range part of the potential $u^{000}(r)$ is often taken to be either the hard-sphere interaction

$$u_{\mathrm{HS}}(r) = \begin{matrix} \infty & \text{for} & r < d \\ 0 & \text{for} & r > d \end{matrix} \tag{3.4}$$

or the Lennard-Jones (LJ) potential

$$u_{\mathrm{LJ}}(r) = 4\varepsilon \left[\left(\frac{\sigma}{r} \right)^{12} - \left(\frac{\sigma}{r} \right)^{6} \right] \tag{3.5}$$

where d is the hard-sphere diameter and ε and σ are parameters characterizing the LJ potential. Two purely dipolar fluids ($Q=0$) have been extensively studied and will occupy a large part of our discussion. These are the dipolar hard-sphere and Stockmayer fluids that correspond to choosing $u^{000}(r) = u_{\mathrm{HS}}(r)$ and $u_{\mathrm{LJ}}(r)$, respectively. There have also been a number of calculations for fluids of hard spheres with dipoles and quadrupoles, but the influence of higher multipole moments on the dielectric constant has not yet been considered.

It is convenient[12, 30, 38, 56, 58, 59] to expand the pair correlation function $h(12)$ in a basis set of rotational invariants. One obtains

$$h(12) = \sum_{mnl} h^{mnl}(r) \Phi^{mnl}(12) \tag{3.6a}$$

where

$$h^{mnl}(r) = \frac{\int h(12) \Phi^{mnl}(12) \, d\Omega_1 d\Omega_2}{\int \left[\Phi^{mnl}(12) \right]^2 d\Omega_1 d\Omega_2} \tag{3.6b}$$

For purely dipolar systems this gives the terms of the Wertheim[12] expansion (2.21) used in Section II, with the projections $h_D(r)$ and $h_\Delta(r)$ becoming $h^{112}(r)$ and $h^{110}(r)$ in the present notation.

For an infinite system, the dielectric constant ε is given by the Kirkwood relationship[4, 5, 10, 17]

$$\frac{(\varepsilon-1)(2\varepsilon+1)}{9\varepsilon}=yg \tag{3.7a}$$

where

$$g=\frac{\langle M^2\rangle}{N\mu^2}=1+\frac{N-1}{\mu^2}\langle\boldsymbol{\mu}_1\cdot\boldsymbol{\mu}_2\rangle$$

$$=1+\frac{\rho}{3}\int h^{110}(r)\,d\mathbf{r} \tag{3.7b}$$

$y=4\pi\beta\mu^2\rho/9$ ($\beta=1/kT$), N is the number of particles and $\mathbf{M}$ the total dipole moment of the system. The dielectric constant can also be obtained through the limiting expression[4, 5, 10]

$$h^{112}(r)\rightarrow\frac{(\varepsilon-1)^2}{4\pi\varepsilon\rho yr^3}\qquad\text{as}\quad r\rightarrow\infty \tag{3.8}$$

These equations relating ε and $h(12)$ were derived in Section II and provide useful routes to ε. If $h(12)$ is exact, both (3.7) and (3.8) must yield the same dielectric constant. In fact, however, self-consistency holds under much weaker conditions,[30] which are satisfied by several approximate theories.

B. Integral Equation Theories

1. The Mean Spherical Approximation

The mean spherical approximation (MSA) is defined by three equations relating the pair correlation function $h(12)$ and the direct correlation function $c(12)$. These are[12, 56]

$$h(12)-c(12)=\frac{\rho}{4\pi}\int c(12)h(32)\,d(3) \tag{3.9a}$$

$$g(12)=0\qquad\text{for}\quad r<d \tag{3.9b}$$

and

$$c(12)=-\beta u(12)\qquad\text{for}\quad r>d \tag{3.9c}$$

where $g(12)=h(12)+1$, ρ is the number density, and $d(3)$ indicates that the integration is to be taken over the position and angular coordinates of particle 3. Equation 3.9a is the Ornstein-Zernike (OZ) relation, (3.9b) is an exact result for hard particles, simply stating that such particles cannot overlap, and (3.9c) is the closure relation characterizing the MSA. The MSA was first applied to polar fluids by Wertheim,[12] who obtained a completely analytical solution for dipolar hard spheres. Formal solutions to the MSA for arbitrary multipolar potentials have been obtained by Blum.[56] In general, however, it is very difficult to find analytic expressions for $h(12)$ and ε, and one must resort to numerical methods.[30, 38, 58, 59] Although the MSA is a rather inaccurate theory, many of the techniques pioneered by Wertheim[12] have proved to be very useful in the solution of more accurate approximations. We shall write in detail the formal MSA results for the general dipole-quadrupole system and indicate how the equations simplify when either Q or μ vanishes.

The exact solution of the MSA for the potential defined by (3.1) has the form[56, 59]

$$h(12) = h^{000}(r) + h^{110}(r)\Phi^{110}(12) + h^{112}(r)\Phi^{112}(12) + h^{121}(r)\Phi^{121}(12)$$
$$+ h^{123}(r)\Phi^{123}(12) + h^{211}(r)\Phi^{211}(12) + h^{213}(r)\Phi^{213}(12)$$
$$+ h^{220}(r)\Phi^{220}(12) + h^{222}(r)\Phi^{222}(12) + h^{224}(r)\Phi^{224}(12) \quad (3.10a)$$

where

$$h^{000}(r) = \frac{1}{(4\pi)^2} \int h(12)\, d\Omega_1\, d\Omega_2 \quad (3.10b)$$

$$h^{110}(r) = \frac{3}{(4\pi)^2} \int h(12)\Phi^{110}(12)\, d\Omega_1\, d\Omega_2 \quad (3.10c)$$

$$h^{112}(r) = \frac{3}{2(4\pi)^2} \int h(12)\Phi^{112}(12)\, d\Omega_1\, d\Omega_2 \quad (3.10d)$$

$$h^{121}(r) = \frac{6}{(4\pi)^2} \int h(12)\Phi^{121}(12)\, d\Omega_1\, d\Omega_2 \quad (3.10e)$$

$$h^{123}(r) = \frac{1}{4(4\pi)^2} \int h(12)\Phi^{123}(12)\, d\Omega_1\, d\Omega_2 \quad (3.10f)$$

$$h^{220}(r) = \frac{5}{4(4\pi)^2} \int h(12)\Phi^{220}(12)\, d\Omega_1\, d\Omega_2 \quad (3.10g)$$

$$h^{222}(r) = \frac{25}{14(4\pi)^2} \int h(12)\Phi^{222}(12)\, d\Omega_1\, d\Omega_2 \quad (3.10h)$$

$$h^{224}(r) = \frac{45}{224(4\pi)^2} \int h(12)\Phi^{224}(12)\, d\Omega_1\, d\Omega_2 \quad (3.10i)$$

and $d\Omega = \sin\theta\, d\theta\, d\phi$. Expressions for $h^{211}(r)$ and $h^{213}(r)$ are not given, since they are identical to (3.10e) and (3.10f), respectively. The direct correlation function $c(12)$ is given by an analogous set of equations, and for a single-component system one has the symmetry relations, $h^{123} = -h^{213}$, $h^{121} = -h^{211}$, $c^{121} = -c^{211}$, and $c^{123} = -c^{213}$. The rotational invariants included in (3.10a) form a closed set under the generalized convolution of the OZ equation (3.9a). This means that when the angular integrations are performed, this particular set of $\Phi^{mnl}(12)$ will generate only themselves and one another. The MSA closure (3.9c) contains only invariants belonging to this set, hence (3.10a) constitutes an exact solution. Fourier transforming (3.9a) and carrying out the angular integrations, it can be shown[56, 59] that the projections $\tilde{h}^{mnl}(k)$ and $\tilde{c}^{mnl}(k)$ must satisfy the equations

$$\frac{1}{\rho}\tilde{\eta}^{000} = \tilde{h}^{000}\tilde{c}^{000} \tag{3.11a}$$

$$\frac{1}{\rho}\tilde{\eta}^{110} = \frac{1}{3}\tilde{h}^{110}\tilde{c}^{110} + \frac{2}{3}\tilde{h}^{112}\tilde{c}^{112} + \frac{1}{6}\tilde{h}^{121}\tilde{c}^{121} + 4\tilde{h}^{123}\tilde{c}^{213} \tag{3.11b}$$

$$\frac{1}{\rho}\tilde{\eta}^{112} = \frac{1}{3}\tilde{h}^{110}\tilde{c}^{112} + \frac{1}{3}\tilde{h}^{112}\tilde{c}^{110} + \frac{1}{3}\tilde{h}^{112}\tilde{c}^{112} + \frac{1}{60}\tilde{h}^{121}\tilde{c}^{211} + \frac{3}{5}\tilde{h}^{121}\tilde{c}^{213}$$

$$+ \frac{3}{5}\tilde{h}^{123}\tilde{c}^{211} + \frac{8}{5}\tilde{h}^{123}\tilde{c}^{213} \tag{3.11c}$$

$$\frac{1}{\rho}\tilde{\eta}^{121} = \frac{1}{3}\tilde{h}^{110}\tilde{c}^{121} + \frac{1}{15}\tilde{h}^{112}\tilde{c}^{121} + \frac{12}{5}\tilde{h}^{121}\tilde{c}^{123} - \frac{2}{5}\tilde{h}^{121}\tilde{c}^{220}$$

$$+ \frac{7}{25}\tilde{h}^{121}\tilde{c}^{222} + \frac{12}{25}\tilde{h}^{123}\tilde{c}^{222} + \frac{32}{5}\tilde{h}^{123}\tilde{c}^{224} \tag{3.11d}$$

$$\frac{1}{\rho}\tilde{\eta}^{123} = \frac{1}{3}\tilde{h}^{110}\tilde{c}^{123} + \frac{1}{10}\tilde{h}^{112}\tilde{c}^{121} + \frac{4}{15}\tilde{h}^{112}\tilde{c}^{123} + \frac{1}{50}\tilde{h}^{121}\tilde{c}^{222}$$

$$+ \frac{4}{15}\tilde{h}^{121}\tilde{c}^{224} - \frac{2}{15}\tilde{h}^{123}\tilde{c}^{220} + \frac{8}{25}\tilde{h}^{123}\tilde{c}^{222} + \frac{8}{15}\tilde{h}^{123}\tilde{c}^{224} \tag{3.11e}$$

$$\frac{1}{\rho}\tilde{\eta}^{211} = \frac{1}{3}\tilde{h}^{211}\tilde{c}^{110} + \frac{1}{15}\tilde{h}^{211}\tilde{c}^{112} + \frac{12}{5}\tilde{h}^{213}\tilde{c}^{112} - \frac{2}{5}\tilde{h}^{220}\tilde{c}^{211}$$

$$+ \frac{7}{25}\tilde{h}^{222}\tilde{c}^{211} + \frac{12}{25}\tilde{h}^{222}\tilde{c}^{213} + \frac{32}{5}\tilde{h}^{224}\tilde{c}^{213} \tag{3.11f}$$

$$\frac{1}{\rho}\tilde{\eta}^{213} = \frac{1}{10}\tilde{h}^{211}\tilde{c}^{122} + \frac{1}{3}\tilde{h}^{213}\tilde{c}^{110} + \frac{4}{5}\tilde{h}^{213}\tilde{c}^{112} - \frac{2}{5}\tilde{h}^{220}\tilde{c}^{213} + \frac{1}{50}\tilde{h}^{222}\tilde{c}^{211}$$

$$+ \frac{8}{25}\tilde{h}^{222}\tilde{c}^{213} + \frac{4}{15}\tilde{h}^{224}\tilde{c}^{211} + \frac{8}{15}\tilde{h}^{224}\tilde{c}^{213} \tag{3.11g}$$

$$\frac{1}{\rho}\tilde{\eta}^{220} = -\frac{1}{12}\tilde{h}^{211}\tilde{c}^{121} - 2\tilde{h}^{213}\tilde{c}^{123} - \frac{2}{5}\tilde{h}^{220}\tilde{c}^{220} - \frac{7}{25}\tilde{h}^{222}\tilde{c}^{222}$$

$$-\frac{112}{45}\tilde{h}^{224}\tilde{c}^{224} \tag{3.11h}$$

$$\frac{1}{\rho}\tilde{\eta}^{222} = \frac{1}{12}\tilde{h}^{211}\tilde{c}^{121} + \frac{1}{7}\tilde{h}^{211}\tilde{c}^{123} + \frac{1}{7}\tilde{h}^{213}\tilde{c}^{121} + \frac{16}{7}\tilde{h}^{213}\tilde{c}^{123}$$

$$-\frac{2}{5}\tilde{h}^{220}\tilde{c}^{222} - \frac{2}{5}\tilde{h}^{222}\tilde{c}^{220} - \frac{3}{35}\tilde{h}^{222}\tilde{c}^{222} + \frac{16}{35}\tilde{h}^{222}\tilde{c}^{224}$$

$$+\frac{16}{35}\tilde{h}^{224}\tilde{c}^{222} + \frac{160}{63}\tilde{h}^{224}\tilde{c}^{224} \tag{3.11i}$$

$$\frac{1}{\rho}\tilde{\eta}^{224} = \frac{3}{14}\tilde{h}^{211}\tilde{c}^{213} + \frac{3}{14}\tilde{h}^{213}\tilde{c}^{121} + \frac{3}{7}\tilde{h}^{213}\tilde{c}^{123} - \frac{2}{5}\tilde{h}^{220}\tilde{c}^{224}$$

$$+\frac{9}{175}\tilde{h}^{222}\tilde{c}^{222} + \frac{2}{7}\tilde{h}^{222}\tilde{c}^{224} - \frac{2}{5}\tilde{h}^{224}\tilde{c}^{220} + \frac{2}{7}\tilde{h}^{224}\tilde{c}^{220}$$

$$+\frac{12}{35}\tilde{h}^{224}\tilde{c}^{224} \tag{3.11j}$$

where $\eta^{mnl} = h^{mnl} - c^{mnl}$, and the tilde denotes the Hankel transform

$$\tilde{h}^{mnl}(k) = 4\pi i^l \int r^2 j_l(kr) h^{mnl}(r)\, dr \tag{3.11k}$$

$j_l(kr)$ being the spherical Bessel function.

Comparing (3.1), (3.2), and (3.9) immediately yields the closure relations

$$c^{000}(r) = -\left[1 + \eta^{000}(r)\right] \tag{3.12a}$$
$$c^{mnl}(r) = -\eta^{mnl}(r) \quad \left[\text{if } (mnl) \neq (000)\right] \qquad \text{for } r < d \tag{3.12b}$$

and

$$c^{mnl}(r) = 0 \quad \left[\text{if}(mnl) = (000),\ (110),\ (121),\ (221),\ (220),\ \text{or } (222)\right] \tag{3.12c}$$

$$\text{for } r > d$$

$$c^{mnl}(r) = -\beta u^{mnl}(r) \quad \left[\text{if } (mnl) = (112),\ (123),\ (213),\ \text{or } (224)\right] \tag{3.12d}$$

The MSA is now reduced to solving (3.11) and (3.12). Equations 3.11a, 3.12a, and 3.12c [with $(mnl) = (000)$] completely decouple from the rest and in fact define the Percus-Yevick (PY) approximation for a hard-sphere fluid at density ρ. Thus in the MSA the angle-averaged or radial distribution function does not depend on the multipolar interactions. The PY approximation for hard spheres was solved by Wertheim[60] some time ago.

For purely dipolar systems the problem is greatly simplified. The expansions for the correlation functions reduce[12] to

$$h(12)=h^{000}(r)+h^{110}(r)\Phi^{110}(12)+h^{112}(r)\Phi^{112}(12) \tag{3.13a}$$

$$c(12)=c^{000}(r)+c^{110}(r)\Phi^{110}(12)+c^{112}(r)\Phi^{112}(12) \tag{3.13b}$$

and only three equations are obtained in Fourier space. These are (3.11a) and the following pair:

$$\frac{1}{\rho}\tilde{\eta}^{110} = \frac{1}{3}\tilde{h}^{110}\tilde{c}^{110} + \frac{2}{3}\tilde{h}^{112}\tilde{c}^{112} \tag{3.14a}$$

$$\frac{1}{\rho}\tilde{\eta}^{112} = \frac{1}{3}\tilde{h}^{110}\tilde{c}^{112} + \frac{1}{3}\tilde{h}^{112}\tilde{c}^{110} + \frac{1}{3}\tilde{h}^{112}\tilde{c}^{112} \tag{3.14b}$$

Wertheim[12] has shown that for dipolar hard spheres (3.14), (3.12b), and (3.12d) can be transformed to yield equations formally analogous to those occurring in the PY approximation for hard spheres. This allows us to obtain analytic expressions for $h(12)$, $c(12)$, the various thermodynamic properties, and the dielectric constant.[12, 61] The dielectric constant of dipolar hard spheres is given by the equations

$$\varepsilon = \frac{q(2\xi)}{q(-\xi)} \tag{3.15a}$$

$$y = \tfrac{1}{3}\left[q(2\xi)-q(-\xi)\right] \tag{3.15b}$$

where

$$q(x)=\frac{1+2x^2}{(1-x)^4} \tag{3.15c}$$

and as before, $y=4\pi\beta\mu^2\rho/9$. We note that the MSA is self-consistent in that both (3.7a) and (3.8) yield the same result of ε. In fact this is true for all approximations that simultaneously satisfy the OZ equation and the asymptotic limit $c(12)\rightarrow\beta\mu^2\Phi^{112}(12)/r^3$ as r goes to infinity. Expanding the MSA result for small y yields[12]

$$\frac{\varepsilon-1}{\varepsilon+2} =y- \frac{15}{16}y^3 +\cdots \tag{3.16}$$

which in the $\rho\rightarrow0$ limit is correct[62, 63] to order y^3. The MSA, however, has one physically unrealistic feature. It is obvious from (3.15) that the MSA ε is determined by the single parameter y, whereas physically we would expect the dipolar hard-sphere result to depend on two parameters, $\beta\mu^2$ and ρ.

For purely quadrupolar and dipole-quadrupole systems, (3.11) and (3.12) must be solved numerically. This is briefly discussed in Section III.B.3. We note that for purely quadrupolar fluids ($\mu=0$), only the four projections (000), (220), (222), and (224) occur in the MSA solution.[38] MSA results for dipole,[31, 47, 49] quadrupole,[38] and dipole-quadrupole[59] fluids have been compared with MC calculations, and in general the agreement is very poor. Several examples of this are given in Section III.D.

2. Beyond the MSA: Cluster Series Expansion and Generalized Mean Spherical Approximations

Two further approximations that build on the MSA solution are derived in Section II. These are the so-called LIN and L3 approximations, which have been investigated by Stell and Weis[45] for dipolar hard spheres. Both theories substantially improve upon the MSA.

The LIN approximation was first obtained by Verlet and Weis,[31] who linearized the EXP approximation of Andersen, Chandler, and Weeks.[64] This leads to the following approximation for the pair correlation function,

$$h(12)=h_{HS}(r)+g_{HS}(r)\mathcal{C}(12) \tag{3.17a}$$

where

$$\begin{aligned}\mathcal{C}(12)&=h_{MSA}(12)-h_{MSA}^{000}(r)\\&=h_{MSA}^{110}(r)\Theta^{110}(12)+h_{MSA}^{112}(r)\Phi^{112}(12)\end{aligned} \tag{3.17b}$$

and the subscripts HS are used to denote "exact" hard-sphere results. The LIN approximation for the projections $h^{110}(r)$ and $h^{112}(r)$ is in much better agreement with MC calculations[45] than the MSA. However, (3.7a) and (3.8) do not yield identical results for ε. The asymptotic behavior of $h^{112}(r)$ is unaltered by (3.17a), hence (3.8) merely gives the MSA result for ε. Equation 3.7a does give a different approximation and for dipolar hard spheres, and ε is a function of two variables, ρ and $\beta\mu^2$, rather than just the single parameter y. This is an obvious improvement on the MSA.

The L3 or "linearized three-node" approximation consists of the equations

$$\begin{aligned}g(12)=g_{HS}(r)\Bigg[1+\mathcal{C}(12)+\rho\int\mathcal{S}(13)\mathcal{S}(23)\,d(3)\\+2\rho\int\mathcal{S}(13)\mathcal{H}(23)\,d(3)\Bigg]\end{aligned} \tag{3.18a}$$

where

$$\mathcal{S}(12) = h_{HS}(r)\mathcal{C}(12) \qquad (3.18b)$$

and

$$\mathcal{K}(12) = h_{HS}(r) + \mathcal{C}(12) \qquad (3.18c)$$

The L3 ε is considerably larger than the LIN value, and numerical results for both approximations are given in Section III.D.

Stell and Weis[45] also consider two other theories for $h(12)$. These are the self-consistent Ornstein-Zernike approximation[16] (SCOZA) and the closely related Padé producing approximation[65] (PPA) of Høye and Stell. The SCOZA is an application of the generalized mean spherical approximation (GMSA) of Høye, Lebowitz, and Stell.[13] The SCOZA and the PPA are not self-contained theories, and they require that several parameters be specified as input. For dipolar hard spheres these include the contact values of the projections $h^{000}(r)$, $h^{110}(r)$, and $h^{112}(r)$, as well as the internal energy and ε. In the PPA, as its name suggests, the internal energy is supplied by the accurate Padé approximant of Rushbrooke, Stell, and Høye.[61] Stell and Weis[45] apply the SCOZA by using MC results for the necessary contact values and internal energy. They then adjust ε until the $h^{000}(r)$, $h^{110}(r)$, and $h^{112}(r)$ obtained are in good agreement with Monte Carlo results. Thus by essentially "fitting" MC data, the SCOZA provides an indirect estimate of ε. Care must be taken, however, to fit only the short-range part of $h^{110}(r)$ and $h^{112}(r)$, since the long-range part is seriously influenced by the MC boundary conditions.[52] This problem is discussed in Section III.D.1.

3. The Linearized and Quadratic Hypernetted-Chain Approximations

The linearized and quadratic hypernetted-chain approximations (LHNC and QHNC, respectively) consist of the OZ equation (3.9a) coupled with closure relations that can be easily obtained[30, 38] by expanding the well-known hypernetted-chain (HNC) approximation.[66] Alternative derivations of the LHNC and QHNC results were given in Section II, and it was noted that the LHNC is essentially equivalent to the single superchain (SSC) approximation of Wertheim[29] and the generalized mean field (GMF) theory of Henderson and Gray.[36] Here we describe the HNC approach and write the explicit equations for dipole-quadrupole systems.[59]

The HNC approximation is defined by the equation

$$c(12) = h(12) - \ln g(12) - \beta u(12) \qquad (3.19)$$

where $g(12) = h(12) + 1$. Inserting the $h(12)$ expansion (3.6a) into (3.19) and rearranging, we obtain

$$c(12) = \left[h^{000}(r) - \ln g^{000}(r) - \beta u^{000}(r) \right]$$
$$- \ln\left[1 + X(12) \right] - \beta\left[u(12) - u^{000}(r) \right] \qquad (3.20a)$$

where

$$X(12) = \frac{h(12) - h^{000}(r)}{g^{000}(r)} \qquad (3.20b)$$

and $g^{000}(r) = h^{000}(r) + 1$. The LHNC closure is found[30, 38] by expanding $\ln[1 + X(12)]$ and retaining only the term that is linear in $X(12)$. The solution to this approximation is of the same general form as the MSA, with only the projections occurring in (3.10a) contributing to $h(12)$ and $c(12)$. The QHNC approximation is obtained[38] by retaining terms to order $X^2(12)$ in the logarithmic expansion, coupled with the additional assumption that we can ignore all terms in the $h(12)$ and $c(12)$ expansions except those that occur in the MSA solution (3.10a). Equation 3.10a obviously contains the minimum subset of projections necessary to treat the $\beta u(12)$ part of (3.19) correctly. The assumption that all other terms in (3.6a) [and the analogous expansion for $c(12)$] can be neglected is very *arbitrary*, and the QHNC approximation is inconsistent in some respects.[38, 58, 59] Nevertheless, the terms included in (3.10a) are sufficient to totally determine the thermodynamic properties and the dielectric constant. In addition, the QHNC has proved to be a very good approximation, particularly for purely dipolar[58] and purely quadrupolar[38] fluids.

The solution of the LHNC and QHNC approximations proceeds much as in the case of the MSA. One must solve (3.11) subject to appropriate closure relations linking the $c^{mnl}(r)$ and $h^{mnl}(r)$ projections. These are found by substituting the expanded HNC equation into the expression (3.10b) through (3.10i) [with $h^{mnl}(r)$ replaced by $c^{mnl}(r)$] defining the $c^{mnl}(r)$. The QHNC approximation for dipole-quadrupole fluids yields[59]

$$c^{000}(r) = \exp\left[\eta^{000}(r) - \beta u^{000}(r) + \frac{B^{000}(r)}{g^{000}(r)^2} \right] - \eta^{000}(r) - 1 \qquad (3.21a)$$

$$c^{mnl}(r) = g^{000}(r)\left[\eta^{mnl}(r) - \beta u^{mnl}(r) \right] - \eta^{mnl}(r) + \frac{B^{mnl}(r)}{g^{000}(r)} \qquad (3.21b)$$

when $(mnl) = (112)$, (123), (213), or (224), and

$$c^{mnl}(r) = g^{000}(r)\eta^{mnl}(r) - \eta^{mnl}(r) + \frac{B^{mnl}(r)}{g^{000}(r)} \qquad (3.21c)$$

for all remaining projections. The $B^{mnl}(r)$ for dipole-quadrupole systems[59] are given in Appendix C. For purely dipolar fluids, $B^{112}(r) = B^{110}(r) = 0$ and $B^{000}(r)/g^{000}(r)^2$ can be written in the relatively simple form

$$\frac{B^{000}(r)}{g^{000}(r)^2} = \frac{1}{3}\left[\eta^{112}(r) - \beta u^{112}(r)\right]^2 + \frac{1}{6}\eta^{110}(r)^2 \qquad (3.22)$$

For hard particles [i.e., $u^{000}(r) = u_{HS}(r)$] (3.21) are applied for $r > d$, and for $r < d$ the closure relations are given by (3.12a) and (3.12b). For systems without hard cores such as the generalized Stockmayer fluid [i.e., $u^{000}(r) = u_{LJ}(r)$] (3.21) completely define the closure. The LHNC closure relations can be obtained[59] by setting all the $B^{mnl}(r)$ to zero in (3.21). This gives a set of equations considerably simpler than the QHNC results.

Equations 3.11, 3.12, and 3.21 are complete and could be solved as they stand. However, at high densities the results would contain some inaccuracy because of the HNC treatment of the spherical part of the interaction potential. This can be avoided to some extent by means of a perturbation treatment first proposed by Lado.[67] The application of Lado's technique to the present problem is straightforward[38, 58, 59] and results in replacing (3.21a) with the relationship

$$c^{000}(r) = g_S(r)\exp\left[\Delta\eta^{000}(r) + \frac{B^{000}(r)}{g^{000}(r)^2}\right] - \eta^{000}(r) - 1 \qquad (3.23)$$

where $\Delta\eta^{000}(r) = \eta^{000}(r) - \eta_S(r)$ and the subscript S is used to denote "exact" results for the fluid defined by the spherically symmetric pair potential $u^{000}(r)$. The radial distribution function $g_S(r)$ is specified as input, and (3.23) ensures that the correct result $g^{000}(r) \rightarrow g_S(r)$ is obtained in the limit of vanishing multipole moments. In practice, all calculations[30, 38, 58, 59] have been done using the more accurate closure (3.23) rather than (3.21a). Equations 3.21b, 3.21c, and 3.23 are sometimes called the "reference" or "corrected" QHNC [or LHNC, if $B^{mnl}(r) = 0$] approximation, but since no ambiguity exists here, we simply refer to the QHNC (or LHNC) approximation, with the use of (3.23) being understood. For hard-core potentials $g_S(r) = g_{HS}(r)$ and is often taken[30, 38, 58, 59] to be Verlet-Weis[68] fit to MC data. For generalized Stockmayer fluids $g_S(r) = g_{LJ}(r)$, and MC or MD results are used.[58]

Two important comments can now be made. First, it is clear on inspection of the equations that the LHNC approximation for $g^{000}(r)$ does not depend on the anisotropic part of $u(12)$ but only on the spherically symmetric term $u^{000}(r)$. This is similar to the MSA situation except that now instead of the PY approximation for the spherically symmetric fluid, one obtains the HNC approximation from (3.21a), or the "exact" result if (3.23) is used. The QHNC approximation is physically more realistic with $h^{000}(r)$ being coupled to the other projections through the $B^{000}(r)/g^{000}(r)^2$ term in (3.21a) and (3.23). This means that the QHNC $g^{000}(r)$ does depend on the multipolar interactions. This attractive feature of the QHNC theory is very important at low density.[58]

Our second comment concerns a problem arising from the arbitrary nature of the QHNC definition. As stated above, the QHNC correlation functions are defined to include only the terms occurring in the MSA solution. This means that for purely dipolar fluids $h(12)$ is given by (3.13a) and by (3.10a) for dipole-quadrupole systems. Now in the MSA and LHNC approximation no ambiguity exists, since in the $Q \to 0$ limit all but three terms in (3.10a) vanish and the dipolar solution (3.13a) is recovered. However, this is not true for the QHNC theory. In this case the (220), (222), and (224) projections do not vanish at $Q = 0$, and the QHNC result (as defined above) for purely dipolar fluids is not recovered. Instead, a different approximation is obtained that includes the projections (220), (222), and (224) in addition to the (000), (110), and (112) terms present in the QHNC. This new approximation gives results very similar to the QHNC for small μ, but no solution appears to exist for large dipole moments.[59] This problem is also encountered in attempting to solve the QHNC for systems characterized by very small quadrupole moments coupled with relatively large dipole moments.[59] Thus in general the QHNC is not as satisfactory for dipole-quadrupole systems[59] as it is for purely dipolar[58] or purely quadrupolar[38] fluids. We note that taking the $\mu \to 0$ limit does give the QHNC approximation for a purely quadrupolar system.

The LHNC and QHNC approximations have not been solved analytically, but numerical solutions can be obtained by iteration. This is also true of the MSA except for the previously discussed dipolar hard-sphere system solved by Wertheim.[12] The details of the numerical solution are described in Refs. 30, 38, 58, and 59. Essentially, (3.11) and the appropriate closure relations are written in terms of c^{mnl} and η^{mnl} and iterated until a solution is obtained. This means that all equations defining a particular approximation are simultaneously satisfied. The present problem is very similar to that encountered in the familiar integral equation theories for simple fluids,[66] but there is one significant difference. Equations 3.11 require the calculation of higher order Hankel transforms, whereas in the simple fluid case only zeroth-order Hankel or Fourier transforms are necessary.

Numerically, the Hankel transforms are best obtained by first calculating integral transforms, which we denote by $\hat{c}^{mnl}(r)$[or $\hat{\eta}^{mnl}(r)$ or $\hat{h}^{mnl}(r)$]. The $\hat{c}^{112}(r)$ was introduced by Wertheim,[12] and generalized results have been obtained by Blum.[56] Blum shows that the calculation of $\tilde{c}^{mnl}(k)$ [defined by (3.11k)] reduces to taking the zeroth-order Hankel (Fourier) transform of $\hat{c}^{mnl}(r)$ if l is even, or the first-order Hankel transform if l is odd. Thus one has

$$\tilde{c}^{mnl}(k) = 4\pi \int_0^\infty r^2 j_0(kr)\hat{c}^{mnl}(r)\, dr \tag{3.24a}$$

$$\hat{c}^{mnl}(r) = \frac{1}{2\pi^2} \int_0^\infty k^2 j_0(kr)\tilde{c}^{mnl}(k)\, dk \tag{3.24b}$$

if l is even and

$$\tilde{c}^{mnl}(k) = 4\pi i \int_0^\infty r^2 j_1(kr)\hat{c}^{mnl}(r)\, dr \tag{3.25a}$$

$$\hat{c}^{mnl}(r) = -\frac{i}{2\pi^2} \int_0^\infty k^2 j_1(kr)\tilde{c}^{mnl}(k)\, dk \tag{3.25b}$$

if l is odd. The integral transforms and their inverses depend on l and are defined as follows:

$$\hat{c}^{mn0}(r) = c^{mn0}(r) \tag{3.26}$$

$$\hat{c}^{mn1}(r) = c^{mn1}(r) \tag{3.27}$$

$$\hat{c}^{mn2}(r) = c^{mn2}(r) - 3\int_r^\infty \frac{c^{mn2}(s)}{s}\, ds \tag{3.28a}$$

$$c^{mn2}(r) = \hat{c}^{mn2}(r) - \frac{3}{r^3}\int_0^r s^2 \hat{c}^{mn2}(s)\, ds \tag{3.28b}$$

$$\hat{c}^{mn3}(r) = c^{mn3}(r) - 5r\int_r^\infty \frac{c^{mn3}(s)}{s^2}\, ds \tag{3.29a}$$

$$c^{mn3}(r) = \hat{c}^{mn3}(r) - \frac{5}{r^4}\int_0^r s^3 \hat{c}^{mn3}(s)\, ds \tag{3.29b}$$

and

$$\hat{c}^{mn4}(r) = c^{mn4}(r) + \frac{15}{2} \int_r^\infty \frac{c^{mn4}(s)}{s} ds - \frac{35}{2} r^2 \int_r^\infty \frac{c^{mn4}(s)}{s^3} ds \quad (3.30a)$$

$$c^{mn4}(r) = \hat{c}^{mn4}(r) + \frac{15}{2r^3} \int_0^r s^2 \hat{c}^{mn4}(s) ds - \frac{35}{2r^5} \int_0^r s^4 \hat{c}^{mn4}(s) ds$$

$$(3.30b)$$

Although we have written these transformations for $c^{mnl}(r)$ only, an analogous set applies to $h^{mnl}(r)$ and $\eta^{mnl}(r)$. Calculating the Hankel transforms in this manner has two distinct advantages. First, it allows the long-range part of certain projections such as the term proportional to $1/r^3$ in $c^{112}(r)$ to be treated exactly, and this is of crucial importance[30] in the calculation of ε. Second, both the Fourier and first-order Hankel transforms can be calculated using fast Fourier transform techniques, and this is enormously advantageous from a computational viewpoint.

The LHNC or QHNC ε can be obtained from either (3.7a) or (3.8), since both routes must give ε consistently.[30] Dipolar hard spheres have been studied extensively using both the LHNC[30, 58] and QHNC[58] theories, and before discussing the results it is convenient to introduce the reduced variables $\rho^* = \rho d^3$ and $\mu^* = (\beta \mu^2/d^3)^{1/2}$, which are sufficient to characterize this system. It is found[58] that for a limited range of density ($\rho^* = 0.2$–0.8) and dipole moment ($\mu^{*2} = 0.5$–2.75) the LHNC results are fitted to within 1% by

$$\varepsilon - 1 = \frac{3y}{1 - a_1 y} \left[1 + (1 - a_1)y + a_2 y^2 + a_3 y^3 \right] \quad (3.31)$$

where $a_1 = 0.4341\rho^{*2}$, $a_2 = -(0.05 + 0.75\rho^{*3})$, and $a_3 = -0.026\rho^{*2} + 0.173\rho^{*4}$. The QHNC approximation for dipolar hard spheres has also been solved[58.] for a range of density and dipole moments, but the results have not been fitted to a simple formula. It is interesting to note that for dipolar hard spheres QHNC solutions cannot be found near the liquid-vapor coexistence region.[69] A number of LHNC and QHNC calculations have also been done for Stockmayer fluids[58] and for hard spheres with both dipoles and quadrupoles.[59] A detailed description of these and other results is given in Section III.D.

An interesting graph theoretical analysis of the LHNC approximation for dipolar systems has been given by Rushbrooke.[63] He concludes that in the limit $\rho^* \to 0$ the LHNC ε reduces to the MSA result for all values of y. This prediction has been confirmed numerically.[58]

4. A Perturbation Expansion of the HNC Equation

Agrofonov, Martinov, and Sarkisov[70] have recently proposed another theory for dipolar hard spheres based on the HNC approximation. Their approach is similar in spirit to that followed in the usual thermodynamic perturbation theory (TPT) of dipolar fluids.[71, 72] It is assumed that $h(12)$ can be expanded in the power series

$$h(12) = \sum_{i=0}^{\infty} \lambda^i h^{(i)}(12) \tag{3.32}$$

where $\lambda = \beta\mu^2/d^3 = \mu^{*2}$. Substituting this expansion together with an analogous one for $\eta(12)$ into the HNC equations and equating terms of equal order in λ leads to a set of coupled integral equations for $h^{(i)}(12)$ and $\eta^{(i)}(12)$. Agrofonov et al.[70] succeed in solving the approximation obtained by considering terms to order λ^2. This yields the pair correlation function

$$h(12) = h^{(0)}(r) + \lambda h^{(1)}(12) + \lambda^2 h^{(2)}(12) + \cdots \tag{3.33a}$$

where

$$h^{(1)}(12) = g^{(0)}(r)\left(\frac{d}{r}\right)^3 \Phi^{112}(12) \tag{3.33b}$$

$$h^{(2)}(12) = g^{(0)}(r)\left[A(r) + B(r)\left(\cos^2\theta_1 + \cos^2\theta_2\right) + C(r)\cos\theta_1\cos\theta_2 \right.$$

$$\left. + D(r)\sin\theta_1\sin\theta_2\cos(\phi_1 - \phi_2) + \tfrac{1}{2}\left(\frac{d}{r}\right)^6\left(\Phi^{112}(12)\right)^2\right] \tag{3.33c}$$

$g^{(0)}(r) = h^{(0)}(r) + 1$ is the HNC approximation for the hard-sphere radial distribution function, and θ and ϕ are the polar and azimuthal angles describing the dipolar orientations with respect to the vector $\mathbf{r}_{12}$ joining their centers; $A(r)$, $B(r)$, $C(r)$, and $D(r)$ are density-dependent functions that must be found numerically.[70] The λ expansions for the projections $h^{000}(r)$, $h^{110}(r)$, and $h^{112}(r)$ can be found by inserting (3.33a) into (3.10b) through (3.10d). One obtains[70]

$$h^{000}(r) = h^{(0)}(r) + \lambda^2 g^{(0)}(r)\left[A(r) + \frac{2}{3}B(r) + \frac{1}{3}\left(\frac{d}{r}\right)^6\right] + \cdots \tag{3.34a}$$

$$h^{110}(r) = \frac{\lambda^2}{3}g^{(0)}(r)\left[2D(r) + C(r)\right] + \cdots \tag{3.34b}$$

and

$$h^{112}(r) = g^{(0)}(r) \left[\lambda \left(\frac{d}{r} \right)^3 + \frac{\lambda^2}{3} (C(r) - D(r)) + \cdots \right] \qquad (3.34c)$$

The similarity of (3.33a) to the TPT expansion[71] is obvious. To order λ, (3.33a) gives

$$h(12) = h^{(0)}(r) + g^{(0)}(r) \frac{\beta \mu^2}{r} \Phi^{112}(12) + \cdots \qquad (3.35)$$

which is identical to the first-order TPT result except for the trivial difference that in TPT the HNC approximation $g^{(0)}(r)$ is replaced by the "exact" hard-sphere result. The problem with the TPT expansions truncated after terms of order λ^2 is well known.[47, 48, 61] At least for the thermodynamic properties, the series does not converge except for values of $\lambda \ll 1$, and indeed the internal energy data given by Agrofonov et al.[70] indicate that the convergence of the present expansion is no better.

Nevertheless, Agrofonov et al.[70] calculate the dielectric constant using (3.7a) and (3.34b) and compare with previous theories. Somewhat surprisingly, they find that at $\rho^* = 0.8$ their values lie quite close to the LHNC results for $\lambda = \mu^{*2} < 1.5$. This suggests that perhaps the λ expansion for ε is more rapidly convergent than the thermodynamic results would imply. Agrofonov et al.[70] correctly point out that some of the difference between their results and the LHNC theory may well be due to their use of the HNC approximation for the hard-sphere radial distribution function, whereas "exact" hard-sphere results are used in the LHNC calculations. This should be investigated further. Also it would be interesting to compare the λ expansions for $h^{110}(r)$ and $h^{112}(r)$ with LHNC or QHNC results. Some work in this direction is already found in Gray and Gubbins (on ε) and Murad et al. (on h^{224}).[63]

C. Computer Simulations

The computer simulation of dipolar fluids and in particular the calculation of accurate dielectric constants has proved to be a very difficult problem for which a completely satisfactory solution has yet to be found.[55] There has, nevertheless, been a good deal of recent progress and the fundamental nature of the problems involved is now recognized and better understood. The difficulties all stem from the long-range nature of the dipolar forces. It is never possible to simulate a truly infinite system, but for fluids with short-range potentials there are approximate methods[73] that give essentially exact results. However, for dipolar fluids this is not the case, and the pair correlation function obtained in a computer simulation will depend on exactly how the long-range dipolar interactions are treated. There have been a number of MC and MD calculations[23, 31, 47–54] for simple dipolar fluids, with the dipolar interactions being handled in different ways, but it is now clear[23, 52] that none of these methods gives the pair correlation function for the truly infinite system described by approximate theories. This means that one must be very careful when attempting to evaluate approximate theories by comparing with computer simulations, and this problem is discussed at length in Section III.D.

There are two basic problems associated with the MC or MD calculation of the dielectric constant. First, the relationship between ε and the mean square moment obtained in the computer calculation will depend on exactly how the dipolar interactions are handled. For example, the Kirkwood formula (3.7a) only holds if $h^{110}(r)$ is that of an infinite sample, and hence does not apply in most computer situations. To find the correct relationship for a given simulation method is not a trivial problem, but for several commonly applied procedures the appropriate formulas are now known.[23, 52, 55] The second, and perhaps more fundamental question, concerns whether the dielectric constant given by a particular simulation is really the true infinite-system value, or whether it is it seriously influenced by the approximate methods used in the calculation. In the absence of exact results, this question is obviously difficult to answer fully, but a detailed and, we hope, useful examination of the problem appears in Section III.D.2.

1. Periodic Boundary Conditions

In most computer simulations of fluids periodic boundary conditions are applied.[73] This means that N particles are placed inside a basic cell of length L, and the configuration so obtained is imagined to be repeated periodically in space to form an infinite array. This procedure succeeds in removing surface effects, but it introduces a periodicity that for fluids is quite artificial.[73] The configurational energy of this periodic system can be calculated in different ways, and these methods are often referred to as different "boundary conditions," with the underlying periodicity being understood. For dipolar fluids the pair correlation function as well as the relationship between ε and $\langle M^2 \rangle$ will depend on how the configurational energy is calculated. The following methods have been applied to dipolar systems.

The Spherical Cutoff and Minimum Image Methods. Two commonly used and easily implemented approximations are the so-called spherical cutoff (SC) and minimum image (MI) methods.[73] Both involve a simple truncation of the interaction potential, and no attempt is made to take long-range interactions into account. In the SC case each particle interacts only with neighbors lying within a sphere of radius R_C, centered on the particle. The MI method is similar, but each particle is considered to be at the center of a cube of length L and to interact with all neighbors lying within the cube. Both the SC and MI methods have been applied to dipolar fluids,[31, 47–49, 52] and $h(12)$ is found[52] to depend strongly on the cutoff radius in the SC calculations and the system size in the MI case. SC and MI results for dipolar hard spheres are shown in Figs. 1 and 2. It is obvious that $h^{110}(r)$ is particularly sensitive to boundary conditions. It has been shown[52] that in SC calculations $h^{110}(r)$ is not only determined by R_C but is also influenced by the underlying periodicity of the system. An interesting analysis of the results shown in Figs. 1 and 2 has recently been given by Neumann and Steinhauser.[74] These authors show that the behavior of $h^{110}(r)$ and $h^{112}(r)$ under SC and MI boundary conditions is consistent with what one would expect from an examination of continuum models.

Despite the obvious problems with SC and MI methods, such calculations serve to give a first, if somewhat crude, evaluation of approximate theories.[31, 47–49, 52] For example, the radial distribution function $g^{000}(r)$, the thermodynamic properties, and to a lesser extent the short-range part of $h^{110}(r)$ and $h^{112}(r)$ are not seriously influenced by the cutoff radius or system size.[52] In addition, SC calculations can provide a relatively unambiguous, if rather indirect, test of some approximate theories. Although it is not

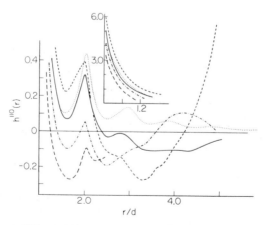

Fig. 1. Values of $h^{110}(r)$ for dipolar hard spheres at $\rho^* = 0.8$ and $\mu^* = 2.75$: long dashes, h_{SC}^{110}, $N = 108$, $R_C = 2.5d$; dots and dashes, h_{SC}^{110}, $N = 256$, $R_C = 3.4d$; solid curve, h_{SC}^{110}, $N = 864$, $R_C = 5.1d$; short dashes, h_{MI}^{110}, $N = 256$; the dotted curve is the LHNC result for an infinite system. (Results from Ref. 52.)

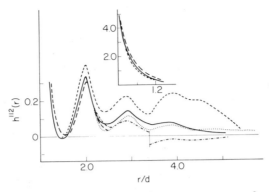

Fig. 2. Values of $h^{112}(r)$ for dipolar hard spheres at $\rho^* = 0.8$ and $\mu^{*2} = 2.75$. The curves are for the calculations described in Fig. 1. (Results from Ref. 52.)

possible to simulate an infinite system, one can sometimes adopt a reverse strategy and solve the approximate theory for a system with spherically truncated dipolar interactions.[30] Results so obtained can then be directly compared with SC calculations. The LHNC and QHNC approximations have been compared with MC calculations[30, 52, 58, 59] in this way, and the results are described in Section III.D.1.

For SC and MI calculations, ε and $\langle M^2 \rangle$ are thought to be related by the formula

$$\frac{\varepsilon - 1}{\varepsilon + 2} = yg \qquad (3.36)$$

where $g = \langle M^2 \rangle / N\mu^2$, M being the *total* dipole moment of the central cube. This relationship was first used in the context of computer calculations by Rahman and Stillinger[92] in their MD simulation of "waterlike" particles, and it has since been applied in MC calculations for simple dipolar systems.[47, 48, 52, 59] Equation 3.36 is of a form usually associated with spherical samples, and this has led to some discussion[47, 48, 75–77] of whether, and if so how, it should be applied in the computer situation. For example, it has been suggested[75–77] that the appropriate mean square moment to use in (3.36) is that of the cutoff sphere rather than that of the total sample. Although (3.36) is not generally correct for cubic geometry, it can be argued[48] that it is valid for an isotropic fluid under SC or MI boundary conditions, and empirically this appears to be true.[47, 48, 52] Also empirically it is found[52] that the mean square moment of the truncation sphere depends on R_C and clearly violates the upper bound $yg \to 1$ as $y \to \infty$. The mean square moment of the cube, on the other hand, always satisfies this condition.[47, 52]

Some insight into (3.36) and its application to the computer situation can be obtained by considering a closely related situation mentioned in Section II.A. This is an infinite system, but one in which the dipolar interactions are spherically truncated. For such a system, (3.36) is an exact result.[4, 5, 30] The computer simulations employ periodic boundary conditions as well as a spherical cutoff, hence do not exactly correspond to the system just described. Nevertheless, the situations are very similar, and we would not expect the periodicity to influence the formal results. It is clear[4, 5, 30] that for the infinite system with a truncated potential $\langle M^2 \rangle$ is the mean square moment of the entire sample, or

$$g = 1 + \frac{\rho}{3} \int h_{SC}^{110}(r) \, d\mathbf{r} \tag{3.37}$$

where the subscripts SC denote the spherical cutoff result, and the integration is over all space. For the periodic system constructed in the computer simulation calculating $\langle M^2 \rangle / N\mu^2$ for the central cell is obviously equivalent to integrating over all space and thus is the appropriate quantity to use in (3.36).

Unfortunately, the discussion above proves to be rather academic, since (3.36) has a serious practical problem.[47, 52] The relationship between ε and g is such that for $\varepsilon \gtrsim 10$ ($\mu^{*2} \gtrsim 1$ for dipolar hard spheres), small uncertainties in g lead to very large errors in ε. Thus except for systems with relatively small dielectric constants, (3.36) is not very useful.

Truly Periodic Boundary Conditions.   It is possible to calculate the configurational energy of a periodic system taking into account all the images of every particle.[73] We refer to this method as truly periodic boundary conditions (TPBC) to distinguish it from other methods such as SC or MI, which employ periodic boundary conditions but calculate the energy in an approximate manner. From our viewpoint, two basic problems must be considered before TPBC can be applied to dipolar systems. The first is fundamental and concerns exactly how, and indeed if, the lattice sums can be done. Second, for a given summation the formula relating the dielectric constant and mean square moment of the periodic system must be found. These questions have been recently considered in important papers by De Leeuw, Perram, and Smith[23, 78] and the following is a brief outline of their results.

If we consider N particles interacting with an arbitrary pair potential $u(\mathbf{r}_{12}, \Omega_1, \Omega_2)$, under TPBC the total configurational energy U is given by

$$U = \frac{1}{2} \sum_{\mathbf{n}}{}' \sum_{i=1}^{N} \sum_{j=1}^{N} u(\mathbf{r}_{ij} + L\mathbf{n}, \Omega_i, \Omega_j) \tag{3.38}$$

where the pair interactions are to be calculated in the MI convention,[73] the sum on $\mathbf{n}$ is over all simple cubic lattice points with integer coordinates, and the prime indicates that if $i=j$ the $\mathbf{n}=0$ terms must be omitted. The sums in (3.38) are absolutely convergent only if for large r the bare potential satisfies the inequality

$$|u(\mathbf{r}, \mathbf{\Omega}_1, \mathbf{\Omega}_2)| \leq A r^{-3-\gamma} \qquad (3.39)$$

where $r=|\mathbf{r}|$ and A and γ are both greater than zero. The dipole-dipole potential (3.2a) is proportional to $1/r^3$, which obviously does not satisfy (3.39). This means that for dipolar systems the sums in (3.38) are only conditionally convergent, hence must be treated very carefully. In fact, the answer obtained depends on precisely how the sums are done.[23] This problem does not appear to have been recognized in earlier work.[51, 77]

De Leeuw et al.[23] show that a result can be found by adding the sums in a sequence of spherical shells. This corresponds to packing together replications of the central cell to form an infinite sphere. The bare potential $u(12)$ is then replaced by the effective interaction $\phi_{\text{TPBC}}(12)$ and the configurational energy is calculated by the usual MI procedure. De Leeuw et al.[23] show that for their particular summation, $\phi_{\text{TPBC}}(12)$ is given by

$$\phi_{\text{TPBC}}(12) = -\frac{1}{L^3}(\mathbf{\mu}_1 \cdot \nabla)(\mathbf{\mu}_2 \cdot \nabla)\psi\left(\frac{\mathbf{r}_{12}}{L}\right) + \frac{4\pi}{3L^3}(\mathbf{\mu}_1 \cdot \mathbf{\mu}_2) \qquad (3.40)$$

where ∇ is the gradient operator and $\psi(\mathbf{r}_{12}/L)$ is defined in Ref. 23. The first term in (3.40) is the effective potential defining what are usually called[51, 53, 54, 77] Ewald or Ewald-Kornfield boundary conditions. Several authors have applied boundary conditions of this type to simple polar fluids,[23, 51, 53, 54] and following tradition we shall refer to these as Ewald calculations.

To obtain appropriate expressions relating the dielectric constant to the mean square moment, De Leeuw et al.[23] consider the periodic sphere formed by replicating the central cell to be embedded in an infinite continuum of dielectric constant ε'. This is reminiscent of techniques first applied by Kirkwood,[17] and the infinite limits must be taken properly. One then finds[23] that the effective pair interaction for this system, $\phi(\varepsilon'; 12)$, can be written as

$$\phi(\varepsilon'; 12) = \phi_{\text{TPBC}}(12) - \frac{8\pi}{3L^3}\frac{\varepsilon'-1}{2\varepsilon'+1}(\mathbf{\mu}_1 \cdot \mathbf{\mu}_2) \qquad (3.41)$$

where $\phi_{\text{TPBC}}(12)$ is defined by (3.40) and the second term represents the interaction of the embedded periodic sphere with the dielectric continuum.

The dielectric constant of the periodic sphere, ε, is given by

$$\frac{(\varepsilon-1)(2\varepsilon'+1)}{3(\varepsilon+2\varepsilon')}=yg(\varepsilon') \tag{3.42}$$

where $g(\varepsilon')=\langle M^2\rangle/N\mu^2$, M being the total moment of the central cube. We note that in (3.42) g is a function of ε', and this suggests that ε might also depend on the value that is chosen for ε'. However, De Leeuw et al.[23, 78] show that this is not the case. All choices of ε' will give the same value of ε, provided the correct effective interaction $\phi(\varepsilon';12)$ is used. It is obvious from (3.42) that the formula relating ε and $g(\varepsilon')$ varies with ε', and the following examples are of particular importance in computer simulations.

For $\varepsilon'=1$, $\phi(1;12)=\phi_{\text{TPBC}}(12)$, and (3.42) becomes

$$\frac{\varepsilon-1}{\varepsilon+2}=yg(1) \tag{3.43}$$

which, as we might expect, is the formula for a sphere in vacuum. Equation 3.43 is analogous to (3.36) and suffers from all the practical problems discussed above. De Leeuw et al.[23] show that the method proposed by Ladd[50] for calculating the configurational energy is in fact equivalent to using the effective interaction $\phi_{\text{TPBC}}(12)$. Thus for Ladd's calculations we would expect (3.43) to be the appropriate formula for ε, rather than (3.7a), which Ladd assumed to be valid. In fact, the Kirkwood formula (3.7a) is obtained by setting $\varepsilon'=\varepsilon$, which again is an expected result.

From the point of view of recent work, the most interesting choice is $\varepsilon'=\infty$. Then from (3.41) and (3.42) one obtains

$$\phi(\infty;12)=\phi_{\text{TPBC}}(12)-\frac{4\pi}{3L^3}(\boldsymbol{\mu}_1\cdot\boldsymbol{\mu}_2) \tag{3.44}$$

and

$$\varepsilon-1=3yg(\infty) \tag{3.45}$$

The $\phi(\infty;12)$ is just the first term in (3.40), or the effective potential used in Ewald calculations. Thus (3.45) is the formula relating ε and g for Ewald boundary conditions. A different derivation of (3.45) has recently been given by Felderhof.[79]

It should be emphasized that although ε is independent of the choice of ε', the pair correlation function is not. De Leeuw et al.[78] show that for two different values, ε' and ε'', the projections h^{000}, h^{110}, and h^{112} are approximately related by the equations

$$h^{000}(\varepsilon'';r)=h^{000}(\varepsilon';r)+\frac{\lambda[g(\varepsilon')]^2}{3N(1-\lambda g(\varepsilon')/3)}h^{110}(\varepsilon';r)+O\left(\frac{1}{N^2}\right)$$

(3.46a)

$$h^{110}(\varepsilon'';r)=h^{110}(\varepsilon';r)+\frac{\lambda[g(\varepsilon')]^2}{N(1-\lambda g(\varepsilon')/3)}\left[1+h^{000}(\varepsilon';r)\right]+O\left(\frac{1}{N^2}\right)$$

(3.46b)

and

$$h^{112}(\varepsilon'';r)=h^{112}(\varepsilon';r)+O\left(\frac{1}{N^2}\right)$$ (3.46c)

where

$$\lambda=6y\left[\frac{\varepsilon''-1}{2\varepsilon''+1}-\frac{\varepsilon'-1}{2\varepsilon'+1}\right]$$ (3.46d)

It can be seen from (3.46b) that even for large r there remains a constant finite difference between $h^{110}(\varepsilon'';r)$ and $h^{110}(\varepsilon';r)$.

Ewald boundary conditions were first applied to dipolar fluids by Jansoone,[51] and Adams and McDonald[77] have studied dipolar lattices. Recently Ewald methods have been applied in MD calculations for Stockmayer fluids by Pollock and Alder[54] and in MC calculations for dipolar hard spheres by Adams[53] and by De Leeuw et al.[23] De Leeuw et al.[23] also report results for the effective potential $\phi(1;12)$, which is just $\phi_{TPBC}(12)$. The details of these recent results are described in Section III.D, but there are two important points worth noting here.

First, it is clear from the discussion above that the Ewald pair correlation function [especially the $h^{110}(r)$ projection] must differ in some respect from that of the infinite system considered in theoretical treatments. If we ignore the periodicity; the system described by the effective potential $\phi(\varepsilon;12)$ or $\varepsilon'=\varepsilon$ is comparable with the theoretical situation, but such calculations have not yet been done. Alternatively, perhaps $h(\varepsilon;12)$ could be obtained from the Ewald result $h(\infty;12)$ by means of the approximate formulas (3.46), but again this has not been attempted.

Second, we remark that since ε does not depend on ε', there is no reason in principle for not comparing Ewald and theoretical estimates of ε. However, such comparisons must be interpreted with caution, since it is not known to what extent the Ewald results are influenced by the periodic boundary conditions. The LHNC and QHNC theories suggest[55] that for dense dipolar hard-sphere systems, ε can be sensitive to correlations much longer in range than even the largest value of $L/2$ yet used in an Ewald simulation. If this is true, the Ewald calculations are likely to seriously underestimate ε for all but small values of μ^{*2}. This possibility is examined in detail in Section III.D.2.

The Mean Reaction Field Method. Another method that attempts to take the long-range dipolar interactions into account is the mean reaction field (MRF) method suggested by Barker[80] and applied by Barker and Watts[75] in their MC calculations on liquid "water." The MRF method is very similar to the SC approach described above, but it seeks to account for the long-range dipolar interactions by surrounding the truncation sphere with a polarizable dielectric continuum. The central particle then interacts directly with all particles within the truncation sphere and with the reaction field **R**. The **R** arises from the polarization of the continuum by all the dipoles within the truncation sphere and is given by Onsager's expression[81]

$$\mathbf{R} = \frac{2(\varepsilon - 1)}{(2\varepsilon + 1)R_C^3} \mathbf{M}_S \qquad (3.47)$$

where $\mathbf{M}_S$ is the total dipole moment of the truncation sphere, and the dielectric constant of the continuum is taken to be that of the sample, ε. The ε is also assumed to obey (3.7a), with g given by

$$g = \frac{\langle M_S^2 \rangle}{N\mu^2} = 1 + \frac{4\pi\rho}{3} \int_0^{R_C} r^2 h_{MRF}^{110}(r)\, dr \qquad (3.48)$$

In principle ε could be obtained by iterating until (3.7a) and (3.48) are simultaneously satisfied. For large ε, however, **R** is nearly independent of ε, so usually the iteration is never carried out.[53, 77]

2. Nonperiodic Boundary Conditions

Two other methods, both of which explicitly avoid the use of periodic boundary conditions, have been proposed for the simulation of dipolar systems. Friedman[82] has proposed that a sample of N particles be enclosed in a spherical cavity within a dielectric continuum and has shown how the energy of such a system can be obtained. This method, however, has been applied in MC calculations by Valleau and Minns[83] and does not appear to work well in practice.

Another approach has been suggested by Bossis[84] and applied by Bossis, Quentrec, and Brot[85] to two-dimensional systems. These authors simply simulate a finite isolated disk of two-dimensional Stockmayer particles and find that apparently reliable estimates of the dielectric constant can be obtained.[85] An adaptation of methods first proposed by Berendsen[86] for finite spheres is used to determine ε. This approach appears to be very encouraging, but since this chapter is primarily concerned with three-dimensional fluids, we give no further details here. To exploit these two-dimensional calculations more fully, it would now be useful to apply some of the theories discussed in Section III.B to two-dimensional systems. Although in principle finite, isolated three-dimensional systems could be studied, such computations would likely prove difficult in practice, since large numbers of particles would be required to construct spherical samples of sufficient size.

D. Quantitative Results

In this section the results given by the various theories are described and compared, insofar as is possible, with MD or MC calculations. Also a qualitative comparison with experimental data for real liquids is made. The computer simulations do not provide as clear an evaluation of the different approximate theories as one would like, since for the reasons discussed in Section III.C, totally convincing estimates of ε have not been obtained. Therefore, to get some idea of the accuracy of the different approximations and to illustrate several of the points made in Section III.C, it is useful to begin by examining the pair correlation function.

The systems we consider are characterized by the reduced density $\rho^* = \rho a^3$, the reduced dipole moment $\mu^* = (\beta\mu^2/a^3)^{1/2}$, and the reduced quadrupole moment $Q^* = (\beta Q^2/a^5)^{1/2}$ where $a = d$ for hard particles and $a = \sigma$ for generalized Stockmayer fluids. In addition, for generalized Stockmayer systems the parameter $T^* = kT/\varepsilon$ must be specified.

1. The Pair Correlation Function

Dipolar Hard Spheres and Stockmayer Particles. As described at length in Section III.C, computer simulations of dipolar systems do not give the pair correlation function of an infinite system, and in general the computer results should not be compared directly with approximate theories. In much of the earlier work[31, 47–49] the severity of the problem was not fully recognized, and theoretical and simulation results were often compared without taking the influence of boundary conditions into account. These comparisons do give at least a rough idea of the accuracy of the various approximations. For example, it is clear that the MSA is very poor[31, 47, 49] and that better results can be expected from the LIN and L3 approximations.[31, 45] It is possible, however, to obtain a much better evaluation of a particular approximation by solving the theory for an infinite system with a spherically

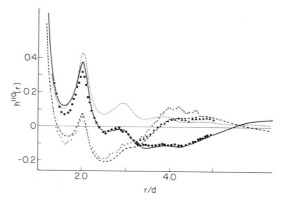

Fig. 3.   Values of $h^{110}(r)$ for dipolar hard spheres at $\rho^* = 0.8$ and $\mu^{*2} = 2.57$: dashed curve, LHNC for an infinite system with the potential truncated at $R_C = 3.4d$; crosses, MC results for $N = 256$, $R_C = 3.4d$; triangles, MC results for $N = 864$, $R_C = 3.4d$; solid curve, LHNC for an infinite system with the potential truncated at $R_C = 5.1d$; big dots, MC results for $N = 864$, $R_C = 5.1d$; dotted curve, LHNC for an infinite system with an untruncated potential. (Results from Ref. 52.)

truncated potential and comparing with SC results.[30, 52, 58, 59] This proce-dure is not entirely free of ambiguity, since the SC results are to some ex-tent influenced by the underlying periodicity not taken into account by the theory.[52] Also of course we are testing the theory for the spherically trun-cated potential, not the full potential. Nevertheless, this method provides a useful test of the pair correlation function.

The LHNC and QHNC approximations have been compared with Monte Carlo SC calculations,[30, 52, 58, 59] and typical results for dipolar hard spheres

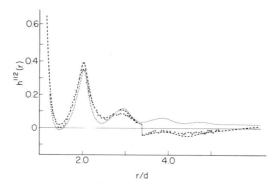

Fig. 4.   Values of $h^{112}(r)$ for dipolar hard spheres at $\rho^* = 0.8$ and $\mu^{*2} = 2.75$. The symbols have the same meaning as in Fig. 3. (Results from Ref. 52.)

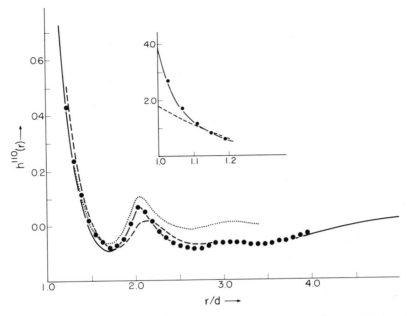

Fig. 5.   Values of $h^{110}(r)$ for dipolar hard spheres at $\rho^* = 0.4$ and $\mu^{*2} = 2.75$. The big dots are MC results for $N = 256$, $R_C = 4.2 d$. The solid and dashed curves are the QHNC and LHNC approximations, respectively, for a spherically truncated potential. The dotted curve is the QHNC result for an infinite system with an untruncated potential. (Results from Ref. 58.)

at high and low density are shown in Figs. 3 to 6. It is found[58] that although the LHNC and QHNC theories are roughly similar in accuracy at high density, the QHNC becomes clearly superior as the density is decreased (Figs. 5 and 6). This is easily understood, since the LHNC approximation $g^{000}(r) = g_{HS}(r)$ is good for dense fluids but poor at low densities, where the dipolar forces exert a greater influence on the spatial structure. The LHNC and QHNC theories have also been applied[58] to dense Stockmayer fluids ($\rho^* = 0.8$, $T^* = 1.35$), and comparable accuracy is obtained. The infinite system results also included in Figs. 3 to 6 serve to emphasize the dramatic effect of the spherical cutoff.

The other approximations described in Section III.C have not been solved for a spherically truncated potential, but an estimate of their accuracy can be obtained[87] by comparing with the LHNC or QHNC theories for an infinite system. We are of course assuming that the LHNC and QHNC approximations remain accurate for the full (untruncated) dipolar interaction and lie close to the true infinite system result. The MSA, LIN, L3, and LHNC theories for a dense dipolar hard-sphere system are compared in

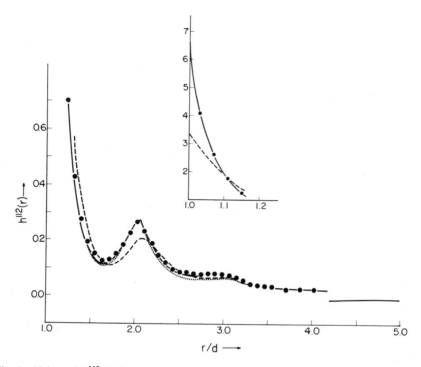

Fig. 6. Values of $h^{112}(r)$ for dipolar hard spheres at $\rho^* = 0.4$ and $\mu^{*2} = 2.75$. The symbols have the same meaning as in Fig. 5. (Results from Ref. 58.)

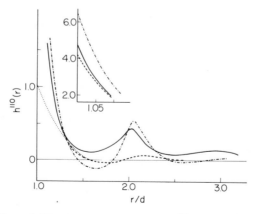

Fig. 7. Comparison of different approximations for $h^{110}(r)$ for dipolar hard spheres at $\rho^* = 0.8$ and $\mu^{*2} = 2.75$: dotted curves, MSA; dashed curves, LIN; dots and dashes, L3; solid curves, LHNC. (Results from Ref. 87.)

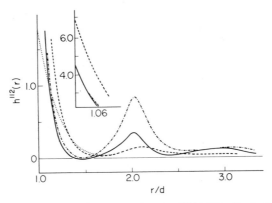

Fig. 8.   Comparison of different approximations for $h^{112}(r)$ for dipolar hard spheres at ρ^* $=0.8$ and $\mu^{*2}=2.75$. The curves represent the same theories as in Fig. 7. (Results from Ref. 87.)

Figs. 7 and 8. The LIN and L3 approximations considerably improve on the MSA near contact, but the second neighbor peak appears to be underestimated by LIN and overestimated by L3.

MRF results for dipolar hard spheres at $\rho^*=0.8$ have been reported by Adams[53] and by Levesque, et al.[52] Adams has carried out a 500-particle ($R_C=3.85d$) MRF calculation at $\mu^{*2}=2.75$ assuming that $\varepsilon=50$ in the reaction-field expression (3.47). The MRF results (Figs. 9 and 10) lie below the LHNC and QHNC results for an infinite system, but well above the SC

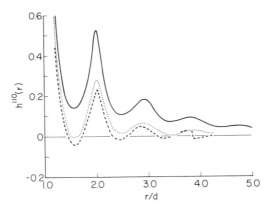

Fig. 9.   Values of $h^{110}(r)$ for dipolar hard spheres at $\rho^*=0.8$ and $\mu^{*2}=2.75$. The solid curve is the QHNC theory for an infinite system; the dotted and dashed curves are, respectively, the Ewald ($N=500$) and MRF ($N=500$, $R_C=3.85d$) results of Adams.[53]

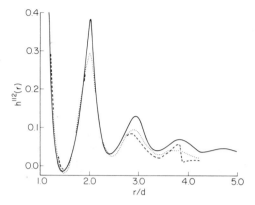

Fig. 10. Values of $h^{112}(r)$ for dipolar hard spheres at $\rho^* = 0.8$ and $\mu^{*2} = 2.75$. The curves represent the same calculations as in Fig. 9. (Results from Ref. 53.)

results for large r. The MRF results reported by Levesque, et al.[52a] show only a slight improvement upon the SC situation. However, it has very recently been learned[52b] that the MRF results reported in ref. [52a] are in error and the corrected results are in qualitative agreement with those of Adams. Further investigations of the MRF method using both MC and integral equation calculations are currently being carried out.[52b]

QHNC and Ewald results for Stockmayer fluids have been compared by Pollock and Alder[54] (Figs. 11 and 12), and Adams[53] has made similar comparisons for dipolar hard spheres (Figs. 9 and 10). We know from the work

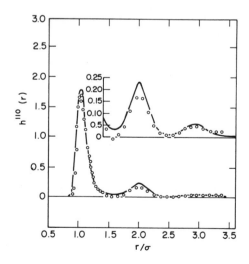

Fig. 11. Values of $h^{110}(r)$ for a Stockmayer fluid at $\rho^* = 0.8$, $T^* = 1.35$, and $\mu^{*2} = 2.269$. The solid curve is the QHNC theory and the circles are the Ewald ($N = 256$) results of Pollock and Alder.[54]

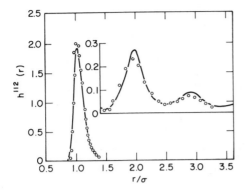

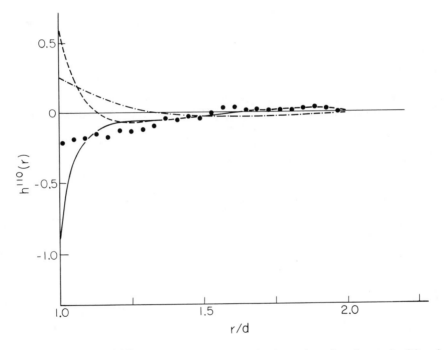

Fig. 12. Values of $h^{112}(r)$ for a Stockmayer fluid at $\rho^* = 0.8$, $T^* = 1.35$, and $\mu^{*2} = 2.269$. The solid curve is the QHNC theory and the circles are the Ewald ($N = 256$) results of Pollock and Alder.[54]

Fig. 13. Values of $h^{110}(r)$ for hard spheres with dipoles and quadrupoles at $\rho^* = 0.8$ and $\mu^* = Q^* = 1.0$. The dots are MC results ($N = 256$, $R_C = 3.4d$), and the solid, dashed, and dash-dot curves represent the QHNC, LHNC, and MSA, respectively, for a spherically truncated potential. (Results from Ref. 59.)

of De Leeuw et al.[23, 78] that the Ewald and QHNC pair correlation functions must be expected to differ in some way. This is especially true of the $h^{110}(r)$ projection. Nevertheless, Pollock and Alder[54] find that for Stockmayer systems the Ewald and QHNC results lie very close together (cf. Figs. 11 and 12), at least for the range of separations possible in the computer calculation. Adams,[53] on the other hand, finds rather large discrepancies between the Ewald and QHNC results for dipolar hard spheres (Figs. 9 and 10). This is interesting and a little surprising since we would have expected[58] both models to give similar results. For both dipolar hard spheres and Stockmayer particles, however, the Ewald method appears to give the closest approximation to a truly infinite system yet obtained in a computer simulation.

Hard Spheres with Dipoles and Quadrupoles. The LHNC, QHNC, and mean spherical approximations have been solved[59] for fluids of hard spheres with both dipole and quadrupole moments. Theoretical results for spherically truncated potentials have been compared with Monte Carlo (SC)

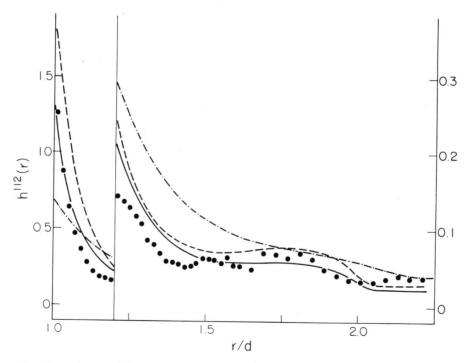

Fig. 14. Values of $h^{112}(r)$ for hard spheres with dipoles and quadrupoles at $\rho^* = 0.8$ and $\mu^* = Q^* = 1$. The symbols have the same meaning as in Fig. 13. (Results from Ref. 59.)

calculations as described above. The accuracy of the LHNC and QHNC approximations is found[59] to vary with the state parameters and with the particular projection considered, but in general the agreement with MC results is not as good as that obtained for purely dipolar systems. Examples of $h^{110}(r)$ and $h^{112}(r)$ for a dipole-quadrupole system are shown in Figs. 13 and 14. All three theories as well as the MC calculations indicate the strong dependence on the quadrupole moment of $h^{110}(r)$, hence ε. For fluids having significant quadrupole moments, $h^{110}(r)$ is found[59] to be nearly structureless and everywhere reduced in magnitude with respect to the purely dipolar or $Q=0$ result. This is extremely important leading to a greatly reduced dielectric constant[59] (cf. Section III.D.2). From the point of view of computer calculations, it is worth noting that even relatively small quadrupole moments tend to dominate the dipolar correlations, and spherically truncating the potential has little effect on $h(12)$ for $r < R_C$. This is of importance in the simulation of realistic liquids (e.g., HCl or NH_3) where substantial quadrupole moments are usually found.

2. The Dielectric Constant

Dipolar Hard Spheres and Stockmayer Particles. A number of theoretical results showing the μ^{*2} dependence for ε for a dense dipolar hard-sphere system are compared in Fig. 15 and in Table II. In addition to the MSA, LIN, L3, LHNC, and QHNC theories discussed in Section III.B, the older approximations of Debye[2] and Onsager,[19] as well as a very recent theory of Berkowitz and Adelman,[88] are also included. Berkowitz and Adelman[88] argue that the principal source of error in Onsager's theory is the neglect of a "two-cavity effect" purely electrostatic in origin. Estimating this effect they obtain the approximate formula

$$\frac{(\varepsilon-1)(2\varepsilon+1)}{9\varepsilon} \geq 3y^2 \tag{3.49}$$

which gives values of ε much larger than those obtained from Onsager's theory. However, there appears to be a serious problem with this result. If we consider the expansion

$$\frac{\varepsilon-1}{\varepsilon+2} = a_1 y + a_2 y^2 + a_3 y^3 + \cdots \tag{3.50}$$

the Berkowitz-Adelman (BA) theory gives $a_1 = 0$, $a_2 = 3$, and $a_3 = 0$. These are to be compared with the exact[62, 63] (at $\rho^* = 0$ in the case of a_3) coefficients $a_1 = 1$, $a_2 = 0$, and $a_3 = -15/16$. It is obvious that the limiting behavior given by the BA theory is incorrect. Equation 3.49 is completely

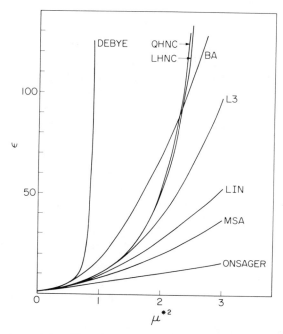

Fig. 15. Comparison of the different approximations for the dielectric constant of dipolar hard spheres at $\rho^* = 0.8$. See text and Table II for references.

TABLE II
Dielectric Constant Results for Dipolar Hard Spheres at $\rho^* = 0.8$

μ^{*2}	Theory						Berkowitz and Adelman	Monte Carlo	
	Onsager	MSA	LIN	L3	LHNC	QHNC		SC[a]	Ewald
0.5	3.17	3.59	3.84	3.76	3.75	3.75	4.81	3.73±0.1	
1.0	5.62	7.80	9.27	9.66	9.62	9.62	17.4	9.0±0.5	7.3[b]
2.0	10.60	20.00	27.06	37.93	50.0	51.7	67.9		22.7[c]
2.75	14.36	31.86	45.62	79.17	250.0	444.9	127.9		90±10[b]

[a] From Ref. 52a.
[b] From Ref. 53.
[c] From Ref. 23.

263

missing the terms in y and y^3, giving instead a term proportional to y^2 that does not occur in the exact result. It is interesting to note that the Onsager approximation gives $a_1 = 1$, $a_2 = 0$, and $a_3 = -2$ and thus, in this respect at least, is clearly superior to the BA theory. Berkowitz and Adelman[88] state that their results are in satisfactory agreement with the LHNC theory. However, it is obvious from Fig. 15 and Table II that this is true only insofar as both approximations predict that ε rises very rapidly with μ^{*2}. From Table II it can be seen that the quantitative agreement is in fact very poor. For example at $\mu^{*2} = 1$ the BA result is nearly twice as large as the LHNC value.

For dipolar hard spheres the LHNC and QHNC theories have been investigated[30, 58] for a wide range of density and dipole moments, and the following observations are of interest. Although for the range of μ^{*2} shown in Fig. 15 the LHNC and QHNC results lie very close together, this is not true for larger values of μ^{*2}. For example at $\rho^* = 0.8$ and $\mu^{*2} = 2.75$ (cf. Table II) the QHNC value is larger by nearly a factor of two! The bulk of this discrepancy can be traced[58] to relatively small differences in the "tail" of $h^{110}(r)$. Indeed, for dipolar fluids at high density the LHNC and QHNC theories suggest that ε is extremely sensitive to rather long-range dipolar correlations.[30, 55] The correlation range important in the LHNC approximation is illustrated in Fig. 16. Here we plot the ratio g_R/g, where

$$g_R = 1 + \frac{4\pi\rho}{3} \int_0^R r^2 h^{110}(r)\, dr \tag{3.51}$$

and $g = g_\infty$ is the Kirkwood g factor occurring in (3.7a). To obtain the LHNC ε from (3.7a), it is obvious that the integration must be continued until $g_R = g$ or $g_R/g = 1$. Thus Fig. 16 shows how the range of correlations contributing to ε becomes larger as μ^{*2} is increased. At large values of μ^{*2} ε is very sensitive to small changes in the long-range (i.e. $r > 8d$) part of $h^{110}(r)$, and this sensitivity gives rise to the large discrepancy between the LHNC and QHNC approximations for dense fluids. We remark that although only the LHNC results for g_R/g are given, the QHNC theory is qualitatively similar. Also, if ε is obtained from (3.8) rather than (3.7a), correlations of roughly comparable range must be considered,[55] or in other words $h^{112}(r)$ does not reach its asymptotic limit before $g_R/g \simeq 1$. At lower densities ($\rho^* \lesssim 0.6$) the significant correlation range is shorter, and both the LHNC and QHNC approximations give very similar results for all values of μ^{*2} considered.[58] For example, at $\rho^* = 0.6$, $\mu^{*2} = 3.0$ the LHNC and QHNC theories give $\varepsilon = 37.1$ and 40.0, respectively. We shall see below that to properly interpret simulation results, it is important to know the range of correlations making a significant contribution to ε.

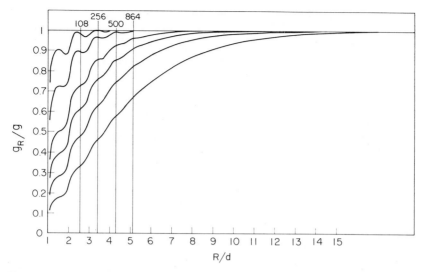

Fig. 16.   Values of g_R/g for dipolar hard spheres at $\rho^* = 0.8$. The vertical lines are drawn at $L/2$ for $N = 108$, 256, 500, and 864. From top to bottom, the curves are for $\mu^{*2} = 1$, 1.5, 2.0, 2.25, 2.5, and 2.75.

The LHNC and QHNC approximations have also been applied[58] to dense Stockmayer fluids, and results for $\rho^* = 0.8$, $T^* = 1.35$ are shown in Fig. 17. It is found[58] that for $\mu^{*2} \lesssim 2.0$ the dielectric constant of a Stockmayer fluid lies very close to the dipolar hard-sphere result, provided the thermodynamic states considered are chosen in accordance with simple thermodynamic perturbation theory. For larger dipole moments, however, the dielectric constant of the Stockmayer fluid is considerably lower than that of the corresponding dipolar hard-sphere system. Also, for Stockmayer particles the LHNC and QHNC theories are in better agreement at large μ^{*2} and the QHNC now gives the *smaller* rather than the *larger* value.

As discussed in Section III.C, computer simulations of dipolar systems provide something less than a definitive test of the approximate theories. Nevertheless, it is possible to get some idea of the accuracy of the various approximations. Several conclusions can be immediately reached by examining the pair correlation functions discussed in Section III.D.1. For example, it is clear that for dipolar systems the Onsager[19] formula

$$\frac{(\varepsilon - 1)(2\varepsilon + 1)}{9\varepsilon} = y \tag{3.52}$$

must underestimate ε, since it merely sets $g = 1$ in (3.7a), completely ignoring the term that is dependent on $h^{110}(r)$. It is also obvious that the MSA is

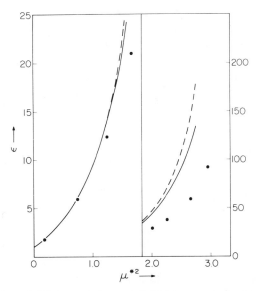

Fig. 17.   The dielectric constant of Stockmayer fluids at $\rho^* = 0.8$ and $T^* = 1.35$. The solid and dashed curves are the QHNC and LHNC approximations, respectively. The dots are the Ewald ($N = 256$) results of Pollock and Alder.[54]

very poor and again in all likelihood underestimates ε. It is a little more difficult to evaluate the remaining approximations. For a limited range of separations ($r \lesssim 5d$), SC calculations have shown[52, 58] that the QHNC theory gives a good approximation to $h^{110}(r)$ for spherically truncated potentials. For dense fluids this is also true of the LHNC theory. Thus if equivalent accuracy is obtained for the full dipolar interaction, we would expect these theories to be good at least for systems where ε is determined by correlations lying within the range $r \lesssim 5d$. Figure 17 indicates that for dipolar hard spheres at $\rho^* = 0.8$ this is the case if $\mu^{*2} \lesssim 2.0$. For larger values of μ^{*2}, ε is influenced by correlations lying outside the range tested in the SC calculations.

The most extensive and numerically reliable estimates of ε itself are provided by the MD (Ewald) calculations of Pollock and Alder[54] for Stockmayer fluids at $\rho^* = 0.8$ and $T^* = 1.35$. These estimates are compared with the LHNC and QHNC approximations in Fig. 17. The agreement between the theoretical and MD results is very good for relatively small values of μ^{*2} but becomes poor at the larger values, with the computer estimates lying well below the QHNC results. Pollock and Alder[54] conclude that the theory is in serious error at the larger values of μ^{*2}. However, in view of Fig. 16 and the discussion given above concerning the range of the correlations determining ε, this conclusion is perhaps premature. In an Ewald (or indeed any) simulation using periodic boundary conditions, correlations greater in range than $L/2$ are not properly taken into account. Thus if for larger dipole moments ε is very sensitive to correlations much greater in range than $L/2$, as the theories suggest, care must be taken in interpreting computer results for "small" periodic systems. It is not at all obvious that the dielectric constant so obtained will be close to that of the nonperiodic infinite system described by the approximate theories. In fact, for Stockmayer systems at $\rho^* = 0.8$, $T^* = 1.35$, the significant correlation range is roughly similar to that shown in Fig. 16, and the contribution to ε from correlations lying outside the central cube could explain much of the discrepancy between the QHNC and Ewald results.[89] Pollock and Alder[54] do report that their calculations show no significant number dependence, which suggests that the neglect of long-range correlations is not the origin of the discrepancy between the QHNC and Ewald results. However, this argument is less than convincing, since for large μ^{*2} varying N from 108 to 500 covers only a relatively small part of the significant correlation range (cf. Fig. 16), and examining the N dependence for systems of this size may not be a very sensitive test. Thus at least for larger values of μ^{*2} the status of the Ewald results for ε is not clear, and further investigation is necessary.

Pollock and Alder[54] also consider polarization fluctuations of finite wavelength and calculate the function

$$g(k) = 1 + \frac{\rho}{3} \int_V h^{110}(\mathbf{r}) e^{i\mathbf{k}\cdot\mathbf{r}} d\mathbf{r} \tag{3.53}$$

where V is the sample volume. For an infinite nonperiodic system, (3.53) simply becomes

$$g(k) = 1 + \frac{\rho}{3} \tilde{h}^{110}(k) \tag{3.54}$$

where $\tilde{h}^{110}(k)$ is defined by (3.11k) and $g(0)$ is just the Kirkwood g factor occurring in (3.7a). For finite k, Ewald and theoretical calculations of $g(k)$ may be compared directly. However, the Ewald results are discontinuous at $k=0$, and it is apparent from (3.45) and (3.7a) that the Ewald $g(0)$ is not the Kirkwood g given by the infinite system theories. Thus Ewald estimates of the Kirkwood g-factor must be obtained by eliminating ε from (3.45) and (3.7a).

The results of Pollock and Alder[54] for Stockmayer fluids are compared with the QHNC theory in Figs. 18 and 19. It can be seen that at $\mu^{*2} = 0.741$

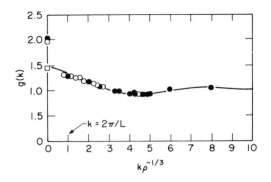

Fig. 18.   Values of $g(k)$ for a Stockmayer fluid at $\rho^* = 0.8$, $T^* = 1.35$, and $\mu^{*2} = 0.741$. The curve is the QHNC theory, and the solid and open circles are the Ewald results of Pollock and Alder[54] for $N = 256$ and 500, respectively. The open square is the Ewald ($N = 500$) result for the Kirkwood g-factor.

excellent agreement is obtained for all k. At $\mu^{*2} = 2.269$ the agreement is good for $k \geq 2\pi/L$ (which is the smallest finite wavelength allowed in the simulation), but there is a large discrepancy at $k = 0$. This discrepancy is of course apparent in the dielectric constant data, and its possible origin is discussed above and in Ref. 55.

Some Monte Carlo estimates of ε for dipolar hard spheres at $\rho^* = 0.8$ are given in Table II. At $\mu^{*2} = 0.5$ and 1.0, the SC results allow at least a rough evaluation of the different approximations. Taking into account the rather large statistical errors and the possible influence of boundary conditions, the LIN, L3, LHNC, and QHNC theories are all more or less in agreement with the MC calculations. The Onsager and MSA results appear to be too low, whereas the BA values are much too large. At $\mu^{*2} = 2.0$ and 2.75 it is not possible to reach any firm conclusions, but several remarks concerning the reported MC results are in order.

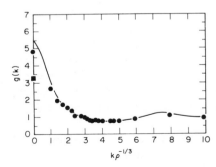

Fig. 19.   Values of $g(k)$ for a Stockmayer fluid at $\rho^* = 0.8$, $T^* = 1.35$, and $\mu^{*2} = 2.269$. The curve is the QHNC theory, the dots are the Ewald results of Pollock and Alder[54] for $N = 256$, and the square is the Ewald result for the Kirkwood g-factor.

At $\mu^{*2} = 2.0$ De Leeuw et al.[23] report two values for ε. These are 22.7 and 14.9, obtained with the effective potentials $\phi(\infty; 12)$ (Ewald) and $\phi(1; 12)$ $= \phi_{TPBC}(12)$, respectively. In principle, both effective potentials should yield the same dielectric constant (cf. Section III.C.1), and De Leeuw et al.[23] claim that their estimates of ε are consistent within statistical error. The discrepancy between these numbers and the QHNC result, $\varepsilon = 51.7$, is much too large to be completely explained by the contribution from correlations lying outside the central cube. (In the QHNC theory the contribution to ε from correlations greater in range than $L/2 \simeq 3.4d$ is about 20%.) Thus if these MC estimates[23] are to be taken seriously, De Leeuw et al. suggest that LHNC and QHNC theories break down even for values of μ^{*2} as low as 2.0. There are, however, two problems with this conclusion.

First of all, essentially the entire range of correlations contributing to ε at $\mu^{*2} = 2.0$ has been tested using SC calculations as described in Section III.D.1, and there is no evidence for such a breakdown. Thus the results of De Leeuw et al.[23] imply that the theories fail drastically for the full dipolar interaction but at the same time remain highly accurate for the spherically truncated case. This result is possible, but it would be very surprising. Second, there appears to be a discrepancy between the result of De Leeuw et al.[23] and that of Pollock and Alder,[54] who find a considerably larger dielectric constant ($\varepsilon = 29.0$) for the "equivalent" Stockmayer fluid. It is unlikely that this is due to the different short-range potentials, and some explanation should be found. In fact, as noted above, the theories predict that the Stockmayer fluid will have a dielectric constant *lower* than that of the corresponding dipolar hard-sphere system.[58]

The Ewald calculations of Adams[53] also bear comment. The estimate $\varepsilon = 90 \pm 10$ given in Table II was not obtained from (3.45) but rather by extrapolating results obtained with an applied electric field. Using (3.45), Adams reports estimates of ε ranging from 23.5 to 54.5, depending on N and the numerical accuracy to which the Ewald sums are calculated. Again these results are not very conclusive, and in all likelihood both the theoretical and the MC results are in error.

Hard Spheres with Dipoles and Quadrupoles. LHNC and MSA results[59] for fluids of hard spheres with dipoles and quadrupoles are shown in Fig. 20. In both approximations ε decreases with increasing quadrupole moment. The LHNC results are particularly dramatic, since the dipolar hard-sphere or $Q^* = 0$ value is very large to begin with. As discussed in Section III.B.3, the QHNC approximation is not very satisfactory for dipole-quadrupole systems, since solutions are not found[59] for some values of μ^* and Q^*. When solutions can be obtained, however, the QHNC ε also decreases sharply with quadrupole moment.

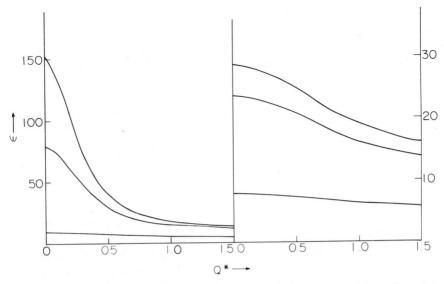

Fig. 20. The variation of ε with Q^* for fluids of hard spheres with dipoles and quadru-poles at $\rho^* = 0.8$. *Left*: LHNC results; *right*: MSA results. From top to bottom, the curves are for $\mu^* = 1.6$, 1.5, and 1.0. (Results from Ref. 59.)

The MSA, LHNC, and QHNC theories have been compared[59] with Monte Carlo (SC) estimates of ε. The results (Table III) are most usefully viewed in the following manner. For dipolar hard spheres at $\rho^* = 0.8$ and $\mu^* = 1.5$, the LHNC ε is 78.5 and the QHNC value is larger. Thus although the different theoretical and MC estimates included in Table III are not everywhere in particularly good agreement, it is clear that the discrepancies are in fact rather small compared with the difference between the large Q^* and $Q^* = 0$ values. Therefore, it appears safe to conclude that ε drops very

TABLE III

Dielectric Constant Results[a] for Fluids of Hard Spheres with Dipoles and Quadrupoles at $\rho^* = 0.8$

		Theory			Monte Carlo
μ^*	Q^*	MSA	LHNC	QHNC	(SC)
1.0	1.0	6.27	5.67	4.61	6.8 ± 0.3
1.5	0.5	20.7	29.9	—[b]	15.6 ± 1.1
1.5	1.0	16.29	12.68	10.33	10.2 ± 0.8

[a] All results are taken from Ref. 59.
[b] A QHNC solution is not obtained[59] at $\mu^* = 1.5$, $Q^* = 0.5$.

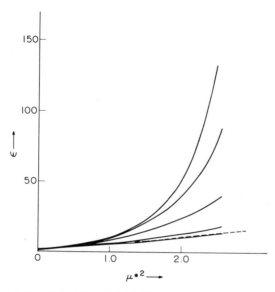

Fig. 21.   The variation of ε with μ^{*2} for fluids of hard spheres with dipoles and quadrupoles at $\rho^*=0.8$. From top to bottom the solid curves represent the LHNC results at $Q^*=0$, 0.25, 0.5, 1.0, and 1.5. The dashed curve is the Onsager approximation. (Results from Ref. 59.)

sharply with increasing Q^*. In fact, the MC calculations suggest that the LHNC *underestimates* both the magnitude and the abruptness of the decrease in ε.

In Fig. 21 ε is plotted as a function of μ^{*2} for several values of Q^*. It can be seen that the dependence of ε on μ^{*2} varies rather drastically with quadrupole moment and at the larger values of Q^* the μ^{*2} dependence is well approximated by the Onsager formula (3.52) for simple dipolar systems. Physically, this can be understood if we recall that the Onsager approximation can be obtained by setting $g=1$ or $\langle \boldsymbol{\mu}_1 \cdot \boldsymbol{\mu}_2 \rangle = 0$ in (3.7b). Quadrupolar interactions increase the probability of finding particles roughly oriented in T-like configurations for which $\boldsymbol{\mu}_1 \cdot \boldsymbol{\mu}_2 = 0$. Consequently we would expect $\langle \boldsymbol{\mu}_1 \cdot \boldsymbol{\mu}_2 \rangle$ to become smaller as Q^* is increased. The effectiveness of quadrupoles in disrupting dipolar orientations is evident in the thermodynamic properties as well,[48, 59] which were first studied by Stell et al.[61], who found great sensitivity to changes in Q^*.

Real Liquids.   Real molecules usually have both dipole and quadrupole moments. Thus the results described above strongly suggest that without the inclusion of higher multipole moments, simple dipolar models will be hopelessly inadequate for most real liquids. This is illustrated qualitatively in Fig. 22, which compares experimental results for a number of common liquids[90] with the LHNC theory for dipolar hard spheres at the typical liquid density, $\rho^* = 0.8$. QHNC results for Stockmayer fluids at $\rho^* = 0.8$ and $T^* = 1.35$ are also shown. If the integral equation theories are even roughly correct, it is obvious from Fig. 22 that the simple dipolar models give dielectric constants that are *much larger* than those usually observed experimentally. This is particularly striking when we note that polarization effects that would make ε larger still (cf. Section IV) are not taken into account by the present models.

The Onsager[19] approximation for polarizable molecules can be written[90] in the form

$$\frac{(\varepsilon - \varepsilon_\infty)(2\varepsilon + \varepsilon_\infty)}{\varepsilon(\varepsilon_\infty + 2)^2} = y \tag{3.55}$$

where ε_∞ is the high-frequency dielectric constant. We note that if $\varepsilon_\infty = 1$,

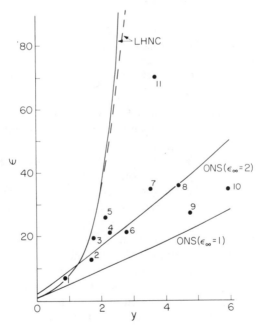

Fig. 22. Comparison with the dielectric constant of real liquids: LHNC results for dipolar hard spheres at $\rho^* = 0.8$ (solid curve), Stockmayer fluids at $\rho^* = 0.8$, $T^* = 1.35$ (dashed curve); ONS = Onsager. The dots are the experimental results[90] for the following liquids: (1) CH_3I, (2) CH_3Cl, (3) NH_3, (4) $CH_3 - C - CH_3$, (5) CH_3F, (6) $CH_3 - C - H$, (7) $C_6H_5NO_2$, (8) $\overset{\|}{O}$ $\overset{\|}{O}$ CH_3NO_2, (9) C_2H_5CN, (10) CH_3CN, (11) H_2O.

(3.55) reduces to (3.52) for nonpolarizable particles. Results for $\varepsilon_\infty = 1$ and for a typical liquid value, $\varepsilon_\infty = 2$, are included in Fig. 22. It can be seen that the Onsager theory, particularly if polarizability is included, is quite a good approximation for real liquids. This of course has been known for many years and is very likely attributable to the quadrupolar interactions, which in real liquids effectively kill the contribution to g that is ignored by the Onsager theory. It is also possible that other forces, such as anisotropic short-range interactions, contribute to the destruction of $\langle \boldsymbol{\mu}_1 \cdot \boldsymbol{\mu}_2 \rangle$ in molecular fluids. Thus although the Onsager theory is in all likelihood very inaccurate for simple dipolar models, it is nevertheless a good approximation for real liquids. This is a fortunate example of approximations made in the theoretical treatment acting to largely cancel the inadequacies of the model.

Finally we observe that the dielectric constant of water lies well above that of the other liquids shown in Fig. 22. This is often attributed to hydrogen bonding effects, but in view of the present discussion it is worth noting that water does not have an axially symmetric quadrupole and, more important, the component of the quadrupole moment tensor along the dipole (Q_{zz}) is practically zero for the water molecule.[91] Thus the axially symmetric models we consider do not even approximately apply to water, and it is interesting to speculate that perhaps the water quadrupole is less effective at destroying the dipolar correlations. This would lead to a larger dielectric constant, and the possibility constitutes one of the important problems remaining to be investigated.

IV. POLAR–POLARIZABLE FLUIDS

A. Models and Formally Exact Expressions

Almost all the formalism and the approximation schemes of Sections II and III have a natural extension to systems of polarizable dipolar particles, but the precise details of the extension depend on the way polarizability is introduced into the Hamiltonian. We refer to the two quite distinct Hamiltonian models that have been most thoroughly developed in this context as the constant-polarizability model and the fluctuating-polarizability model. The dielectric behavior of the former was first systematically investigated from a statistical mechanical viewpoint by Kirkwood[93] and by Yvon,[94] who considered the model almost exclusively in the absence of permanent dipole moments. (Kirkwood[17] subsequently pioneered an exact formulation of the statistical mechanics of polar molecules, but largely as a separate enterprise that did not attempt to treat the polarizability exactly.) The general case of polar-polarizable particles remained only very partially developed[95] until the recent fundamental work of Wertheim.[29, 96–99] A systematic statistical mechanical investigation of the fluctuating-polarizability model was initiated by van Vleck[100] and extended somewhat by W. F. Brown,[101] but it enjoyed little further development until the very recent work of Høye and Stell,[102] who treat the general polar-polarizable case, and Pratt,[103] who considers the nonpolar case. Pratt's work is the outgrowth of an approach developed by Chandler[104] and by Chandler and Pratt[105] on the atom-atom correlation function formalism, in terms of which Chandler[104] had earlier generalized the Høye-Stell equation (2.40) to the case of polarizable polar molecules. We rederive Chandler's generalization of that equation directly in terms of the Høye-Stell formalism below. As indicated there, the underlying model is one of fluctuating polarizability rather than constant polarizability (although the derivations do not address themselves to the details of the model). The result provides an as-yet-unexploited link between ε and molecular parameters.

In the constant-polarizability model, each particle is assumed to carry an induced polarization **p** that is instantaneously proportional to the local electric field $\mathbf{E}_l$ that acts on it, with a fixed tensor constant of proportionality, so that

$$\mathbf{p} = \boldsymbol{\alpha} \cdot \mathbf{E}_l \qquad (4.1\mathrm{a})$$

where $\boldsymbol{\alpha}$ is the polarizability. For particles that each carry a permanent moment $\boldsymbol{\mu}$ as well, the total dipole moment **m** of a particle is given in this model by

$$\mathbf{m} = \mathbf{p} + \boldsymbol{\mu} \qquad (4.1\mathrm{b})$$

For simplicity, we treat here only a system of identical particles.

If we define a molecular coordinate system by taking the z-axis along the permanent dipole, then we shall consider here $\boldsymbol{\alpha}$ of the form

$$\boldsymbol{\alpha} = \begin{bmatrix} \alpha_{xx} & 0 & 0 \\ 0 & \alpha_{yy} & 0 \\ 0 & 0 & \alpha_{zz} \end{bmatrix} \qquad (4.2)$$

Thus α_{zz} is the polarizability in the direction of the permanent dipole. The local field $(\mathbf{E}_l)_i$ acting on the ith particle in a system of particles is a fluctuating quantity, given at any instant by

$$(\mathbf{E}_l)_i = - \sum_{j \neq 1} \mathbf{T}_{ij} \cdot \mathbf{m}_j \qquad (4.3)$$

where $\mathbf{T}_{ij}$ is the dipole tensor.

In the fluctuating-polarizability model, the polarization carried by a particle is regarded as the amplitude of a quantity that oscillates (i.e., an internal motion of the molecule representing the relative displacement of positive and negative charge), even in the presence of fixed E_l. In the purely classical version of the model, the fluctuation of internal coordinates is purely thermal, the internal degrees of freedom of an isolated particle fluctuating when it is in a heat bath according to the dictates of classical statistical mechanics. Since the polarization per particle represents a potential energy—call it $\phi(\mathbf{p})$—even when $E_l = 0$, the probability of an isolated particle in a heat bath having induced polarization $\mathbf{p}$ is proportional to $\exp[-\beta\phi(\mathbf{p})]$ when $E_l = 0$. More generally, it is proportional to $\exp[-\beta\phi_1(\mathbf{p}, \mathbf{m})]$, where

$$\phi_1(\mathbf{p}, \mathbf{m}) = \phi(\mathbf{p}) - \mathbf{m} \cdot \mathbf{E}_l \qquad (4.4)$$

The instantaneous polarization is still given by $\mathbf{m} = \mathbf{p} + \boldsymbol{\mu}$, but a more important quantity is now the mean polarization

$$\langle \mathbf{m} \rangle = \int \mathbf{m} g(\mathbf{m}) \, d\mathbf{m} \qquad (4.5)$$

where

$$g(\mathbf{m}) = \frac{\exp[-\beta\phi_1]}{\int \exp[-\beta\phi_1] \, d\mathbf{m}} \qquad (4.6)$$

When $\phi(\mathbf{p})$ is a harmonic potential,

$$\phi(\mathbf{p}) = c_x p_x^2 + c_y p_y^2 + c_z p_z^2 \qquad (4.7)$$

then there is a linear relation between $\langle \mathbf{m} \rangle - \boldsymbol{\mu}$ and E_l for E_l of arbitrary strength:

$$\langle \mathbf{m} \rangle = \boldsymbol{\mu} + \boldsymbol{\alpha} \cdot \mathbf{E}_l \qquad (4.8a)$$

with $\boldsymbol{\alpha}$ given by (4.2), where

$$\alpha_{xx} = \frac{1}{2c_x}, \qquad \alpha_{yy} = \frac{1}{2c_y}, \qquad \alpha_{zz} = \frac{1}{2c_z} \qquad (4.8b)$$

In the fluctuating-polarizability model with harmonic $\phi(\mathbf{p})$, the $\boldsymbol{\alpha}$ of (4.8) is identified as the polarizability of an isolated particle. With terms of higher order in the components of **p** added to the right-hand side of (4.7), $\langle \mathbf{m} \rangle - \boldsymbol{\mu}$ will no longer be linear in $\mathbf{E}_I$ except in the limit of $\mathbf{E}_I \to 0$, and the field-independent polarizability of isolated particle is defined by (4.8a) only in that limit. More generally, one has higher-order terms in $\mathbf{E}_I$,

$$\langle \mathbf{m} \rangle = \boldsymbol{\mu} + \boldsymbol{\alpha} \cdot \mathbf{E}_I + \frac{1}{2} \boldsymbol{\beta} : \mathbf{E}_I \mathbf{E}_I + \frac{1}{6} \boldsymbol{\gamma} : \mathbf{E}_I \mathbf{E}_I \mathbf{E}_I + \cdots \qquad (4.9)$$

A few words concerning terminology are in order here. In the absence of permanent moments, the constant-polarizability model is sometimes referred to as the dipole–induced-dipole (DID) model, but the phrase "DID result" sometimes seems to refer to the use of the model along with certain approximations that have become standard in its connection—primarily, those that involve dropping all but the leading terms in polarizability in the virial coefficients of the Clausius-Mossotti function and related quantities. The term "instantaneous approximation" is used by Brown[101] to refer to the constant-polarizability model. [The fluctuating-polarizability model might alternatively be called the fluctuating-polarization model, since the primary variable of interest that is fluctuating is **m**, the mean and variance of which we typically consider. However, except when the ith particle is an isolated one, its $\mathbf{m}_i$ and $\mathbf{p}_i$ will be fluctuating even in the constant-polarizability model, simply because $(\mathbf{E}_I)_i$ is. This manifestation of fluctuation is sometimes[93] referred to as "translational fluctuation," to distinguish it from the fluctuation associated with internal coordinates.]

1. Constant-Polarizability Results

Wertheim has extended many of the exact results of Section II to the case of dipolar molecules with constant polarizability. (Higher permanent multipoles are assumed to be absent.) For example, the direct extensions of (2.30a) and (2.30b) were obtained by him[96] in terms of the function

$$\overline{\mathbf{A}}(\mathbf{r}_1, \mathbf{r}_2) = \beta^{-1} \left. \frac{\delta \ln \Xi[E_0]}{\delta E_0(\mathbf{r}_1) \delta E_0(\mathbf{r}_2)} \right|_{E_0 = 0} \qquad (4.10)$$

which takes over the role played by $h_\delta(12)$ in Section II. The $\ln \Xi[E_0]$ is the grand partition function of the system in the presence of applied field $E_0(\mathbf{r})$. (In this subsection, IV.A.1, we follow Wertheim's notation except where otherwise noted.)

In the thermodynamic limit, $\overline{\mathbf{A}}(\mathbf{r}_1, \mathbf{r}_2) \to \overline{\mathbf{A}}_\infty(\mathbf{r}_{12})$, a function of $\mathbf{r}_1$ and $\mathbf{r}_2$ only through $\mathbf{r}_{12}$. Any second-rank tensor function $\mathbf{a}(\mathbf{r}_{12})$ must have the form

$$\mathbf{a}(\mathbf{r}_{12}) = a_I(r_{12}) \mathbf{I} + a_T(r_{12}) \mathbf{T}_0(\mathbf{r}_{12}) \qquad (4.11)$$

where $\mathbf{I}$ is the unit tensor and $r^{-3}\mathbf{T}_0(\mathbf{r}_{12})$ is the dipole tensor

$$\mathbf{T}(\mathbf{r})=r^{-3}\mathbf{T}_0(\mathbf{r})= \lim_{d\to 0} \begin{cases} 3r^{-5}\mathbf{r}\cdot\mathbf{r}-r^{-3}\mathbf{I} & \text{for } r>d \\ 0 & \text{for } r<d \end{cases} \qquad (4.12)$$

For such tensors the special role of the Δ and D components of the orientationally dependent correlation functions discussed in Section II is taken over by the analogous I and T components here. The Fourier transform of $\mathbf{a}(r_{12})$ given by (4.11) is

$$\tilde{\mathbf{a}}(k)=\tilde{a}_I(k)\mathbf{I}+\bar{a}_T(k)\mathbf{T}_0(\mathbf{k}) \qquad (4.13)$$

where $\tilde{a}_I(k)$ and $\bar{a}_T(k)$ are the Fourier and Hankel transforms of $a_I(r)$ and $a_T(r)$ given by (2.22b) and (2.22c), respectively. [Wertheim denotes $\bar{a}_T(k)$ as $\tilde{a}_T(k)$.] Letting

$$\bar{\mathbf{A}}_\infty(\mathbf{r}_{12})=\bar{A}_I(r_{12})\mathbf{I}+\bar{A}_T(r_{12})\mathbf{T}_0(\mathbf{r}_{12}) \qquad (4.14)$$

Wertheim finds,[29] as extensions of (2.30a) and (2.30b), respectively,

$$\frac{(\varepsilon-1)(2\varepsilon+1)}{3\varepsilon}=4\pi\int\bar{A}_I(r)\,d\mathbf{r} \qquad (4.15)$$

$$\frac{(\varepsilon-1)^2}{\varepsilon}=(4\pi)^2\lim_{r\to\infty}r^3\bar{A}_T(r) \qquad (4.16)$$

Here G, the generalization of our z, is given by

$$G=\left(\frac{4\pi}{3}\right)\tilde{G}_I(0) \qquad (4.17)$$

where $\tilde{G}_I(0)$ is the $\mathbf{I}$ component of the function $\tilde{G}(\mathbf{k})$, related to $\tilde{A}_\infty(\mathbf{k})$, the transform of $\bar{\mathbf{A}}_\infty(\mathbf{r}_{12})$, by the OZ-type relation

$$\tilde{A}_\infty(\mathbf{k})=\tilde{G}(\mathbf{k})-\left(\frac{4\pi}{3}\right)\tilde{G}(\mathbf{k})\cdot\mathbf{T}_0(\mathbf{k})\cdot\tilde{A}_\infty(\mathbf{k}) \qquad (4.18)$$

Thus $\tilde{G}_I(k)$ is playing the role played by $\tilde{\Sigma}_\Delta(k)$ in (2.25a). Equation 4.15 can also be written in a form that directly generalizes (2.15) or (2.26) through the introduction of the functional inverse of $\bar{A}_\infty(r)$:

$$\int\bar{\mathbf{Y}}(r_{13})\bar{\mathbf{A}}_\infty(r_{32})\,d\mathbf{r}_3=\delta(r_{12})\mathbf{I} \qquad (4.19)$$

As shown in Ref. 96,

$$\frac{\varepsilon-1}{\varepsilon+2} = \frac{4\pi}{\int \overline{Y}_l(r)\,d\mathbf{r}} \tag{4.20}$$

Thus $\overline{Y}(\mathbf{r})$ takes over the role played in (2.15d) by $c_\delta(12)$, the functional inverse of $\rho(1)\rho(2)h_\delta(12)$. A generalization of (2.37) can also be derived as outlined briefly in Ref. 29 and considered in detail in Ref. 98. For a finite spherical sample of macroscopic volume V in a vacuum, one obtains

$$\int_V \overline{A}(\mathbf{r}_1,\mathbf{r}_2)\,d\mathbf{r}_2 = \frac{3(\varepsilon-1)}{4\pi(\varepsilon+2)}\mathbf{I} \tag{4.21}$$

In addition to the direct analogs of the nonpolarizable-particle expressions considered above, Wertheim has also derived a family of expressions involving ε_∞ as well as ε, where ε_∞ is the high-frequency dielectric constant that corresponds to ε computed in a static system in which molecular reorientation is suppressed. In the model under consideration, the relevant aspect of such reorientation is reorientation of the permanent dipole vector, so the computation of ε_∞ effectively reduces to the computation of ε in the absence of the permanent dipole moment; the expressions involving ε_∞ then follow trivially from the expressions above by subtraction. For example, corresponding to (4.21), there is the expression

$$\int_V \overline{A}_\infty(\mathbf{r}_1,\mathbf{r}_2)\,d\mathbf{r}_2 = \frac{3(\varepsilon_\infty-1)}{4\pi(\varepsilon_\infty+2)}\mathbf{I} \tag{4.22}$$

hence

$$\int_V \left[\overline{A}(\mathbf{r}_1,\mathbf{r}_2)-\overline{A}_\infty(\mathbf{r}_1,\mathbf{r}_2)\right]d\mathbf{r}_2 = \left[\frac{3(\varepsilon-1)}{4\pi(\varepsilon+2)} - \frac{3(\varepsilon_\infty-1)}{4\pi(\varepsilon_\infty+2)}\right]\mathbf{I} \tag{4.23}$$

Equation 4.23 is in some ways a more natural expression than (4.21) because for a spherical sample of macroscopic volume, it can be shown[98] that

$$\beta\langle\mathbf{M}_V^2\rangle = \int_V\int_V \left[\overline{A}(\mathbf{r}_1,\mathbf{r}_2)-\overline{A}_\infty(\mathbf{r}_1,\mathbf{r}_2)\right]d\mathbf{r}_1\,d\mathbf{r}_2 \tag{4.24}$$

where $\mathbf{M}_V$ is the total dipole moment of the sample.[106] Similarly, for a macroscopic spherical region of volume V in a uniform infinite system, one again has (4.24), along with (4.15) used for both ε and ε_∞, to give[98]

$$\beta\langle M_V^2\rangle = \frac{V}{4\pi}\left[\frac{(\varepsilon-1)(2\varepsilon+1)}{3\varepsilon} - \frac{(\varepsilon_\infty-1)(2\varepsilon_\infty+1)}{3\varepsilon_\infty}\right]\mathbf{I} \qquad (4.25)$$

This is to be distinguished from an earlier expression of Fröhlich,[95]

$$\langle M_V^2\rangle = \frac{V}{4\pi}\frac{(\varepsilon-\varepsilon_\infty)(2\varepsilon+1)^2}{3\varepsilon(2\varepsilon+\varepsilon_\infty)}\mathbf{I} \qquad (4.26)$$

which refers instead to a macroscopic spherical sample embedded in a passive dielectric continuum, for which ε and ε_∞ are equal [and equal to the ε of the sample exhibited in (4.26)]. Felderhof rederives (4.25) and (4.26) from a macroscopic viewpoint and discuss the relation between them and related expressions in great detail in Ref. 107.

2. Fluctuating-Polarizability Model

The work by Høye and Stell on the fluctuating-polarizability model[102] regards molecules with different $\mathbf{m}$ and $\mathbf{p}$ as being molecules of different species. The route to ε that has been exploited by Høye and Stell in this connection involves (2.26) and the appropriate generalizations of (2.25c) to (2.25e). When fluctuating polarizability is added, we have the probability density $\rho_{\mu,\mathbf{p}}$ that gives the probable distribution of particles with permanent moment μ and instantaneous induced moment $\mathbf{p}$. This gives a distribution for fixed $\mathbf{m}$ of

$$\rho_\mathbf{m} = \int \rho_{\mu,\mathbf{p}}\,d\mu, \qquad |\mu| \text{ fixed at } \mu \qquad (4.27)$$

The $\rho_\mathbf{m}$ is independent of orientation of $\mathbf{m}$ and thus corresponds to a density ρ_m of polar molecules of moment magnitude m. This corresponds in turn to the continuum limit of a mixture of dipolar particles, all of which have the same intermolecular potential except for dipolar moment magnitudes. Thus in (2.25d) $\rho\mu^2$ is replaced by

$$\rho\langle\mathbf{m}^2\rangle = \int \rho_{\mu,\mathbf{p}}m^2\,d\mu\,d\mathbf{p} = \int \rho_m m^2\,dm \qquad (4.28)$$

The $\langle m^2\rangle$ can thus be thought of as the square of the moment m_e of an equivalent single-component nonpolarizable polar system, and we can write

$$3y_e = \frac{4\pi}{3} \beta \rho m_e^2 = q_1 - q_2 \tag{4.29}$$

in place of (2.25d), retaining the expression (2.26)

$$\varepsilon = \frac{q_1}{q_2} \tag{4.30}$$

Thus in the Høye-Stell formalism, the central problem reduces to the computation of the $\rho_{\mu,\mathbf{p}}$ or its equivalent, corresponding in Wertheim's formalism to the computation of $\bar{A}_\infty(r_{12})$ or its equivalent.

The generalization of (2.40) to molecules with fluctuating polarizability can be best understood in terms of an Ornstein-Zernike equation, which can be written in Fourier space as

$$\tilde{h}(k) = \tilde{\omega}(k)\tilde{c}(k)[1 - \rho\tilde{\omega}(k)\tilde{c}(k)]^{-1}\tilde{\omega}(k) \tag{4.31}$$

where the $\tilde{h}(k)$, $\tilde{\omega}(k)$, and $\tilde{c}(k)$ are matrices, the elements of which are the Fourier transforms, $\tilde{h}_{ij}(k)$, $\tilde{\omega}_{ij}(k)$, and $\tilde{c}_{ij}(k)$ of the site-site correlation functions $h_{ij}(r)$, $c_{ij}(r)$, and $\omega_{ij}(r)$ that refer to the interaction of pairs of atoms of species i and j. The ρ is the molecular density. The $\tilde{h}_{ij}(k)$ and $h_{ij}(r)$ refer to the pair correlation between atoms *in different molecules*, and $\tilde{\omega}_{ij}(k)$ and $\omega_{ij}(r)$ refer to the pair correlation between atoms within the same molecule (as well as including the delta-function "self-correlation" term that always arises when one considers the probability of simultaneously finding one particle centered at $\mathbf{r}_i$ and one particle—possibly the same one—centered at $\mathbf{r}_j$). Thus the relation among $\tilde{\omega}$, $\tilde{h}$, and the usual particle-particle correlation function for the atoms, viewed as particles in a mixture, is given by

$$\delta_{ij} + \rho\tilde{H}_{ij}(k) = \tilde{\omega}_{ij}(k) + \rho\tilde{h}_{ij}(k) \tag{4.32a}$$

or, in matrix notation,

$$I + \rho\tilde{H} = \tilde{\omega} + \rho\tilde{h} \tag{4.32b}$$

where I is the unit matrix and we are using H_{ij} to denote the usual pair correlation function between particles of species i and j in a mixture. [We would normally use h_{ij} for this quantity, but in this subsection we must work with both the H_{ij} and h_{ij} in (4.32a).] Now the usual OZ equation for a mixture can be written, in matrix notation as

$$(I + \rho \tilde{H})(I - \rho \tilde{C}) = I \tag{4.33}$$

where $\tilde{C}$ is the direct correlation matrix in Fourier space, with elements $\tilde{C}_{ij}(k)$. [We would normally use $\tilde{c}_{ij}(k)$ to denote this quantity.] We now note that (4.31) and (4.33) are the same equation, when we define $\tilde{c}$ in terms of $\tilde{C}$ by the equation

$$I - \rho \tilde{C} = \tilde{\omega}^{-1} - \rho \tilde{c} \tag{4.34}$$

This follows immediately from some elementary algebra, when (4.33) is rewritten in terms of $\tilde{\omega}$, $\tilde{h}$, and $\tilde{c}$. Chandler and his colleagues[104, 105] refer to (4.31) as an "Ornstein-Zernike-like" equation; we see in fact that if (4.34) is used to define $\tilde{c}$, (4.31) is nothing but the usual OZ equation for the mixture of atoms that make up the molecules. The $\tilde{\omega}^{-1}$ is the contribution to $\tilde{C}$ from the *intra*molecular correlations, whereas $\tilde{c}$ is the contribution from correlations between atoms in different molecules, that is, direct *inter*molecular correlations.

In the case of rigid molecules, for a distance between atom i and atom j of d_{ij}, one has

$$\tilde{\omega}_{ij}(k) = j_0(kd_{ij}); \qquad \omega_{ij}(r) = \delta(r - d_{ij})/4\pi d_{ij}^2 \tag{4.35}$$

so that expanding $\tilde{\omega}_{ij}(k)$ about $k = 0$,

$$\tilde{\omega}_{ij}(k) = \omega_{ij}^{(0)} + k^2 \omega_{ij}^{(2)} + \cdots \tag{4.36}$$

one finds

$$\omega_{ij}^{(2)} = -\tfrac{1}{6} d_{ij}^2 \tag{4.37}$$

In this case, if atoms i and j carry charges q_i and q_j, respectively, we have

$$\sum_{i,j} q_i q_j \omega_{ij}^{(2)} = \frac{m^2}{3} \tag{4.38}$$

since

$$\sum_{i,j} q_i q_j d_{ij}^2 = -2m^2 \tag{4.39}$$

In the more general case of $\omega_{ij}(k)$ associated with nonrigid molecules, one has instead of (4.38)

$$\sum_{i,j} q_i q_j \omega_{ij}^{(2)} = \frac{(m')^2}{3} + \frac{\alpha'}{\beta} \tag{4.40}$$

which follows from the usual definitions[104] of molecular dipole moment, the magnitude of which we have denoted m', and molecular polarizability, the trace of which we have denoted $3\alpha'$. We note that for nonrigid molecules, $\tilde{\omega}_{ij}(k)$ is state dependent (i.e., ρ and β dependent), and thus m' and α' are too. The m' is just $|\langle \mathbf{m}(1) \rangle|$, the magnitude of the mean polarization of the molecule (measured in the coordinate system of that molecule), and $3\alpha'$ is just $\beta[\langle \mathbf{m}^2 \rangle - |\langle \mathbf{m}(1) \rangle|^2]$. One has, in terms of the m_e of (4.29),

$$m_e^2 = (m')^2 + \frac{3\alpha'}{\beta} \tag{4.41}$$

The generalization of (4.38) to (4.40) is the sole change necessary in the discussions given in Refs. 24 and 25 of (2.40), which can be written in the form [see (7) of Ref. 25]

$$\frac{\varepsilon - 1}{\varepsilon} = \lim_{k \to 0} \frac{2\pi\beta\rho}{3} \nabla_k^2 \left[\sum_{i,j} \rho q_i q_j \tilde{h}_{ij}(k) + \frac{m^2}{3} k^2 \right] \tag{4.42}$$

which thus generalizes to

$$\frac{\varepsilon - 1}{\varepsilon} = \lim_{k \to 0} \frac{2\pi\beta\rho}{3} \nabla_k^2 \left[\sum_{i,j} \rho q_i q_j \tilde{h}_{ij}(k) + \left\{ \frac{(m')^2}{3} + \frac{\alpha'}{\beta} \right\} k^2 \right] \tag{4.43}$$

Expanding $\tilde{h}_{ij}$ about $k = 0$,

$$\tilde{h}_{ij}(k) = h_{ij}^{(0)} + k^2 h_{ij}^{(2)} \tag{4.44}$$

we can rewrite these expressions as

$$\frac{\varepsilon - 1}{\varepsilon} = 4\pi\beta\rho^2 \sum_{i,j} q_i q_j h_{ij}^{(2)} + \frac{4\pi}{3} \beta\rho m^2 \tag{4.45}$$

and

$$\frac{\varepsilon - 1}{\varepsilon} = 4\pi\beta\rho^2 \sum_{i,j} q_i q_j h_{ij}^{(2)} + 4\pi\beta\rho \left\{ \frac{(m')^2}{3} + \frac{\alpha'}{\beta} \right\} \tag{4.46}$$

respectively. Equation 4.45 is the form of (4.42) originally given as (35) in Ref. 24 except for minor notational differences. Equation 4.46 is the form given by Chandler.[104] We note that the site-site intermolecular correlation function $h_{ij}(r)$ whose transform appears in (4.43) is defined in a way that automatically embodies averaging over all intramolecular site-site charge displacements (i.e., over all internal molecular states).

B. Implementation of Wertheim's SSCA for Polar-Polarizable Fluids with Constant Polarizability

Wertheim's formulation of his SSC approximation, which we have already discussed in the context of nonpolarizable fluids in Sections II and III, applies to the more general case of polar-polarizable fluids. In describing this case we use his notation. For polarizable dipolar hard spheres, the approximation is defined by the integral equations[29]

$$\hat{\mathfrak{N}}_T(r) = \beta^{-1} B \big(\hat{W}_T * \hat{H}_T + \hat{W}_T * H_I + W_I * \hat{H}_T \big) \tag{4.47a}$$

$$\mathfrak{N}_I(r) = \beta^{-1} B \big(2\hat{W}_T * \hat{H}_T + W_I * H_I \big) \tag{4.47b}$$

subject to the closure relations

$$H_T(r) = H_I(r) = 0 \qquad \text{for} \quad r < d \tag{4.47c}$$

$$W_I(r) = g_{\text{HS}}(r)\mathfrak{N}_I(r) - \mathfrak{N}_I(r) \qquad \text{for} \quad r > d \tag{4.47d}$$

$$W_T(r) = g_{\text{HS}}(r)\left[\mathfrak{N}_T(r) + \frac{\beta}{r^3} \right] - \mathfrak{N}_T(r) \tag{4.47e}$$

where, with $\alpha' = \text{Tr}\,\boldsymbol{\alpha}'$

$$B = \rho \left(\frac{1}{3}\beta m'^2 + \alpha' \right) \tag{4.47f}$$

$\mathfrak{N}_T = H_T - W_T$, $\mathfrak{N}_I = H_I - W_I$. The functions H_T, H_I, W_T, and W_I are defined by Wertheim,[29] and the caret denotes an integral transform of the type defined in (3.28a). The m' and α' are the renormalized dipole moment and polarizability of Wertheim.[29] We discuss their physical significance below. In Wertheim's "first renormalization" or 1-R approximation $\mathbf{m}'$ and $\boldsymbol{\alpha}'$ are related to the permanent dipole moment, $\boldsymbol{\mu}$, and polarizability $\boldsymbol{\alpha}$ by the equations[29]

$$\boldsymbol{\alpha}' = \boldsymbol{\alpha} + 8\pi K_H \beta^{-1} B \boldsymbol{\alpha}' \cdot \boldsymbol{\alpha} \tag{4.48a}$$

and

$$\mathbf{m}' = \boldsymbol{\mu} + 8\pi K_H \beta^{-1} B \boldsymbol{\alpha}' \cdot \boldsymbol{\mu} \tag{4.48b}$$

where

$$K_H = \int_d^\infty \frac{H_T(r)}{r} \, dr \tag{4.48c}$$

(Wertheim[97] has also introduced a "second renormalization" or 2-R approximation, but this development does not directly enter the computation of ε, although it is relevant to ε_∞. Wertheim[99] has shown that for key thermodynamic properties, the 2-R results differ only slightly from those given by the 1-R theory.)

Equations 4.47 can be written in a more familiar form[108] if we introduce the dimensionless functions

$$h_T(r) = \beta^{-1} b H_T(r) \tag{4.49a}$$

$$h_I(r) = \beta^{-1} b H_I(r) \tag{4.49b}$$

$$w_T(r) = \beta^{-1} b W_T(r) \tag{4.49c}$$

$$w_I(r) = \beta^{-1} b W_I(r) \tag{4.49d}$$

where

$$b = \frac{3B}{\rho} = \beta m'^2 + 3\alpha' \tag{4.49e}$$

Writing (4.47a) and (4.47b) in terms of these new functions and taking the Fourier transform, we obtain

$$\tilde{\eta}_T(k|2) = \frac{1}{3}\rho \Big[\tilde{w}_T(k|2)\tilde{h}_T(k|2) + \tilde{w}_T(k|2)\tilde{h}_I(k|0) $$
$$+ \tilde{w}_I(k|0)\tilde{h}_T(k|2) \Big] \tag{4.50a}$$

and

$$\tilde{\eta}_I(k|0) = \frac{1}{3}\rho \Big[2\tilde{w}_T(k|2)\tilde{h}_T(k|2) + \tilde{w}_I(k|0)\tilde{h}_I(k|0) \Big] \tag{4.50b}$$

where $\tilde{f}(k|l)$ denotes the Hankel transform

$$\tilde{f}(k|l) = 4\pi i^l \int_0^\infty r^2 j_l(kr) f(r) \, dr \tag{4.50c}$$

The closure relations (4.47c) to (4.47e) become

$$h_I(r) = h_T(r) = 0 \qquad \text{for} \quad r < d \qquad (4.50d)$$

$$w_I(r) = g_{HS}(r)\eta_I(r) - \eta_I(r) \qquad \text{for} \quad r > d \qquad (4.50e)$$

and

$$w_T(r) = g_{HS}(r)\left[\eta_T(r) + \frac{b}{r^3}\right] - \eta_T(r) \qquad \text{for} \quad r > d \qquad (4.50f)$$

where $\eta_T = h_T - w_T$ and $\eta_I = h_I - w_I$. The 1-R equations (4.48a) and (4.48b) reduce to

$$\alpha' = \alpha + \frac{8\pi}{3} K_h \rho \alpha' \cdot \alpha \qquad (4.51a)$$

and

$$\mathbf{m}' = \boldsymbol{\mu} + \frac{8\pi}{3} K_h \rho \alpha' \cdot \boldsymbol{\mu} \qquad (4.51b)$$

where

$$K_h = \int_d^\infty \frac{h_T(r)}{r} dr \qquad (4.51c)$$

It is now obvious that (4.50a) to (4.50f) are exactly analogous to (3.14a), (3.14b), (3.12b), (3.21b), and (3.21c) [with $B^{mnl}(r)$ set to zero in (3.21b) and (3.21c)] defining the SSC or LHNC (reference version) approximation for nonpolarizable dipolar hard spheres. The effective dipole moment m_e is given by

$$\beta m_e^2 = b = \beta m'^2 + 3\alpha' \qquad (4.52)$$

In the limit $\alpha = 0$, $h_T(r)$, $h_I(r)$, $w_T(r)$, and $w_I(r)$ become $h^{112}(r)$, $h^{110}(r)$, $c^{112}(r)$, and $c^{110}(r)$ respectively. Equations 4.50 and 4.51 can be readily solved by iteration.[108] The situation is very similar to that described in Section III.B.3 for nonpolarizable systems except that in addition to the usual integral equations, the relationships (4.51a) and (4.51b) must also be satisfied.

Physically, the quantities $\mathbf{m}'$, m_e^2, and α' can be identified[96, 97] as follows:

$$\mathbf{m}' = \langle \mathbf{m} \rangle = \langle \boldsymbol{\mu} + \mathbf{p} \rangle \qquad (4.53a)$$

$$m_e^2 = \langle m^2 \rangle \qquad (4.53b)$$

and

$$\alpha' = \frac{\beta}{3}(\langle m^2 \rangle - |\langle \mathbf{m} \rangle|^2) \qquad (4.53c)$$

where $\mathbf{m}$ is the total and $\mathbf{p}$ the induced dipole moment. Thus $\mathbf{m}'$ is the average dipole moment (as measured in the molecular coordinate system) of a particle, m_e^2, is the mean square dipole moment, and α' is related to the fluctuation in the mean moment.

Wertheim[29] shows that the dielectric constant of the polar-polarizable fluid is given by

$$\varepsilon = \frac{q_1}{q_2} \qquad (4.54a)$$

where

$$q_1 = 1 - \frac{1}{3}\rho\left[\tilde{w}_I(0|0) + 2\tilde{w}_T(0|2) \right] \qquad (4.54b)$$

and

$$q_2 = 1 - \frac{1}{3}\rho\left[\tilde{w}_I(0|0) - \tilde{w}_T(0|2) \right] \qquad (4.54c)$$

It is also easy to show that for the present approximation the formulas

$$q_1 - q_2 = -\rho\tilde{w}_T(0|2) = 3y_e \qquad (4.55a)$$

$$\frac{(\varepsilon - 1)(2\varepsilon + 1)}{9\varepsilon} = y_e\left[1 + \frac{1}{3}\rho\tilde{h}_I(0|0) \right] \qquad (4.55b)$$

and

$$\tilde{h}_T(0|2) = \frac{-(\varepsilon - 1)^2}{3\varepsilon y_e \rho} \qquad (4.55c)$$

where $y_e = 4\pi\beta m_e^2\rho/9$ must hold. Equations 4.55b and 4.55c are essentially analogous to (2.30c) and (2.30b) giving the dielectric constant for non-polarizable polar systems with the role of h_Δ and h_D now being played by h_I and h_T, respectively. Thus in the present approximation the dielectric constant of the polar-polarizable system is just that of a rigid dipolar fluid characterized by the dipole moment, m_e.

Wertheim[29] shows that if one makes the approximation

$$g_{HS}(r) = \begin{matrix} 0 & r<d \\ 1 & r>d \end{matrix} \qquad (4.56)$$

then (4.50a) to (4.50f) become just the MSA for dipolar hard spheres having the dipole moment m_e. Thus ε can be found analytically[12] and is given by the equations[12, 29]

$$\varepsilon = \frac{q(2\xi)}{q(-\xi)} \qquad (4.57a)$$

$$y_e = \frac{1}{3}\left[q(2\xi) - q(-\xi) \right] \qquad (4.57b)$$

$$\alpha' = \alpha + 16\xi d^{-3}\alpha' \cdot \alpha \qquad (4.57c)$$

and

$$\mathbf{m}' = \mu + 16\xi d^{-3}\alpha' \cdot \mu \qquad (4.57d)$$

where $q(x)$ is defined by (3.15c) and we have used the relationship $8\pi K_h \rho d^3 = 16\xi$.

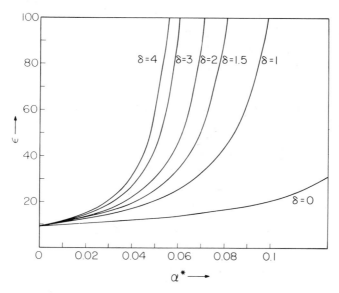

Fig. 23.   SSCA results for the dielectric constant of polarizable dipolar hard spheres at $\rho^* = 0.8$ and $\mu^{*2} = 1$; $\delta = \alpha_{zz}/\alpha_{xx}$.

Some results obtained[108] by solving the SSC (1-R) approximation [(4.50) and (4.51)] for polarizable dipolar hard spheres at $\mu^{*2} = 1$ and $\rho^* = 0.8$ are shown in Fig. 23. The ε is plotted as a function of the averaged reduced polarizability $\alpha^* = \frac{1}{3} \text{Tr}\, \boldsymbol{\alpha}^*$ where $\boldsymbol{\alpha}^* = \boldsymbol{\alpha}/d^3$. For all curves $\alpha_{xx} = \alpha_{yy}$ and the degree of anisotropy $\delta = \alpha_{zz}/\alpha_{xx}$, where α_{zz} is the component of the polarizability tensor along the dipole vector. It can be seen that for $\delta \geq 1$, ε rises very rapidly with increasing α^*. Furthermore, ε is very sensitive to the anisotropy of $\boldsymbol{\alpha}$ increasing with δ for fixed α^*. We note that for larger values of μ^{*2} the ε is found[108] to increase even more sharply with α^*. In all likelihood the very large dielectric constants obtained are an artifact of the polarizable dipolar hard-sphere model. We would expect the inclusion of a linear quadrupole moment to greatly reduce ε, just as it does for non-polarizable polar systems.

C. Implementation of the Høye-Stell Results for Polar-Polarizable Fluids with Fluctuating Polarizability

When $\phi(\mathbf{p})$ is harmonic, as in (4.7) — the model of strictly field-independent ε — it has been shown by Høye and Stell[102] that the probability density $\rho_{\mu,\mathbf{p}}$ is of Gaussian form for a certain important class of approximations that include the MSA, the LIN, the ADCA, and the SSCA –LHNCA. Within this class of approximations, ε as a function of α and m for a polar-polarizable fluid with harmonic $\phi(\mathbf{p})$ can be obtained by relatively simple transcription from ε as a function of m in the nonpolarizable case. One must replace m by m_e (the m of a dielectrically equivalent nonpolarizable system), where

$$m_e^2 = (m')^2 + \frac{3\alpha'}{\beta}$$

as given by (4.41). Moreover, Høye and Stell have shown that in the context of each one of this class of approximations, the relation among m, $\alpha\; [= \frac{1}{3}\text{Tr}\,\boldsymbol{\alpha}]$, m', and α' is precisely the same as it is for the constant-polarizability model. One immediate result of this is that all the MSA and SSCA results for ε discussed in Section IV.B are also the MSA and SSCA results in the fluctuating-polarizability model with harmonic $\phi(\mathbf{p})$. The LIN and ADC results for ε are equally easy to describe in this model. Using (2.30c) to find the LIN ε, we get

$$\frac{(\varepsilon - 1)(2\varepsilon + 1)}{9\varepsilon} = y_e g^{\text{LIN}}(\rho, y_e) \tag{4.58a}$$

$$y_e = \frac{4\pi\beta m_e^2 \rho}{9} \tag{4.58b}$$

$$m_e^2 = (m'_{\text{MSA}})^2 + \frac{3\alpha'_{\text{MSA}}}{\beta} \tag{4.58c}$$

Here $g^{\text{LIN}}(\rho, y_e)$ is the same function of ρ and y_e that $g^{\text{LIN}}(\rho, y)$ is of ρ and y in the nonpolarizable case. This is easy to compute. From (2.145) and (2.30c)

$$g^{\text{LIN}}(\rho, y) = 1 + \frac{1}{3}\rho \int g_0(r) \mathcal{C}_\Delta(r)\, d\mathbf{r} \qquad (4.58\text{d})$$

$$g^{\text{LIN}}(\rho, y) = g^{\text{MSA}}(\rho, y) + \Delta g(\rho, y) \qquad (4.58\text{e})$$

$$\Delta g(\rho, y) = \frac{\rho}{3} \int h_0(r) \mathcal{C}_\Delta(r)\, d\mathbf{r} = \frac{\rho}{3(2\pi)^3} \int \tilde{h}_0(k)\tilde{\mathcal{C}}_\Delta(k)\, d\mathbf{k} \qquad (4.58\text{f})$$

where $\mathcal{C}_\Delta(r)$ is the Δ component of the chain function $\mathcal{C}(r)$ of (2.141). The $\tilde{\mathcal{C}}_\Delta(k)$ is a closed-form function of k, as is $\tilde{h}_0(k)$ in the Percus-Yevick approximation for hard spheres. In (4.58), m' and α' (hence m_e and y_e) are evaluated for given m and α according to the same approximation that yields the polar-polarizable MSA results; hence the use of the MSA label in that equation.

The ε in the ADCA is no harder to compute; it involves exactly the same ingredients. Because the ADCA (2.107a) involves $c(12)$ in a simpler way than it involves $h(12)$, the extension of (2.15) to the polarizable case is the most natural route to ε in its connection. Using it we obtain

$$\frac{\varepsilon - 1}{\varepsilon + 2} = \frac{y_e}{1 - \frac{1}{3}\rho\tilde{c}_\Delta(0)^{\text{MSA}} - \Delta g(\rho, y_e)} \qquad (4.59\text{a})$$

Here $\rho\tilde{c}_\Delta(0)^{\text{MSA}}$ is a function only of y_e; it is the same function of y_e that the nonpolarizable $\rho\tilde{c}_\Delta(0)^{\text{MSA}}$ is of y. It can be related to the core parameter Θ of Section II according to the equation

$$-\frac{1}{3}\rho\tilde{c}_\Delta(0)^{\text{MSA}} = y\Theta^{\text{MSA}}(y) \qquad (4.59\text{b})$$

The $\Theta^{\text{MSA}}(y)$ is given by (2.79b). We can rewrite (4.59b) as

$$\frac{\varepsilon - 1}{\varepsilon + 2} = \frac{y_e}{1 + y_e\Theta^{\text{MSA}}(y_e) - \Delta g(\rho, y_e)} \qquad (4.59\text{c})$$

In the ADCA, y_e is again given by (4.58c). For the general case of polarizable-polar particles, the LIN approximation for ε has not yet been quantitatively examined. Preliminary results of ours[108] for the ADCA indicate that it is highly successful for nonpolar particles but rapidly becomes inaccurate as m is increased to values of interest for polar particles, yielding far too large an ε.

D. The Case of Nonpolar Polarizable Fluids

All the results given in Section IV.A to IV.C hold for the special case of zero permanent dipole moment. This case is special quantitatively as well as qualitatively in the sense that for values of α/σ^3 (σ = molecular diameter) typical of real nonpolar particles such as the noble-gas particles, the Høye-Stell parameter y_e, equivalent to Wertheim's B of our (4.47f), is very much smaller than for values of $\beta m^2/\sigma^3$ and α/σ^3 typical of real polar molecules of interest. In fact, for typical α/σ^3 in the nonpolar case, the Clausius-Mossotti equation

$$\frac{\varepsilon-1}{\varepsilon+2} = \frac{4\pi}{9}\rho\, Tr\,\alpha \tag{4.60}$$

is already a quite accurate approximation in the case of both the models we have been considering, just as it has long been known to be for most real nonpolar molecules. For many decades, in fact, the key theoretical question with regard to ε for nonpolar fluids was not how to improve (4.60), but why it is such a satisfactory approximation as it stands, since its original derivations[109] involved assumptions that are clearly not satisfied even in the noble-gas fluids. The work of Kirkwood,[93] Yvon,[94] Brown,[110] and the Dutch school of dielectric theoreticians[111] went a long way toward clarifying this point. The Dutch workers first reexpressed[111a] the Kirkwood-Yvon results slightly (in a form that lends itself especially well to correcting the Clausius-Mossotti result in powers of ρ or α) and then generalized[111b] those results from the classical constant-polarizability case to the case of fluctuating polarizability associated with a quantum mechanical rather than classical description of internal states. These studies revealed that in both the constant-polarizability and quantum fluctuating-polarizability models (4.60) represents the exact $(\varepsilon-1)/(\varepsilon+2)$ through second order in α $[=\frac{1}{3}\mathrm{Tr}\,\alpha]$. Writing

$$\frac{\varepsilon-1}{\varepsilon+2} = \frac{4\pi\rho}{3}\alpha[1+S] \tag{4.61}$$

one finds that S is of order α^2. The studies that incorporated quantum averaging in a fundamental way further revealed that formidable technical difficulties stand in the way of making accurate quantitative estimates of the deviation from the Clausius-Mossotti result for even the simplest real molecules by means of a first-principles study.[111b, 112]

The parameter α/σ^3 is so small for most nonpolar particles ($\alpha/\sigma^3 \approx 0.04$ for argon and ≈ 0.06 for xenon) that the approximation

$$S \approx S_2 \tag{4.62}$$

can be expected to be a very good one at typical α/σ^3 values, where S_2 is defined by

$$S_2 = \alpha^2 \lim_{\alpha \to 0} (S/\alpha^2) \qquad (4.63a)$$

Thus S_2 can be regarded as the lowest-order term in a formal expansion in α

$$S = \sum_{i=2}^{\infty} \alpha^i S_{(i)}, \qquad S_2 = \alpha^2 S_{(2)} \qquad (4.63b)$$

except that there is no good evidence, to our knowledge, that S is analytic in α about $\alpha = 0$, so that one should perhaps be just as prepared to find (at the critical point, at least)

$$S = S_2 + O(\alpha^3 \ln \alpha) \qquad (4.63c)$$

as a simple power series. Nevertheless α is clearly an appropriate parameter of smallness in considering S.

For simplicity we now focus on the constant-polarizability model for orientation-independent potentials and scalar α. Here one has [with $\mathbf{T}_{ij} = \mathbf{T}(\mathbf{r}_{ij})$ given by (4.12)]

$$S_2 = \frac{1}{3}\alpha^2 \mathrm{Tr}\left[\sum_k \langle \mathbf{T}_{ik}\cdot\mathbf{T}_{ki}\rangle + \sum_k \sum_{l \neq k} \langle \mathbf{T}_{ik}\cdot\mathbf{T}_{kl}\rangle \right]$$

$$= \frac{1}{3}\alpha^2\left\{ \rho \int g(r)\mathrm{Tr}[\mathbf{T}(r)]^2\,d\mathbf{r} \right.$$

$$\left. + \rho^2 \iint [g_3(0,\mathbf{r}_1,\mathbf{r}_2) - g(r_1)g(r_{12})]\,\mathrm{Tr}[\mathbf{T}(\mathbf{r}_1)\cdot\mathbf{T}(\mathbf{r}_{12})]\,d\mathbf{r}_1\,d\mathbf{r}_2 \right\}$$

$$= 2\alpha^2\left\{ \rho \int g(r)r^{-6}\,d\mathbf{r} \right.$$

$$\left. + \rho^2 \int [g_3(0,\mathbf{r}_1,\mathbf{r}_2) - g(r_1)g(r_{12})]r_1^{-3}r_{12}^{-3}P_2(\cos\theta)\,d\mathbf{r}_1\,d\mathbf{r}_2 \right\} \qquad (4.64)$$

where θ is the angle between $\mathbf{r}_1$ and $\mathbf{r}_{12}$ and P_2 is the second Legendre polynomial. If one expands the S_2 function in this model in ρ instead of α, one finds

$$S = 2\alpha^2\rho \int g(r)r^{-6}\left(1 + \frac{\alpha}{r^3}\right)^{-1}\left(1 - \frac{2\alpha}{r^3}\right)^{-1}\,d\mathbf{r} + O(\rho^2)$$

a result that essentially goes back to Silberstein.[113] It has been rigorously proved in the constant-scalar-polarizability case that

$$S \geq 0 \tag{4.65}$$

and also that

$$S_2 \geq 0 \tag{4.66}$$

The first case follows from any one of several arguments[114]; the latter is trivial, since[115]

$$S_2 = \frac{1}{3} \alpha^2 \left\langle \left(\sum_j \mathbf{T}_{ij} \right) : \left(\sum_k \mathbf{T}_{ik} \right) \right\rangle \geq 0$$

Generalizations of such inequalities to the polar-polarizable case can be found in an elegant study by Wertheim.[98]

Computer simulation results for S_2 are somewhat sparse and involve the usual uncertainties involved in extrapolating results for a truncated $\mathbf{T}(\mathbf{r})$ used in a periodic box to untruncated $\mathbf{T}(\mathbf{r})$ in an infinite system.[116] Nevertheless for polarizable hard-sphere and Lennard-Jones particles, it is probably safe to say that the estimates currently available from the combined use of analytic and simulation input are enough to provide a reliable guide to the ρ and β dependence of S_2 over the full fluid range of those variables. The most comprehensive studies of S_2 have been made by Stell and Rushbrooke[115] and by Graben, Rushbrooke, and Stell,[117] for the hard-sphere and Lennard-Jones cases, respectively. Both these works utilize the simulation results of Alder, Weis, and Strauss,[116] as well as exact density-expansion results, and numerical results of the Kirkwood superposition approximation

$$g_3(0, \mathbf{r}_1, \mathbf{r}_2) = g(r_1)g(r_2)g(r_{12}) \tag{4.67}$$

The use of (4.67) in (4.64) appears to yield accurate S_2 up to densities of around $\rho\sigma^3 \approx \frac{1}{2}$ at typical liquid temperatures. As ρ increases beyond this, S_2 rapidly begins to be overestimated by the use of (4.67), as shown in Fig. 24 for a hard-sphere fluid. (As seen in the figure, however, for realistic values of α/σ^3, S_2 is extremely small in the first place.) An approximation that is far better at typical liquid densities (and nearly as good at lower densities) is one introduced by Stell and Høye[118] in their study of the critical behavior of S_2:

$$h_3(0, \mathbf{r}_1, \mathbf{r}_2)_\delta = 0 \tag{4.68}$$

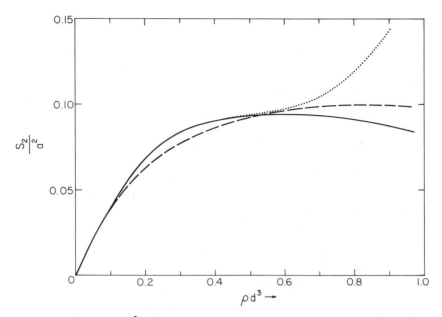

Fig. 24. Values of S_2/a^2, where $a = 4\pi\alpha/3$ for a system of polarizable hard spheres, given by the SSCA (dotted curve), the ADCA (dashed curve), and the best estimate of the exact value (solid curve). The best estimate is based on the results of computer simulations[116] as well as analytic estimates.

Here $h_3(0,\mathbf{r}_1,\mathbf{r}_2)_\delta$ is the three-particle analog of $h_\delta(r) = h(r) + \delta(r)/\rho$. It is the "modified" correlation function that occurs naturally in fluctuation theory,

$$h_3(0,\mathbf{r}_1,\mathbf{r}_2)_\delta = \rho^{-3}\langle\Delta\rho(0)\Delta\rho(\mathbf{r}_1)\Delta\rho(\mathbf{r}_2)\rangle \qquad (4.69)$$

For a lattice gas on its critical isochore, (4.68) is exact, and it is for this reason that Stell and Høye chose to use it in their study of the critical behavior of S_2, which we summarize in the next section.

In the original treatment of de Boer et al.[111a] and of Mazur and Jansen,[111b] S was expressed in terms of g, g_3, and so on [as in (4.64)]. Bedeaux and Mazur[119] subsequently reexpressed S (and its frequency-dependent generalization) in an elegant representation involving

$$h_{n,\delta} = \rho^{-n}\langle\Delta\rho(0)\Delta\rho(\mathbf{r}_1)\cdots\Delta\rho(\mathbf{r}_{n-1})\rangle \qquad (4.70)$$

Their choice of $\mathbf{T}(\mathbf{r})$ (and its frequency-dependent generalization) was given by (4.12) (but with d identified as a hard-core diameter that remains finite). Felderhof[120] then derived and discussed a result equivalent to the Bedeaux-Mazur representation from a somewhat different point of view. As Høye and Stell pointed out in the context of problems involving a permanent dipole, one can introduce a core parameter Θ in treating $T(r)$ for $r < d$, as discussed in detail in Section II. Sullivan and Deutch[121] generalized the use of Θ to the case of polarizable nonpolar particles, showing that depending on one's choice of Θ, one will get as a lowest-order result at zero frequency the Clausius-Mossotti approximation, the Onsager approximation, or the polarizable analog of the MSA developed by Wertheim (labeled the PMSA by him) in close correspondence to the three $\gamma \to 0$ results considered here in Sections II.C and II.D. Finally Høye and Bedeaux[122] extended to all orders the use of the core parameter that yields the PMSA result as a lowest-order result, and explicitly evaluated ε in next highest order. The PMSA is given by (4.57), and the next-order Høye-Bedeaux result is essentially the constant-polarizability version of the ADCA result given by (4.59c).

In the PMSA, S_2 (hence S) is very poorly approximated, with $g(r)$ and $g_3(0, r_1, r_2)$ both replaced by their zero-density limits in (4.64). In the SSCA, S_2 (hence S) is improved greatly, with S_2 given by (4.64) in which (4.67) is used. In the ADCA, S_2 (hence S) is still better, with S_2 given by (4.64) in which g_3 is expressed in terms of $g(r)$ by means of (4.68) and $g(r)$ is given its $\alpha = 0$ value. [In the constant-polarizability case, $g(r)$ and g_3 are rigorously independent of α, but in the classical fluctuating-polarizability model they are only so independent in the context of the particular class of approximations discussed in Section II.C.] The ADCA form of S_2 so obtained is extremely simple. It is given by

$$S_2 = \alpha^2 \rho^2 \int \left[h(r) + \frac{\delta(\mathbf{r})}{\rho} \right] O_2(r) \, d\mathbf{r}$$

$$= \alpha^2 \rho^2 \int h(r) O_2(r) \, d\mathbf{r} + \frac{8\pi\rho\alpha^2}{3d^3} \tag{4.71a}$$

where

$$O_2(r) = \begin{cases} 0 & \text{for } r > 2d \\ \dfrac{4\pi}{3d^3} \left[2 - \dfrac{3}{2}\dfrac{r}{d} + \dfrac{1}{8}\left(\dfrac{r}{d}\right)^3 \right] & \text{for } r < 2d \end{cases} \tag{4.71b}$$

This is also the S_2 associated with the lowest-order correction to the Clausius-Mossotti result that comes out of the original Bedeaux-Mazur expansion. (This lowest-order Bedeaux-Mazur correction yields a result somewhat different from the ADCA for arbitrary α, however.) If one writes the SSCA and ADCA approximations for S as

$$S = S_2 + \Delta S \qquad (4.72)$$

one finds that ΔS is indeed completely negligible compared to S_2 for $\alpha/\sigma^3 \leq 0.1$ in both approximations. For a hard-sphere system, S_2 in the ADCA and SSCA is shown in Fig. 24 compared to the best available estimate of S_2.

How do these model results compare with the S defined by (4.61) for real molecules? They are very different! The S for noble-gas fluids does not satisfy (4.65). For example, for helium S seems to be always negative,[123, 124] whereas for argon it is positive for lower densities but becomes negative at liquid densities.[125] Thus although the polarizability of real nonpolar molecules is related to ε much as it is in the constant-polarizability model in an overall way [i.e., in both cases, (4.60) holds to good approximation], corrections to (4.60) are quite different in the two cases. The difference is best understood not in terms of the polarizability $\boldsymbol{\alpha}$ associated with an isolated particle defined by (4.8) or (4.9), but by the polarizability $\boldsymbol{\alpha}(ij)$ associated with an isolated pair of particles, $\boldsymbol{\alpha}(ijk)$ associated with an isolated triplet of particles, and so on. The dependence of $\boldsymbol{\alpha}(ij)$ on small r_{ij} (i.e., in the neighborhood of σ) appears to differ strikingly in the models we have been considering here from the dependence found in real monatomic molecules (and presumably, most other molecules as well). This difference [and presumably corresponding differences in $\boldsymbol{\alpha}(ijk)$] appears to be the primary reason for the difference in the behavior of S in the models we have considered compared to the experimental S.[126]

One can estimate $\boldsymbol{\alpha}(ij)$ for atoms such as helium and argon with reasonable accuracy through a combination of experimental and theoretical input[127] and also approximate (with much less assurance) $\boldsymbol{\alpha}(ijk)$ for such atoms in terms of $\boldsymbol{\alpha}(ij)$ using superposition-type approximations. Using such results, one can then generalize our S_2 to yield approximate expressions of greater relevance to real particles. The greater the degree of experimental input in quantitatively parameterizing the pair polarizabilities that go into such expressions, the more accurate such expressions can be made; but they do not yet provide an independent means of predicting ε.

E. Critical Properties of ε

1. The Nonpolar Case

The specific heat of a fluid at critical density ρ_c is believed to behave, as the critical temperature T_c is approached, according to the relation

$$c_v = A t^{-\alpha} + B \qquad (4.73)$$

where $t=|T-T_c|$ and $\alpha\approx\frac{1}{8}$. [There is a notational dilemma associated with (4.73) for us, since the α there represents standard notation for a critical exponent that has nothing to do with the α that stands for $\frac{1}{3}\mathrm{Tr}\,\alpha$, also in standard notation. Stell and Høye,[118] in discussing critical behavior of ε, used θ to stand for $\frac{1}{3}\mathrm{Tr}\,\alpha$, and we do likewise in this subsection, IV.E, but *only* here.]

Here $c_v=(\partial u_{\mathrm{TOT}}/\partial T)_\rho$ where u_{TOT} is total internal energy per particle, which is the sum of a nonsingular kinetic contribution u_{KIN} plus a configuration contribution u, which can be written in a simple fluid as

$$u=\frac{\rho}{2}\int g(12)\phi(12)\,d\mathbf{r}_{12} \qquad (4.74)$$

with $g(12)$ the pair distribution function and $\phi(12)$ the pair potential. Equation 4.73 is also believed to hold for the lattice gas, as does (4.74), with the integral appropriately interpreted as a lattice sum.

We recall now that at $\rho=\rho_c$, (4.68) is exact for a lattice gas; hence, on the basis of widely held notions of universality, ρ can also be assumed in a continuum-fluid computation without doing violence to the structure of the dominant singularities that emerge at the fluid critical point as $t\to 0$, $\rho=\rho_c$. Since (4.71) follows without further assumptions from (4.64) and (4.68), (4.71) appears to be an appropriate expression for the study of critical behavior of S_2 in the constant-polarizability model. It is especially useful if we note that there is gross similarity between the function O_2 in (4.71) and the attractive part of a typical pair potential. To exploit this, consider the potential

$$\phi(r)=\begin{cases} \infty & \text{for} \quad r<d \\ \phi_0 O_2(r) & \text{for} \quad r>d \end{cases} \qquad (4.75)$$

which approximates the overall form of the noble-gas potentials reasonably well if we make judicious choices of d and ϕ_0. For such a model we can make two strong statements:

1. First, the quantity

$$S_2+(\theta\rho)^2\left[4\pi\int_d^{2d}O_2(r)r^2\,dr-\frac{8\pi}{3\rho d^3}\right] \qquad (4.76)$$

and u are literally proportional to each other as functions of t at ρ_c. They thus share the same $t^{1-\alpha}$ singularity, and $\partial S_2/\partial t$ and c_v share the same $t^{-\alpha}$ singularity. Moreover, the coefficients of the $\partial S_2/\partial t$ and c_v singularity must be proportional, with proportionality constant given by

$$\lim_{t\to 0}\left(\frac{\partial S_2/\partial T}{c_v}\right)=\frac{2(\theta\rho_c)^2}{\rho_c\phi_0}\tag{4.77}$$

2. Second, in the context of (4.62), this means that $\partial\varepsilon/\partial T$ and c_v similarly share their singularity with proportionality constant

$$\lim_{t\to 0}\left(\frac{\partial\varepsilon/\partial T}{c_v}\right)=\frac{8\pi(\varepsilon_c+2)^2(\theta\rho_c)^3}{9\rho_c\phi_0}\tag{4.78}$$

where ε_c is the critical value of ε. For typical values of $\theta\rho_c$, (4.62) indeed offers an appropriate context, since $\theta\rho_c\le 0.02$ for nonpolar fluids of interest, and the terms omitted in (4.78) are $O(\theta\rho_c)^4$. The same order-of-magnitude estimate illuminates why critical anomalies in $\partial\varepsilon/\partial T$ are so hard to detect experimentally—they come with a coefficient of order $(\theta\rho_c)^3$.

From the observations above, we can see that if real molecules had constant polarizability, we should expect to have great difficulty experimentally observing a critical anomaly in $\partial\varepsilon/\partial T$ because of its small magnitude, but no great difficulty in making theoretical predictions concerning its form or its order of magnitude relative to that of c_v.

Since nonpolar molecules do not have constant polarizability, what can we say about their dielectric anomaly? The difference in the pair and triplet polarizabilities in real noble-gas molecules and our model molecules can be summarized in terms of S_2 by saying that we would expect a similar sort of integral to provide a reasonable approximation to S_2 at ρ_c, but with an $O_2(r)$ quite different from that given by (4.71b). In the case of helium, for example, where the linearity[124] of S_2 in ρ suggests that three-body effects are of less importance than in argon, S_2 might be reasonably approximated at ρ_c by the expression

$$S_2=\frac{\rho_c}{2\theta}\int g(r)\alpha(r)\,d\mathbf{r}\tag{4.79}$$

where $\alpha(r)$ is the variation in the trace of the pair polarizability tensor, which becomes strongly enough negative for small r to give rise to a negative S_2 and for large r has the functional form

$$\alpha(r)=\frac{\text{const }\theta^3}{r^6}\tag{4.80}$$

(In the constant-polarizability model, const$=4$.) Such an S_2 will also share the singularity of u as would an S_2 of the more general form

$$S_2 = (\theta \rho_c)^2 \left[\int g(r) A(r) \, d\mathbf{r} + \text{const} \right] \tag{4.81}$$

as long as $A(r)$ is of short enough range for the integral to exist. There seems little doubt, therefore, that $\partial S_2 / \partial T$ for real molecules shares the specific-heat singularity. What seems less easy to estimate in general is its magnitude. To see why [and to confirm that (4.81) will yield $\partial S_2 / \partial T \sim t^{-\alpha}$], we note that for (4.73) to be satisfied via (4.74) for the variety of $\phi(12)$ for which it is expected to hold—the lattice gas nearest-neighbor (or few-neighbor) interaction, the hard-sphere plus square-well potential, the Lennard-Jones potential, etc.—one expects $g(\mathbf{r}_{12})$ to have in all such cases the form, for fixed $\mathbf{r}_{12}$ and $t \to 0$, $\rho = \rho_c$, $T \ge T_c$,

$$g(\mathbf{r}_{12}) = g_c(\mathbf{r}_{12}) + t^{1-\alpha} g_{1-\alpha}(\mathbf{r}_{12}) + t g_1(\mathbf{r}_{12}) + \cdots \tag{4.82}$$

for $r_{12} > \sigma$, [with $g_x(\mathbf{r}_{12}) \approx 0$ for $r_{12} < \sigma$, all x.] When inserted into (4.74), (4.82) will yield (4.73). When (4.82) is inserted into (4.81) we find similarly

$$S_2 = (\theta \rho_c)^2 \left[\text{const} + t^{1-\alpha} \int g_{1-\alpha}(r_{12}) A(r) \, d\mathbf{r} \right.$$

$$\left. + t \int g_1(r) \overset{\bullet}{A}(r) \, d\mathbf{r} + \cdots \right] \tag{4.83}$$

hence

$$\frac{\partial S_2}{\partial t} = (\theta \rho_c)^2 \left[t^{-\alpha} (1-\alpha) \int g_{1-\alpha}(r) A(r) \, d\mathbf{r} \right.$$

$$\left. + \int g_1(r) A(r) \, d\mathbf{r} + \cdots \right] \tag{4.84}$$

Thus the $t^{-\alpha}$ singularity persists, but in general, we must know something about both $A(r)$ and $g_{1-\alpha}(r)$ to directly estimate the magnitude of the $t^{-\alpha}$ singularity, or compare it with the magnitude of the c_v singularity. In this more general case, we see that we still expect equations of the form given in (4.77) and (4.78), with ϕ_0 now emerging as a useful measure of the ratio of integrated magnitude of $\phi(12)$ and $A(r)$ weighted by $g_{1-\alpha}(r)$, since we recover (4.77) from (4.82), using (4.84) and (4.74), with ϕ_0 defined by

$$\phi_0 = \frac{\int g_{1-\alpha}(r)\phi(12)\,d\mathbf{r}}{\int g_{1-\alpha}(r)A(r)\,d\mathbf{r}} \tag{4.85}$$

The definition of ϕ_0 given by (4.75) is a special case of (4.85) applying to $\phi(r) \sim \text{const } O_2(r)$ with $A(r)$ given by $O_2(r)$. Here $\phi_0 < 0$. For helium, if (4.79) were used, we would have

$$\phi_0 = 2\theta^3 \rho_c \frac{\int g_{1-\alpha}(r)\phi(12)\,d\mathbf{r}}{\int g_{1-\alpha}(r)\alpha(r)\,d\mathbf{r}} \tag{4.86}$$

which might well be very large in magnitude (and *positive*) because of the small-r negativity of $\alpha(r)$ that could contribute a bit more to the denominator of (4.86) than the positive tail of $\alpha(r)$. The $\partial\varepsilon/\partial T$ anomaly of helium might therefore be expected to be unmeasurably small (and anomalous in sign, with $\phi_0 > 0$) as a result of this, taken together with the extremely small $\theta\rho_c$ for helium.

Can we estimate $g_{1-\alpha}(r)$ in the small-r region, where the $\phi(12)$ and $A(r)$ will make their biggest contribution to u and S_2, respectively? There are various ways of getting an order-of-magnitude estimate. For example, for a hard-core potential of core diameter d, the virial theorem gives

$$\beta p = \rho + \frac{2\pi}{3}\rho^2 d^3 g(d+) - \frac{\rho^2}{6}\int \beta \frac{\partial\phi(r)}{\partial r} r g(r)\,d\mathbf{r} \tag{4.87a}$$

If

$$\phi(r) = \begin{cases} \infty & \text{for } r < d \\ \phi_1 r^{-6} & \text{for } r > d \end{cases} \tag{4.87b}$$

the last term can be easily written in terms of the internal energy u to give

$$\beta p = \rho + \frac{2\pi}{3}\rho^2 d^3 g(d+) + 2\rho\beta u \tag{4.87c}$$

Since we expect $\partial p/\partial t \sim t^{1-\alpha}$ with $u \sim t^{1-\alpha}$, the $t^{1-\alpha}$ contribution from the

$g(d+)$ term in (4.87) must exactly cancel that of the u term, so

$$-\frac{2\pi}{3}\rho d^3\frac{\partial g(d+)}{\partial t}\approx 2\beta_c c_v \qquad \text{as} \quad t\to 0 \qquad (4.88a)$$

or [from (4.74) and (4.82)]

$$-\frac{2\pi}{3}\rho d^3 t^{-\alpha}(1-\alpha)g_{1-\alpha}(d+)\approx 2\beta_c c_v \qquad \text{as} \quad t\to 0 \qquad (4.88b)$$

This gives us precise knowledge of $g_{1-\alpha}(d+)$. If we were to assume that $g_{1-\alpha}(r)$ is essentially constant throughout the small-r region over which $A(r)$ makes its dominant contribution, (4.88b) with (4.74) would yield the order-of-magnitude estimate

$$2\beta_c\phi_1\approx d^6 \qquad (4.89)$$

for a potential given by (4.87b). For such a potential one expects $\beta_c\phi_1$ to be perhaps two or three times as large as this, suggesting that $g_{1-\alpha}(r)$ decays fairly rapidly from its contact value $g_{1-\alpha}(d+)$ as r increases from d, which is reasonable. This information can be used with (4.84) and (4.88b) to make order-of-magnitude estimates relating $\partial S_2/\partial t$, c_v, and $\int_{r>\sigma}A(r)d\mathbf{r}$.

Arguments based on quite different scaling theory assumptions yield much the same order-of-magnitude estimates. For $r_{12}\to\infty$ and fixed $x=\kappa r_{12}$, where κ is an inverse correlation length, $\kappa\propto t^\nu$, we can use (4.82) with

$$g_{1-\alpha}(r)\approx G_{1-\alpha}r^{(1-\alpha)/\nu-1-\eta} \qquad (4.90)$$

according to scaling assumptions, where $G_{1-\alpha}$ is a constant and η a critical exponent ($\eta\approx 0.06$). This is an asymptotic limit different from the fixed-r, $\kappa\to 0$ limit most obviously relevant to u and S_2. However, if we assume a strong form of matching of asymptotic regimes, we are led to use (4.90) for all $r>\sigma$ in (4.84) and (4.74), which gives as $T\to T_c$,

$$\frac{\partial S_2}{\partial t}\to(\theta\rho_c)^2 t^{-\alpha}(1-\alpha)G_{1-\alpha}\int_{r>\sigma} r^{(1-\alpha)/\nu-1-\eta}A(r)d\mathbf{r} \qquad (4.91a)$$

$$c_v\to\left(\frac{\rho_c}{2}\right)t^{-\alpha}(1-\alpha)G_{1-\alpha}\int_{r>\sigma} r^{(1-\alpha)/\nu-1-\eta}\phi(12)d\mathbf{r} \qquad (4.91b)$$

hence

$$\phi_0=\frac{\displaystyle\int_{r>\sigma} r^{(1-\alpha)/\nu-1-\eta}\phi(12)d\mathbf{r}}{\displaystyle\int_{r>\sigma} r^{(1-\alpha)/\nu-1-\eta}A(r)d\mathbf{r}} \qquad (4.91c)$$

Because of the strong matching assumption, however, (4.91) is perhaps a less reliable result than the estimates based on (4.87).

At the time of writing, the experimental status of the anomaly in $\partial\varepsilon/\partial t$ is ambiguous. It had been thought that the anomaly had been observed in several fluids,[128–130] but subsequent experiments[131, 132] have raised a possibility that the detection reported earlier might have been only apparent. (The newer experiments[131, 132] also appear to put reliable upper bounds on the anomaly for ^{3}He, Ne, and SF$_6$.)

Among the approximations we have considered, the MSA will have no anomaly, but the LIN and ADC approximations will give the $t^{-\alpha}$ anomaly for $\partial\varepsilon/\partial T$. This is clear from the form of Δg given by (4.58f). The $\int h_0(r)\mathcal{C}_\Delta(r)\,d\mathbf{r}$ there is an integral of the type $\int h(r)A(r)\,d\mathbf{r}$ we have been considering here. In fact, the S_2 of the ADCA given by (4.71) is the form that we have taken as our starting point in Section IV.E.

2. The Dipolar Case

The general expressions for ε in terms of $h(12)$ do not seem to lend themselves as directly in the case of dipolar particles to the sort of analysis that we have just given for the nonpolar case. We shall therefore content ourselves to making a few observations concerning approximations that can be expected to exhibit the correct critical exponent of ε in the dipolar case. The arguments of Ref. 37 suggest that this behavior should be the same in the polar and nonpolar cases. In agreement with our own statistical mechanical nonpolar result, they further suggest $\partial\varepsilon/\partial T \sim t^{-\alpha}$.

A key ingredient in our understanding of the critical behavior of dipolar systems lies in Stell's[133] observation that one can establish a correspondence in structure between

$$h_S(r) = \frac{\int h(12)\,d\Omega_1\,d\Omega_2}{\Omega^2}$$

for such systems and $h(r)$ for simple nonpolar (e.g., monatomic) polarizable systems. The correspondence implies that the liquid-gas critical point of dipolar systems will lie in the same universality class as that of the noble-gas fluids, despite the enormous difference in range and symmetry of the dipolar and nonpolar potentials. It also asserts that $h_S(r)$ will have the same sort of t and ρ dependence in the critical region as the noble-gas $h(r)$, hence as the $h(r)$ described by (4.82). [This correspondence between polar and nonpolar $h_S(r)$ hinges on arguments very similar to the one that enables Høye and Stell[102, 134] to use the same formalism to describe correlations in polar- and nonpolar-polarizable systems.]

Perhaps the most useful approximation for our purposes here is a renormalized ADCA

$$c_\Delta(r) = c_\Delta^{MSA}(r) + h_S(r)h_\Delta^{MSA}(r) \tag{4.92}$$

which is the ADC result for $c_\Delta(r)$, "renormalized" by the use of $h_S(r)$ instead of $h_0(r)$ in (2.103b), in keeping with the discussion between (2.112) and (2.113). Alternatively we can use the similarly "renormalized" LIN result,

$$h_\Delta(r) = h_\Delta^{MSA}(r) + h_S(r)h_\Delta^{MSA}(r) \tag{4.93}$$

Using (2.25b) with (4.92) or (2.30c) with (4.93), we shall obtain an ε that has a critical anomaly, with $\partial\varepsilon/\partial T \sim t^{-\alpha}$.

For dipolar molecules in which the quantity $\beta\mu^2$ is appreciably greater than 3α, the whole problem concerning the sensitivity of the anomaly magnitude on the precise forms of $\alpha(ij)$ and $\alpha(ijk)$ is absent, since the anomaly magnitude will depend entirely on the dipole moment magnitude in the absence of polarizability. Let us take the special case of nonpolarizable dipolar spheres for simplicity. Then the virial theorem yields

$$\beta p = \rho + \frac{2\pi}{3}\rho^2 d^3 g_S(d+) + \rho\beta u \tag{4.94}$$

Since we expect $\partial p/\partial T \sim t^{1-\alpha}$ with $u \sim t^{1-\alpha},$ the $t^{1-\alpha}$ contribution from the $g_S(d+)$ term must exactly cancel that of the u term, so

$$-\frac{2\pi}{3}\rho d^3 \frac{\partial g_S(d+)}{\partial t} \approx \beta_c c_v \qquad \text{as} \qquad t \to 0 \tag{4.95}$$

a result analogous to (4.88).

Let us consider the implications of assuming that the value of $\partial g_S(r)/\partial t$ at $r = d+$ is representative of its value in the small-r range over which the short-range function $h_\Delta^{MSA}(r)$ makes its greatest contribution to the integral

$$\Delta_S = \frac{\rho}{3}\int h_S(r)h_\Delta^{MSA}(r)\,d\mathbf{r} \tag{4.96}$$

This is the integral that determines the singular behavior of ε in the renormalized ADC and LIN approximations. It is the same function as the Δg appearing in (4.58) and (4.59), with $h_0(r)$ there replaced by $h_S(r)$. We have then the order-of-magnitude estimate

$$\lim_{t \to 0} \left(\frac{\partial \Delta_S / \partial t}{c_v} \right) \approx \frac{-\beta_c \tilde{h}_\Delta(0)^{\text{MSA}}}{2 \pi d^3} \tag{4.97}$$

The absolute magnitude of this estimate will bound the true absolute magnitude if $\partial h_S(r)/\partial t$ decays smoothly with increasing r, as is reasonable. In the renormalized ADCA

$$\frac{\varepsilon + 2}{\varepsilon - 1} = \left(\frac{\varepsilon + 2}{\varepsilon - 1} \right)^{\text{MSA}} - \frac{\Delta_S}{y} \tag{4.98}$$

and in the renormalized LIN

$$\frac{(\varepsilon - 1)(2\varepsilon + 1)}{9\varepsilon} = \left[\frac{(\varepsilon - 1)(2\varepsilon + 1)}{9\varepsilon} \right]^{\text{MSA}} + y\Delta_S \tag{4.99}$$

The critical anomaly of $\partial \varepsilon / \partial t$ appears to have been recently observed[135] in carbon monoxide, a weakly polar substance.

V. ELECTROLYTE SOLUTIONS: THE SOLUTE–DEPENDENT DIELECTRIC CONSTANT

The frequency-dependent dielectric constant of a conducting medium $\varepsilon(\omega)$, diverges as ω approaches zero. Ionic solutions conduct, hence the static dielectric constant $\varepsilon(0)$, at least as it is usually defined, becomes infinite. Thus if we refer to the static dielectric constant of an ionic system, it is necessary to define that term precisely. An excellent discussion of this question has been recently given by Hubbard, Colonomos, and Wolynes.[136]

For ionic solutions it is possible to define[136] an apparent dielectric constant,

$$\varepsilon_{\text{SOL}} = \lim_{\omega \to 0} \left(\varepsilon(\omega) - \frac{4 \pi \sigma}{i \omega} \right) \tag{5.1}$$

where σ is the static conductivity. The ε_{SOL} is the part of $\varepsilon(\omega)$ remaining finite as $\omega \to 0$, and it can be measured experimentally.[136, 137] However, ε_{SOL} is not a true equilibrium property, since it contains dynamic as well as equilibrium contributions.[136, 138] One can write

$$\varepsilon_{\text{SOL}} = \varepsilon_E + \varepsilon_D \tag{5.2}$$

where ε_E represents the equilibrium part and ε_D the various dynamic effects. We are concerned here only with the equilibrium contribution, ε_E. Interesting discussions of the dynamic contributions are given in Refs. 136 and 138.

The ε_E is essentially analogous to the static dielectric constant of nonconducting fluids discussed in Sections I to IV. It can be written in terms of the equilibrium correlation functions, and in recent papers[139-145] on the equilibrium theory of ionic solutions it is often referred to as the "solute-dependent dielectric constant." It is important to note, however, that ε_E constitutes only a part of the experimentally measured quantity, ε_{SOL}.

In the remainder of Section V we discuss simple models for ionic solutions and give formal expressions, approximate theories, and quantitative results for ε_E.

A. Simple Models for Ionic Systems

Consider a mixture in which all species carry both a point charge and a permanent point dipole moment. We are interested only in uncharged solvents and ions that are not dipolar, but it is convenient to write the equations for the general case and then obtain the desired ionic solution results by setting appropriate charges and dipole moments to zero. Thus species α is characterized by a charge q_α, a dipole moment μ_α, and a number density ρ_α. The system must be electrically neutral, so

$$\sum_\alpha \rho_\alpha q_\alpha = 0 \qquad (5.3)$$

where the sum on α is over all species.

The pair potential $u_{\alpha\beta}(12)$, for two particles of species α and β separated by a distance r, can be written in the form

$$u_{\alpha\beta}(12) = u_{\alpha\beta}^{000}(r) + u_{\alpha\beta}^{101}(r)\Phi^{101}(12) + u_{\alpha\beta}^{011}(r)\Phi^{011}(12) + u_{\alpha\beta}^{112}(r)\Phi^{112}(12)$$

$$(5.4a)$$

where

$$u_{\alpha\beta}^{000}(r) = u_{\alpha\beta}^{S}(r) + \frac{q_\alpha q_\beta}{r} \qquad (5.4b)$$

$$u_{\alpha\beta}^{101}(r) = \frac{\mu_\alpha q_\beta}{r^2} \qquad (5.4c)$$

$$u_{\alpha\beta}^{011}(r) = \frac{-q_\alpha \mu_\beta}{r^2} \qquad (5.4d)$$

$$u_{\alpha\beta}^{112}(r) = \frac{-\mu_\alpha \mu_\beta}{r^3} \qquad (5.4e)$$

The $u_{\alpha\beta}^S(r)$ can be any spherically symmetric short-range potential, and the rotational invariants, $\Phi^{mnl}(12)$ are defined by (3.3a) and given in Appendix B. The charge-charge, charge-dipole, and dipole-dipole interactions are easily recognized in (5.4a).

Quantitative results have been obtained[143, 145] for charged hard spheres in dipolar hard-sphere solvents. For this model $u_{\alpha\beta}^S(r)$ is the hard-sphere interaction

$$u_{\alpha\beta}^{HS}(r) = \begin{array}{ll} \infty & \text{for} \quad r < d_{\alpha\beta} \\ 0 & \text{for} \quad r > d_{\alpha\beta} \end{array} \qquad (5.5)$$

where $d_{\alpha\beta} = (d_\alpha + d_\beta)/2$, d_α being the diameter of species α.

B. Integral Equation Theories

1. The Mean Spherical Approximation

The MSA for a mixture is defined by the equations

$$h_{\alpha\beta}(12) - c_{\alpha\beta}(12) = \frac{1}{4\pi} \sum_\gamma \rho_\gamma \int h_{\alpha\gamma}(13) c_{\gamma\beta}(32) d(3) \qquad (5.6a)$$

$$g_{\alpha\beta}(12) = 0 \qquad \text{for} \quad r < d_{\alpha\beta} \qquad (5.6b)$$

and

$$c_{\alpha\beta}(12) = -\beta u_{\alpha\beta}(12) \qquad \text{for} \quad r > d_{\alpha\beta} \qquad (5.6c)$$

where $g_{\alpha\beta}(12) = h_{\alpha\beta}(12) + 1$. This definition is exactly analogous to the MSA for a single-component system. Equation 5.6a is the OZ relation for a mixture, and $h_{\alpha\beta}(12)$ and $c_{\alpha\beta}(12)$ are, respectively, the total and direct correlation functions for species α and β. For an m-component system (5.6a) yields a set of m^2 equations that are conveniently written in the matrix notation

$$\mathbf{h}(12) - \mathbf{c}(12) = \frac{1}{4\pi} \int \mathbf{h}(13) \boldsymbol{\rho} \mathbf{c}(32) \, d(3) \qquad (5.7a)$$

where

$$\boldsymbol{\rho} = \begin{bmatrix} \rho_1 & 0 & \cdots & 0 \\ 0 & \rho_2 & \cdots & 0 \\ \vdots & & & \\ 0 & 0 & \cdots & \rho_m \end{bmatrix} \qquad (5.7b)$$

and the remaining matrices are of the form

$$
\mathbf{A} = \begin{bmatrix} A_{11} & \cdots & \cdots & A_{1m} \\ A_{21} & A_{22} & \cdots & A_{2m} \\ \vdots & & & \vdots \\ A_{m1} & \cdots & \cdots & A_{mm} \end{bmatrix} \tag{5.7c}
$$

The MSA for charge-dipole systems has been solved by Blum[56, 146] and also by Adelman and Deutch.[147] The formal solution is

$$
\mathbf{h}(12) = \mathbf{h}^{000}(r) + \mathbf{h}^{101}(r)\Phi^{101}(12) + \mathbf{h}^{011}(r)\Phi^{011}(12) + \mathbf{h}^{110}(r)\Phi^{110}(12)
$$

$$
+ \mathbf{h}^{112}(r)\Phi^{112}(12) \tag{5.8a}
$$

where

$$
\mathbf{h}^{mnl}(r) = \frac{3}{(4\pi)^2} \int \mathbf{h}(12)\Phi^{mnl}(12)\, d\Omega_1 d\Omega_2 \tag{5.8b}
$$

if $(mnl) = (101)$ or (011), with $\mathbf{h}^{110}(r)$ and $\mathbf{h}^{112}(r)$ defined by (3.10c) and (3.10d), respectively. The direct correlation function $\mathbf{c}(12)$ is given by a similar set of relationships. Equation 5.7a can be Fourier transformed and combined with the $\mathbf{h}(12)$ and $\mathbf{c}(12)$ expansions to yield[56, 143, 146, 147]

$$
\tilde{\eta}^{000} = \tilde{h}^{000}\rho\tilde{c}^{000} + \frac{1}{3}\tilde{h}^{011}\rho\tilde{c}^{101} \tag{5.9a}
$$

$$
\tilde{\eta}^{011} = \tilde{h}^{000}\rho\tilde{c}^{011} + \frac{1}{3}\tilde{h}^{011}\rho\tilde{c}^{110} + \frac{2}{3}\tilde{h}^{011}\rho\tilde{c}^{112} \tag{5.9b}
$$

$$
\tilde{\eta}^{101} = \tilde{h}^{101}\rho\tilde{c}^{000} + \frac{1}{3}\tilde{h}^{110}\rho\tilde{c}^{101} + \frac{2}{3}\tilde{h}^{112}\rho\tilde{c}^{101} \tag{5.9c}
$$

$$
\tilde{\eta}^{110} = \frac{1}{3}\tilde{h}^{101}\rho\tilde{c}^{011} + \frac{1}{3}\tilde{h}^{110}\rho\tilde{c}^{110} + \frac{2}{3}\tilde{h}^{112}\rho\tilde{c}^{112} \tag{5.9d}
$$

$$
\tilde{\eta}^{112} = \frac{1}{3}\tilde{h}^{101}\rho\tilde{c}^{011} + \frac{1}{3}\tilde{h}^{110}\rho\tilde{c}^{112} + \frac{1}{3}\tilde{h}^{112}\rho\tilde{c}^{110} + \frac{1}{3}\tilde{h}^{112}\rho\tilde{c}^{112} \tag{5.9e}
$$

where $\eta^{mnl} = h^{mnl} - c^{mnl}$, and the tildes denote the Hankel transforms defined by (3.11k). Comparing (5.4), (5.6b), (5.6c), and (5.8a), one obtains the

MSA closure relations

$$c_{\alpha\beta}^{000}(r)=-\left[1+\eta_{\alpha\beta}^{000}(r)\right] \quad\quad\quad\quad\quad\quad\quad\quad (5.10a)$$

$$c_{\alpha\beta}^{mnl}(r)=-\eta_{\alpha\beta}^{mnl}(r) \quad \left[\text{if } (mnl)\neq(000)\right] \quad\Bigg\}\text{ for }\; r<d_{\alpha\beta} \quad (5.10b)$$

$$c_{\alpha\beta}^{mnl}(r)=0 \quad \left[\text{if}(mnl)=(000)\text{ or }(110)\right] \quad\quad\quad\quad (5.10c)$$

$$c_{\alpha\beta}^{mnl}(r)=-\beta u_{\alpha\beta}^{mnl}(r) \quad \left[\begin{array}{l}\text{if }(mnl)=(101),\\ (011),\text{ or }(112)\end{array}\right. \Bigg\}\text{ for }\; r>d_{\alpha\beta} \quad (5.10d)$$

All equations written from (5.6) are valid in the MSA in the case of all species having both a charge and a dipole moment. The MSA equations describing an m-component electrolyte consisting of a single solvent and $m-1$ ionic species are obtained by setting $\mu_1=\mu_2=\cdots\mu_{m-1}=0$, $\mu_m=\mu$, and $q_m=0$. Then the pair correlation functions for the different species are of the form

$$h_{ij}(r)=h_{ij}^{000}(r) \quad\quad\quad\quad (5.11a)$$

$$h_{i\mu}(12)=h_{i\mu}^{000}(r)+h_{i\mu}^{011}(r)\Phi^{011}(12) \quad\quad\quad\quad (5.11b)$$

$$h_{\mu i}(12)=h_{\mu i}^{000}(r)+h_{\mu i}^{101}(r)\Phi^{101}(12) \quad\quad\quad\quad (5.11c)$$

and

$$h_{\mu\mu}(12)=h_{\mu\mu}^{000}(r)+h_{\mu\mu}^{110}(r)\Phi^{110}(12)+h_{\mu\mu}^{112}(r)\Phi^{112}(12) \quad (5.11d)$$

where the subscripts i and j indicate ionic species and μ denotes the dipolar solvent. Again of course there is a corresponding set of equations for the $c_{\alpha\beta}(12)$. For the ionic solutions the matrix equations (5.9) reduce[143] to

$$\tilde{\eta}_{ij}=\sum_n \rho_n \tilde{h}_{in}\tilde{c}_{nj}+\rho_\mu \tilde{h}_{i\mu}^{000}\tilde{c}_{j\mu}^{000}-\tfrac{1}{3}\rho_\mu \tilde{h}_{i\mu}^{011}c_{j\mu}^{011} \quad\quad (5.12a)$$

$$\tilde{\eta}_{i\mu}^{000}=\sum_n \rho_n \tilde{h}_{in}\tilde{c}_{n\mu}^{000}+\rho_\mu \tilde{h}_{i\mu}^{000}\tilde{c}_{\mu\mu}^{000} \quad\quad (5.12b)$$

$$\tilde{\eta}_{\mu\mu}^{000}=\sum_n \rho_n \tilde{h}_{n\mu}^{000}\tilde{c}_{n\mu}^{000}+\rho_\mu \tilde{h}_{\mu\mu}^{000}\tilde{c}_{\mu\mu}^{000} \quad\quad (5.12c)$$

$$\tilde{\eta}_{i\mu}^{011}=\sum_n \rho_n \tilde{h}_{in}\tilde{c}_{n\mu}^{011}+\tfrac{1}{3}\rho_\mu \tilde{h}_{i\mu}^{011}\tilde{c}_{\mu\mu}^{0} \quad\quad (5.12d)$$

$$\tilde{\eta}_{\mu\mu}^{0}=-\sum_n \rho_n \tilde{h}_{n\mu}^{011}\tilde{c}_{n\mu}^{011}+\tfrac{1}{3}\rho_\mu \tilde{h}_{\mu\mu}^{0}\tilde{c}_{\mu\mu}^{0} \quad\quad (5.12e)$$

and

$$\tilde{\eta}_{\mu\mu}^1 = \tfrac{1}{3}\rho_\mu \tilde{h}_{\mu\mu}^1 \tilde{c}_{\mu\mu}^1 \tag{5.12f}$$

where the sum on n is over all ionic species. We have used the symmetry relations $h_{ij} = h_{ji}$, $h_{i\mu}^{000} = h_{\mu i}^{000}$, and $h_{i\mu}^{011} = -h_{\mu i}^{101}$, and have introduced the combinations

$$f_{\mu\mu}^0 = f_{\mu\mu}^{110} + 2f_{\mu\mu}^{112} \tag{5.12g}$$

$$f_{\mu\mu}^1 = f_{\mu\mu}^{110} - f_{\mu\mu}^{112} \tag{5.12h}$$

where f represents h, c, or η. We note that if there are $m - 1$ ionic species, (5.12a) results in $m(m-1)/2$ equations, and (5.12b) and (5.12d) give $m-1$ each, for a total of $\tfrac{1}{2}(m-1)(m+4)+3$. Equations 5.12 together with the closure relations (5.10) constitute the MSA for an ionic solution. The equations determining $h_{i\mu}^{000}(r)$ and $h_{\mu\mu}^{000}(r)$ completely decouple, and one obtains PY results for hard-sphere mixtures. The remaining equations can be solved numerically.[143]

2. The LHNC and QHNC Approximations

The LHNC and QHNC closures are defined and discussed in Section III.B.3. For charge-dipole systems one obtains[143]

$$c_{\alpha\beta}^{000}(r) = \exp\left[\eta_{\alpha\beta}^{000}(r) - \beta u_{\alpha\beta}^{000}(r) + \frac{B_{\alpha\beta}^{000}(r)}{g_{\alpha\beta}^{000}(r)^2}\right] - \eta_{\alpha\beta}^{000}(r) - 1 \tag{5.13a}$$

$$c_{\alpha\beta}^{mnl}(r) = g_{\alpha\beta}^{000}(r)\left[\eta_{\alpha\beta}^{mnl}(r) - \beta u_{\alpha\beta}^{mnl}(r)\right] - \eta_{\alpha\beta}^{mnl}(r) \tag{5.13b}$$

if $(mnl) = (011)$, (101), or (112), and

$$c_{\alpha\beta}^{110}(r) = g_{\alpha\beta}^{000}(r)\eta_{\alpha\beta}^{110}(r) - \eta_{\alpha\beta}^{110}(r) \tag{5.13c}$$

where

$$\frac{B_{\alpha\beta}^{000}(r)}{g^{000}(r)^2} = \tfrac{1}{6}\left[\left(\eta_{\alpha\beta}^{011}(r) - \beta u_{\alpha\beta}^{011}(r)\right)^2 + \left(\eta_{\alpha\beta}^{101}(r) - \beta u_{\alpha\beta}^{101}(r)\right)^2\right.$$

$$\left. + \left(\eta_{\alpha\beta}^{110}(r)\right)^2 + 2\left(\eta_{\alpha\beta}^{112}(r) - \beta u_{\alpha\beta}^{112}(r)\right)^2\right] \tag{5.13d}$$

For hard particles, (5.13a) to (5.13d) are applied for $r > d_{\alpha\beta}$, and (5.10a) and

(5.10b) constitute the closure for $r < d_{\alpha\beta}$. As described in Section III.B.3, errors arising from the HNC treatment of $u_{\alpha\beta}^{000}(r)$ are partially removed by applying Lado's technique,[67] which replaces (5.13a) with an expression analogous to (3.23). Here again of course the LHNC and QHNC theories must be solved numerically.[143]

C. The Solute-Dependent Dielectric Constant

1. Formal Relationships

The OZ equation (5.6a), together with the plausible assumption

$$c_{\alpha\beta}(12) \rightarrow -\beta u_{\alpha\beta}(12) \qquad \text{as} \quad r_{\alpha\beta} \rightarrow \infty \qquad (5.14)$$

is sufficient to determine formal expressions relating ε_E and the various correlation functions. Such expressions have been obtained by Adelman,[139] by Chan et al.,[141] and recently by Levesque et al.[143] The following is a derivation and discussion of these results in a unified notation.

Adelman[139] has shown that (5.6a), the OZ equation for mixtures, can be transformed into an effective OZ equation that can be written in the form

$$\tilde{h}_{ij}(k) - \tilde{c}_{ij}^{\text{eff}}(k) = \sum_n \rho_n \tilde{h}_{in}(k) \tilde{c}_{nj}^{\text{eff}}(k) \qquad (5.15)$$

where the summation is over solute species only and the tildes denote Fourier transforms. Thus (5.15) explicitly involves only the solute species, with all solvent effects being by definition exactly included in the effective direct correlation function $c_{ij}^{\text{eff}}(r)$. If we introduce an effective solute-solute pair potential $w_{ij}^{\text{eff}}(r)$, then $c_{ij}^{\text{eff}}(r)$ is related to $w_{ij}^{\text{eff}}(r)$ in exactly the same way that $c(12)$ is related to the usual pair potential.[139] For ionic solutions one finds[139, 143] that

$$c_{ij}^{\text{eff}}(r) \rightarrow -\beta w_{ij}^{\text{eff}}(r) \rightarrow -\frac{\beta q_i q_j}{\varepsilon_E r} \qquad \text{as} \quad r \rightarrow \infty \qquad (5.16)$$

which essentially defines the solute-dependent dielectric constant ε_E. It is obvious that ε_E plays a role very similar to that of the usual dielectric constant of a nonconducting fluid. In the limit of infinite ion dilution, $w_{ij}^{\text{eff}}(r)$ becomes the usual ion-ion potential of mean force, and ε_E is just the dielectric constant of the solvent, ε.

Equation 5.16 is obtained[139, 143] by considering the asymptotic behavior of $c_{ij}^{\text{eff}}(r)$ as r goes to infinity, or, equivalently, the form of $\tilde{c}_{ij}^{\text{eff}}(k)$ as k goes to zero. Equations 5.12 and 5.14 are sufficient for this purpose. Eliminating

$\tilde{h}_{i\mu}^{000}$ and $\tilde{h}_{i\mu}^{011}$ from (5.12a) through the use of (5.12b) and (5.12d), and comparing with (5.15) yields[143]

$$\tilde{c}_{ij}^{\text{eff}} = \tilde{c}_{ij} + \rho_\mu \frac{\tilde{c}_{i\mu}^{000} \tilde{c}_{j\mu}^{000}}{1 - \rho_\mu \tilde{c}_{\mu\mu}^{000}} - \frac{1}{3}\rho_\mu \frac{\tilde{c}_{i\mu}^{011} \tilde{c}_{j\mu}^{011}}{1 - \frac{1}{3}\rho_\mu \tilde{c}_{\mu\mu}^{0}} \tag{5.17}$$

Equations 5.4 and 5.14 imply that as r goes to infinity,

$$c_{ij}(r) \rightarrow \frac{-\beta q_i q_j}{r} \tag{5.18a}$$

$$c_{i\mu}^{011}(r) \rightarrow \frac{\beta q_i \mu}{r^2} \tag{5.18b}$$

and

$$c_{\mu\mu}^{112}(r) \rightarrow \frac{\beta \mu^2}{r^3} \tag{5.18c}$$

which in turn give[139, 143]

$$\tilde{c}_{ij}(k) \rightarrow \frac{-4\pi\beta q_i q_j}{k^2} \tag{5.19a}$$

$$\tilde{c}_{i\mu}^{011}(k) \rightarrow \frac{4\pi\beta q_i \mu i}{k} \tag{5.19b}$$

and

$$c_{\mu\mu}^{112}(k) \rightarrow \frac{-4\pi\beta\mu^2}{3} \tag{5.19c}$$

as k goes to zero. Combining (5.17) and (5.19), one obtains that

$$\tilde{c}_{ij}^{\text{eff}}(k) \rightarrow \frac{-4\pi\beta q_i q_j}{k^2}\left(\frac{a_1}{a_0}\right) = \frac{-4\pi\beta q_i q_j}{\varepsilon_E k^2} \qquad \text{as} \quad k \rightarrow 0 \tag{5.20a}$$

where

$$\varepsilon_E = \frac{a_0}{a_1} \tag{5.20b}$$

$$a_0 = 1 - \frac{1}{3}\rho_\mu \tilde{c}_{\mu\mu}^{0(0)} \tag{5.20c}$$

$$a_1 = 1 - \frac{1}{3}\rho_\mu \tilde{c}_{\mu\mu}^{1(0)} \tag{5.20d}$$

$\tilde{c}_{\mu\mu}^{0(0)} = \tilde{c}_{\mu\mu}^{0}(k=0)$ and $\tilde{c}_{\mu\mu}^{1(0)} = c_{\mu\mu}^{1}(k=0)$ (for notation see Appendix D). Equation 5.20b relating ε_E and the direct correlation function was given by Adelman.[139] Although for finite k, $\tilde{c}_{ij}^{\text{eff}}(k)$ will depend on the components neglected in the expansion (5.8a), the $k=0$ result does not. Thus at least for the present model equation (5.20b) is formally exact. Furthermore, (5.20b) is easily identified as (2.26), which is thus generally valid for rigid particles of cylindrical symmetry, as already discussed in Sec. II for uncharged particles.

It is possible[143] to relate ε_E and the dipole-dipole correlation function $h_{\mu\mu}(12)$. If we first consider the infinite dilution case ($\rho_1 = \rho_2 = \cdots \rho_{m-1} = 0$), it follows immediately from (5.12e) and (5.12f) that

$$a_0\left(1 + \tfrac{1}{3}\rho_\mu \tilde{h}_{\mu\mu}^{0(0)}\right) = 1 \tag{5.21a}$$

and

$$a_1\left(1 + \frac{1}{3}\rho_\mu \tilde{h}_{\mu\mu}^{1(0)}\right) = 1 \tag{5.21b}$$

where $\tilde{h}_{\mu\mu}^{0(0)} = \tilde{h}_{\mu\mu}^{0}(k=0)$ and $\tilde{h}_{\mu\mu}^{1(0)} = \tilde{h}_{\mu\mu}^{1}(k=0)$. Combining (5.20) to (5.21b) gives

$$\frac{(\varepsilon_E - 1)(2\varepsilon_E + 1)}{9\varepsilon_E} = y\left(1 + \frac{1}{3}\rho_\mu \tilde{h}_{\mu\mu}^{110(0)}\right) = yg \tag{5.22a}$$

where we have used the relationship

$$a_0 - a_1 = -\rho_\mu \tilde{c}_{\mu\mu}^{112(0)} = \frac{4\pi\rho_\mu \beta\mu^2}{3} = 3y \tag{5.22b}$$

Equation 5.22a is just the Kirkwood expression (3.7a), and thus it is obvious that at infinite dilution $\varepsilon_E = \varepsilon$, the dielectric constant of the pure solvent. For ion-dipole mixtures, (5.21b) remains valid; but it is shown in Appendix D that at finite ion concentration (5.21a) is replaced by

$$a_0\left(1 + \tfrac{1}{3}\rho_\mu \tilde{h}_{\mu\mu}^{0(0)}\right) = 1 + 3y\left(1 + \tfrac{1}{3}\rho_\mu \tilde{h}_{\mu\mu}^{0(0)}\right) \tag{5.23}$$

Also for ion-dipole mixtures it is found[143] (Appendix D) that when $\rho_i > 0$

$$\tilde{h}^{112(0)} = \tilde{h}^{112}(k=0) = 0 \tag{5.24}$$

This differs from the dipolar fluid case [$\tilde{h}_{\mu\mu}^{112}(0) = -(\varepsilon - 1)^2/3y\varepsilon\rho_\mu$ (cf. Section II.A)] and results from the Debye screening of the dipole-dipole interactions. Combining (5.20b), (5.21b), (5.23), and (5.24), we obtain, when

$\rho_i > 0$

$$\varepsilon_E - 1 = 3y\left(1 + \tfrac{1}{3}\rho_\mu h_{\mu\mu}^{110(0)}\right) = 3yg \qquad (5.25)$$

where as before, $g = \langle M^2 \rangle / N\mu^2$. This is the fluctuation formula valid for ion-dipole mixtures. Equation 5.25 is similar in form to the "tin-foil theorem" first attributed[136] to Onsager. In the context of ε_E, this formula has been given by Fulton[148] and suggested as a conjecture by Hubbard et al.[136] The derivation outlined above is essentially that of Levesque et al.[143]

The ε_E can also be related[141, 143] to the ion-ion correlation functions $h_{ij}(r)$. One obtains[141, 143] (Appendix D)

$$\varepsilon_E = 4\pi\beta \sum_{ij} \rho_i q_i \rho_j q_j \tilde{h}_{ij}^{(2)} \qquad (5.26a)$$

where

$$\tilde{h}_{ij}^{(2)} = -\tfrac{1}{6} \int h_{ij}(r) r^2 \, d\mathbf{r} \qquad (5.26b)$$

This result was obtained by Chan et al.[141] and is similar to the second moment condition first derived by Stillinger and Lovett.[149] However, it should be noted that Stillinger and Lovett[149] consider only primitive model electrolytes, and the dielectric constant occurring in their formula is that of the pure solvent. For the mean spherical, LHNC, and QHNC approximations, all three formulas [(5.20b), (5.25), and (5.26)] must give ε_E consistently.

2. Quantitative Results

The LHNC and QHNC approximations have been solved[143] for three-component mixtures consisting of charged hard spheres in dipolar hard-sphere solvents. Only 1:1 electrolytes with ions of equal diameter have been considered. Thus denoting the positive and negative ionic species with the obvious subscripts, $+$ and $-$, one has $q_+ = -q_- = q$, $d_+ = d_-$, and the system is characterized by the reduced variables $\rho_+^* = \rho_+ d_+^3$, $\rho_\mu^* = \rho_\mu d_\mu^3$, $\mu^{*2} = \beta\mu^2/d_\mu^3$, and $q^{*2} = \beta q^2/d_+$. In addition, the solvent-ion diameter ratio, d_μ/d_+, must be specified. For some values of these parameters, convergence problems are encountered in the QHNC theory, and these difficulties are discussed in Ref. 143. However, for many ionic solutions numerical results can be obtained.

The dependence of ε_E and g on ionic concentration is shown in Fig. 25. Results for several solutions are included. It is useful to note that if $d_+ \simeq 4$ Å, then $\rho_+^* = 0.0019$ and 0.038 correspond to 0.05 and $1M$ (M = moles/liter)

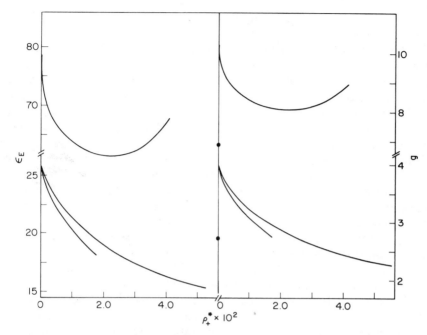

Fig. 25. The solute-dependent dielectric constant and associated g-values. From top to bottom the curves are for the following parameters: $\mu^{*2} = 2.25$, $\rho_\mu^* = 0.8$, $q^{*2} = 86$, and $d_\mu/d_+ = 0.68$; $\mu^{*2} = 2.5$, $\rho_\mu^* = 0.6$, $q^{*2} = 40$, and $d_\mu/d_+ = 1$; $\mu^{*2} = 2.5$, $\rho_\mu^* = 0.6$, $q^{*2} = 136$, and $d_\mu/d_+ = 0.68$. The dots are the g-values for the pure dipolar solvent. The discontinuity in g at $\rho_+ = 0$ can be clearly seen. (Results from Ref. 143.)

solutions, respectively. It follows from (5.22a) and (5.25) that if ε_E is continuous, g must be discontinuous at $\rho_+ = 0$. The discontinuity in g is obvious in Fig. 25.

At the solvent density, $\rho_\mu^* = 0.6$, we see that ε_E decreases continuously with increasing ρ_+^*. This behavior is qualitatively consistent with experimental results for aqueous solutions,[137] but the present model is much too simple to permit meaningful quantitative comparison. Also it is important to recall the ε_E is but one of the contributions to the apparent dielectric constant ε_{SOL}, measured experimentally. The behavior of ε_E at $\rho_\mu^* = 0.8$ is somewhat different. Initially, ε_E decreases as before, but it now passes through a minimum and increases again at the higher concentrations. This is not observed experimentally, for a very simple reason.[143] In the theoretical calculations the ions are added at constant volume rather than at constant pressure. Thus as ions are added, ρ_μ remains constant but the total number density, $\rho = \rho_+ + \rho_- + \rho_\mu$, increases. At high density ε_E is very sensitive to small changes in ρ, and it is the increase in ρ that causes ε_E to increase with ρ_+ at the higher concentrations. This is not a surprising observation, since it is known[58] that at high density ($\rho^* \sim 0.8$) the QHNC dielectric constant of dipolar spheres varies rapidly with density. Also, since ρ_μ is held fixed and there is no loss of dielectric material, the decrease in ε_E at the lower concentrations must be totally due to the influence of the ions on the dipolar correlations. The experiments are carried out at constant pressure where ρ, ρ_μ, and ρ_+ are related in a nontrivial manner that depends on the equation of state. This relationship is not known for the present model. Perhaps holding ρ, rather than ρ_μ, fixed as ρ_+ is increased would better approximate the experimental situation, but this has not been investigated. At $\rho_\mu^* = 0.6$, ε_E is not very sensitive to small changes in ρ and no minimum occurs in the range of ρ_+ examined.

Pérez-Hernández and Blum[144] have recently obtained essentially analytic expressions for ε_E in the MSA, and quantitative results are given by Vericat et al.[145] The MSA results are found[145] to be qualitatively similar to those described above, with ε_E decreasing as the ionic concentration is increased. Vericat et al.[145] also compare the MSA results for ε_E with experimental values of ε_{SOL} for aqueous solutions. They show that if one is willing to treat the dipole moment of water as an adjustable parameter, rough agreement can be obtained.[145] However, in view of the simplicity of the model, the likely inaccuracy of the MSA (cf. Section III.D), and the failure to account for the dynamic contributions to ε_{SOL}, it is difficult to reach any conclusions from such comparisons.

Another theory for ε_E has been proposed by Adelman and Chen.[140] Their approach differs from the theories described above in that they do not rely completely on statistical mechanical methods, but attempt to solve generalized Poisson-Boltzmann equations for ion-solvent mixtures. The approximation obtained[140] predicts that ε_E *increases* with ρ_+ for all concentrations. This is obviously in serious disagreement with the integral-equation theories, and it probably means that the approximation of Adelman and Chen[140] is rather poor. However, this has not been proved and further investigation is necessary.

VI. THINGS TO COME

The sensitivity of ε to the magnitude of the quadrupole moment for "linear" quadrupoles in dipolar-fluid model calculations was discussed in Section III. This property points the way to two questions that must be explored in future work.

1. How quantitatively sensitive is ε to other forms of anisotropy that characterize the pair potential in real polar molecules?
2. What are the full generalizations of the formally exact expressions for ε we have exhibited in terms of correlation functions when one generalizes the models we have been considering to simultaneously include quadrupoles and higher multipoles, as well as short-range anisotropy of arbitrary symmetry along with a dipolar term and polarizability?

The first question has hardly been touched. One of the few preliminary results available, an exploratory calculation of Martina and Stell[150] on a dipolar system with hard diatomic core, uses approximations of untested accuracy. Much more work is needed in regard to core anisotropy as well as the effect of quadrupole terms of general symmetry.

Quite probably the answer to the second question will look not too much different from the expressions for the models that have been thoroughly analyzed here, but the establishment of this result may turn out to be tedious. We have seen in Section II how to handle rigid nonpolarizable particles of arbitrary symmetry using the formalism of Høye and Stell. The addition of fluctuating polarizability has been considered by those authors only for molecules of cylindrical symmetry, but its extension to molecules of arbitrary symmetry is unlikely to raise fundamental problems. On the other hand, particles lacking cylindrical symmetry even in the nonpolarizable case are substantially more awkward to deal with than cylindrically symmetric particles. In treating the constant-polarizability case, Wertheim excludes all permanent multipoles beyond the dipole; clearly the quadrupole at least must also be included to provide a realistic model for many real fluids of interest.

Significant progress in the nonpolar case will be, if anything, *more* difficult than in the polar case, because in the former case small deviations from the Clausius-Mossotti equation represent the whole story of interest, and they are crucially sensitive to the form of the pair-polarizability $\alpha(12)$ [and presumably the triplet term $\alpha(123)$ as well]. Neither the constant-polarizability model or the classical harmonically fluctuating polarizability model considered in Section IV gives rise to an $\alpha(12)$ that looks like that of real molecules. There are a number of challenging problems to be met in this regard. Can one find a classical non-harmonically fluctuating polarizability model that yields realistic $\alpha(12)$ [and $\alpha(123)$]? Or can one efficiently use the Høye-Stell formalism in a semiclassical way simply by inserting the true quantum mechanical functions for isolated molecules in place of the $g(\mathbf{m})$ of (4.6)? (This seems promising.) Can we ultimately do good enough first-principle calculations to get $\alpha(12)$ and $\alpha(123)$ without working backward from experimental input? (This is a hard problem in quantum chemistry rather than one of statistical mechanics.) And once we have reasonable $\alpha(12)$ and $\alpha(123)$, by hook or by crook, will there be any analytically tractable approximation of high accuracy that will emerge for the S of (4.61) as a functional of these quantities?

A quite different sort of extension of the work considered here is already well under way and promises to yield new and fundamentally important results, namely, work on a fluid in the presence of a nonzero electric field. Nienhuis and Deutch considered certain aspects of such a system in one of their papers,[20] showing one way of formally characterizing the polarizability of such a system as a function of full Maxwell field E rather than applied field E_0. More recently, Høye and Stell,[134] Martina and Stell,[151] Rasaiah and Stell,[152] and Ramshaw[153] have been investigating this area. Høye and Stell were able to recover the thermodynamic relations that describe such a system (as given, e.g., by Landau and Lifshitz[154]) from the statistical mechanics of a Hamiltonian model of molecular interaction. Martina and Stell have found a systematic cluster expansion method for directly implementing (1.1) to obtain ε, and they consider dielectric saturation and other nonlinear effects. Rasaiah and Stell have used an integral equation approach to study such effects, especially electrostriction. Ramshaw has obtained a very interesting closed-form functional representation of the one-particle distribution function $\rho(1)$ in the presence of an external field of arbitrary strength.

We do not go into any of the above mentioned work here, except to show that the ε that comes out of (1.1) for a system in a nonzero field E is indeed the same ε that we have been considering in Sections I to V in terms of correlations in systems free of external fields. To see this, we use the standard equation for P in terms of the one-particle density in the presence of a field $\rho(\Omega, E)$

$$P = \int \mu(\Omega)\rho(\Omega, E)\, d\Omega \tag{6.1}$$

We can compute $\rho(\Omega, E)$ by starting with $\rho(12)$ for a system with a tagged dipolar particle 1 in a sea of its fellow particles and letting particle 1 grow bigger and bigger (with appropriately scaled bigger and bigger dipole moment as well as diameter) until it becomes a macrosphere or wall (or infinite radius on the scale set by the microscopic radii) through which an external field E_0 is passing.[155] We then back far enough from the wall for the resulting $\rho(12)$ to be independent of distance from the wall (on the microscale). We now have $\rho(\Omega, E)$, which—to linear order in E—proves to be of the form[155]

$$\rho(\Omega, \mathbf{E}) = \left(\frac{\rho}{\Omega}\right)\left(1 + \frac{3\beta E_0 \mu \cos\theta}{2q_1 + q_2}\right) \tag{6.2}$$

where θ is the angle between $\mathbf{E}_0$ and the μ of a fluid particle, and q_1 and q_2, which characterize the dielectric properties of the fluid, are given by (2.25). The $\mathbf{E}_0$ is related to the Maxwell field $\mathbf{E}$ at the point of interest by the simple equation

$$\mathbf{E} = \frac{3}{2\varepsilon + 1}\mathbf{E}_0 \tag{6.3}$$

Putting (6.2) and (6.3) into (6.1) and putting that result in turn into (1.1) yields

$$\varepsilon = \frac{q_1}{q_2} \tag{6.4}$$

which is exactly (2.26). It should be a reassuring result for those who may have felt some unease over whether the vast formal scaffolding we have had to erect has gone up on the right plot—the ε we have been talking about is, after all, the same conceptual object that Clausius, Mossotti, and Debye were talking about.

APPENDIX A

Use of Test Charges

Another way to obtain (2.55) and (2.56) is to compute the effective interaction between two test charges. This is easily done by use of (21) of ref. 4 with a and b taken as the coefficients of the D_k and Δ terms, respectively, in (2.54). We get

$$\frac{1}{\varepsilon} = \frac{1-z}{1+2z} = 1 - 3y - 2ya - yb$$

$$= 1 - 3y - ya(2+\Theta)(1-(1-\Theta)y) = \frac{1-(1-\Theta)y}{1+(2+\Theta)y} \tag{A.1}$$

which immediately leads to the results (2.55) and (2.56) for ε and z.

We also can make the corresponding computation using our test charge and dipole or two test dipoles defined in Section II of Ref. 4. Putting $K = \hat{\mathbf{k}}\cdot\hat{\mathbf{s}}$, we have by convolution (integration over $\hat{\mathbf{s}}$) $KD_k = \frac{2}{3}K$; $K\Delta = \frac{1}{3}K$. In addition, we have the convolution rules [Ref. 4, (B.8)]. For our test charge and dipole we then get [using (17) from Ref. 4]:

$$\frac{\varepsilon+2}{3\varepsilon}K = \frac{1}{1+2z}K = K\left[1 - 3yD_k(1 + aD_k + b\Delta)\right]$$

$$= \left[1 - 2y\left(1 + \tfrac{2}{3}ay + \tfrac{1}{3}by\right)\right]K$$

$$= \left[1 - 2y - \tfrac{2}{3}ay(2+\Theta)(1-(1-\Theta)y)\right]K$$

$$= \frac{1+\Theta y}{1+(2+\Theta)y}K \qquad\qquad (A.2)$$

For our two test dipoles we similarly get [using (18) from Ref. 4 for ε],

$$\left(\frac{\varepsilon+2}{3}\right)^2 \frac{1}{\varepsilon}[D_k - 2z\Delta] = \frac{1}{(1+2z)(1-z)}[D_k - 2z\Delta]$$

$$= D_k\left[1 - 3yD_k(1 + aD_k + b\Delta)\right]$$

$$= D_k\left[1 - 3y\left((1 + \tfrac{1}{3}a + \tfrac{1}{3}b)D_k + \tfrac{2}{3}a\Delta\right)\right]$$

$$= \left[1 - y\left(1 + a + \tfrac{1}{3}b\right)\right]D_k + \left[-2y\left(1 + \tfrac{1}{3}a + \tfrac{1}{3}b\right)\right]\Delta$$

$$= \frac{(1+\Theta y)^2}{(1+(2+\Theta)y)(1-(1-\Theta)y)}\left[D_k - 2\frac{y}{1+\Theta y}\Delta\right] \qquad (A.3)$$

One sees that (A.2) and (A.3) both immediately lead to result (2.55) for z and by that to result (2.56) for ε.

APPENDIX B

Explicit Expressions for the Rotational Invariants

The $\Phi^{mnl}(12)$ used in this chapter can be conveniently written in the form

$$\Phi^{000}(12) = 1 \qquad\qquad (B.1)$$
$$\Phi^{101}(12) = T_1 \qquad\qquad (B.2)$$
$$\Phi^{011}(12) = T_2 \qquad\qquad (B.3)$$
$$\Phi^{110}(12) = T_1T_2 + T_3 \qquad\qquad (B.4)$$
$$\Phi^{112}(12) = 2T_1T_2 - T_3 \qquad\qquad (B.5)$$
$$\Phi^{121}(12) = \tfrac{1}{2}[T_1(3T_2^2 - 1) + 3T_2T_3] \qquad\qquad (B.6)$$
$$\Phi^{211}(12) = \tfrac{1}{2}[T_2(3T_1^2 - 1) + 3T_1T_3] \qquad\qquad (B.7)$$

$$\Phi^{123}(12) = 3T_1(3T_2^2 - 1) - 6T_2T_3 \tag{B.8}$$
$$\Phi^{213}(12) = 3T_2(3T_1^2 - 1) - 6T_1T_3 \tag{B.9}$$
$$\Phi^{220}(12) = 1 - 3T_1^2T_2^2 - 3T_3^2 - 6T_1T_2T_3 \tag{B.10}$$
$$\Phi^{222}(12) = 2 - 3T_1^2 - 3T_2^2 + 6T_1^2T_2^2 - 3T_3^2 + 3T_1T_2T_3 \tag{B.11}$$

and

$$\Phi^{224}(12) = 1 - 5T_1^2 - 5T_2^2 + 17T_1^2T_2^2 + 2T_3^2 - 16T_1T_2T_3 \tag{B.12}$$

where

$$T_1 = \hat{\boldsymbol{\mu}}_1 \cdot \hat{\mathbf{r}}_{12} = \cos\theta_1 \tag{B.13}$$
$$T_2 = \hat{\boldsymbol{\mu}}_2 \cdot \hat{\mathbf{r}}_{12} = \cos\theta_2 \tag{B.14}$$
$$T_3 = \hat{\boldsymbol{\mu}}_1 \cdot \hat{\boldsymbol{\mu}}_2 - (\hat{\boldsymbol{\mu}}_1 \cdot \hat{\mathbf{r}}_{12})(\hat{\boldsymbol{\mu}}_2 \cdot \hat{\mathbf{r}}_{12}) = \sin\theta_1 \sin\theta_2 \cos(\phi_1 - \phi_2) \tag{B.15}$$

and $\hat{\boldsymbol{\mu}}_1$ and $\hat{\boldsymbol{\mu}}_2$ are unit vectors directed along dipoles 1 and 2, $\hat{\mathbf{r}}_{12} = \mathbf{r}_{12}/|\mathbf{r}_{12}|$, and θ and ϕ are, respectively, polar and azimuthal angles defined with respect to the vector $\mathbf{r}_{12} = \mathbf{r}_2 - \mathbf{r}_1$.

APPENDIX C

The $B^{mnl}(r)$ for Dipole-Quadrupole Systems

The functions $B^{mnl}(r)$ occurring in the QHNC equations (3.21) for dipole-quadrupole systems are defined as follows:

$$
\begin{aligned}
B^{000}(r) = \frac{1}{2}\Bigg[& \frac{1}{3}(110)^2 + \frac{2}{3}(112)^2 + \frac{1}{6}(121)^2 + 4(123)^2 \\
& + \frac{1}{6}(211)^2 + 4(213)^2 + \frac{4}{5}(220)^2 \\
& + \frac{14}{25}(222)^2 + \frac{224}{45}(224)^2 \Bigg]
\end{aligned} \tag{C.1}
$$

$$
\begin{aligned}
B^{110}(r) = \frac{3}{2}\Bigg[& -\frac{8}{15}(110)(220) + \frac{28}{75}(112)(222) + \frac{1}{45}(121)(211) \\
& + \frac{16}{5}(121)(213) \Bigg]
\end{aligned} \tag{C.2}
$$

$$
\begin{aligned}
B^{112}(r) = \frac{3}{4}\Bigg[& \frac{28}{75}(110)(222) - \frac{8}{75}(112)(220) - \frac{8}{75}(112)(222) \\
& + \frac{64}{25}(112)(224) \\
& + \frac{46}{225}(121)(211) + \frac{4}{25}(121)(213) \\
& + \frac{4}{25}(123)(211) + \frac{64}{25}(123)(213) \Bigg]
\end{aligned} \tag{C.3}
$$

$$B^{121}(r) = 3\left[\frac{1}{45}(110)(211) + \frac{46}{225}(112)(211)\right.$$

$$+ \frac{4}{25}(112)(213) - \frac{2}{15}(121)(220)$$

$$- \frac{2}{75}(121)(222)$$

$$\left. + \frac{92}{175}(123)(222) + \frac{64}{105}(123)(224)\right] \qquad (C.4)$$

$$B^{123}(r) = \frac{1}{8}\left[\frac{16}{5}(110)(213) + \frac{4}{25}(112)(211) + \frac{64}{25}(112)(213)\right.$$

$$+ \frac{92}{175}(121)(222)$$

$$+ \frac{64}{105}(121)(224) - \frac{32}{35}(123)(220)$$

$$\left. - \frac{128}{175}(123)(222) + \frac{256}{35}(123)(224)\right] \qquad (C.5)$$

$$B^{211}(r) = 3\left[\frac{1}{45}(110)(121) + \frac{46}{225}(112)(121) + \frac{4}{25}(112)(123) - \frac{2}{15}(211)(220)\right.$$

$$\left. - \frac{2}{75}(211)(222) + \frac{92}{175}(213)(222) + \frac{64}{105}(213)(224)\right] \qquad (C.6)$$

$$B^{213}(r) = \frac{1}{8}\left[\frac{16}{5}(110)(123) + \frac{4}{25}(112)(121) + \frac{64}{25}(112)(123)\right.$$

$$+ \frac{92}{175}(211)(222)$$

$$+ \frac{64}{105}(211)(224) - \frac{32}{35}(213)(220)$$

$$\left. - \frac{128}{175}(213)(222) + \frac{256}{35}(213)(224)\right] \qquad (C.7)$$

$$B^{220}(r) = \frac{5}{8}\left[-\frac{4}{15}(110)^2 - \frac{4}{75}(112)^2 - \frac{1}{15}(121)^2 - \frac{16}{35}(123)^2\right.$$

$$- \frac{1}{15}(211)^2 - \frac{16}{35}(213)^2$$

$$\left. - \frac{16}{35}(220)^2 + \frac{12}{175}(222)^2 - \frac{256}{315}(224)^2\right] \qquad (C.8)$$

$$B^{222}(r) = \frac{25}{28}\left[-\frac{4}{75}(112)^2 - \frac{1}{75}(121)^2 - \frac{64}{175}(123)^2 - \frac{1}{75}(211)^2\right.$$

$$- \frac{64}{175}(213)^2$$

$$+ \frac{164}{1225}(222)^2 - \frac{256}{441}(224)^2 + \frac{28}{75}(110)(112) + \frac{92}{175}(121)(123)$$

$$+ \frac{92}{175}(211)(213) + \frac{24}{175}(220)(222) + \frac{1088}{1225}(222)(224) \Bigg] \quad (C.9)$$

$$B^{224}(r) = \frac{45}{448} \Bigg[\frac{32}{25}(112)^2 + \frac{128}{35}(123)^2 + \frac{128}{35}(213)^2 + \frac{544}{1225}(222)^2$$

$$+ \frac{1024}{245}(224)^2$$

$$+ \frac{64}{105}(121)(123) + \frac{64}{105}(211)(213) - \frac{512}{315}(220)(224)$$

$$- \frac{512}{441}(222)(224) \Bigg] \quad (C.10)$$

where (mnl) denotes $h^{mnl}(r)$.

APPENDIX D

The Derivation of (5.23), (5.24), and (5.25)

In the vicinity of $k=0$ the correlation functions can be expanded in the form[141, 143]

$$\tilde{c}_{ij} = \tilde{c}_{ij}^{(-2)}k^{-2} + \tilde{c}_{ij}^{(0)} + \cdots \quad (D.1a)$$

$$\tilde{c}_{i\mu}^{011} = -i\left(\tilde{c}_{i\mu}^{011(-1)}k^{-1} + \tilde{c}_{i\mu}^{011(1)}k + \cdots \right) \quad (D.1b)$$

$$\tilde{h}_{ij} = \tilde{h}_{ij}^{(0)} + \tilde{h}_{ij}^{(2)}k^2 + \cdots \quad (D.1c)$$

and

$$\tilde{h}_{i\mu}^{011} = -ik\tilde{h}_{i\mu}^{011(1)} + \cdots \quad (D.1d)$$

where it is assumed that the pair correlation functions are finite at $k=0$, or that the charge-charge and charge-dipole interactions are screened at long range.[22, 25] Then from (5.19a) and (5.19b) it is obvious that

$$\tilde{c}_{ij}^{(-2)} = -4\pi\beta q_i q_j \quad (D.2a)$$

and

$$\tilde{c}_{i\mu}^{011(-1)} = -4\pi\beta q_i \mu \quad (D.2b)$$

Inserting (D.1a) to (D.1d) into (5.12a) and comparing terms of equal order in k yields two results. The k^{-2} term gives the local charge neutrality condition

$$\sum_n \rho_n q_n \tilde{h}_{in}^{(0)} = -q_i \tag{D.3}$$

and collecting terms of order k^0, one obtains[143]

$$\tilde{h}_{ij}^{(0)} - \tilde{c}_{ij}^{(0)} = \sum_n \rho_n \left(\tilde{h}_{in}^{(0)} \tilde{c}_{nj}^{(0)} + \tilde{h}_{in}^{(2)} \tilde{c}_{nj}^{(-2)} \right) + \rho_\mu \tilde{h}_{i\mu}^{000(0)} \tilde{c}_{j\mu}^{000(0)}$$
$$+ \frac{1}{3} \rho_\mu \tilde{h}_{i\mu}^{011(1)} \tilde{c}_{j\mu}^{011(-1)} \tag{D.4}$$

Multiplying (D.4) by $\rho_i q_i$ and summing over i yields

$$4\pi\beta \sum_{ij} \rho_i q_i \rho_j q_j \tilde{h}_{ij}^{(2)} + \frac{1}{3} 4\pi\beta\rho_\mu\mu \sum_i \rho_i q_i \tilde{h}_{i\mu}^{011(1)} = 1 \tag{D.5}$$

where we have used (D.3) and the total charge neutrality condition,

$$\sum_i \rho_i q_i \tilde{h}_{i\mu}^{000(0)} = 0 \tag{D.6}$$

The coefficients of k^{-1} in (5.12d) give

$$\sum_i \rho_i q_i \tilde{h}_{i\mu}^{011(1)} + \mu\left(1 + \frac{1}{3}\rho_\mu \tilde{h}_{\mu\mu}^{0(0)}\right) = 0 \tag{D.7}$$

and collecting the terms of order k^0 in (5.12e), we obtain

$$1 + \frac{1}{3}\rho_\mu \tilde{h}_{\mu\mu}^{0(0)} = \frac{1}{1 - \frac{1}{3}\rho_\mu\left(\tilde{c}_{\mu\mu}^{0(0)} + 4\pi\beta\mu^2\right)} \tag{D.8a}$$

$$= \frac{1}{1 - \frac{1}{3}\rho_\mu \tilde{c}_{\mu\mu}^{1(0)}} \tag{D.8b}$$

where (D.8b) follows from the use of (D.7), (5.12g), (5.12h), and (5.19). Equation 5.23 is easily obtained by rearranging (D.8a). Also (5.12f) allows (D.8b) to be rewritten in the form

$$1 + \frac{1}{3}\rho_\mu \tilde{h}_{\mu\mu}^{0(0)} = 1 + \frac{1}{3}\rho_\mu \tilde{h}_{\mu\mu}^{1(0)} \tag{D.8c}$$

and from (D.8c), (5.12g), and (5.12h), it is immediately apparent that

$$\tilde{h}_{\mu\mu}^{112(0)} = \tilde{h}_{\mu\mu}^{112}(k=0) = 0 \tag{D.9}$$

This is the relationship (5.24).

Finally, combining (D.5), (D.7), (D.9), and (5.25), we obtain

$$4\pi\beta \sum_{ij} \rho_i q_i \rho_j q_j \tilde{h}_{ij}^{(2)} = 1 + 3y\left(1 + \frac{1}{3}\rho_\mu \tilde{h}_{\mu\mu}^{110(0)}\right) = \varepsilon_E \tag{D.10a}$$

where

$$\tilde{h}_{ij}^{(2)} = \frac{d^2}{dk^2}\tilde{h}_{ij}(k)\bigg|_{k=0} = -\frac{1}{6}\int h_{ij}(r)r^2\,d\mathbf{r} \tag{D.10b}$$

These are the formulas (5.26).

Acknowledgments

G. S. acknowledges the National Science Foundation, and the Donors of the Petroleum Research Fund, administered by the American Chemical Society, for support of this research. G. N. P. acknowledges the financial support of the National Research Council of Canada. We are indebted to Harold Friedman for helpful comments on the manuscript of the article.

References

1. M. S. Wertheim, *Annu. Rev. Phys. Chem.*, **30**, 471 (1979).
2. P. Debye, *Phys. Z.*, **13**, 97 (1912).
3. J. S. Høye and G. Stell, *J. Chem. Phys.*, **70**, 2894 (1979).
4. J. S. Høye and G. Stell, *J. Chem. Phys.*, **61**, 562 (1974).
5. J. S. Høye and G. Stell, *J. Chem. Phys.*, **64**, 1952 (1976).
6. J. L. Lebowitz, G. Stell, and S. Baer, *J. Math. Phys.*, **6**, 1282 (1965).
7. G. Stell, J. L. Lebowitz, S. Baer, and W. K. Theumann, *J. Math. Phys.*, **7**, 1532 (1966); see also G. Stell and W. K. Theumann, *Phys. Rev.*, **186**, 581 (1969).
8. J. Ramshaw, *J. Chem. Phys.*, **57**, 2684 (1972).
9. J. D. Ramshaw, *J. Chem. Phys.*, **68**, 5199 (1978).
10. G. Nienhuis and J. M. Deutch, *J. Chem. Phys.*, **55**, 4213 (1971); **56**, 235 (1972).
11. J. D. Ramshaw, *J. Chem. Phys.*, **70**, 1577 (1979).
12. M. S. Wertheim, *J. Chem. Phys.*, **55**, 4291 (1971).
13. J. S. Høye, J. L. Lebowitz, and G. Stell, *J. Chem. Phys.*, **61**, 3252 (1974).
14. See (A1) to (A4) of Ref. 13, but note that our Hankel transform $\bar{a}(k)$ is denoted $\tilde{a}(k)$ there. See also Appendix C of J. S. Høye and G. Stell, *J. Chem. Phys.*, **63**, 5342 (1975), where the notation is consistent with that used here.
15. M. S. Wertheim, *Mol. Phys.*, **25**, 211 (1973).
16. J. S. Høye and G. Stell, *J. Chem. Phys.*, **67**, 524 (1977).
17. J. G. Kirkwood, *J. Chem. Phys.*, **7**, 911 (1939).
18. G. Oster and J. G. Kirkwood, *J. Chem. Phys.*, **11**, 175 (1943).

19. L. Onsager, *J. Am. Chem. Soc.*, **58**, 1486 (1936).
20. G. Nienhuis and J. M. Deutch, *J. Chem. Phys.*, **56**, 1819 (1972).
21. D. W. Jepsen and H. L. Friedman, *J. Chem. Phys.*, **38**, 846 (1963).
22. J. S. Høye and G. Stell, *J. Chem. Phys.*, **68**, 4145 (1978); **71**, 1985 (1979). See also J. S. Høye and G. Stell, in *Faraday Discuss. Chemical Soc.*, No. 64, 16 (1978).
23. S. De Leeuw, J. W. Perram, and E. R. Smith, *Proc R. Soc. Lond.*, A **373**, 27 (1980).
24. J. S. Høye and G. Stell, *J. Chem. Phys.*, **65**, 18 (1976). See also J. S. Høye and G. Stell, *J. Chem. Phys.*, **66**, 795 (1977).
25. J. S. Høye and G. Stell, *J. Chem. Phys.*, **67**, 1776 (1977).
26. P. C. Hemmer, *J. Math. Phys.*, **5**, 75 (1964).
27. S. A. Adelman and J. M. Deutch, *J. Chem. Phys.*, **59**, 3971 (1973); W. Sutherland, G. Nienhuis, and J. M. Deutch, *Mol. Phys.*, **27**, 721 (1974); E. Martina and J. M. Deutch, *Chem. Phys.*, **27**, 183 (1978); D. Isbister and R. J. Bearman, *Mol. Phys.*, **28**, 1297 (1974); **32**, 597 (1976).
28. J. D. Ramshaw, *J. Chem. Phys.*, **64**, 3666 (1976).
29. M. S. Wertheim, *Mol. Phys.*, **26**, 1425 (1973).
30. G. N. Patey, *Mol. Phys.*, **34**, 427 (1977).
31. L. Verlet and J. J. Weis, *Mol. Phys.*, **28**, 665 (1974).
32. J. S. Høye and G. Stell, *J. Chem. Phys.*, **73**, 461 (1980).
33. G. Stell, *J. Chem. Phys.*, **55**, 1485 (1971).
34. G. Stell, in C. Domb and M. S. Green, Eds., *Phase Transitions and Critical Phenomena*, Vol. 5B, Academic Press, London, 1976.
35. G. Stell, in B. Berne, Ed., *Modern Theoretical Chemistry*, Vol. 5, Plenum Press, New York, 1977.
36. R. L. Henderson and C. G. Gray, *Can. J. Phys.*, **56**, 571 (1978); C. G. Gray and R. L. Henderson, *Can. J. Phys.*, **57**, 1605 (1979).
37. L. Mistura, *J. Chem. Phys.*, **59**, 4563 (1973). See also J. Goulon, J. L. Greffe, and D. W. Oxtoby, *J. Chem. Phys.*, **70**, 4742 (1979), for a similar prediction.
38. G. N. Patey, *Mol. Phys.*, **35**, 1413 (1978).
39. H. C. Andersen and D. Chandler, *J. Chem. Phys.*, **55**, 1497 (1971); H. C. Andersen, D. Chandler, and J. D. Weeks, *J. Chem. Phys.*, **56**, 3812 (1972).
40. J. A. Barker and D. Henderson, *J. Chem. Phys.*, **47**, 4714 (1967).
41. J. D. Weeks, D. Chandler, and H. C. Andersen, *J. Chem. Phys.*, **54**, 5237 (1971). See also L. Verlet and J. J. Weis, *Phys. Rev. A*, **5**, 939 (1972).
42. G. A. Mansoori and F. B. Canfield, *J. Chem. Phys.*, **51**, 4958 (1969); J. Rasaiah and G. Stell, *Mol. Phys.*, **18**, 249 (1970).
43. G. Stell, College of Engineering Report No. 228, State University of New York at Stony Brook, July 1972. This appeared subsequently as ref. 34, above.
44. H. C. Andersen and D. Chandler, *J. Chem. Phys.*, **57**, 1918 (1972).
45. G. Stell and J. J. Weis, *Phys. Rev. A*, **16**, 757 (1977).
46. The general structure of this representation was discussed in a somewhat different connection by one of us in D. Henderson, G. Stell, and E. Waisman, *J. Chem. Phys.*, **62**, 4247 (1975). It is an immediate extension of that discussed in G. Stell, *Physica*, **29**, 517 (1963).
47. G. N. Patey and J. P. Valleau, *J. Chem. Phys.*, **61**, 534 (1974); *Chem. Phys. Lett.*, **21**, 297 (1973).
48. G. N. Patey and J. P. Valleau, *J. Chem. Phys.*, **64**, 170 (1976).
49. I. R. McDonald, *J. Phys. C*, **7**, 1225 (1974).
50. A. J. C. Ladd, *Mol. Phys.*, **36**, 463 (1978); **33**, 1039 (1977).
51. V. M. Jansoone, *Chem. Phys.*, **3**, 78 (1974).

52. D. Levesque, G. N. Patey, and J. J. Weis, (a) *Mol. Phys.*, **34**, 1077 (1977); (b) *Mol. Phys.* (to appear, 1981).
53. D. J. Adams, *Mol. Phys.* **40**, 1261 (1980).
54. E. L. Pollock and B. J. Alder, *Physica*, **102A**, 1 (1980).
55. G. N. Patey, in D. Ceperley, Ed., *The Problem of Long Range Forces in Computer Simulation of Condensed Media*, NRCC Proceedings No. 9, 1980.
56. L. Blum and A. J. Torruella, *J. Chem. Phys.*, **56**, 303 (1971); L. Blum, *ibid.*, **57**, 1862 (1972); **58**, 3295 (1973). Rotational-invariant representations had been used earlier for pair potentials (but not in the context of solving the OZ equation). See, e.g. C. G. Gray and J. Van Kranendonk, *Can. J. Phys.*, **44**, 2411 (1966), *ibid*, **46**, 135 (1968), R. L. Armstrong, *ibid*, **46**, 1131 (1968).
57. A. R. Edmonds, *Angular Momentum in Quantum Mechanics*, Princeton University Press, Princeton, NJ, 1960.
58. G. N. Patey, D. Levesque, and J. J. Weis, *Mol. Phys.*, **38**, 219 (1979).
59. G. N. Patey, D. Levesque, and J. J. Weis, *Mol. Phys.*, **38**, 1635 (1979).
60. M. S. Wertheim, *Phys. Rev. Lett.*, **10**, 321 (1963).
61. G. S. Rushbrooke, G. Stell, and J. S. Høye, *Mol. Phys.*, **26**, 1119 (1973). For further Padé results see G. Stell, J. C. Rasaiah, and H. Narang, *Mol. Phys.*, **27**, 1393 (1974), J. C. Rasaiah, B. Larsen, and G. Stell, *J. Chem. Phys.*, **63**, 722 (1975); **64**, 913 (1976).
62. D. W. Jepsen, *J. Chem. Phys.*, **44**, 774 (1966).
63. G. S. Rushbrooke, *Mol. Phys.*, **37**, 761 (1979). See also the work of Gray and Gubbins, *Mol. Phys.*, **30**, 1481 (1975), for expansions of the Kirkwood g-factor for simple models that include quadrupolar terms and Murad et al., *Chem. Phys. Lett.*, **65**, 187 (1979) for h^{224}.
64. H. C. Andersen, D. Chandler, and J. D. Weeks, *J. Chem. Phys.*, **57**, 2626 (1972).
65. J. S. Høye and G. Stell, *J. Chem. Phys.*, **63**, 5342 (1975).
66. R. O. Watts, in K. Singer, Ed., *Statistical Mechanics*, Vol. 1, *Specialist Periodical Reports*, Chemical Society, London, 1973, Ch. 1.
67. F. Lado, *Mol. Phys.*, **31**, 1117 (1976).
68. L. Verlet and J. J. Weis, *Phys. Rev. A*, **5**, 939 (1972).
69. K.-C. Ng, J. P. Valleau, G. M. Torrie, and G. N. Patey, *Mol. Phys.*, **38**, 781 (1979).
70. Yu. V. Agrofonov, G. A. Martinov, and G. N. Sarkisov, *Mol. Phys.*, **39**, 963 (1980).
71. K. E. Gubbins and C. G. Gray, *Mol. Phys.*, **23**, 187 (1972).
72. W. R. Smith, in K. Singer, Ed., *Statistical Mechanics*, Vol. 1, *Specialist Periodical Reports*, Chemical Society, London, 1973, Ch. 2.
73. J. P. Valleau and S. G. Whittington, in B. Berne, Ed., *Modern Theoretical Chemistry*, Vol. V, *Statistical Mechanics*, Part A, Plenum Press, New York, 1977.
74. M. Neumann and O. Steinhauser, *Mol. Phys.*, **39**, 437 (1980).
75. J. A. Barker and R. O. Watts, *Chem. Phys. Lett.*, **3**, 144 (1969); *Mol. Phys.*, **26**, 789 (1973).
76. R. O. Watts, *Mol. Phys.*, **28**, 1069 (1974).
77. D. J. Adams and I. R. McDonald, *Mol. Phys.*, **32**, 931 (1976).
78. S. De Leeuw, J. W. Perram, and E. R. Smith, *Proc. R. Soc. Lond.* A **373**, 57 (1980).
79. B. U. Felderhof, *Physica*, **101A**, 275 (1980).
80. J. A. Barker, *Molecular Dynamics and Monte Carlo Calculations on Water*, CECAM Report, 1972, p. 21.
81. H. Frölich, *Theory of Dielectrics*, 2nd ed., Oxford University Press, Oxford, 1958.
82. H. L. Friedman, *Mol. Phys.*, **29**, 1533 (1975).
83. J. P. Valleau and C. Minns, private communication; also see C. Minns, M. Sc. thesis, University of Toronto, 1977.
84. G. Bossis, *Mol. Phys.*, **38**, 2023 (1979).

85. G. Bossis, B. Quentrec, and C. Brot, *Mol. Phys.*, **39**, 1233 (1980).

86. H. J. C. Berendsen, *Molecular Dynamics and Monte Carlo Calculations on Water*, CECAM Report, 1972, p. 29.

87. J. J. Weis, in J. Dupuy and A. J. Dianoux, Eds., *Microscopic Structure and Dynamics of Liquids*, Plenum Press, New York, 1978, p. 381.

88. M. Berkowitz and S. A. Adelman, *J. Chem Phys.*, **72**, 4795 (1980).

89. D. Levesque, J. J. Weis, and G. N. Patey, unpublished results.

90. C. J. F. Böttcher, *Theory of Electric Polarization*, 2nd ed., Elsevier, Amsterdam, 1973.

91. F. H. Stillinger, *Adv. Chem. Phys.*, **31**, 1 (1975).

92. A. Rahman and F. H. Stillinger, *J. Chem. Phys.*, **55**, 3336 (1971).

93. J. G. Kirkwood, *J. Chem. Phys.*, **4**, 592 (1936).

94. J. Yvon, *Actualités Scientifiques et Industrielles*, Nos. 542 and 543, Hermann & Cie., Paris, 1937.

95. The most comprehensive development prior to Wertheim's work is due to M. Mandel and P. Mazur, *Physica*, **24**, 116 (1958). It is an exact formal development that elucidated the status of a number of earlier less complete studies and avoided the errors that plagued some of those. It did not lend itself to the formulation of a systematic tractable approximation scheme in the way that Wertheim's work does, however, and does not as directly yield the variety of exact expressions for ε that come out of Wertheim's work. The work preceding that of Mandel and Mazur includes that of F. E. Harris and B. J. Alder, *J. Chem. Phys.*, **21**, 1031, 1351 (1953) and **22**, 1806 (1954); F. E. Harris, *J. Chem. Phys.*, **23**, 1663 (1955); A. D. Buckingham, *Proc. R. Soc. London, Ser. A*, **238**, 235 (1956); A. D. Buckingham and J. A. Pople, *Trans. Faraday Soc.*, **51**, 1173 (1955); and H. Fröhlich, *Theory of Dielectrics*, 2nd ed., Oxford University Press, Oxford, 1958. See also H. Fröhlich, *J. Chem. Phys.*, **22**, 1804 (1954); *Physica*, **22**, 898 (1956); and B. K. P. Scaife, *Proc. Phys. Soc.*, **70B**, 314 (1957).

96. M. S. Wertheim, *Mol. Phys.*, **33**, 95 (1977).

97. M. S. Wertheim, *Mol. Phys.*, **34**, 1109 (1977).

98. M. S. Wertheim, *Mol. Phys.*, **36**, 1217 (1978).

99. M. S. Wertheim, *Mol. Phys.*, **37**, 83 (1979).

100. J. H. van Vleck, *J. Chem. Phys.*, **5**, 556, 991 (1937).

101. W. F. Brown, Jr., in *Handbuch der Physik*, Vol. 17, Springer-Verlag, Berlin, 1956.

102. J. S. Høye and G. Stell, *J. Chem. Phys.*, **73**, 461 (1980). See also the Appendix of J. S. Høye and G. Stell, *J. Chem. Phys.*, **72**, 1597 (1980).

103. L. R. Pratt, *Mol. Phys.*, **40**, 347 (1980).

104. D. Chandler, *J..Chem. Phys.*, **67**, 1113 (1977); L. R. Pratt and D. Chandler, *J. Chem. Phys.*, **66**, 147 (1977).

105. D. Chandler and L. R. Pratt, *J. Chem. Phys.*, **65**, 2625 (1976), and references therein to earlier work.

106. Equations 4.24 and 4.23 yield an expression for $\langle M_v^2 \rangle$ in terms of ε first obtained by Harris and Alder, cited in Ref. 95.

107. B. U. Felderhof, *J. Phys. C*, **12**, 2423 (1979). This work includes a valuable discussion of the relation among various results derived in the references cited in Ref. 95 and the precise significance of those results.

108. G. N. Patey and G. Stell, to be published.

109. H. A. Lorentz, *Theory of Electrons*, Teubner, Leipzig, 1909 (reprinted by Dover, New York, 1952). O. F. Mossotti, *Bibl. Univ. Modena*, **6**, 193 (1847); *Mem. Math. Fis. Modena*, **24**:II, 49 (1850). R. Clausius, *Die mechanische Wärmetheorie* Vol. II, Braunschweig, 1879. In *Ann. Phy.*, **49**, 1 (1916), Ewald showed that for a lattice of polarizable atoms of cubic symmetry, the local field is essentially that of the continuum considered by Lorentz.

110. W. F. Brown, Jr., *J. Chem. Phys.*, **18**, 1193, 1200 (1950).
111a. J. de Boer, F. van der Maesen, and C. A. ten Seldam, *Physica*, **19**, 265 (1953).
111b. L. Jansen and P. Mazur, *Physica*, **21**, 193 (1955); P. Mazur and L. Jansen, *Physica*, **21**, 208 (1955). See also closely related work of A. D. Buckingham and J. A. Pople, *Trans. Faraday Soc.*, **51**, 1029 (1955); 1035 (1956).
112. L. Jansen and A. D. Solem, *Phys. Rev.*, **104**, 1291 (1956); L. Jansen, *ibid.*, **112**, 434 (1958).
113. L. Silberstein, *Phil. Mag.*, **33**, 92, 521 (1917).
114. R. W. Hellwarth, *Phys. Rev.*, **152**, 156 (1966); **163**, 205 (1967); S. Prager, W. Kunken, and H. L. Frisch, *J. Chem. Phys.*, **52**, 4925 (1970).
115. G. Stell and G. S. Rushbrooke, *Chem. Phys. Lett.*, **24**, 531 (1974).
116. B. J. Alder, J. J. Weis, and H. L. Strauss, *Phys. Rev. A*, **7**, 281 (1973); B. J. Alder, H. L. Strauss, and J. J. Weis, *J. Chem. Phys.*, **59**, 1002 (1973); **62**, 2328 (1975).
117. H. W. Graben, G. S. Rushbrooke, and G. Stell, *Mol. Phys.*, **30**, 373 (1975).
118. G. Stell and J. S. Høye, *Phys. Rev. Lett.*, **33**, 1268 (1974).
119. D. Bedeaux and P. Mazur, *Physica*, **67**, 23 (1973). See S. K. Kim and P. Mazur, *Physica*, **71**, 579 (1974) for the extension to mixtures.
120. B. U. Felderhof, *Physica*, **76**, 786 (1974).
121. D. E. Sullivan and J. M. Deutch, *J. Chem. Phys.*, **64**, 3870 (1976).
122. J. S. Høye and D. Bedeaux, *Physica*, **87A**, 288 (1977).
123. R. H. Orcutt and R. H. Cole, *Physica*, **31**, 1779 (1965), *J. Chem. Phys.*, **46**, 697 (1967).
124. E. C. Kerr and R. H. Sherman, *J. Low Temp. Phys.*, **3**, 451 (1970).
125. A. Michels, C. A. ten Seldam, and D. J. Overdijk, *Physica*, **17**, 781 (1951).
126. The relevant literature on this subject is substantial, and we do not attempt to give it here. A good review was given by W. M. Gelbart, *Adv. Chem. Phys.*, **26**, 1 (1974), but there has been much work done subsequently. See D. W. Oxtoby, *J. Chem. Phys.*, **69**, 1184 (1978); B. J. Alder et al., *J. Chem. Phys.*, **70**, 4091 (1979); and L. Frommhold, *Adv. Chem. Phys.*, (1980), for representative results.
127. See, for example, B. J. Alder, J. C. Beers II, H. L. Strauss, and J. J. Weis, preprint.
128. R. Hocken and L. R. Wilcox, *Bull Am. Phys. Soc.*, **17**, 614 (1972).
129. C. L. Hartley, D. T. Jacobs, R. C. Mockler, and W. J. O'Sullivan, *Phys. Rev. Lett.*, **33**, 1129 (1974).
130. R. Hocken, M. A. Horowitz, and S. C. Greer, *Phys. Rev. Lett.*, **37**, 964 (1976).
131. T. Doiron and H. Meyer, *Phys. Rev. B*, **17**, 2141 (1978), B. J. Thijsse, T. Dorian, and J. M. H. Levelt Sengers, *Chem. Phys. Lett.* **72**, 546 (1980).
132. M. H. W. Chan, *Phys. Rev. B*, **21**, 1187 (1980).
133. G. Stell, *Phys. Rev. Lett.*, **32**, 286 (1974).
134. J. S. Høye and G. Stell, *J. Chem. Phys.*, **72**, 1597 (1980).
135. M. Chan, private communication.
136. J. B. Hubbard, P. Colonamos, and P. Wolynes, *J. Chem. Phys.*, **71**, 2652 (1979).
137. J. B. Hasted, *Aqueous Dielectrics*, Chapman and Hall, London, 1973; J. B. Hasted, D. M. Ritson, and C. H. Collie, *J. Chem. Phys.*, **16**, 1 (1948).
138. J. B. Hubbard and L. Onsager, *J. Chem. Phys.*, **67**, 4850 (1977).
139. S. A. Adelman, *J. Chem. Phys.*, **64**, 724 (1976); *Chem. Phys. Lett.*, **38**, 567 (1976).
140. S. A. Adelman and J.-H. Chen, *J. Chem. Phys.*, **70**, 4291 (1979).
141. D. Y. C. Chan, D. J. Mitchell, B. W. Ninham, and B. A. Pailthorpe, *J. Chem. Phys.*, **69**, 691 (1978).
142. D. Y. C. Chan, D. J. Mitchell, and B. W. Ninham, *J. Chem. Phys.*, **70**, 2946 (1979).
143. D. Levesque, J. J. Weis, and G. N. Patey, *J. Chem. Phys.*, **72**, 1887 (1980).
144. W. Pérez-Hernández and L. Blum, private communication.
145. F. Vericat, L. Blum, and J. R. Grigera, private communication.

146. L. Blum, *J. Chem. Phys.*, **61**, 2129 (1974); *Chem. Phys. Lett.*, **26**, 200 (1974); *J. Stat. Phys.*, **18**, 451 (1978).
147. S. A. Adelman and J. M. Deutch, *J. Chem. Phys.*, **60**, 3935 (1974).
148. R. L. Fulton, *J. Chem. Phys.*, **68**, 3095 (1978).
149. F. H. Stillinger and R. Lovett, *J. Chem. Phys.*, **48**, 3858 (1968).
150. E. Martina and G. Stell, *J. Chem. Phys.*, **69**, 931 (1978).
151. E. Martina and G. Stell, to be published.
152. J. C. Rasaiah and G. Stell, to be published.
153. J. D. Ramshaw, *J. Chem. Phys.*, **73**, 5294 (1980).
154. L. D. Landau and E. M. Lifshitz, *Electrodynamics of Continuous Media*, Pergamon Press, London, 1960.
155. In the context of the MSA, this procedure to obtain $\rho(\Omega, E_0)$ (but not ε) was followed by D. Isbister and B. Freasier, *J. Stat. Phys.*, **20**, 331 (1979), who obtained (6.2) and (6.3) with the approximate q_1 and q_2 that characterize the MSA result. The same result, and its exact generalization given here, was independently obtained by Martina and Stell. Equation 6.2 is also consistent with the linear-order result for $\rho(\Omega, E)$ obtained by a different method by Nienhuis and Deutch.[20]

LIGHT SCATTERING FROM THIN LIQUID FILMS

A. VRIJ, J. G. H. JOOSTEN, AND H. M. FIJNAUT

Van't Hoff Laboratory University of Utrecht
Utrecht, The Netherlands

CONTENTS

I. INTRODUCTION

Everyone is familiar with soap films: those thin, liquid structures that form soap bubbles and foam. It is much less widely known, however, that soap films can serve as model systems for the study of long-range colloidal interaction forces between suspended particles. Some of the characteristics of soap films can be observed easily when films are formed in a looped wire frame, which is dipped into and withdrawn from an aqueous solution containing some soap or detergent (Fig. 1.1). After a short time beautiful interference colors appear in reflected light, showing that the thickness of the film is comparable to the wavelength of visible light. When the frame is held vertical, colored horizontal bands are observed, slowly descending, which gradually become broader and brighter.[1] This implies that in the vertical direction the films at first have large gradients in thickness, which become smaller when drainage of the film liquid due to gravity causes the film as a whole to thin progressively. In time one may observe "black spots," mostly in the upper parts of the film bordering the supporting frame. The "black spots" are extremely thin film patches with thickness in the order of 10 nm.

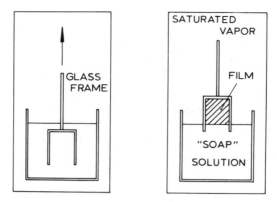

Fig. 1.1.   Preparation of a soap film by raising a glass frame from an aqueous soap solution.

At about this stage, but sometimes much earlier, the film ruptures if it is not protected against evaporation.

These phenomena can be studied in a controlled way when the disturbing effect of air currents and evaporation are minimized by placing both film and solution in a closed vessel. The film thinning appears to proceed more gradually and regularly, especially near the end of the drainage period. At the top of the frame the colored bands make room for a band with a silvery appearance, corresponding to a film thickness of around 120 nm (~one-quarter of the wavelength of light). Then, slowly, the whole film becomes more and more gray and finally black, starting from the top. The black film has a thickness of a few tens of nanometers or less. If appropriate precautions are taken, these black films can remain stable for several months or longer.

A. Interaction Forces

The phenomena described above have been known for a long time; indeed, Newton[2] reported on the "black spots" in soap films. In the past 25 years, however, these thin, liquid structures have become a subject of intensive scientific studies.[3] One of the main reasons is that the interaction forces between colloidal particles suspended in a liquid are of the same nature as those operating in soap films. Because the film geometry is well defined (i.e., a thin, flat liquid sheet, macroscopic in lateral extension), it is an attractive experimental subject for studying these forces, in particular with optical means.

The presence of attractive "colloidal" forces becomes apparent in the black spots.[4] The black film appears to be in a sort of metastable equilibrium state with a finite thickness in the colloidal size range. What are these colloidal forces? Clearly they must have a range much larger than atomic dimensions. Relevant for us are two main types: long-range van der Waals–London forces and long-range electrostatic forces.

1. Van der Waals–London Forces

The van der Waals–London forces find their origin in (quantum) fluctuations in the electric polarizability of matter. In first approximation they may be calculated by a simple summation of the attractive van der Waals–London forces between atoms. The distance on which these forces are felt may be up to several hundreds of nanometers.

2. Electrostatic Forces

The electrostatic forces are caused by soap ions adsorbed at the two opposing air-water interfaces that sandwich the aqueous core. The core contains water, a few nonadsorbed soap ions, and other ions for example, Na^+ and Cl^- from added sodium chloride (common salt) or from other impurities present in the solution from which the film was made. Except for the thinnest films, the percentage of water is overwhelming. For instance, a film with a core thickness of 50 nm contains a weight-percentage of, say, 97% water, 2.5% adsorbed soap, and 0.5% NaCl. The repulsive Coulomb forces between the opposing film surfaces are screened by a diffuse "charge cloud" built up by ions in the core. The range of the screened forces, or the "thickness" of the charge cloud (called the electrical double layer in colloid science), is given by the Debye parameter $1/\kappa$, which is proportional to the electrolyte concentration to the power $-\frac{1}{2}$. This parameter is defined by (2.6). One calculates, for instance, that $1/\kappa$ ranges from 3 to 30 nm for aqueous NaCl solutions at 25°C containing 10^{-2} to 10^{-4} mol/dm^3 (1 mol/dm^3 NaCl solution contains 58 g of NaCl and water added to a volume of 1 dm^3). Since both van der Waals–London and electrostatic forces are long-range with respect to atomic dimensions, they can be described without taking into account the detailed atomic structure of the film liquid.

B. Thin Films as They Occur in Nature or Practice

Thin, liquid films as such are not only systems of interest, they also occur frequently in nature and in laboratory practice, for example, in *foams*. In *emulsions*, to name another case, thin water films between "oil" (instead of "air") phases are present in a concentrated emulsion of small oil droplets dispersed in water. The properties of the thin water film determine the interaction forces between the oil droplets and will determine, for example, whether the emulsion in stable. Also the reverse case, oil films in water, occurs. Extremely thin ($\sim$5 nm) "oil-in-water" films are prototypes of lipid bilayers occurring in biological membranes. Thin liquid films on a solid support[5] are present in all kinds of surface wetting phenomena[6] and are important in lubrication and in transport processes in chemical engineering.

Also in these cases colloidal forces play a role in the behavior of the films. In practice, however, films occur often in rather complex situations that are difficult to disentangle. Fundamental studies on certain prototypes of films may help in the understanding of these phenomena.

C. Experimental Methods

Several experimental methods exist to study the interaction forces in films. Five such approaches are described next.

1. The thickness of equilibrium films is measured as a function of electrolyte concentration.[7] The attractive van der Waals–London forces, which tend to thin the film, are balanced by repulsive electrostatic forces, which tend to thicken the film. Adding electrolyte has its main influence on the electrostatic forces, which become more effectively screened, so that the films thickness becomes smaller. The film thickness is calculated from the optical reflection coefficient.

2. Small, circular films, a few millimeters in diameter, can be made in a circular glass ring[3, 8–10] (Fig. 1.2). The film thickness can be changed by applying a small variable hydrostatic suction at the film edge. Larger suctions, up to 1 atm, can be reached when the film is made in a hole drilled in a porous glass plate.[11] This method has the advantage that the equilibrium of forces mentioned in item 1 can be changed by superimposing a known (suction) force, without changing the electrolyte concentration.

3. A (less satisfactory) variant of method 2 is to study the kinetics of the film thinning process.[8, 12] To extract information on the interaction forces, the dynamics of film thinning must be known. Formation of "dimples" complicates the situation.

4. At the place where the film contacts the bulk liquid, a curved surface profile (sometimes called a Plateau border) is present. For a plane vertical film this is sketched in Fig. 1.3. In a thick film this profile is smooth, but in a thin film, where the interaction forces become perceptible, a discontinuity emerges that can be measured as a so-called contact angle[13–16] (e.g., from a Fresnel diffraction pattern[17, 18] of visible light). A similar phenomenon can be observed in small circular

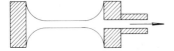

Fig. 1.2. Formation of a soap film in a circular glass ring having a diameter of a few millimeters.

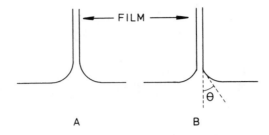

Fig. 1.3.   Vertical soap film in contact with soap solution: *A*, film without contact angle; *B*, film with contact angle θ.

films.[19, 20] Contact angles θ range from, say, 10 minutes to 20 degrees, depending on the contribution of the interaction forces to the force balance governing the surface profile.

5. One of us proposed to study the *scattering of light* by the film surfaces.[21] The surface of a liquid is not perfectly flat but always contains small ripples or corrugations because of the thermal motion of the molecules. Surface ripples having a wavelength comparable to the wavelength of light give a measurable scattering (diffuse reflection) if they are illuminated by visible light. According to Einstein,[22] these large-scale fluctuations may be calculated with the law of equipartition of energy. For the surface of a bulk liquid, this leads to an inverse proportionality of the scattering intensity and the surface tension.

In thin, liquid films two additional phenomena must be considered.

1. The total scattering from *both* film surfaces is modified, compared with the scattering from the two single surfaces, by interference from the light in the film.
2. Not only the surface tension but also the long-range interaction forces in the film will determine the scattering of the light. This makes it possible to determine these forces from the intensity of the scattered light.

More recently we have found[23] that *fluctuations* in the scattered light intensity around its mean value give information about dynamics of the ripple motions in the film (intensity fluctuation spectroscopy). The rest of this chapter treats the light-scattering method in more detail.

For the interested reader, we suggest some books[24–26] and review papers in the field of soap films.[27–33]

II. THIN LIQUID FILMS

This section provides more detail about thin films and the governing interaction forces.

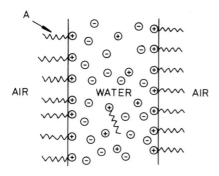

Fig. 2.1. Cross section of a soap film in air: A is the tail of a soap ion. The surfaces contain adsorbed soap ions, and the aqueous core contains electrolyte. The film is electrically neutral.

A. Soap Film Structure

In Section I we introduced the reader to some notions about liquid soap films that can be observed in everyday life. Such films are what we call *free liquid films*, the term "free" meaning that the film is supported only at its periphery, with both film surfaces in contact with gas (or liquid). Figure 2.1 is a schematic picture of a cross section of such a film. At both air-water surfaces, surface active molecules (A) are adsorbed. Molecule A is a soap molecule,* but it could be another molecule that is "surface active."

Surface active molecules (surfactants) invariably contain two parts: one (hydrophilic) part that prefers to be surrounded by water (the positive charged *head* of molecule A), and another (hydrophobic) part that prefers to be surrounded by air (or any nonpolar liquid immiscible with water) and shown as a wavy line, called *tail*. This constitution of the molecule makes it surface active because, clearly, it prefers to be situated at the interface. An example is hexadecyl trimethylammonium bromide (HDTAB); see Fig. 2.2. Not all the soap molecules are adsorbed at the film surfaces; a few are present in the core of the film, this number being smaller, the larger the surface "activity" of the molecule is. The surface activity increases with the length of the tail. The adsorbed soap molecules form a monolayer behaving as a kind of two-dimensional fluid with a certain "surface pressure." The (positive) surface pressure is equal to the surface tension of the solution without surfactant minus the surface tension of the solution with surfactant. Between the two surfaces one has the aqueous core containing ions from the added electrolyte (here, K^+, Br^-) and from the soap.

One could ask whether the presence of an adsorbed surfactant is really necessary. The answer is yes. The soap layer, although only a monolayer thick, has a profound influence on the flow properties of the film liquid be-

---

*For convenience, we adhere to the term "soap," although the substance may not be a soap in chemical terms.

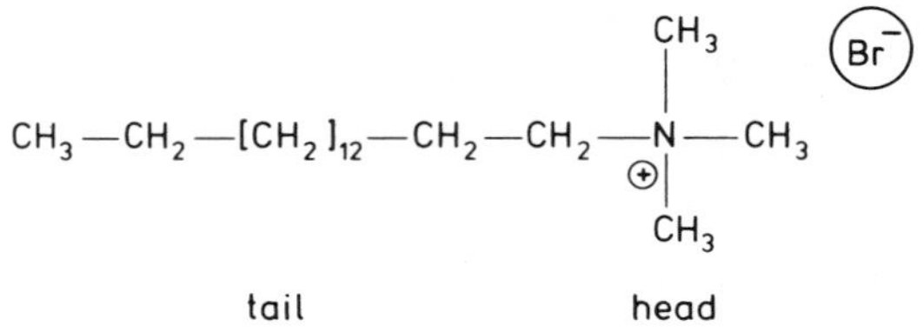

$$CH_3-CH_2-[CH_2]_{12}-CH_2-CH_2-\overset{\overset{\displaystyle CH_3}{|}}{\underset{\underset{\displaystyle CH_3}{\oplus|}}{N}}-CH_3 \qquad \overset{\frown}{Br^-}$$

tail   head

Fig. 2.2.   Chemical formula of a hexadecyl trimethylammonium bromide (HDTAB) molecule.

cause it sensitively affects the flow boundary conditions. During its formation, the film goes through stages of turbulent flow. The surface of a pure liquid (with a constant surface tension) would stretch indefinitely under the action of a surface stress, and a film of pure liquid would rupture immediately. Liquid surfaces covered with a monolayer, however, have a variable surface tension that depends on the surface density of the surfactant. Local stretching of a surface element diminishes the surface pressure and thus increases the surface tension, which tends to contract the surface again. This "healing effect" is known as the Marangoni effect. It prevents the film of becoming dangerously thin upon stretching.

B. Electric Double Layer Repulsion Forces

The ionized soap molecules adsorbed at the film surfaces give them a positive charge. The average density is given by

$$\sigma = N_+^{\sigma} \nu e \qquad (2.1)$$

Here N_+^{σ} is the surface number density of the soap ions, e is the elementary charge, and ν the number of elementary charges per ion (for HDTAB, $\nu = 1$). The surface charge is compensated by an equal amount of opposite charge present on ions in the aqueous film core to make the system electroneutral. The positive ions in the core will be repelled and the negative ions will be attracted to the surfaces. Both effects, however, are counteracted by the thermal motion of the ions, which results in a "diffuse" cloud of ions that screens the effect of the surface charges. Treating the surface charge as continuous and the ions in the core as point charges, distributed according to Boltzmann's law in a continuous dielectric ($\varepsilon \sim 80$), one has:

$$n_+(z) = n_+^0 \exp\left[-\frac{\nu e \psi(z)}{k_B T}\right]$$

$$n_-(z) = n_-^0 \exp\left[\frac{\nu e \psi(z)}{k_B T}\right] \qquad (2.2)$$

where $n_+(z)$ is the average number density of positive ions in the core with valence ν (the same for positive, negative, and soap ions), as a function of the position z along an axis perpendicular to the surface. Furthermore, $\psi(z)$ is the average electrostatic potential at z, and k_B, T, and ε are Boltzmann's constant, the absolute temperature, and the dielectric constant of the solvent. The terms $n_i(z)$ and $\psi(z)$ are related by Poisson's law through the charge density $\rho(z)$

$$\frac{d^2\psi(z)}{dz^2} = -\frac{4\pi\rho(z)}{\varepsilon} \tag{2.3}$$

where $\rho(z)$ is given by

$$\rho(z) = n_+ \nu e - n_- \nu e \tag{2.4}$$

Combining (2.2) through (2.4) results in the so-called Poisson-Boltzmann equation[34]

$$\frac{d^2\psi(z)}{dz^2} = \kappa^2 \sinh \frac{\nu e \psi(z)}{k_B T} \tag{2.5}$$

where κ is the reciprocal Debye length, mentioned in Section I, defined by

$$\kappa^2 = \frac{4\pi(n_+^0 + n_-^0)\nu e^2}{\varepsilon k_B T} \tag{2.6}$$

Because of electroneutrality, $n_+^0 = n_-^0 \equiv n^0$ is the bulk ion density. Furthermore, σ is found to be

$$\sigma = -\frac{\varepsilon}{4\pi}\left(\frac{d\psi}{dz}\right)_{surface} = \left[\frac{n^0 \varepsilon k_B T}{2\pi}\left(2\cosh\frac{\nu e \psi_0}{k_B T} - 2\cosh\frac{\nu e \psi_m}{k_B T}\right)\right]^{1/2} \tag{2.7}$$

Here ψ_0 is the potential at the surface and ψ_m the potential in the middle of the film. For $e\psi/k_B T = 1$, $\psi \simeq 25$ mV at 25°C. Equation 2.5, subject to the boundary condition (2.7), can be solved for a flat film geometry in the representation of elliptic functions.[34]

Now the double layer repulsion force per unit surface area (the double layer pressure) $\Pi_{DL}(h)$ is found to be:

$$\Pi_{DL}(h) = 2n^0 k_B T\left[\cosh\left(\frac{\nu e \psi_m}{k_B T}\right) - 1\right] \tag{2.8}$$

where h is the thickness of the aqueous core. If h is large with respect to the Debye length ($\kappa h \gg 1$), the following approximate solution is found:

$$\Pi_{DL}(h) \simeq 64 n^0 k_B T \mu^2 e^{-\kappa h} \tag{2.9}$$

where

$$\mu = \tanh \frac{\nu e \psi_0}{4 k_B T} \tag{2.10}$$

From the double layer pressure another useful quantity can be found:

$$V_{DL}(h) = \int_\infty^h \Pi_{DL}(h)\, dh \tag{2.11}$$

It is a (potential) free energy of interaction, being the work (at constant T) of bringing the surfaces from $h = \infty$ to $h = h$, letting the aqueous solution in the core leak to a reservoir with electrolyte concentration n^0 at atmospheric pressure.

Many attempts have been made to improve the classical Poisson-Boltzmann equation to include discrete charge effects, finite ion size, and so on (see, e.g., Refs. 35–37). At present some fundamental progress is being made on the basis of certain models in modern fluid-state theory, in which the hard-core repulsions of the ions are incorporated in a consistent way.[38-41] The Poisson-Boltzmann equation was found to be a limiting case of the hypernetted chain approximation at low densities.[42] Also a computer simulation was reported.[43]

One aim of our experiments is to check the applicability of (2.9).

C. Long-Range van der Waals–London Forces

1. Van der Waals Forces Between Molecules

In his thesis van der Waals expressed his desire to determine "a quantity that plays a peculiar role in Laplace's theory of capillarity."[44, 45] He was referring to a molecular pressure, "a measure for the cohesion of matter." He was—in the Newtonian tradition—looking for a way of grasping intermolecular forces, the forces that would appear in his own equation of state.

Attractive forces between neutral molecules may be partially explained with electric dipole interactions. Keesom[46] explained attractive forces between permanent dipoles in thermal motion, and Debye[47] investigated permanent and induced dipoles. The induced-induced dipole interaction could be understood adequately only after the advent of quantum theory. The general character of these forces was explained by London[48] as a perturbation of zero-point energies. He also made a connection with the dispersion of electric oscillators. Therefore these forces are also called "dispersion forces." All contributions give the following form for the attractive potential:

$$V_w(r) = -\frac{\lambda_w}{r^6} \qquad (2.12)$$

For permanent dipole–permanent dipole interaction, $\lambda_w \sim \mu_p^4 / k_B T$; for permanent dipole–induced-dipole interaction, $\lambda_w \sim \alpha \mu_p^2$, and for induced-dipole–induced-dipole interaction, $\lambda_w \sim \alpha^2 \hbar \omega$, where μ_p is the permanent dipole moment, α the polarizability, $\hbar$ Planck's constant, and ω the excitation frequency of the molecules.

2. Van der Waals–London Forces Between Macroscopic Particles

In 1932 Kallmann and Willstätter[49] and also Bradley[50] recognized that attractive forces between *colloidal particles* would emerge from a pairwise summation of dispersion forces between atoms. This was further investigated theoretically by Hamaker[51] and de Boer.[52] For a flat geometry, two half-spaces separated by a gap of thickness h, this leads to the form

$$V_w(h) = -\frac{A_H}{12 \pi h^2} \qquad (2.13)$$

where $A_H = \pi^2 \lambda_w \rho^2$ is the so-called Hamaker constant, with ρ the number density of the atoms. Expression 2.13 shows clearly the long-range character of these forces. This can be extended to half-spaces in a medium.

It may seem somewhat puzzling that in the case of a slab of material surrounded by vacuum—a model for a soap film—the film "feels" a tendency to become thinner. This can be explained by the notion that a molecule in the film is surrounded by fewer other molecules than a molecule in the bulk. It, therefore, has a higher potential energy in the film than in the bulk and tends to move from the film to the bulk liquid.

Equation 2.13 is valid only for small values of h. For larger distances the propagation speed of the electromagnetic field has to be taken into account (so-called retardation effect), which gives weaker attractive forces. We avoid these complications here, also because the (approximate) summation procedures described in this section are now superseded by a macroscopic theory of dispersion forces in which these effects are treated in a natural way.

3. Macroscopic Theory of Dispersion Forces

An early approach to calculating interaction forces between conducting plates from the change in the zero-point energy of the electromagnetic field modes is due to Casimir.[53] Lifshitz[54] and Dzyaloshinskii et al.[55] derived equations for dispersion forces in the more general case of dielectric slabs with quantum field theory. The equations are very complicated. As an illustration, for the simplest case—two half-spaces separated by a vacuum of thickness h in the nonretarded limit—the following equation is given;

$$V_w(h) = -\frac{\hbar}{32\pi^2 h^2} \int_0^\infty d\xi \int_0^\infty x^2 \, dx \left[\left(\frac{\varepsilon(i\xi)+1}{\varepsilon(i\xi)-1} \right)^2 e^x - 1 \right]^{-1} \quad (2.14)$$

with $i = \sqrt{-1}$. Here $\varepsilon(\omega)$ is the frequency-dependent dielectric "constant" of the material needed only on the imaginary axis: $\omega = i\xi$, where ε is real. Note that the dependence on h is the same as with pairwise summation (2.13). For further details and other cases we refer to books in this field.[56, 57]

In most cases $\varepsilon(\omega)$ is not known over the whole electromagnetic spectrum. Fairly accurate interpolation formulas can be used for several systems.[56] Numerical calculations have been made for soap films by Ninham and Parsegian.[58] Their formula for $\varepsilon(\omega)$ was used to calculate the dispersion forces for our type of film.[59] Results are given in Section VI. It is noteworthy that $V_w(h)$ is found to be not simply proportional to h^{-2}, so that retardation effects cannot be neglected in our soap films.

In analogy to the double layer pressure, one can introduce a van der Waals–London pressure: $\Pi_w = -\partial V_w/\partial h$.

D. Combination of Electrical Double Layer and van der Waals–London Forces

The total interaction free energy of a film follows from a combination of electrical double layer and van der Waals–London interactions.

$$V(h) = V_{DL}(h) + V_w(h) \quad (2.15)$$

[However, $V_w(h)$ also contains contributions from $\varepsilon(\omega)$ at zero frequency that are due to motion of the charged ions. This may give a small interdependency of V_{DL} and V_w, which will be neglected.[60]]

To show the general behavior of $V(h)$, let us use (2.9) and (2.11) with (2.13) and write

$$V(h) = B\kappa e^{-\kappa h} - \frac{D}{h^2} \quad (2.16)$$

with $B = (8/\pi)\varepsilon[(k_B T/e)\tanh(\nu e\psi_0/4k_B T)]^2$ and $D = A_H/12\pi$.

Expression 2.16 has a maximum and a minimum (Fig. 2.3). This situation occurs in a stable film. The minimum describes the situation of an

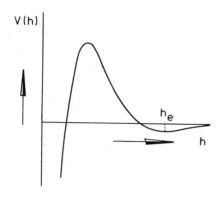

Fig. 2.3. Scheme of the interaction free energy of a soap film as a function of thickness h. The thickness $h = h_e$ (equilibrium). Influence of gravity is not shown.

equilibrium film in which the electrical double layer and the van der Waals–London forces equilibrate. We will see later that d^2V/dh^2 can be obtained from light scattering for $h \gg h$ (equilibrium). For larger values of κ and (D/B) (i.e. with larger electrolyte concentration, larger van der Waals–London forces and/or smaller double layer forces), the extrema disappear and no stable film is predicted (see Section IV.D.3 on stability). This general picture is not changed when more accurate expressions are used for $V_w(h)$.

We further note that (2.15) has to be supplemented with a hydrostatic pressure term. For a vertical film in the field of gravity, it is equal to $\rho_m hgH$, where ρ_m is the mass density difference between liquid and air, H is the height in the film above the surface of the solution, and g is the gravitation constant.

The total interaction between the interfaces is also denoted by the so-called disjoining pressure[61]

$$\Pi = -\frac{\partial V}{\partial h} = \Pi_{DL} + \Pi_w \qquad (2.17)$$

In the equilibrium situation we therefore have $\Pi = \rho_m gH$.

Finally we mention that the colloidal interactions have also been measured directly as a force between solid plates (see, e.g., Refs. 35 and 56). Recently, forces between mica plates in an electrolyte solution have been measured.[62]

III. LIGHT SCATTERING FROM LIQUID SURFACES

That liquid surfaces scatter light was first predicted by von Smoluchowski[63] in 1908. He expected the phenomenon to be visible near the critical point where the surface tension of the liquid is small. A quantitative theory was developed by Mandelstamm,[64] who described the thermal roughness of the surface as a spectrum of surface waves. The average square amplitude of each mode he obtained by invoking the equipartition theorem as applied by Einstein[22] to density fluctuations, to calculate the light scattering of (bulk) liquids. The theory was extended by Andronov and Leontovicz[65] and by Gans,[66] who also considered the scattering outside the plane of incidence and the states of polarization.

Raman and Ramdas,[67] using the sun as a light source, reported a great number of measurements of some 60 liquids on the intensity (measured by photographic photometry) and the state of polarization. Care was taken to measure on clean surfaces obtained in a closed bulb of a distillation apparatus in which the liquid was distilled back and forth several times. For the common liquids where the scattering intensity is small, the surface tension could be calculated only within a factor of 2. It is noteworthy that they could measure the surface tension of carbon dioxide near the critical point and found it to vary from 0.050 to 0.0015 mN/m for temperatures ranging from 30.00 to 31.10°C within an accuracy of $\sim$30%. With a mercury surface they found a much smaller angular dependency of the scattering than with other liquids, a result that deserves further consideration. It is interesting to note that they already did some investigations on surface layers of oleic acid and some fluorescent dyes on water. The dye methyl violet produced a faint yellowish-orange surface opalescence like the surface of the dry crystals.

Further papers are scarce. The state of the art until about 1967 is given in a review.[68] In this paper also the very complicated equations[65, 66] for the nonsymmetrical geometries are transformed into a more tractable form with the help of Fresnel coefficients.

A. Description of Surface Corrugations

We now outline Mandelstamm's theory for the surface of a pure liquid. The average position of the horizontal surface is taken in the XY-plane at $z = 0$, and the vertical elevation of the surface from its average position (in the positive Z-direction) at a certain instant is given by $\zeta = \zeta(x, y)$. The thickness of the surface layer is assumed small with respect to the wavelength of light used, and the liquid is assumed to be incompressible. Both approximations are good for liquids removed from the critical point. The function $\zeta(x, y)$ is expanded in Fourier components[64, 68] in a square with a side length a.

$$\zeta = \sum_{\rho = -\infty}^{+\infty} \sum_{\sigma = -\infty}^{+\infty} \zeta_{\rho\sigma} \exp\left[\frac{2\pi i}{a} (\rho x + \sigma y) \right] = \sum_{\mathbf{K}} \zeta_{\mathbf{K}} e^{i\mathbf{K}\cdot\mathbf{s}} \quad (3.1)$$

where $\mathbf{s} = (x, y)$; $\mathbf{K} = (2\pi/a)(\rho, \sigma)$; $|\mathbf{K}| = K = (2\pi/a)(\rho^2 + \sigma^2)^{1/2} = 2\pi/\Lambda$

with K and Λ the wave number and wavelength of the surface wave with wave-vector $\mathbf{K}$, and with $\zeta_{\mathbf{K}} = \zeta^*_{-\mathbf{K}}$, when ζ^* is the complex conjugate of ζ. Because only surface wavelengths Λ in the order of the wavelength of light or larger are of importance here, the system can be treated on a quasi-macroscopic scale. Furthermore, $|\zeta| \ll \Lambda$, which implies that the surface fluctuations can be linearized, resulting in uncoupled Fourier components.

The work (at constant T) to create a surface fluctuation can be written as follows;

$$\Delta F = \Delta F_1 + \Delta F_2 \tag{3.2}$$

where $\Delta F_1 = \gamma \Delta O$ is due to an increase in surface area ΔO with surface tension γ. For small ζ one has

$$\Delta F_1 = \tfrac{1}{2}\gamma \int_{-(1/2)a}^{+(1/2)a} \int \left[\left(\frac{\partial \zeta}{\partial x} \right)^2 + \left(\frac{\partial \zeta}{\partial y} \right)^2 \right] dx \, dy \tag{3.3}$$

The contribution ΔF_2 is due to gravity and can be written as

$$\Delta F_2 = \tfrac{1}{2}\rho_m \int_{-(1/2)a}^{+(1/2)a} \int \zeta(x, y)^2 \, dx \, dy \tag{3.4}$$

On substituting (3.1) into (3.3) and (3.4), it is found that ΔF consists of a sum $\Sigma_{\rho, \sigma} \Delta F_{\rho\sigma}$ of squares, with

$$\Delta F_{\rho\sigma} = \tfrac{1}{2}\zeta_{\rho\sigma}\zeta^*_{\rho\sigma}a^2\left(\gamma K^2 + \rho_m g \right) \tag{3.5}$$

The probability of finding a value $|\zeta_{\rho\sigma}|$ is proportional to $\exp(-\Delta F_{\rho\sigma}/k_{\mathbf{B}}T)$; that is, the modes fluctuate independently with a Gaussian probability distribution. Application of the equipartition theorem then yields

$$\langle \zeta_{\mathbf{K}}\zeta^*_{\mathbf{K}} \rangle = \frac{\left(k_{\mathbf{B}}T/a^2 \right)}{\gamma K^2 + \rho_m g} \simeq \frac{\left(k_{\mathbf{B}}T/a^2 \right)}{\gamma K^2} \tag{3.6}$$

Because only waves, say, in the range $\Lambda = 4\text{--}30\lambda$ can be detected, the gravity contribution is often negligible, except when γ is very small. For normal liquids the root-mean-square value of the surface elevations is in the order of 0.5 nm.

It may be of interest to note that Mandelstamm's analysis forms the basis of the capillary wave theory of surface tension.[69, 70]

It is also possible to calculate the spatial correlations between surface elevations, a distance s apart,

$$\langle \zeta(s=0)\zeta(s)\rangle \sim \frac{1}{s^{1/2}}\exp\left[-s\left(\frac{\rho_m g}{\gamma} \right)^{1/2} \right] \quad \text{for } s \text{ large}$$

which shows that the surface elevations have a very long correlation length, given by what is called in surface science the capillary constant $[\gamma/(\rho_m g)]^{1/2}$ and having a value in the order of 1 mm.

The surface elevations bear a (formal) resemblance to (bulk) density fluctuations of a fluid near the critical point, which also have a large range correlation, and for which Ornstein and Zernike[71] found the same form as (3.6).

For more recent statistical mechanical theories on interfaces, see, for example, Refs. 72 to 75.

B. Surface Light Scattering

Consider a plane light wave with wave vector $\mathbf{k}$, which falls on the liquid-air surface containing thermal ripples with a small amplitude. Rayleigh[76] found that if the illuminated surface area is much larger than Λ^2 and $\zeta \ll \Lambda$, each component (ρ, σ) gives a first-order diffracted light wave $\mathbf{k}_1$; the direction of which is given by the polar and azimuthal angles θ and ϕ, as follows:

$$\sin\theta\cos\phi - \sin\theta_0 = \frac{(2\pi/a)\rho}{k_1} = \left(\frac{\lambda_1}{a} \right)\rho$$

$$\sin\theta\sin\phi = \frac{(2\pi/a)\sigma}{k_1} = \left(\frac{\lambda_1}{a} \right)\sigma \tag{3.7}$$

This applies to light waves (of wavelength λ_1) scattered (or diffracted) in the reflection half-space. A similar equation[65, 66, 68] applies to scattering in the transmission half-space.

Relations 3.7 may be visualized as follows. Consider a large sphere with radius unity, centered around the illuminated spot in the liquid surface (XY-plane) (Fig. 3.1). The intersection with the XY-plane is a circle with unit radius. The incident, reflected, and scattering directions are defined by the points A, B, and C on the surface of the sphere. The projections on the XY-plane are given by A', B', and C'. The incident and reflected wave vectors are in the XZ-plane. The reflected wave vector $\mathbf{k}_r$ makes an angle θ_0 with the positive Z-axis. The scattered (diffracted) wave vector $\mathbf{k}_1$ makes an angle θ with the positive Z-axis and an azimuthal angle ϕ (in the XY-plane) with the positive X-axis.

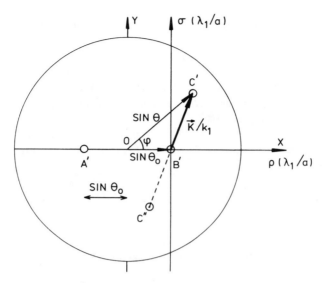

Fig. 3.1. "Ewald circle" showing the relation of the angles between the directions of incident, reflected, and scattered light to the reciprocal lattice parameters ρ and σ.

Relations 3.7 express the conservation of momentum in the surface, and $k \simeq k_1$ expresses the conservation of energy.

For each pair of integral values (ρ, σ), the direction of the diffracted wave can be read from Fig. 3.1. When λ/a is small, the points are narrowly spread. Each (real) surface wave component gives *two* diffracted waves (in the reflected half-space), one direction given by the point C' and a complementary one given by the point C'', the mirrored point of C' with respect to B'; $[(\rho, \sigma) \rightarrow (-\rho, -\sigma)]$.

For an incident light wave with unit amplitude and with the electrical vector perpendicular to the plane of incidence (XZ), the amplitude of the diffracted waves can be written as follows:

$$A_{\rho\sigma} = \frac{ik_1 \zeta_{\rho\sigma}}{2 \cos \theta} (n^2 - 1) t_s t_{0s} \cos \phi \qquad (3.8)$$

$$t_{0s} = \frac{2 \cos \theta_0}{\cos \theta_0 + n \cos \theta_0'} \qquad (3.9)$$

$$t_s = \frac{2 \cos \theta}{\cos \theta + n \cos \theta'}$$

Here $n = n_2/n_1$ is the ratio of the refractive indices of liquid (n_2) and air (n_1). The θ' is defined by Snell's law: $\sin \theta / \sin \theta' = n$; t_{0s} and t_s are Fresnel transmission coefficients, and $i = \sqrt{-1}$. Results for other states of polarization can be found elsewhere.[68]

The intensity of the scattered light is proportional to $|A_{\rho\sigma}|^2$ times the number of diffracted waves $\Delta\rho\,\Delta\sigma$ falling into a solid angle $d\Omega$. This number can be found easily from Fig. 3.1. On the surface of the unit sphere $d\Omega$ corresponds to a surface area $d\Omega$ around C and in the XY-plane with a surface area $\cos\theta\,d\Omega$ around C'.

Thus $\Delta\rho\,\Delta\sigma = (a/\lambda_1)^2 \cos\theta\,d\Omega$. The average (light) energy flow falling within $d\Omega$ per unit of illuminated surface area and for unit incident intensity is then, using (3.6) and (3.9):

$$S(\theta; \phi; \theta_0) = \frac{\frac{1}{4}(k_B T/\lambda_1^2)(n^2-1)^2 t_s^2 t_{0s}^2 \cos^2\phi}{\gamma(\sin^2\theta_0 + \sin^2\theta - 2\sin\theta_0 \sin\theta\cos\phi)} \tag{3.10}$$

Note that the length a of (3.6) just drops out as it should, and that (see Fig. 3.1)

$$K^2 = k_1^2(\sin^2\theta_0 + \sin^2\theta - 2\sin\theta_0\sin\theta\cos\phi) \tag{3.11}$$

As an illustration of the order of magnitude of surface light scattering, one calculates for $n = 1.36$; $\theta_0 = 60°$; $\theta = 45°$; $\gamma = 30$ mN/m; $\lambda = 546$ nm; $S = 7.5 \times 10^{-7}$.

The predicted angular dependence of the scattered intensity is very pronounced. It strongly increases on approaching the reflected beam ($\phi \simeq 0$; $\theta \simeq \theta_0$). This also applies for the scattering in the transmission half-space on approaching the transmitted beam. It may be of interest to note that a similar situation occurs in bulk fluids near a critical point, where large density fluctuations (with large spatial gradients) create a kind of "internal surface" giving rise to pronounced scattering in the forward direction.

The Mandelstamm theory takes only surface elevations into account. Recently a more fundamental treatment by Bouchiat and Langevin[77] gives the same result found by Mandelstamm, Andronov and Leontovicz, and Gans, but in addition allows the calculation of contributions to the surface polarization caused by surface density and orientation fluctuations of adsorbed layers. These contributions, however, are small except near two-dimensional critical points.

IV. LIGHT SCATTERING FROM THIN LIQUID FILMS

For thin liquid (soap) films one has to deal with three complications compared with single interfaces.[21, 78]

1. Interference of the light waves reflected and refracted by both film surfaces.
2. The necessary presence of a monolayer of soap molecules.
3. Long-range interaction forces across the film.

Let us start with the optical problem, 1.

A. Interference of Light Waves in the Film

First we will confine ourselves to an (optically) homogeneous film with refractive index n and an (average) thickness h. Instead of corrugations on one surface, one has to deal with corrugations on two film surfaces

$$\zeta_I(x, y) = \sum_K \zeta_{I,K} e^{iK \cdot s} \tag{4.1}$$

$$\zeta_{II}(x, y) = \sum_K \zeta_{II,K} e^{iK \cdot s}$$

The diffraction of light waves in such a film can be treated in the same way[78] used by Rayleigh,[76] that is, applying (macroscopic) Maxwell equations with the appropriate boundary conditions at both surfaces. Afterward[78] it was found that the same answers are obtained when the interference of the reflected, refracted, and diffracted waves are treated in an elementary fashion by adding the diffraction of the single surfaces with appropriate phase factors.

First, consider the interference of the reflected and the refracted (or transmitted) waves. We treat only the case $\phi = 0$ and the electrical vector perpendicular to the plane of incidence. In the diagram of Fig. 4.1, 1 and 3

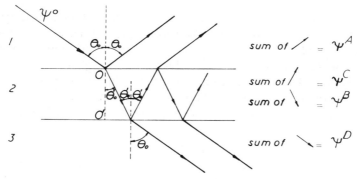

Fig. 4.1. Interference after multiple reflection and refraction of the incident beam. (From Ref. 78, courtesy of Academic Press Inc., New York.)

are air and 2 the film. The incident light wave with amplitude $\Psi^0 = 1$ undergoes multiple reflection and refraction at the upper and the lower surfaces, thus giving the following four amplitudes:

$$\Psi^0 = 1 \tag{4.2}$$

$$\Psi^A = r_0\left(1 - t_0 t_0' U_0 e^{-2i\beta}\right) \tag{4.3}$$

$$\Psi^B = t_0 U_0 e^{-i\beta} \tag{4.4}$$

$$\Psi^C = -t_0 r_0 U_0 e^{-2i\beta} \tag{4.5}$$

$$\Psi^D = t_0 t_0' U_0 e^{-i\beta} \tag{4.6}$$

where

$$\beta = \left(\frac{2\pi}{\lambda_2}\right) h \cos \theta_0' \tag{4.7}$$

with λ_2 the wavelength of light in the film, and

$\theta_0' = $ angle of refraction, given by Snell's law

$$U_0 = \left[1 - r_0^2 e^{-2i\beta}\right]^{-1} \tag{4.8}$$

$$t_0 = \frac{2\cos\theta_0}{\cos\theta_0 + n\cos\theta_0'} \tag{4.9}$$

$$t_0' = \frac{2n\cos\theta_0'}{\cos\theta_0 + n\cos\theta_0'} \tag{4.10}$$

$$r_0 = t_0 - 1 = 1 - t_0' \tag{4.11}$$

Now the incident light will give rise to diffraction at the boundary (1–2); Ψ^C at (2–1), and Ψ^B at (2–3). The waves that diffract into the film also undergo multiple reflection and refraction; multiple diffraction is neglected because the amplitudes would be of higher order in ζ/λ.

The amplitudes of the diffracted waves (including interference) are obtained as follows. The wave Ψ^0 gives rise to one wave diffracted into medium 1 and another one diffracted into medium 2 (Fig. 4.2, dashed lines). The amplitudes are

$$\Psi^0 A_K^{12} \tag{4.12}$$

where A_K^{12} is given by (3.8).

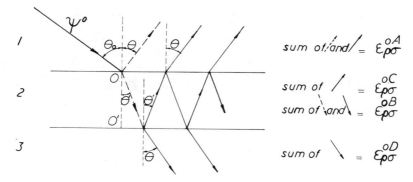

Fig. 4.2. Interference after multiple reflection and refraction of the diffracted light beam. (From Ref. 78, courtesy of Academic Press Inc., New York.)

The wave diffracted into medium 2 gives after interference a wave in medium 1,

$$-\Psi^0 A_{\mathbf{K}}^{12} rt'e^{-2i\alpha}U \tag{4.13}$$

where

$$\alpha = \left(\frac{2\pi}{\lambda_2}\right)h\cos\theta' \tag{4.14}$$

$$U = \left[1 - r^2 e^{-2i\alpha}\right]^{-1} \tag{4.15}$$

and t, t', and r defined by (4.9) to (4.11), with θ_0 and θ_0' replaced by θ and θ'. A relation similar to (4.13) can be found for a wave diffracted into medium 3. The total amplitude, $\varepsilon_{\mathbf{K}}^{0A}$, of the wave diffracted into medium 1, produced by Ψ^0, is the sum of (4.12) and (4.13).

$$\varepsilon_{\mathbf{K}}^{0A} = \Psi^0 A_{\mathbf{K}}^{12}\left[1 - rt'Ue^{-2i\alpha}\right] \tag{4.16}$$

In a similar way the diffracted amplitudes produced by Ψ^B and Ψ^C follow

$$\varepsilon_{\mathbf{K}}^{BA} = \Psi^B A_{\mathbf{K}}^{23} t'Ue^{-i\alpha} \tag{4.17}$$

$$\varepsilon_{\mathbf{K}}^{CA} = \Psi^C A_{\mathbf{K}}^{21}\left[1 - rt'Ue^{-2i\alpha}\right] \tag{4.18}$$

The flow patterns of the dynamic modes

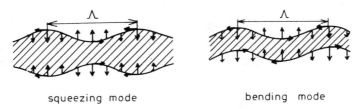

squeezing mode　　　　　　　bending mode

Fig. 4.3.　Schemes of squeezing and bending modes.

where we write

$$A_{\mathbf{K}}^{12} = Pt_0 t\zeta_{\mathrm{I},\mathbf{K}} \tag{4.19}$$

$$A_{\mathbf{K}}^{21} = Pt_0' t\zeta_{\mathrm{I},\mathbf{K}} \tag{4.20}$$

$$A_{\mathbf{K}}^{23} = -Pt_0' t\zeta_{\mathrm{II},\mathbf{K}} \tag{4.21}$$

The total amplitude diffracted into medium 1 (the reflection half-space) is the sum of (4.16), (4.17), and (4.18) and can be written as follows:

$$A_{\mathbf{K}} = M\zeta_{\mathrm{I},\mathbf{K}} + N\zeta_{\mathrm{II},\mathbf{K}} \tag{4.22}$$

or

$$A_{\mathbf{K}} = \tfrac{1}{2}(M+N)(\zeta_{\mathrm{I},\mathbf{K}} + \zeta_{\mathrm{II},\mathbf{K}}) + \tfrac{1}{2}(M-N)(\zeta_{\mathrm{I},\mathbf{K}} - \zeta_{\mathrm{II},\mathbf{K}}) \tag{4.23}$$

with

$$M = L\left(1 - r_0 e^{-2i\beta}\right)\left(1 - re^{-2i\alpha}\right) \tag{4.24}$$

$$N = -Lt_0' t' e^{-i\beta} e^{-i\alpha} \tag{4.25}$$

$$L = \frac{i\pi(n^2 - 1)t_0 t}{\lambda_1 \cos\theta} U_0 U; \quad P = \frac{ik_1(n^2 - 1)}{2\cos\theta} \tag{4.26}$$

Equation 4.23 is written because it is more natural to consider $(\zeta_{\mathrm{I},\mathbf{K}} + \zeta_{\mathrm{II},\mathbf{K}})$ and $(\zeta_{\mathrm{I},\mathbf{K}} - \zeta_{\mathrm{II},\mathbf{K}})$ as new, separate, uncoupled modes, which we call bending and squeezing mode, respectively (see Fig. 4.3).

B. Scattered Intensity for Films without Long-Range Interaction Forces: Bending and Squeezing Modes

The scattered intensity follows from the squared amplitudes as explained in Section III,

$$S^R = (\theta; \phi = 0; \theta_0) = \tfrac{1}{2}HG_0G_3 + \tfrac{1}{2}H(2G_0G_1 - G_0G_3) = HG_0G_1 \tag{4.27}$$

where

$$H = \frac{\frac{1}{4}k_B T}{\lambda_1^2 \gamma}(n^2-1)^2(\sin\theta_0 - \sin\theta)^{-2} \tag{4.28}$$

$$G_0 = \frac{t^2 t_0^2}{R(r^2,\alpha)R(r_0^2,\beta)}$$

$$G_1 = R(r,\alpha)R(r_0,\beta) + t'^2 t_0'^2$$

$$G_3 = G_1 - 2t'^2 t_0'^2 \cos\alpha\cos\beta + 2tt't_0 t_0'\sin\alpha\sin\beta$$

$$R(r,\alpha) = 1 + r^2 - 2r\cos 2\alpha$$

$$R(r_0,\beta) = 1 + r_0^2 - 2r_0\cos 2\beta \tag{4.29}$$

$$R(r^2,\alpha) = 1 + r^4 - 2r^2\cos 2\alpha$$

$$R(r_0^2,\beta) = 1 + r_0^4 - 2r_0^2\cos 2\beta$$

The contributions of the bending and squeezing modes are given by $G_0 G_3$ and $(2G_0 G_1 - G_0 G_3)$, respectively. The superscript R in S^R indicates

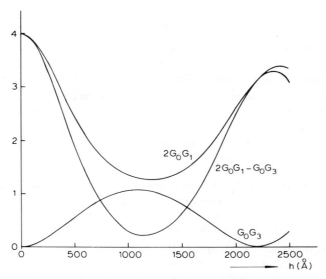

Fig. 4.4. Intensity (reflection side) of the light scattered from fluctuations $(\xi_I + \xi_{II})$: $G_0 G_3$ (bending mode) and from fluctuations $(\xi_I - \xi_{II})$: $2G_0 G_1 - G_0 G_3$ (squeezing mode), where $\theta = \theta_0 = 60°$, $n = 1.36$, and $\lambda_1 = 546$ nm. (From Ref. 68, courtesy of Elsevier Scientific Publishing Company, Amsterdam.)

scattering on the reflection side. An equation similar to (4.27) applies to scattering on the transmission side, S^T.

Examples are shown in Fig. 4.4. Note that the light-scattering curves of the bending and squeezing modes are very different functions of the (average) film thickness h. The curve for the bending mode has a shape (i.e., a minimum at $h \to 0$ and a maximum around $h = 120$ nm) similar to that of the *reflected* intensity, just opposite to that of the squeezing mode. The curve of their sum still shows an opposite trend with the reflected intensity. This effect was indeed found in some of our experiments[78] and was studied in more detail by Mann et al.[79]

C. Influence on the Scattering of a Monolayer of Soap Molecules and of the Bulk of the Film

Let us now turn our attention to the effect of a thin monolayer of soap molecules present on both film surfaces. In the first place it was found (see Section III.B) by Bouchiat and Langevin[77] that density and orientation fluctuations of the soap molecules (in the case of a single interface) make only a negligible contribution to the scattering process, except possibly at a (two-dimensional) critical point of the monolayer, or by addition of a few fluorescent soap molecules. Thus we are left with the optical effect of the monolayers on the interference of the light waves. It turns out (see Section VI) that this interference effect can be taken into account by redefining the (optical) film thickness.

Let us now consider the effect of the scattering of the *bulk* of the film due to density and concentration fluctuations of dissolved components.

Although this contribution often cannot be neglected when the scattering of a *single* surface of a bulk liquid is measured, it is negligible for our type of thin film. To be more precise, the scattering of 1 cm^3 of a 10% solution of the soap sodium dodecyl sulfate is 25×10^{-5} cm^{-1} (for pure water it is 7.5×10^{-7} cm^{-1}). For 1 cm^2 soap film, 300 nm thick, this is less than 1% of the surface scattering.

D. Influence of Long-Range Interaction Forces on the Surface Scattering in Thin Films

1. Surface Fluctuations in Thin Liquid Films

The work to create a surface fluctuation in a film involves the same components as that for a single interface, that is, contributions from surface tension and gravity: $\Delta F = \Delta F_1 + \Delta F_2$ [see (3.2)]. The gravity contribution ΔF_2 will be omitted hereafter because it is always small in comparison to ΔF_1 and because we work in vertical films where ΔF_2 is absent anyway. When the film is thick enough, the corrugations ζ_I and ζ_{II} are independent, so we are left with

$$\Delta F_1 = \frac{1}{2}\gamma\int_{-(1/2)a}^{+(1/2)a}\int\left[\left(\frac{\partial\zeta_{\mathrm{I}}}{\partial x}\right)^2+\left(\frac{\partial\zeta_{\mathrm{I}}}{\partial y}\right)^2+\left(\frac{\partial\zeta_{\mathrm{II}}}{\partial x}\right)^2+\left(\frac{\partial\zeta_{\mathrm{II}}}{\partial y}\right)^2\right]dx\,dy$$

(4.30)

and which leads to (4.27) given above.

When, however, the film thickness is small enough (say, $h=5$ to 100 nm) to make the long-range interaction forces perceptible, this is no longer true because the work to create a surface fluctuation will explicitly depend on $\Delta h=\zeta_{\mathrm{I}}-\zeta_{\mathrm{II}}$ through $V(h)$. Thus one has to take into account a third contribution

$$\Delta F_3 = \frac{dV}{dh}\int_{-(1/2)a}^{+(1/2)a}\int(\zeta_{\mathrm{I}}-\zeta_{\mathrm{II}})\,dx\,dy+\frac{1}{2}\frac{d^2V}{dh^2}\int_{-(1/2)a}^{+(1/2)a}\int(\zeta_{\mathrm{I}}-\zeta_{\mathrm{II}})^2\,dx\,dy$$

(4.31)

where $V(h)$ is given by (2.15). The first integral is zero, the second one is positive. For equilibrium films (see Fig. 2.3) d^2V/dh^2 is positive. The form of (4.31) makes apparent the advantage of the mode separation in (4.23).

It is only the squeezing mode in our representation of ΔF that contains a contribution of the interaction free energy $V(h)$. Thus we write, instead,

$$\Delta F=\Delta F_b+\Delta F_s$$

(4.32)

with

$$\Delta F_b = \frac{1}{4}\gamma\int_{-(1/2)a}^{+(1/2)a}\int\left\{\left[\frac{\partial(\zeta_{\mathrm{I}}+\zeta_{\mathrm{II}})}{\partial x}\right]^2+\left[\frac{\partial(\zeta_{\mathrm{I}}+\zeta_{\mathrm{II}})}{\partial y}\right]^2\right\}dx\,dy$$

(4.33)

$$\Delta F_s = \frac{1}{4}\gamma\int_{-(1/2)a}^{+(1/2)a}\int\left\{\left[\frac{\partial(\zeta_{\mathrm{I}}-\zeta_{\mathrm{II}})}{\partial x}\right]^2\right.$$

$$\left.+\left[\frac{\partial(\zeta_{\mathrm{I}}-\zeta_{\mathrm{II}})}{\partial y}\right]^2+\frac{1}{2}V''(\zeta_{\mathrm{I}}-\zeta_{\mathrm{II}})^2\right\}dx\,dy$$

(4.34)

where we have abbreviated d^2V/dh^2 by V''. Analogously as before, this leads to

$$\langle|\zeta_{\mathrm{I},\mathbf{K}}+\zeta_{\mathrm{II},\mathbf{K}}|^2\rangle = \frac{2k_{\mathrm{B}}T/a^2}{\gamma K^2}$$

(4.35)

$$\langle|\zeta_{\mathrm{I},\mathbf{K}}-\zeta_{\mathrm{II},\mathbf{K}}|^2\rangle = \frac{2k_{\mathrm{B}}T/a^2}{\gamma K^2+2V''}$$

(4.36)

Then the scattering intensity becomes

$$S^R = \frac{\pi^2 k_B T}{2\lambda_1^4}(n^2 - 1)^2\left(\frac{G_0 G_3}{\gamma K^2} + \frac{2G_1 G_0 - G_0 G_3}{\gamma K^2 + 2V''}\right) \tag{4.37}$$

The first term is due to the bending mode and the second one to the squeezing mode.

2. Some Early Experimental Observations

Recalling the contribution of the bending and squeezing modes to the scattering power, we note from Fig. 4.4 that in the thinner regions, say $h < 50$ nm, the contribution from the squeezing mode ($\zeta_I - \zeta_{II}$) given by ($2G_0 G_1 - G_0 G_3$) is much larger than that of the bending mode ($\zeta_I + \zeta_{II}$) given by $G_0 G_3$. Thus a change in the denominator $\gamma K^2 + 2V''$ in (4.37) because of increasing interaction forces in the thinner films, should be observable indeed. The following early experiment[78] clearly shows the expected behavior.

A film was drawn from an aqueous solution containing a nonionic soap and some ionic soap (to give charge stabilization), and glycerol to slow down the drainage speed. The film was observed for one week during which it thinned from 53 to 13 nm. The intensity S^R was measured at angles θ between 20° and 44° and the incident angle was $\theta_0 = 60°$. In Fig. 4.5, S^R is plotted as a function of $\Theta = (\sin\theta_0 - \sin\theta)^{-2} = (k_1/K)^2 = (\Lambda/\lambda_1)^2$. The film thickness was obtained from the reflected intensity R with the equation

$$R = |\Psi^A|^2 = \frac{4r_0^2 \sin^2\beta}{1 + r_0^4 - 2r_0^2 \cos 2\beta} \tag{4.38}$$

with β given by (4.7).

Starting with the thickest film ($h = 530$ Å $= 53$ nm), one notes that S^R is proportional to Θ or K^{-2}, which implies that V'' is negligible. This is also the case for $h = 470$ and 320 Å, although the slope of the plot has increased, implying that $G_0 G_3 + (2G_1 G_0 - G_0 G_3) = 2G_1 G_0$ increases with decreasing h, just as predicted. From $h = 260$ to 130 Å, the plots show an increasing, downward curvature, which can be explained by an increase of V'' with decreasing h, also as expected. The values of γ obtained were 32 ± 5 mN/m, whereas the surface tension of the solution was 33 mN/m. The values of $V''(h)$ thus found were of the expected order of magnitude, although the (approximately exponential) decrease with h was much steeper than expected from theory. More recent experiments are discussed in Section VI.

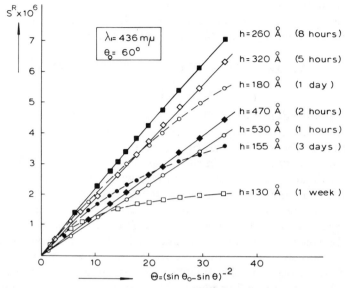

Fig. 4.5. Light scattering of a soap film as a function of $\Theta = (\sin\theta_0 - \sin\theta)^{-2} = (\Lambda/\lambda_1)^2$. (From Ref. 68, courtesy of Elsevier Scientific Publishing Company, Amsterdam.)

3. Instability Regions in Films

Until now it has been (implicitly) assumed that V'' is (always) positive. A positive, increasing V'' gives a decrease in the amplitude of $(\zeta_I - \zeta_{II})$; see (4.36). When V'' is negative, however, ΔF_s turns from positive to negative for some K, that is, at

$$K_c = \left[\frac{-2V''}{\gamma} \right]^{1/2} \tag{4.39}$$

For $K > K_c$ the fluctuations $(\zeta_I - \zeta_{II})$ will die out. For $K < K_c$, however, these fluctuations will tend to grow! Figure 2.3 for $V(h)$ indeed shows a thickness range with $V'' < 0$ [i.e., at the right-hand side of the minimum in $V(h)$, where V'' turns from positive to negative]. We think that this may indeed occur in practice, perhaps sometimes leading to instabilities that cause rupturing of the film. This is discussed further in Section V. A mechanism for film rupture based on this idea was first proposed by Scheludko,[12] who derived an equation similar to (4.39). One of us proposed a more complete theory,[81, 82] which shows the equivalence of this process with what is called "spinodal decomposition."[83] For further discussion and refinements, see, for example, Refs. 80 and 84. Also amenable to this explanation is the sudden appearance of (stable) black holes that sometimes is observed in much thicker, draining films. See also Section VI.

V. DYNAMICS OF FILM FLUCTUATIONS

A. Introduction

Interest in the dynamics of film fluctuations was prompted by the search for a theory of film stability.[81] At that time (1966) no experiments on thermal excited surfaces waves had been performed.

From the theory of waves on single surfaces (for a review, see Ref. 85) it was known that for small values of surface wavelength Λ (i.e., large values of K), the relaxation of the waves was dominated by the action of surface tension. In the case of thin films, also, the effect of interaction forces was expected to influence the relaxation of the film to its mean thickness. Especially, the squeezing mode relaxation, where local thickness fluctuations appear, should show up the influence of colloidal interactions.

The first theory for film fluctuations resembles Cahn's theory[83] of "spinodal decomposition" of unstable bulk systems. A very simple mechanism was adopted for the liquid flow. It was assumed that because of the presence of the soap monolayers, the film surfaces were stagnant (see Section II) and that the film liquid was pumped back and forth through a slab with thickness h according to Reynolds's law:

$$Q = - \frac{h^3}{12\eta} \frac{d\Delta P}{dx} \tag{5.1}$$

when Q is the volume flow rate and ΔP the variation in pressure along the film, which is given by

$$\Delta P = - \frac{1}{2} \gamma \frac{d^2 h}{dx^2} + V'' \Delta h. \tag{5.2}$$

By assuming the existence of squeezing modes, it was found from the conservation of film volume that the amplitude of the mode with wavenumber K relaxes according to

$$\exp\left(\frac{-t}{\tau}\right) \tag{5.3}$$

with

$$\tau^{-1} = \frac{h^3 K^4 \gamma}{24\eta} \left[1 + \frac{2V''}{\gamma K^2}\right] \tag{5.4}$$

A similar equation was found by deGennes.[86] For $V'' > 0$, τ is positive, giving rise to an overdamped relaxation. For $V'' < 0$ and $K < (-2V''/\gamma)^{1/2}$, τ becomes negative, and the squeezing mode will grow in accordance with (4.34), potentially producing film rupture. The maximum value of τ is found at $K_m = (-V''/\gamma)^{1/2}$ with $\tau^{-1} = \tau_m^{-1} = -(h^3\gamma/24\eta)K_m^4$.

Next we review the more refined theories. The connection between the hydrodynamics of film motion and the light scattering experiments from thermal fluctuations is based on Onsager's regression hypothesis,[87] namely, the relaxations of the surface elevations derived from macroscopic theories also pertain to the relaxation of thermally excited fluctuations.

B. Hydrodynamics of Free Films

There are two trends in the existing hydrodynamic theories of free films, which differ in the way the (colloidal) interaction forces are taken into account. We mention first the method of Felderhof,[88] who developed a systematic and consistent electrohydrodynamic theory for a nonviscous liquid. His theory was extended to include viscous behavior by Sche and Fijnaut.[89] In these theories the interaction forces are included in the momentum equations. The other theoretical approach[90–94] considers hydrodynamic equations without the interaction forces. The influence of these interactions are considered only in the normal stress boundary conditions.

Although the Felderhof approach seems to be the more fundamental one, we describe the film hydrodynamics on the basis of the second method mentioned, since the dynamics are more surveyable and the method can more easily be extended to include different types of surface motion. However, the final equations found from the different approaches are compared in Section V.B.5.

1. Hydrodynamic Equations

As before, the film is considered to consist of a volume part, containing an electrolytic solution, bounded by surface layers in which ions may be adsorbed. The surface layers are idealized as Gibbs dividing (mathematical) surfaces[95] having a surface charge density, a surface tension, and a surface elasticity. The liquid satisfies the hydrodynamic equations of motion supplemented with boundary conditions at the surfaces. In writing down the equations, the following assumptions are made: the liquid is assumed to be incompressible; the amplitude of the surface waves is small compared with the wavelength and the thickness of the film, so that the Navier-Stokes equation can be linearized[96]; and we consider capillary waves, which means that the only waves considered are of such small wavelength that the effects of gravity can be neglected[96] (see, however,

Section III.A). The relevant equations are the continuity condition

$$\nabla \cdot \mathbf{v} = 0 \tag{5.5}$$

and the linearized Navier-Stokes equations

$$\rho \frac{\partial \mathbf{v}}{\partial t} = \nabla \cdot \boldsymbol{\sigma} \tag{5.6}$$

where $\mathbf{v} = \mathbf{v}(\mathbf{r}, t)$ is the flow velocity, ρ the mass density, and $\boldsymbol{\sigma}$ the mechanical stress tensor defined by

$$\boldsymbol{\sigma} = -P\mathbf{I} + \eta \nabla \mathbf{v} \tag{5.7}$$

where P is the hydrostatic pressure, $\mathbf{I}$ the unit tensor, and η the shear viscosity of the film liquid.

The procedure for solving (5.5) and (5.6) consists of the introduction of a so-called stream function Ψ and a potential function Φ in such a way that:

$$v_x = -\frac{\partial \Phi}{\partial x} - \frac{\partial \Psi}{\partial z}$$

$$v_z = -\frac{\partial \Phi}{\partial z} + \frac{\partial \Psi}{\partial x} \tag{5.8}$$

where for simplicity only x and z components of the liquid flow are considered. In the case of a nonviscous liquid, the stream function is identically zero, which means that the flow is irrotational. Substitution of (5.8) into (5.5) and (5.6) and integration of the result from the substitution of (5.8) into (5.6) gives:

$$\Delta \Phi = 0 \tag{5.9}$$

$$-\rho \frac{\partial \Psi}{\partial t} + \eta \Delta \Psi = c_1 \tag{5.10}$$

$$-\rho \frac{\partial \Phi}{\partial t} + P = c_2 \tag{5.11}$$

The integrating constants c_1 and c_2 can be found[90] from the conditions at zero flow. The constant c_1 is evidently zero because both the temporal and the space derivatives of Ψ vanish at zero flow. The constant c_2 can be defined from the continuity of pressure in a flat film at the interface. The difference between the surrounding pressure P_0 and the hydrostatic pressure at $z = \pm h/2$ for an equilibrium flat film is given simply by the disjoining pressure $\Pi(h)$ (see Section II.D), where h is the film thickness:

$$P_0 = P\left(z = \pm \frac{h}{2}\right) + \Pi(h) \tag{5.12}$$

Inserting this expression in (5.11) gives:

$$c_2 = P_0 - \Pi(h) \tag{5.13}$$

Since we are finally interested in the Fourier components of the surface displacement, we insert solutions for Φ and Ψ, consisting of plane waves in the x-direction with a z-dependent amplitude, into (5.9) and (5.10). This results in the following expressions[90] for Φ and Ψ inside the film:

$$\Phi = (A \cosh Kz + B \sinh Kz)\exp[i(Kx - \omega t)] \tag{5.14}$$

$$\Psi = (C \cosh mz + D \sinh mz)\exp[i(Kx - \omega t)] \tag{5.15}$$

where m is defined by

$$m^2 \equiv K^2 - \frac{i\omega\rho}{\eta} \tag{5.16}$$

The constants A to D have to be found from the boundary conditions. As before, K is the wave number and ω is the complex frequency. The relation between Φ, Ψ, and the surface element displacement ζ (in the z-direction) and ξ (in the x-direction) is found through the velocity $\mathbf{v}$. Under the previously mentioned condition of surface wave amplitudes small compared to the wavelength of the wave, we have:

$$v_x \approx \frac{\partial \xi}{\partial t} \tag{5.17}$$

$$v_z \approx \frac{\partial \zeta}{\partial t} \tag{5.18}$$

By inspection of (5.8), (5.14), (5.15), (5.17), and (5.18) it is clear that the surface displacements show the same wave character as Φ and Ψ. This means that as far as the dynamics of the surface waves are concerned, we are interested only in the dependence of ω on the system variables, since ω completely determines the time dependence of the surface ripples.

2. Boundary Conditions

To evaluate the boundary conditions, we consider the influence of the interaction forces and the properties of the adsorbed monolayers of soap molecules at both air-liquid interfaces. First we discuss the occurrence of the interaction forces in the boundary conditions. From (5.12) it is seen that the equilibrium condition for a flat, free film can be formulated in terms of the disjoining pressure. This way of describing the influence of interaction forces is followed in the continuity of the normal stress across the liquid interfaces.[90, 94]

$$-\sigma_{zz} + \gamma \frac{\partial^2 \zeta_I}{\partial x^2} + \Pi(h + \zeta_I - \zeta_{II}) = P_0 \qquad \text{at} \quad z = \frac{h}{2} + \zeta_I \qquad (5.19)$$

and

$$-\sigma_{zz} - \gamma \frac{\partial^2 \zeta_{II}}{\partial x^2} + \Pi(h + \zeta_I - \zeta_{II}) = P_0 \qquad \text{at} \quad z = \frac{-h}{2} + \zeta_{II} \qquad (5.20)$$

where ζ_I and ζ_{II} are the elevations of upper and lower interfaces, respectively. In (5.19) and (5.20) the viscosity of the surrounding air has been put zero, as also will be done in the tangential stress boundary conditions, which are:[90, 106, 107]

$$-\sigma_{zx} + \varepsilon_s \frac{\partial^2 \xi_I}{\partial x^2} = 0 \qquad \text{at} \quad z = \frac{h}{2} + \zeta_I \qquad (5.21)$$

and

$$\sigma_{zx} + \varepsilon_s \frac{\partial^2 \xi_{II}}{\partial x^2} = 0 \qquad \text{at} \quad z = \frac{-h}{2} + \zeta_{II} \qquad (5.22)$$

The properties of the adsorbed monolayers are reflected in the surface tension γ and the surface dilatational modulus ε_s. The surface concentration Γ of the soap molecules in subjected to changes through the changes in surface area by the surface dynamics. This means that the surface tension also may vary as function of Γ (see also Section II.A); this now results in a surface stress $(\partial\gamma/\partial\Gamma)\nabla\Gamma$. By the introduction of the dilatational modulus

$$\varepsilon_s \equiv -\Gamma \frac{\partial\gamma}{\partial\Gamma} \qquad (5.23)$$

the tangential stress boundary conditions become as in (5.21) and (5.22), under the assumption that the adsorbed monolayer is insoluble, and thus no bulk-surface diffusion processes occur among surface active molecules. The second terms in (5.19) and (5.20) are expressions for the Laplace pressure caused by surface curvature in the limit of values of ζ_I and ζ_{II} small compared with Λ (the ripple wavelength).

3. Dispersion Relations

The boundary conditions (5.19) to (5.22) give us four equations in which the expressions for Φ and Ψ are inserted through the relations (5.7), (5.8), (5.11), (5.17), and (5.18). The resulting four equations have nontrivial solutions for A to D, if the determinant of coefficients of A to D is zero. In the case considered here, a film surrounded at both sides by the same medium, it appears[90] that zero equating the determinant gives two relations between ω, K, and the system variables: the so-called dispersion relation[94]:

$$\left(\omega^2 + \frac{2i\eta K^2\omega}{\rho}\right)^2 - \omega^2 \left\{\omega_1^2 + \frac{\varepsilon_s mK^2}{\rho}\coth\frac{mh}{2} - \frac{4\eta^2 K^3 m}{\rho^2}\tanh\frac{Kh}{2}\coth\frac{mh}{2}\right\}$$

$$+ \frac{\varepsilon_s K^2\omega_1^2}{\rho}\left(m\coth\frac{mh}{2} - K\coth\frac{Kh}{2}\right) = 0 \qquad (5.24)$$

$$\left(\omega^2 + \frac{2i\eta K^2\omega}{\rho}\right)^2 - \omega^2 \left\{\omega_2^2 + \frac{\varepsilon_s mK^2}{\rho}\tanh\frac{mh}{2} - \frac{4\eta^2 K^3 m}{\rho^2}\coth\frac{Kh}{2}\tanh\frac{mh}{2}\right\}$$

$$+ \frac{\varepsilon_s K^2\omega_2^2}{\rho}\left(m\tanh\frac{mh}{2} - K\tanh\frac{Kh}{2}\right) = 0 \qquad (5.25)$$

where

$$\omega_1^2 = \left\{\frac{\gamma K^3 + 2KV''}{\rho}\right\}\tanh\frac{Kh}{2} \qquad (5.26)$$

and

$$\omega_2^2 = \frac{\gamma K^3}{\rho}\coth\frac{Kh}{2} \qquad (5.27)$$

The dispersion relations (5.24) and (5.25) are too complicated to show simply what the explicit solution for ω is. Therefore we will consider some limiting, experimentally relevant, situations. First, however, we consider the conditions under which (5.24) and (5.25) are obtained. One can observe that the relation (5.24), found from the equations for A to D, corresponds to the situation $(B,C)=0$ and $(A,D)\neq 0$. Inserting (5.14) and (5.15) into (5.8) shows that the situation $(B,C)=0$ and $(A,D)\neq 0$ corresponds to $v_x(x,z)$ $= v_x(x,-z)$, which means that (5.24) is the dispersion relation for the squeezing mode. Alternatively it appears that (5.25) is the dispersion relation for the bending mode.

The results (5.24) and (5.25) can easily be transformed into the dispersion equation for single interfaces,[32, 97, 98, 108] just by taking the limit $h\to\infty$ and bearing in mind that $V''=0$ in this case. The result is:

$$\left(\omega^2+\frac{2i\eta K^2\omega}{\rho}\right)^2-\omega^2\left(\omega_0^2+\frac{\varepsilon_s mK^2}{\rho}-\frac{4\eta^2K^3m}{\rho^2}\right)+\varepsilon_s K^2\omega_0^2(m-K)=0$$

(5.28)

with

$$\omega_0^2=\lim_{h\to\infty}\omega_1^2=\lim_{h\to\infty}\omega_2^2=\frac{\gamma K^3}{\rho}$$

(5.29)

The result (5.28) cannot be solved in an explicit expression for ω. That can be done only numerically.[117]

4. The Long-Wavelength Limit

In the experiments we always deal with film having thickness much less than the wavelength of the observable surface waves (see Section III.A). This means that in the experiments the long-wavelength limit (lwl) is observed. Since the lwl corresponds with the conditions $Kh\ll 1$ and $|mh|\ll 1$ (this condition should be checked afterward), the lwl of squeezing and bending mode can be found from (5.24) and (5.25) by series expansion of the hyperbolic functions. The result for the squeezing mode is[91]:

$$\omega^3+i\left(\frac{4\eta K^2}{\rho}\right)\omega^2-\left(\omega_3^2+\frac{2\varepsilon_s K^2}{\rho h}\right)\omega-\frac{i\varepsilon_s K^2 h\omega_3^2}{6\eta}=0$$

(5.30)

with

$$\omega_3^2\equiv\lim_{Kh\ll 1}\omega_1^2=\frac{h(\gamma K^4+2K^2V'')}{2\rho}$$

(5.31)

Expansion of the hyperbolic terms of (5.25) gives for the lwl of the bending mode:

$$\omega^3 + i\left(\frac{\varepsilon_s K^2 h}{2\eta}\right)\omega^2 - \left(\omega_4^2 + \frac{\varepsilon_s K^4 h}{2\rho}\right)\omega - i\frac{\varepsilon_s K^2 h}{2\eta}\omega_4^2 = 0 \qquad (5.32)$$

with

$$\omega_4^2 \equiv \lim_{Kh \ll 1} \omega_2^2 = \frac{2\gamma K^2}{\rho h} \qquad (5.33)$$

The above-described dispersion relations for the squeezing and bending modes are found under the conditions of zero density and viscosity of the medium surrounding the film. The influence of the air's viscosity and density changes the coefficients in the dispersion relations (5.30) and (5.32). In the case of the squeezing mode, the corrected results are given in the paper by Vrij and others,[91] but for light-scattering experiments on free films, the influence of the surrounding air on liquid motion can be neglected in the experimentally accessible range of surface waves. In the case of the bending mode the air's viscosity and density cannot be neglected.[99]

Both dispersion relations (5.30) and (5.32) permit three solutions for ω, which means that each mode might be composed of three components. One component is purely imaginary, the two others are complex conjugate. The complex roots correspond to propagating damped or growing waves, whereas the purely imaginary root corresponds to an overdamped wave. The spectrum of the solutions of the dispersion relations is therefore very similar to the spectrum of light scattered from a pure liquid, where a central Rayleigh line is surrounded by two Brillouin peaks.

It is interesting to consider two limiting cases for the dispersion relation by considering the limits $\varepsilon_s \to \infty$ and $\varepsilon_s \to 0$. The limit $\varepsilon_s \to \infty$ corresponds with complete rigidity of the adsorbed layer and consequently with the condition of no slip at the surfaces.

The dispersion relations become now:

$$\omega = -i\frac{h^3 K^2}{24\eta}\left(\gamma K^2 + 2V''\right) \qquad (5.34)$$

for the squeezing mode, and

$$\omega = -\frac{iK^2\eta}{2\rho} \pm K\left(\frac{2\gamma}{\rho h} - \frac{K^2\eta^2}{4\rho^2}\right)^{1/2} \qquad (5.35)$$

for the bending mode. Equation 5.34 corresponds to (5.4). From (5.34) we see that the squeezing mode will be aperiodically damped, as long as $\gamma K^2 > -2V''$, and that the mode will grow if $\gamma K^2 < -2V''$ (also, see Section V.A.).

The bending mode shows behavior a bit more complex depending on the sign of the expression between brackets in (5.35). If this expression is positive (as is the case in liquid film experiments reported thus far) the bending mode is a periodically damped wave with frequency $K(2\gamma/\rho h - K^2\eta^2/4\rho^2)^{1/2}$ and with damping constant $K^2\eta/2\rho$. In the other cases, critical damping, aperiodic damping, and even growth of the mode may be expected.

Precise inspection of (5.30) and (5.32) reveals that the limit $\varepsilon_s \to \infty$ in practice is obtained when, say $\varepsilon_s > 10$ mN/m for the usual experimental conditions.[91] Direct measurements of ε_s on surfaces, performed by Prins et al.,[100] revealed that $\varepsilon_s \approx 80$ mN/m for adsorbed layers. This means that the limit $\varepsilon_s \to \infty$ is a good approximation in the theory. From (5.34) and (5.35) it follows that dynamic experiments on surface waves will give information about V'', γ, and η.

5. Comparison with Other Theories

The theoretical approach of Felderhof,[88] Sche and Fijnaut,[89] and Sche[101] is based on (5.5) and a modified momentum equation:

$$\rho \frac{\partial \mathbf{v}}{\partial t} = \nabla \cdot (\boldsymbol{\sigma} + \mathbf{T}) - \rho \nabla W \qquad (5.36)$$

where the Maxwell stress tensor $\mathbf{T}$ contains the local electric fields and the van der Waals potential W reflects the local result of the van der Waals two-body attraction forces between the film molecules. Now the interaction forces are included in the basic equations. The film is considered to consist of a volume part, containing an electrolytic solution, bounded by surface layers having a surface charge density. In addition to the assumption made in Section V.B.1, the ion distribution is considered to be at each instant in thermal equilibrium. This means that the theory is limited to surface waves of frequencies below about 10^7 Hz. The electric potential in the film satisfies the Poisson-Boltzmann equation (see also Section II.B) as a result of the governing quasi-static Maxwell equations for the electric field. The boundary conditions in the Felderhof theory do not contain the disjoining pressure [see (5.19) and (5.20)], but the appropriate element of the Maxwell stress tensor. However, the tangential boundary conditions now also contain the influence of electrical interaction forces. Apart from the stress boundary conditions, the Felderhof approach also contains boundary conditions for the electric field and the electric potential. The van der Waals potential now is a function not only of the thickness, but also of the curvature of the surface.

In his theory, Felderhof also considers the effect of surface charge density fluctuations.[102] He takes into account fluctuations from the change in electrostatic potential in the surface and fluctuations due to the change in surface area. He considers two regimes in the surface motion: the fast regime shows only charge density fluctuations that are due to surface area variations, whereas in the slow regime only surface potential fluctuations remain. The charge density fluctuations due to the change in surface area are in the slow regime, compensated by completion of the adsorbed surface molecules from the film bulk. The more extensive (and more fundamental) basic equations, together with the refinement of the behavior of (surface) charges and the diffusion of surface molecules, makes the theory much more laborious. Comparison of the dispersion relations with the results of the theory in the foregoing sections can easily be done only in the lwl with condition $\varepsilon_s \rightarrow \infty$. In that case the surface tension occurring in the (5.34) and (5.35) must be corrected by terms arising from the interaction forces and the electric field. The numerical value of these corrections[89, 101] are of order of 1 mN/m or less, and they must be compared with usual surface tensions of about 30 mN/m. This means that in this case the simplified hydrodynamic theory is a useful approximation.

6. Discussion

We have confined ourselves to a description of the dynamics of surface roughness and the influence of the interaction forces on these dynamics. In reality, however, there are many more dynamic processes in the film and especially in the adsorbed monolayers that should be considered to describe in full detail the film dynamics. Apart from dynamics of the film surfaces parallel to the normal of the interfaces, motions of the adsorbed surface molecules in the interface must be considered. According to Lucassen-Reynders and Lucassen,[85] the actual stresses in an interface are described by four rheological coefficients, reflecting the viscoelastic properties of the interface. Two of these, the surface dilatational elasticity and the surface dilatational viscosity, measure the surface's resistance against changes in area. The dilatational module ε_s, considered before, expresses the dilatational elasticity. In our description of the film system, we neglected the viscous behavior of the interface, which implies that no diffusion of surface active molecules between bulk and interface was considered. If, however, surface-to-bulk diffusion is taken into account, the expression

(5.23) for ε_τ becomes more complicated[103]:

$$\varepsilon_s = \frac{-\Gamma \dfrac{d\gamma}{d\Gamma}}{1 + \dfrac{i\mu D}{\omega} \dfrac{dc_s}{d\Gamma} \tanh \dfrac{\mu h}{2}} \tag{5.37}$$

where D is the bulk diffusion coefficient of the surface active molecules, and μ is defined by

$$\mu^2 = K^2 - \frac{i\omega}{D} \tag{5.38}$$

and c_s is the surfactant bulk concentration just below the interfaces. Under the condition $\mu h \ll 1$, which seems realistic, since $\omega \simeq 10^3 \ \text{sec}^{-1}$, $h \simeq 5 \times 10^{-8}$ m, $D \simeq 10^{-9} \ \text{m}^2/\text{sec}$, and $K \simeq 10^6/\text{m}$, (5.37) becomes:

$$\varepsilon_s = \frac{-\Gamma \dfrac{d\gamma}{d\Gamma}}{1 + \dfrac{dc_s}{d\Gamma} \dfrac{h}{2} \left(1 + i \dfrac{K^2 D}{\omega} \right)} \tag{5.39}$$

Since in practice the term $(dc_s/d\Gamma)h/2$ for high surface active molecules is very small with respect to 1, we expect to be able to neglect bulk diffusion. The case of surface diffusion is treated in the case of films in the same way as in the case of single interfaces. One now finds for ε_s:

$$\varepsilon_s = \frac{-\Gamma \dfrac{d\gamma}{d\Gamma}}{1 + \dfrac{iK^2 D_s}{\omega}} \tag{5.40}$$

It has been shown[104], however, that for values of $-\Gamma(d\gamma/d\Gamma)$ larger than about $+20$ mN/m, no effect of the surface diffusion coefficient D_s can be observed in the dispersion relation.

The two other rheological coefficients, the surface shear elasticity and the surface shear viscosity, describe the resistance against changes in shape of a surface element. The usually made assumption is that the shear coefficients are negligible compared to the dilatational coefficients.[85] Up to now it appeared to be impossible to derive information about the rheological behavior of the interface from light scattering from liquid films. Single-interface light scattering experiments have yielded some results of viscoelastic behavior of adsorbed monolayers.[105]

The combination of the laws of reflection and refraction by surface waves with the hydrodynamics of these surface deformations, facilitates the calculation of the (dynamic) light scattering from the squeezing and bending modes. However, the shear and dilatational motions in the adsorbed monolayers, which in fact are due to surface density fluctuations, require another method for calculating the scattered intensity.[77] From those calculations, it appears that the contribution to the light scattering from the surface modes is negligible compared to the contribution from the surface deformation (squeezing and bending) modes. Moreover, the theory of Ref. 77 applied to deformation modes gives the same light-scattering results as found from the combination of refraction of light with hydrodynamics.

The hydrodynamics described in the preceding sections can also be applied to the description of the dynamics of liquids films on a substrate. The procedure is simple but rather involved.

VI. TIME–AVERAGED LIGHT–SCATTERING EXPERIMENTS ON LIQUID FILMS

A. Introduction

Time-averaged light-scattering intensities of liquid interfaces are rather difficult to measure. The scattered intensity is weak and strongly angle dependent. Therefore, great care must be taken to avoid unwanted stray light. The absolute calibration of the (average) intensity is laborious, and not much has been reported on single interfaces. We mention here experiments on superfluid helium films.[109] Above all in soap films, a careful conditioning of the film itself is essential to obtain reproducible results from different film samples.

The first experiments performed by one of us with a high-pressure mercury lamp as a light source clearly showed the influence of the colloidal interaction forces on the intensity.[78] The film conditioning was poor, however. Further experiments were reported by Mann and co-workers.[79, 110] This section describes the recent measurements of Donners and Rijnbout.[111, 112]

1. Apparatus

The experiments were performed in a vessel designed by Rijnbout. Special care was taken to obtain a well-defined optical geometry with a minimum of unwanted stray light and good temperature control ($\pm 0.002\,^\circ$C). The vessel was placed in a cage with air thermostat ($\pm 0.2\,^\circ$C) and the temperature of the room varied only within $\pm 1\,^\circ$C. The light source was a He–Ne laser. The film could be set to fixed positions such that $\theta_0 = 30^\circ$, 60°, and 75°, whereas the scattering angle θ could be varied continuously in the plane of incidence. The films were drawn in a circular brass frame (hole diameter 2 cm) heavily plated with silver and were in permanent contact with the solution from which the film was drawn. For a further description of the apparatus and the data collection, we refer to the original papers.[111, 112] An analogous light-scattering vessel is described in more detail in Section VII.

2. Materials

Films were drawn from solutions containing 8.2×10^{-4} mol/dm^3 of purified hexadecyl trimethylammonium bromide (HDTAB) and 8.4 wt% glycerol (added to gain better control of the water vapor pressure equilibrium) in twice-destilled water. The refractive index of the solution was taken ($n = 1.337$); the surface tension was $\gamma = 39$ mN/m ($= 39$ dynes/cm), and the dielectric constant ε was 76.2. All measurements were carried out at 25°C. To avoid irregular drainage of the film upon refreshment of the surface inside the vessel, the temperature of the stock solution (see Section VII.B.1) was carefully controlled.

3. Measurement of the Film Thickness

Film thicknesses were obtained from the reflected intensity, using (4.38) for a homogeneous film with a refractive index equal to that of the surfactant solution. The value thus obtained is the so-called equivalent water layer thickness h_w, which is also used in the light-scattering interference formulas [see (4.29)].

In the equation for the interaction forces, the thickness of the aqueous core h is needed. To obtain h from h_w, slightly different procedures and equations exist.[10, 113–115] Rijnbout's equation is

$$h = h_w - \frac{2(n_{hc}^2 + 2)}{n^2 - 1} R_{hc} \Gamma \tag{6.1}$$

where n and n_{hc} are the refractice indices of the soap solution and the soap layers, respectively, and R_{hc} is the molar refraction of the surfactant. In this study the investigators used $R_{hc} = 1.08 \times 10^{-4}$ m^3/mol, which is the value for dodecyl trimethylammonium bromide corrected for the presence of four extra CH$_2$ groups; $\Gamma = 3.26 \times 10^{-6}$ mol/m^2, and $n_{hc} = 1.435$ (the value of bulk hexadecane). Then one obtains $h = h_w - 3.6$ nm.

4. Scattering Measurement

The light-scattering intensity S^R or S^T was obtained from the photocurrent $i = fS$. The factor f was obtained within a few percent from absolute calibration[111] and was checked frequently by a relative calibration on thicker films (where interaction forces are still negligible). Taking for γ the surface tension of the solution, it was found that the optical thickness dependence and the angular dependence of the scattering were internally consistent within the accuracy ($\sim$2%) of the measurement, which proves the correctness of the scattering equations.

B. Electrical Double Layer Forces

This section gives results obtained for films with a low electrolyte concentration, which give information about electrical double layer forces; the contribution of van der Waals–London forces is very small in this case.

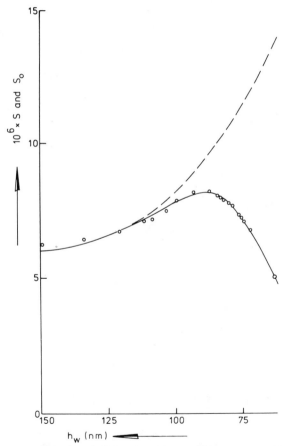

Fig. 6.1. Plot of S_0 and S versus thickness of the film. The dashed curve gives S_0, scattering without interaction between the film surfaces. The solid curve gives S, scattering calculated from (6.2) using the results of Fig. 6.3. Circles indicate experimental results; $\theta_0 = 60°$, $\theta = 54°$. (From Ref. 111, courtesy of Academic Press Inc., New York.)

Two series of measurements were performed: on draining films at constant scattering angle θ, and on (nearly) stationary films at varying θ. The angle of reflection $\theta_0 = 60°$. The results of a draining film are shown in Fig. 6.1, where the scattering intensity S at $\theta = 54°$ is plotted for film thicknesses h_w between 60 and 150 nm. The dashed curve gives S_0, the scattered intensity calculated assuming absence of interaction forces. The curves were analyzed with (4.37) rewritten in the following form:

$$\frac{S_0}{S} = \frac{1 + \mathcal{V}}{1 + \chi \mathcal{V}} \tag{6.2}$$

$$\mathcal{V} = \frac{2}{K^2 \gamma} \frac{d^2 V}{dh^2} \tag{6.3}$$

and

$$\chi = \frac{G_0}{2 G_1}, \qquad \text{scattering on reflection side} \tag{6.4}$$

Some curves of χ versus θ (at $\theta_0 = 60°$) are given in Fig. 6.2 for three film thicknesses. The curves at the left are for the scattering on the reflection

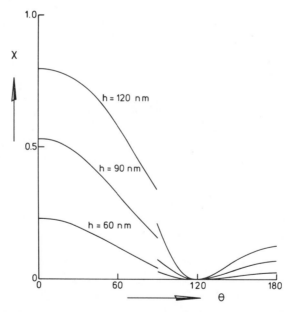

Fig. 6.2.   Dependence of χ on the scattering angle θ for three film thicknesses; $\theta_0 = 60°$, $n = 1.337$, $\lambda = 632.8$ nm. The discontinuity in the curves at $\theta = 90°$ is caused by the differences in the scattering formulas for the reflection and transmission sides of the film. (From Ref. 111, courtesy of Academic Press Inc., New York.)

side. They show that in the thickness range where $\nabla \neq 0$, χ varies between 0.1 and 0.3 The curves on the right-hand side are for scattering on the transmission side.

Values of d^2V/dh^2 as a function of h calculated from the scattering equation are shown in Fig. 6.3. Because the van der Waals–London contribution as calculated from (6.6) (see below) was always smaller than 2%, this contribution was neglected here. The linear plot in Fig. 6.3 was used to calculate ψ_0 and κ from the differentiated form of the left-hand side of (2.16),

$$\frac{d^2V_{DL}}{dh^2} = B\kappa^3 e^{-\kappa h} \tag{6.5}$$

which, indeed, predicts linearity. The slope gives $\kappa = 9.68 \times 10^7 \text{ m}^{-1}$. A large

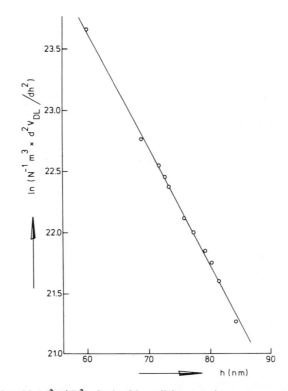

Fig. 6.3. Plot of $\ln(d^2V/dh^2)$ obtained from light-scattering measurements versus the film thickness. The solid curve shows the least-squares fit of the experimental points; $\theta_0 = 60°$, $\theta = 54°$. From slope and intercept one finds, respectively, $\kappa = 9.68 \times 10^7 \text{ m}^{-1}$ and $\ln B\kappa^3 = 29.81$. (From Ref. 111, courtesy of Academic Press Inc., New York.)

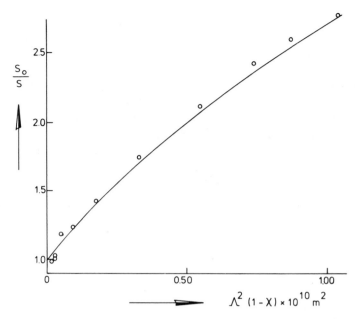

Fig. 6.4. Angle dependence of the light scattering: S_0/S is plotted versus $\Lambda^2(1-\chi)$. The solid curve shows S_0/S calculated from (6.2) using the results of Fig. 6.3; $\theta_0 = 60°$. (From Ref. 111, courtesy of Academic Press Inc., New York.)

number of measurements from different films treated in this way gives values of κ between 9.0 and 10.0×10^7 m^{-1}, in very good agreement with the value calculated from (2.6), which is 9.5×10^7 m^{-1}. The ψ_0's calculated from the intercept scatter more but are mostly between 80 and 100 mV. These numbers for κ and ψ_0 were used to calculate the solid curve in Fig. 6.1. It gives a good fit with the measured points.

After the film had been draining to a stationary thickness of 60.0 ± 2.5 nm, the scattering angle θ was varied from 54° to 26°. In this range S decreased from 5.02×10^{-6} to 0.27×10^{-6} and S_0 from 13.7×10^{-6} to 0.27×10^{-6}. The ratio S_0/S is plotted as a function of $\Lambda^2(1-\chi)$ in Fig. 6.4, where $\Lambda = 2\pi/K$ is the wavelength of the observed scattering mode and is connected with the scattering angle through (3.11). The curve was calculated from the known values of κ and ψ_0. This again shows that the experiments are wholly consistent with the theoretical equations. The ψ_0's calculated from the equilibrium thicknesses of the films were in the range of 60–90 mV.

C. Van der Waals–London Interactions

1. Measurements

The measurements on van der Waals–London forces are less straightforward than those on the electrical double layer forces. First, the double layer forces must be suppressed: this was achieved by adding so much electrolyte (0.1 M KBr) that $1/\kappa \sim 1$ nm. Then, however, the films do not drain regularly to the new equilibrium thickness but spontaneously form black spots at $h \simeq 100$ nm, which makes it impossible to perform light-scattering measurements on these (inhomogeneous) films.

The black spot formation was discussed in Section IV.D.3. It can be shifted[108] to smaller thicknesses by a faster drainage. By using a narrow rectangular frame (2×1 cm), however, the investigators speeded up the draining by a factor of 5; the black spot formation occurred around $\sim$80 nm, however, which is still too large. To obtain even faster thinning, a more forceful method had to be chosen. It was found that by blowing dry air into the vessel, thinning by evaporation also occurs. This speeded up the thinning of the film by another factor of 5 (see Fig. 6.5) and reduced the thickness of black spot formation to $\sim$50 nm, sufficiently low to make van der Waals force measurements feasible.

The evaporation may have several disturbing effects on the film. First, it serves to increase the concentration of electrolyte, soap, and glycerol (say,

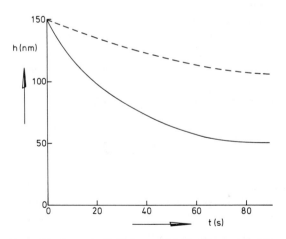

Fig. 6.5. Experimentally determined thickness of the aqueous core of the film versus time for an evaporating film (solid curve) and a normally draining film (dashed curve). (From Ref. 112, courtesy of Academic Press, Inc., New York.)

by a factor of 2). The increased electrolyte concentration compresses the already thin double layer even more, which should have no effect. The increased soap concentration decreases the surface tension, but because the original soap concentration is already close to the critical micel concentration of the soap solution, this effect should be small. The increase in glycerol concentration may change the refractive index of the aqueous core by a few parts per thousand, which is negligible. The evaporation may also lead to some cooling of the film, with a concommittant change in γ. It can be shown, however, that the heat supply from the surroundings is sufficiently fast to make the effect on γ negligibly small.

Also some changes in the measurement procedure were made to increase the response of the interaction forces on S. The measurements were performed in the transmission regime (Fig. 6.2) and very close to the transmitted beam, $\theta \sim 3°$, where the optical factor χ is negligible. This increases the amount of stray light from the diaphragm edges for which, however, a special correction and calibration procedure was developed (for details, see Ref. 112.) Furthermore, a photocounting technique with pulse height discrimination was used to measure the photocurrent, which brought the measuring time of the intensity down below one second, as is necessary for fast-draining films. The outputs of the scattering and thickness measurements were fed into a paper-tape punch.

2. Results and Discussion

Scattering intensities of some hundred films, drawn on the same day, were superimposed and averaged in groups of small h ranges, and processed as indicated above. The results are shown in Fig. 6.6. The measured values were compared with calculations of van der Waals–London forces (see Section II.C.2). The calculations were done for an aqueous film sandwiched by two hexadecane layers using the bulk properties of the hydrocarbon and were cast into the following interpolation formula[59]:

$$\frac{d^2 V_w}{dh^2} = \frac{1}{h^4} \left(\frac{b+ch}{1+dh+eh^2} + q \right) \tag{6.6}$$

The numerical values of b, c, d, e, and q are given elsewhere. The constant q represents the contribution of the zero-frequency term in Dzyaloshinskii's formula, which according to Mitchell and Richmond[116] must be taken equal to zero for films with large values of κh. The values of $d^2 V_w/dh^2$ thus obtained were used to calculate S as shown in Fig. 6.6 for $q=0$ (dashed curve) and $q = -5.4 \times 10^{-22}$ (solid curve). They show a reasonable agreement between theory and experiment within the accuracy of the measurements.

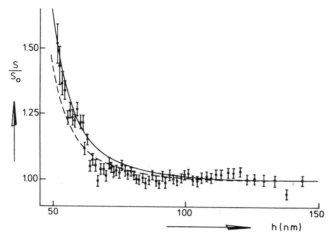

Fig. 6.6. Experimental results for S/S_0 versus h. For comparison, two theoretical curves are shown, based on (6.6): $q = 0$ (dashed), and $q = -5.4 \times 10^{-22}$ (solid). (From Ref. 112, courtesy of Academic Press Inc., New York.)

Here we are measuring attractive forces that tend to induce growing fluctuations of wavelength surpassing a certain limit, as discussed in Section V.A, and some points connected with this circumstance remain to be discussed. From (5.4), here written in a slightly different form,

$$\frac{1}{\tau} = \frac{h^3 \gamma}{24 \eta} K^4 (1 + \mathcal{V})$$

(6.7)

it follows that growing fluctuations (with $\tau < 0$) can occur when $\mathcal{V} < -1$ or $K^2 < -2V''/\gamma$. The most rapidly growing mode exists for $\mathcal{V} = -2$. In the experimental situation ($\theta = 117°$) values down to $\mathcal{V} = -0.3$ could be measured. This implies that ripples with $\mathcal{V} = -2$ will scatter light at an angle of about 1.1° from the transmitted beam, which is outside the range of the detector. It is further of interest to compare τ_d, the relaxation time of the ripples observed by the detector, with τ_m, the time constant of the most rapidly growing ripple. From (6.7) one finds the formula

$$\frac{\tau_d}{\tau_m} = \frac{\mathcal{V}^2}{4(1 + \mathcal{V})}$$

(6.8)

which is depicted in Fig. 6.7 for $\mathcal{V} < -1$. For $\mathcal{V} \simeq -0.3$, this ratio is about 0.03. Thus the observed ripples can still fully relax before the growing mode gains appreciably in amplitude.

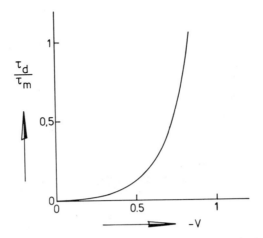

Fig. 6.7.   Ratio of τ_d, the relaxation time of the detected mode, and τ_m, the growing time of the most rapidly growing mode, versus $-\mathcal{V}$ [cf. (6.8)]. (From Ref. 112, courtesy of Academic Press Inc., New York.)

VII. DYNAMIC LIGHT–SCATTERING EXPERIMENTS FROM FREE FILMS

A. Introduction

The light-scattering experiments described in Section VI used the mean intensity, scattered by a thin film, to find the mean square amplitude of the ripples. Now we summarize light-scattering experiments that deal with the dynamics of the ripples.

Because of relaxation of surface elevations, the scattered light has a broadened spectral distribution compared with the incident light. The broadening is too small to be analyzed by the conventional Fabry-Pérot interferometry,[117] however, so the more recent technique of light beating[118, 119] must be used.[120–123] We call this technique intensity fluctuation spectroscopy (IFS).

Basically, in IFS experiments, one can distinguish between two limiting configurations.[121] First, the so-called heterodyne detection scheme, in which the scattered light is "mixed" with a coherent local oscillator signal on the cathode of the photoelectric detector. The local oscillator is often a part of the incident laser light, but this is not necessary. In this configuration the output of the detector contains an exact replica of the optical spectrum (spectrum of the EM *field*). Second, the so-called homodyne or self-beat detection scheme, in which only the scattered light is detected. In this configuration the detector output contains the spectrum of the scattered *intensity*.

For comprehensive reviews of this light-scattering technique, together with a number of applications in different fields, see the monographs of Chu[124] and of Berne and Pecora,[125] and the already cited review article of Cummins and Swinney.[121] For further details, see also Refs. 126 and 127.

The homodyne scheme is the most frequently utilized form of light-beating spectroscopy. In the field of interfaces, however, the heterodyne detection scheme is most often applied. To give some examples of studies on interfaces that have been carried out during the last 15 years, we mention some papers that we believe to be representative.

After the pioneering work of Katyl and Ingard[117, 120] and Bouchiat et al.[122, 123] the study on surfaces of pure liquids was continued and extended by Langevin[128] and Hård et al.[129] These studies suggest that after careful deconvolution of the spectra, the light-scattering results can be interpreted in terms of the bulk values for the surface tension and shear viscosity. No high-viscosity surface region was found, as was claimed by McQueen and Lundström.[130] There is only one exception, in the experiments of Bird and Hills[131] on mercury. A large discrepancy between the viscosity from the light-scattering experiments and the literature value is reported. It is worthwhile to mention also the IFS work on liquid-vapor interfaces near the critical point,[132–135] the work on liquid interfaces covered by mono-layers,[105, 130, 136] and the experiments on the surfaces of liquid crystals[137] in search of viscoelastic behavior.

Intensity fluctuation spectroscopy was used in our laboratory to study the dynamic behavior of surface ripples on thin liquid films. Both squeez-ing[23, 138] and bending[139, 140] modes were examined. To our knowledge one other group[141] of researchers has obtained dynamic light-scattering data from thin soap films; but as far as we know, nothing has been published in the official literature. Also some experiments were reported on lipid bi-layers in water.[143]

In the next section we give a brief summary of the relevant formulas.

1. Correlation Functions and Power Spectra

In Section V.B dispersion relations were given for the squeezing and bending modes, assuming a plane wave solution for the amplitude $\zeta(K, t)$ of both modes:

$$\zeta(K, t) = \hat{\zeta} \exp[i(Kx - \omega t)] \tag{7.1}$$

The dispersion relation can be checked experimentally by IFS experiments. The optical geometry determines a certain K, whereas the relaxation of the mode is given by ω.

It will be clear that for thermal ripples, only statistical properties for the amplitude $\hat{\zeta}$ can be derived. For that purpose we define an autocorrelation function $\phi_s(K, \tau)$ for the squeezing mode and a similar function $\phi_b(K, \tau)$ for the bending mode. These correlation functions are for stationary processes given by

$$\phi_s(K, \tau) = \langle \zeta_s^*(K, 0) \zeta_s(K, \tau) \rangle$$

and

$$\phi_b(K, \tau) = \langle \zeta_b^*(K, 0) \zeta_b(K, \tau) \rangle \tag{7.2}$$

where $\zeta_s(K, t) = \zeta_I(K, t) - \zeta_{II}(K, t)$ for the squeezing mode, $\zeta_b(K, t) = \zeta_I(K, t) + \zeta_{II}(K, t)$ for the bending mode, the asterisk means the complex conjugate quantity, and the angular brackets denote an ensemble average, as before.

These correlation functions can be calculated by using the Onsager regression hypothesis, together with the hydrodynamics as presented in Section V.B. Using (5.34) and (5.35), derived for the long-wavelength limit and the no-slip condition at the interfaces, we have for the squeezing mode:

$$\phi_s(K, \tau) = \langle |\zeta_s(K, 0)|^2 \rangle \exp[-\Gamma_s \tau] \tag{7.3}$$

and for the bending mode:

$$\phi_b(K, \tau) = \langle |\zeta_b(K, 0)|^2 \rangle \exp[-\Gamma_b \tau] \cos \omega_b \tau \tag{7.4}$$

where

$$\Gamma_s = -\operatorname{Im}(\omega) = \frac{K^2 h^3}{24\eta} (\gamma K^2 + 2V''(h)), \tag{7.5}$$

$$\Gamma_b = \frac{K^2 \eta}{2\rho} \quad \text{and} \quad \omega_b = K \left(\frac{2\gamma}{\rho h} - \frac{K^2 \eta^2}{4\rho^2} \right)^{1/2} \tag{7.6}$$

From (7.3) and (7.4) one sees that the squeezing mode is purely diffusive, whereas the bending mode is propagating.

The correlation functions for the amplitude fluctuations can now be connected with the autocorrelation function for the scattered field A (see Section IV.A). This function $g^{(1)}(K, \tau)$ is defined as:

$$g^{(1)}(K, \tau) = \langle A^*(K, 0) A(K, \tau) \rangle \tag{7.7}$$

The time-dependent part of the scattered field $A(K, \tau)$ is proportional to the actual amplitudes of the squeezing mode and bending mode, as given by (4.23). Using ϕ_s and ϕ_b as given by (7.3) and (7.4), and given the independence of the modes, one derives for the intermediate scattering function $g^{(1)}$ from (7.7)

$$g^{(1)}(K, \tau) = \langle I_s(t) \rangle \exp[-\Gamma_s \tau] + \langle I_b(t) \rangle \exp[-\Gamma_b \tau] \cos \omega_b \tau \quad (7.8)$$

where $\langle I_s \rangle$ and $\langle I_b \rangle$ are the mean scattered intensities as calculated in Section IV. The correlation function in the heterodyne detection scheme is proportional to $g^{(1)}(K, \tau)$.

To calculate the autocorrelation function $g^{(2)}(K, \tau)$ of the scattered *intensity*, one must assume that the scattering process obeys Gaussian statistics. If this assumption holds, one has[121]

$$g^{(2)}(K, \tau) = 1 + \alpha |g^{(1)}(K, \tau)|^2, \quad (7.9)$$

where α depends on the optical detection configuration. If only the fluctuating component ΔI of the scattered intensity is measured (as in our case), only the time-dependent part of $g^{(2)}$ is of interest, and this is given by

$$\phi_{\Delta I}(K, \tau) = \alpha \left\{ \langle I_s(t) \rangle^2 \exp[-2\Gamma_s \tau] + 2\langle I_s(t) \rangle \langle I_b(t) \rangle \exp[-(\Gamma_b + \Gamma_s)\tau] \right.$$
$$\left. \times \cos \omega_b \tau + \frac{\langle I_b(t) \rangle^2}{2} \exp[-2\Gamma_b \tau](1 + \cos 2\omega_b \tau) \right\} \quad (7.10)$$

As was shown in Section IV, the scattered intensities of the squeezing and bending modes are of the same order of magnitude in the reflection region and thickness range of interest. This means, as (7.10) indicates, that also in the homodyne scheme the frequency ω_b can be measured. That is, the squeezing mode acts as a local oscillator for the bending mode in this case. As the experimental section (Section VII.B) shows, the detection of the bending mode is performed in the frequency domain, that is, by measuring the power spectrum of the photomultiplier voltage fluctuations instead of time-autocorrelation function.

According to the Wiener-Khinchine theorem, the power spectrum of a fluctuating quantity is given by the (time) Fourier transform of the auto-correlation function. So for the bending mode one derives from (7.4), for the power spectral density $P_b(K, \omega)$:

$$P_b(K, \omega) = \frac{\langle |\zeta_b(K,0)|^2 \rangle}{2\pi} \left\{ \frac{\Gamma_b}{\Gamma_b^2 + (\omega - \omega_b)^2} + \frac{\Gamma_b}{\Gamma_b^2 + (\omega + \omega_b)^2} \right\} \quad (7.11)$$

B. Experiments

1. Experimental Setup

The apparatus used for the dynamic scattering experiments is an improved version of the Rijnbout apparatus, mentioned in Section VI.A.1. The main parts of the apparatus appear in Fig. 7.1, and it is described in detail in Ref. 138. The apparatus depicted in Fig. 7.1 is really isolated from the surroundings, whereas the Rijnbout apparatus has an open connection with the surroundings. The way of isolating the system, and the mechanism for

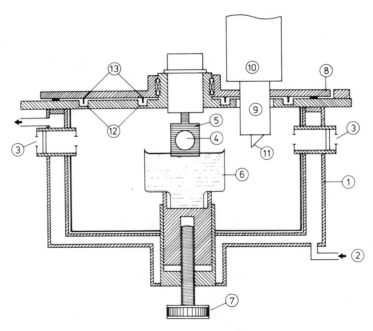

Fig. 7.1. Schematic view of the light-scattering apparatus: 1, stainless steel vessel; 2, inlet for thermostatting fluid; 3, windows to illuminate the film; 4, film; 5, ground glass frame; 6, glass vessel with "soap" solution; 7, mechanism to move vessel up and down; 8, turnable part of the cover; 9, tube; 10, photomultiplier; 11, prism; 12, concentric grooves filled with high-viscosity paraffinic oil; 13, concentric rims. (From Ref. 138, courtesy of American Institute of Physics.)

moving the reservoir with the soap solution up and down without disturbing the pressure inside the apparatus, are the main differences between the apparatus described here and the one mentioned in Section VI.A.1. In general, for thin film studies it is very important that the system be vapor tight.

For the thermostatting of the light-scattering apparatus, see Section VI.A.1. The resulting long-term temperature stability inside the apparatus is $\pm 0.002°C$.

In Fig. 7.2 the optical and detection part of the experimental setup is schematically drawn. The film is illuminated by an argon-ion laser about 10 mm above the surface of the soap solution in the reservoir. The laser operates in the TEM_{00} mode (multimode) at 514.5 or 488 nm, at a power between 100 and 300 mW. The light is polarized perpendicular to the plane of incidence. The lens L_1, with a focal distance of 60 cm, is placed about 40 cm from the film so that the incident beam is slightly focused on the film. The diaphragm system D_1, which is positioned after the incoming window (see Fig. 7.1), is used to remove unwanted stray light. The specularly reflected light is measured by means of a photodiode, which can be set at fixed positions such that $\theta_0 = 45°, 60°$, or $75°$.

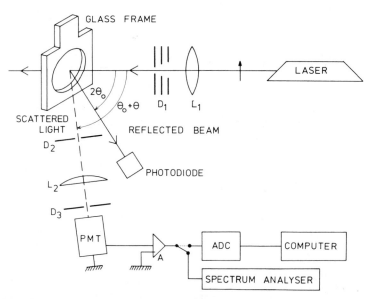

Fig. 7.2.   Optical and detection part of the experimental setup: L_1, lens ($f = 600$ mm); D_1, diaphragm system; D_2, pinhole ($\phi = 1.5$ mm); L_2, lens ($f = 48$ mm); D_3, pinhole ($\phi = 0.2$ mm); A, preamplifier; ADC, analogue-to-digital converter (200 kHz); PMT, photomultiplier tube.

The scattered light is detected by a photomultipler tube (PMT). The area that is "seen" by the PMT encloses the spot of the laser beam on the film. The PMT signal is fed either to a correlator or to a spectrum analyzer. The correlator consists of a fast analogue-to-digital converter and a minicomputer, programmed to calculate the time autocorrelation function of the signal. The correlation function, calculated in this way, is proportional to the time autocorrelation function of the fluctuations in the scattered intensity.

The output of the spectrum analyzer is proportional to the power spectrum of the fluctuations in the scattered intensity. Although the power spectrum of the fluctuations contains the same information as the time autocorrelation function, each has its own merits, making it necessary to use them both to cover a broad spectral range (see Section VII.C.4).

To conclude this section, we remark that the whole optical system, including the laser and light-scattering apparatus, is placed on a vibration isolation system to eliminate exterior vibrations.

2. Materials and Methods

The solutions used contain as standard components hexadecyl trimethylammonium bromide (9.5×10^{-4} mol/dm^3) plus glycerol (8.4 wt%). To increase the ionic strength, KBr was added. The refractive index of the solution appeared to be $n = 1.345$. The surface tension γ_0 of the solutions was measured with a stalagmometer (drop weight method), and the shear viscosity η_0 by an Ubelohde viscosimeter at the same temperature that was used for the light-scattering experiments. The results are given in Table I.

TABLE I

Bulk Solution Properties and Equilibrium Thicknesses (h_2) and Results of Dynamic Light-Scattering Experiments for the Two Systems Described in Text

Ionic Strength[a] (10^{-3} mol/dm^3)	T (°K)	γ_0 (mN/m)	η_0 (mPa-sec)	h_2 (nm)	$\frac{\gamma}{\eta}$ (m/sec)			γ (mN/m)
					Bulk	Drainage	Equilibrium	Bending mode
4.8	297.6	33.8	1.12	35.1	30.2 ± 0.2	30 ± 2[b]	20 ± 1[c]	—
1.9	300.0	35.7	1.07	44.4	33.3	30.8 ± 1.2[d]	20 ± 1[e]	32

[a] Concentration of HDTAB plus the concentration of KBr.
[b] Average value of 10 different films.
[c] Average value of 14 experiments.
[d] Average value of four different films.
[e] Average value of five experiments.

The scattering apparatus is made dustfree by blowing through dustfree air. Then the vapor equilibrium inside the apparatus is established by waiting at least 12 hr. Because of the need to have a dustfree system plus a well-defined vapor equilibrium, the system should be tightly closed. Just before a film is drawn, the surface of the solution in the reservoir is refreshed by supplying a few milliliters of stock solution. Once a film is made, the drainage process starts and continues until the equilibrium thickness is reached. This takes about 12 hr. During the drainage process the reflected intensity is recorded, to find the equivalent water thickness h_w.

The spectral distribution of the scattered light was measured during the drainage of the film at constant angles θ and θ_0 at thicknesses below about 110 nm. Once the film has reached a constant thickness, so that h is constant, the scattering angle θ is varied.

C. Experimental Results

A calculation of the relaxation times Γ_s^{-1} and Γ_b^{-1} and the frequency ω_b, from (7.5) and (7.6), and realistic values for the different parameters, shows that the squeezing mode exhibits its relevant features in the millisecond region, whereas for the bending mode this display occurs in the microsecond region. This very fact makes it possible to study the squeezing and bending modes separately. The properties of the squeezing mode can be measured either at the transmission side or at the reflection side of the film (see Section VI). The bending mode scattering, however, only has enough intensity in the reflection region.

The experimental results for the squeezing mode are mainly taken from Ref. 138.

1. Squeezing Mode

The squeezing mode experiments were carried out using the homodyne detection scheme with scattering angles within $\pm 20°$ of either the reflected beam ($\theta_0 = 60°$) or the transmitted beam. A small region of $\pm 4°$ around either the transmitted or reflected beam was excluded because here a gradual, but not exactly reproducible, transition from homodyne to heterodyne detection is observed.

In the plane of incidence the wave number K of a Fourier component is given, according to (3.11), by

$$K = \frac{2\pi}{\lambda_1} |\sin \theta_0 - \sin \theta|^{-1} \tag{7.12}$$

The observed correlation functions were analyzed according to a single

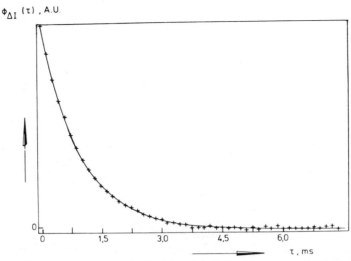

$\phi_{\Delta I}(\tau)$, A.U.

Fig. 7.3. Measured autocorrelation function (+) of the photocurrent fluctuations. Wavelength of the thickness fluctuation (squeezing mode) $\Lambda = 4.90\ \mu$m. The curve indicates nonlinear least-squares fit to the experimental data according to (7.13). Film was drawn from solution with an ionic strength of 1.9×10^{-3} mol/dm^3.

exponential plus background by minimizing the function

$$\sum_{i=1}^{n}\left[A \exp\left(\frac{-\tau_i}{\tau_R}\right) + B - \phi_{\Delta I}(\tau_i)\right]^2 \tag{7.13}$$

where $n = 25$ or 50, $\tau_i = i\tau$, τ is the sample interval time, and $\phi_{\Delta I}(\tau_i)$ is the (measured) correlation function at τ_i. A typical correlation function, with its computer fit, is found in Fig. 7.3. According to (7.10) τ_R in (7.13) must be interpreted by

$$(2\tau_R)^{-1} = \Gamma_s = \frac{K^2 h^3}{24\eta}\left(\gamma K^2 + 2V''\right) \tag{7.14}$$

The other three terms on the right-hand side of (7.10) can be neglected in the analysis of the relaxation of the squeezing mode because $\Gamma_b \gg \Gamma_s$ (Γ_s is typically 100 to 5000 sec^{-1}, whereas Γ_b is in the order of 10^{+6} sec^{-1}; see Section VII.C.4).

2. Squeezing Mode: Drainage Experiments

As already has been pointed out, it takes about 12 hr for a film to reach its equilibrium thickness. At first, the thinning process goes very fast, but below a thickness of about 120 nm the drainage process is slow enough to permit measuring correlation functions under "constant" thickness conditions. Typically, the relative thickness changes, during the measurement of one correlation function, were 10% at $h \simeq 100$ nm and 1% near the equilibrium thickness.

In the thickness region where $\gamma K^2 \gg -2V''$, the quantity $(2\tau_R)^{-1/3}$ should, according to (7.14), behave as a linear function of h at constant K. In Fig. 7.4 some results are depicted for three different scattering angles, showing that the measured data can be fitted to a straight line above a thickness of about 55 nm. According to (7.14) the slope of the lines should be equal to $[\gamma K^4/(24\eta)]^{1/3}$ when $\gamma K^2 \gg -2V''$. Thus knowing K, a value for the observed ratio γ/η can be calculated. The resulting value of γ/η obtained from different films, and the value of this ratio for the bulk solution, are given in Table I. Below about 55 nm the relation between $(2\tau_R)^{-1/3}$ and h is no longer linear, as can be seen in Fig. 7.4 especially for the smallest K. This implies that the value of $2V''$ can no more be neglected with respect to γK^2: the influence of the interactions shows up. Since the curve increases with decreasing h, the value of V'' must be positive. We also describe here a series of experiments carried out on films drawn from a solution of hexadecyl trimethylammonium bromide with an ionic strength

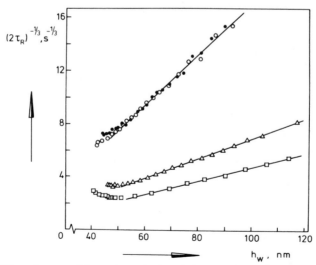

Fig. 7.4. The cube root of the reciprocal relaxation time of the squeezing mode for draining soap films as a function of their thickness h_w at three different wavelengths: circles, $\Lambda = 4.14$ μm; triangles, $\Lambda = 8.22$ μm; squares, $\Lambda = 11.18$ μm. Solid curves are the least-squares fits through the measured data above a thickness of 55 nm. Film was drawn from solution with an ionic strength of 4.8×10^{-3} mol/dm^3.

of 1.9×10^{-3} mol/dm³. Both squeezing and bending mode experiments have been performed on these films. The results of the drainage experiments and the bulk properties of this solution are also given in Table I.

3. Squeezing Mode: Equilibrium Experiments

If the film has reached a constant thickness, the wave number dependence of the relaxation time can be studied. From a rearrangement of (7.14), one finds that if the (experimental) value of $12/(\tau_R h^3 K^2)$ is plotted against K^2, a straight line is expected. The slope of this line equals γ/η and the intercept with the $K^2 = 0$ axis is $2V''/\eta$. Figure 7.5 gives a typical example of such an experiment at constant thickness. For h, the thickness h_2 of the aqueous core is taken. The correction term amounts 3.5 nm (i.e., $h_2 = h_w - 3.5$ nm). The averaged value of γ/η, obtained from five measurements on three different films is also given in Table I. As one sees, the ratio is lower than for the bulk quantities. It is important, however, to know the quantities γ and η separately because otherwise no reliable data concerning the interaction term V'' can be extracted from the experiments at constant thickness.

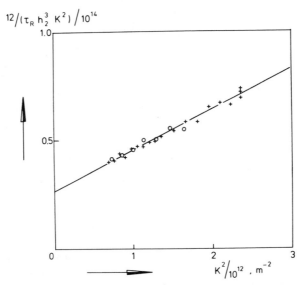

Fig. 7.5. Values of $12(\tau_R h_2^3 K^2)^{-1}$ versus K^2 for a film at constant thickness; $h_2 = 46.0$ nm; circles and crosses denote experimental data measured at angles $\theta > \theta_0$ and $\theta < \theta_0$, respectively. The solid curve is a least-squares fit through all data points. Film was drawn from solution with an ionic strength of 1.9×10^{-3} mol/dm³.

As the next section indicates, the bending mode experiments provide us with a value for γ. In Section VII.D a combination of squeezing and bending mode results on the same films is discussed to obtain values for V''.

4. Bending Mode

Recently we succeeded in detecting the bending mode by means of IFS.[139, 140] The properties of the bending mode have been studied by measuring the power spectrum of the PMT voltage fluctuations. There are several reasons for working in the frequency domain instead of in the time domain. First, as a calculation of the oscillation frequency $\omega_b/2\pi$ shows, we have to deal with rather high frequencies. Most commercially available correlators, however, do not have a high enough resolution. Second, the determination of a peak in the power spectrum is more accurate than the determination of a period of a cosine in the time autocorrelation function.

The frequency shift with respect to $\omega=0$ in the power spectrum is approximately* given by $f_b(K)=\omega_b(K)/2\pi$, whereas the full linewidth at half-maximum is given by $K^2\eta/2\pi\rho$, when the no-slip condition and the long-wavelength limit applies [see (7.6)]. Using bulk values for γ, η, and ρ (Table I), one expects shifts in the megahertz region and linewidths in the kilohertz region for the K and h values under study.

Though the scattered intensity of the bending mode at least at the reflection side is comparable to the intensity scattered by the squeezing mode, the signal-to-noise ratio becomes poor for large scattering angles because of the high-frequencies involved. The range of scattering angles is confined about 2° to about 6° at both sides of the specularly reflected beam. This means that we gradually move over from the heterodyne to the homodyne detection scheme. Very close to the reflected beam, the scattered light is mixed with part of the specularly reflected light that acts as the local oscillator. Figure 7.6 shows an example of a heterodyne spectrum. As already has been remarked, the frequency shift (actually a Doppler shift) of the bending mode can be measured using the homodyne scheme, also. Figure 7.7 shows an example of such a spectrum.

Similar spectra have been measured at different scattering angles in the range $2° \leq |\theta-\theta_0| \leq 6°$, corresponding to wavelengths 30.4 μm $\leq \Lambda \leq 10.3$

---

*In fact the frequency shift is

$$\frac{\omega_b}{2\pi}\left(1-\frac{K^4\eta^2}{4\rho^2\omega_b^2}\right)^{1/2}$$

but in the experimental situation we always have $\omega_b^2 \gg K^4\eta^2/4\rho^2$.

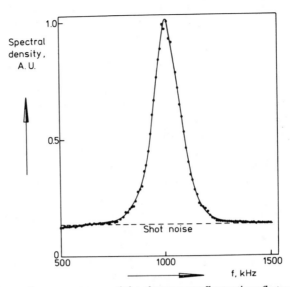

Fig. 7.6. Measured power spectrum of the photocurrent fluctuations (heterodyne scheme): wavelength of the surface wave (bending mode) $\Lambda = 30.9\ \mu$m, and thickness of the film ($h_2 + 2h_1$)=47.5 nm. The composition of the solution is the same as in Fig. 7.5. (From Ref. 139, courtesy of North Holland Publishing Company, Amsterdam.)

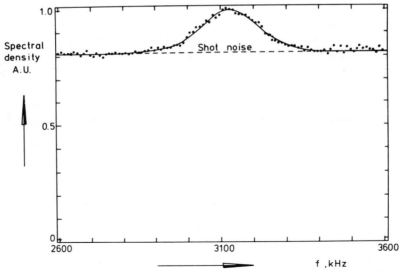

Fig. 7.7. Measured power spectrum of the photocurrent fluctuations (homodyne scheme): wavelength of the surface wave (bending mode) $\Lambda = 11.1\ \mu$m, and the thickness of the film ($h_2 + 2h_1$)=47.5 nm. The composition of the solution is the same as in Figs. 7.5 and 7.6.

μm when $\theta_0 = 60°$. In Fig. 7.8 the resulting frequency shifts, $f_b(K)$, measured on two films with the same equilibrium thickness, are plotted as a
function of the wave number $K(=2\pi/\Lambda)$. It is clear that a linear relationship between $f_b(K)$ and K exists in the range of K values investigated.
According to

$$f_b(K) = \frac{K}{2\pi}\left(\frac{2\gamma}{\rho h}\right)^{1/2} \tag{7.15}$$

the slope of the line in Fig. 7.8 should be equal to $[\gamma/(2\pi^2\rho h)]^{1/2}$. A linear
least-squares fit through the measured data yields a slope of 5.8 m/sec.
Taking for ρ a value of 1018 kg/m^3 (a $1M$ glycerol solution) and for h a
mean value of 47.5 nm $(=h_2 + 2h_1;$ h_1 is the thickness of the monolayer),
one finds $\gamma = 32$ mN/m.

No attempt was made to compare the widths of the peaks (which are in
the order of 100 kHz) with the theory. The reason for this was twofold. First,
since one is dealing with a propagating mode, so-called instrumental
broadening effects are important.[128] The divergence of the laser beam, the
uncertainty of the incident k_1, and the finite solid angle of observation lead
to the detection of many modes, with different K's, that all have different

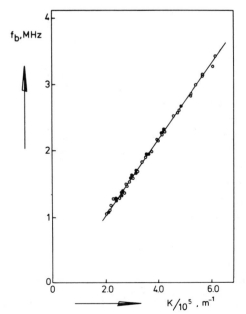

Fig. 7.8. Measured values of frequency shifts f_b against K for two different films drawn from a solution with the
same composition as in Figs. 7.5, 7.6, and
7.7. The mean film thickness $h_2 + 2h_1 =$
47.5 nm. The solid line is a least-squares
fit through all data points. (From Ref.
139, courtesy of North Holland Publishing Company, Amsterdam.)

frequency shifts. To extract reliable information from the widths of the peaks requires a very tedious and difficult deconvolution procedure. Second, as a future paper[99] will demonstrate, the widths of the peaks mainly result from the properties of the vapor phase surrounding the films. Although the frequency shifts are also affected by this effect, it is nearly negligible for films with thicknesses $\gtrsim 50$ nm and wave numbers K as described here.

D. Discussion

As far as the drainage experiments on the squeezing mode are concerned, we may conclude that the ration γ/η obtained in the light-scattering experiments shows, within the experimental error, a reasonable agreement with the value of this ratio found from the bulk quantities γ and η (see Table I). The application of a three-parameter fit [see (7.13)] instead of a two-parameter fit[138] revealed that the correlation function could be analyzed more precisely. It appeared that a small, but systematic background was present in the measured correlation functions. The effect of this background turned out to be more pronounced at shorter interval times when the two-parameter fit was used. Thus the difference of γ/η for larger K's compared to smaller K's is no longer obvious. We may also conclude that although a net flow of liquid out of the film occurs during the drainage process, this flow seems to have no influence on the relaxation of the surface ripples.

The value for γ/η obtained in the squeezing mode experiments at constant thickness, however, shows a significant discrepancy with the bulk value of γ/η (see Table I). As we have seen in the preceding section, the bending mode experiments yield a value for the surface tension of the film. Although it is not obvious that a film tension measured in megahertz frequency range (bending mode) equals the value of this quantity obtained in the kilohertz frequency region (squeezing mode), it could give us at least an estimate for the value to be used in the interpretation of the squeezing mode results. Using the value for γ and the bulk value for η, one finds γ/η = 30 m/sec, so we have just a partial explanation for the difference found in the experiments at constant thickness.

There are a number of possible reasons for the observed differences. The theory used to interpret the experimental results may have been incomplete, or perhaps the assumptions made in the derivations were not met in the experimental situation. As remarked in Section V.B, the more general and systematic electrohydrodynamic theory[89] gives about the same expression for the relaxation time τ_R as (7.5), in the long-wavelength limit (lwl) and for no-slip conditions. Moreover, as can be seen clearly from Fig. 7.5, the K-dependence of the experimental data shows the behavior as predicted by (7.5), which is in favor of the use of this equation.

A theoretical explanation for the increase of the viscosity η could be found in the so-called electroviscous effect. It is well known that an electrolyte solution streaming between two charged walls shows an increase of the apparent shear viscosity. Assuming that the results obtained for plane parallel channels in a steady state by Levine et al.[142] may be used for the film situation, it was found that a maximum increase of about 20% can be expected for the viscosity of the solution inside the film compared to the bulk value. This electroviscous effect is expected to be important only in equilibrium films where an overlap of the electrical double layers occurs; but nevertheless this phenomenon cannot explain the full discrepancy between theory and experiment.

Concerning experimental problems, we may conclude that there is no good reason to believe that an unexpected mixing occurs between reflected light or unwanted stray light and the scattered light.[138] So the experiments, carried out in the K region described here, are really free of homodyne-heterodyne problems. Another possible source for the discrepancy between theory and experiment may be the use of the aqueous core thickness h_2 in the interpretation of the equilibrium experiments. Whether this is permitted is not clear at the moment.

Now we discuss briefly the results obtained from the relaxation of the bending mode. As a comparison with the bulk value for the surface indicates (Table I), there is a discrepancy between this surface tension and the surface tension obtained from the bending mode experiments. It should be realized, however, that the film tension value is obtained from experiments on a time scale totally different from the time scale encountered in obtaining the bulk value. This implies the possibility of processes that influence both values in a completely different way, like adsorption and desorption phenomena of surface active materials at the interface.

A trivial explanation can be found in the presence of contamination. It is very well known in the field of surface and interface research that traces of highly active components, other than the added surfactant, can cause serious problems. It is very hard to rule them out fully, especially when one uses quite aged interfaces. The films described here were always older than 12 hr; thus the results could be obscured by aging effects. Whether this is the case is now under investigation.

To conclude, we present the results for the interaction term V'' as a function of h. Taking for the γ the value as found in the bending mode experiments, we can calculate from the slope of the squeezing mode experiments at constant thickness a value for the shear viscosity η. Now we are

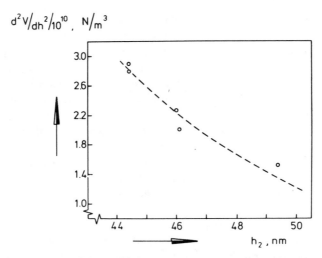

Fig. 7.9. Plot of d^2V/dh^2 versus thickness of the aqueous core h_2, showing theoretical curve according to (6.5) and (6.6). For the surface potential ψ_0 a value of 83 mV was taken. Composition of the solution from which the films were drawn is the same as in Figs. 7.5 to 7.8.

able to find a numerical value for V'' from the intercept that equals $2V''/\eta$ [see (7.5)]. In Fig. 7.9 the resulting values for V'' are plotted as a function of h_2, the thickness of the aqueous core. Also the theoretical curve for V'' is given in Fig. 7.8 as calculated from (6.5) and (6.6) for the electrical double layer and the van der Waals–London contribution to V'', respectively. As one can see, a value of $\psi_0 = 83$ mV yields a reasonable agreement between theory and experiment.

Acknowledgments

We thank Mrs. M. Uit de Bulten and Miss S. Abrams for typing the manuscript and Mr. W. den Hartog for drawing the illustrations.

One of us (J. G. H. Joosten) acknowledges the Foundation for Fundamental Research of Matter (F.O.M.) for supporting in part this research with financial support from the Netherlands Organization for Pure Research (Z.W.O.).

References

1. Color pictures are to be found, for example, in K. J. Mysels, K. Shinoda, and S. Frankel, *Soap Films*, Pergamon Press, New York, 1959; J. Th. G. Overbeek, *Proc. Fourth Int. Congr. Surface Activity*, Vol. 2, Gordon and Breach, New York, 1967, p. 19.
2. I. Newton, *Opticks*, Book II, Part I, obs. 17–21, 1704.
3. B. V. Derjaguin, A. S. Titijevskaja, I. I. Abricossova, and A. D. Malkina, *Discuss. Faraday Soc.*, **18**, 24 (1954).

4. A. J. de Vries, *Rec. Trav. Chim.*, **77**, 383 (1958).
5. H. J. Schulze and C. Cichos, *Z. Phys. Chem.*, **251**, 145 (1972).
6. F. Renk, P. C. Wayner, Jr., and G. M. Homsy, *J. Colloid Interface Sci.*, **67**, 408 (1978).
7. J. Lyklema and K. J. Mysels, *J. Am. Chem. Soc.*, **87**, 2539 (1965).
8. A. Scheludko and D. Exerowa, *Kolloid Z.*, **168**, 24 (1960).
9. H. Sonntag, J. Netzel, and H. Klare, *Kolloid Z.*, **211**, 121 (1966).
10. E. M. Duyvis, *Thesis*, University of Utrecht, the Netherlands, 1962.
11. K. J. Mysels and M. N. Jones, *Discuss. Faraday Soc.*, **42**, 42 (1966).
12. A. Scheludko, *Proc. K. Ned. Akad. Wet.*, Ser. B, **65**, 76 (1962).
13. H. M. Princen and S. G. Mason, *J. Colloid Interface Sci.*, **29**, 156 (1965).
14. K. J. Mysels, H. F. Huisman, and R. I. Razouk, *J. Phys. Chem.*, **70**, 1339 (1966).
15. J. H. Clint, J. S. Clunie, J. R. Tate, and J. F. Goodman, *Nature* (*London*), **223**, 291 (1969).
16. A. Prins, *J. Colloid Interface Sci.*, **29**, 177 (1969).
17. H. M. Princen, *J. Phys. Chem.*, **72**, 3342 (1968).
18. J. A. de Feijter and A. Vrij, *J. Colloid Interface Sci.*, **64**, 269 (1978).
19. A. Scheludko, B. Radoev, and T. Kolarov, *Trans. Faraday Soc.*, **64**, 2213 (1968).
20. D. A. Haydon and J. L. Taylor, *Nature* (*London*), **317**, 739 (1968).
21. A. Vrij, in M. Kerker, Ed., *Interdisciplinary Conference on Electromagnetic Scattering*, Pergamon Press, London 1963, p. 387.
22. A. Einstein, *Ann. Phys.*, **33**, 1275 (1910).
23. H. M. Fijnaut and A. Vrij, *Nature* (*London*), **246**, 118 (1973).
24. C. V. Boys, *Soap Bubbles and the Forces Which Mould Them*, Society for Promoting Christian Knowledge, London; E. and J. B. Young, New York, 1890. Reprinted in Doubleday Anchor Books, Garden City, NY, 1959.
25. K. J. Mysels, K. Shinoda, and S. Frankel, *Soap Films*, Pergamon Press, New York, 1959.
26. C. Isenberg, *The Science of Soap Films and Soap Bubbles*, Tieto Ltd. Clevedon, Avon, England, 1978.
27. J. T. G. Overbeek, *J. Phys. Chem.*, **64**, 1178 (1960).
28. A. Scheludko, *Adv. Colloid Interface Sci.*, **1**, 391 (1967).
29. B. V. Derjaguin, Ed., *Research in Surface Forces*, Vols. 2–4, Consultants Bureau, New York, 1972.
30. J. A. Kitchener, "*Foams and Free Liquid Films*," in J. F. Danielli, K. G. A. Pankhurst, and A. C. Riddiford, Eds., *Recent Progress in Surface Science*, Vol. 1, Academic Press, New York, 1964, p. 51.
31. J. S. Clunie, J. F. Goodman, and B. T. Ingram, "*Thin Liquid Films*," in E. Matijević, Ed., *Surface and Colloid Science*, Vol. 3, Wiley-Interscience, New York, 1971, p. 167.
32. J. A. Mann, Jr., and K. C. Porzio, in *International Review of Science, Physical Chemistry*, Ser. Two, Vol. 7, M. Kerker, Ed., *Surface Chemistry and Colloids*, Butterworths, London, 1975, p. 47.
33. R. Buscall and R. H. Ottewill, "*Thin Films*," in D. H. Everett, Ed., *Specialist Periodical Reports*, *Colloid Science*, Vol. 2, Chemical Society, London, 1975, pp. 191–245.
34. E. J. Verwey and J. T. G. Overbeek, *Theory of the Stability of Lyophobic Colloids*, Elsevier, Amsterdam, 1948.
35. M. J. Sparnaay, *The Electrical Double Layer*, Pergamon Press, Oxford, 1972.
36. B. V. Derjaguin, Ed., *Research in Surface Forces*, Consultants Bureau, New York, for example, Vol. 2, 1972, pp. 75, 84, 94.
37. A. Sanfeld, *Introduction to the Thermodynamics of Charged and Polarized Layers*, Wiley-Interscience, London, 1968.
38. D. Henderson and L. Blum, *J. Chem. Phys.*, **69**, 5441 (1978).

39. S. Levine and C. W. Outhwaite, *J. Chem. Soc., Faraday Trans.* 2, **74**, 1670 (1978).
40. D. Henderson, L. Blum, and W. R. Smith, *Chem. Phys. Lett.*, **63**, 381 (1979).
41. M. Fixman, *J. Chem. Phys.*, **70**, 4995 (1979).
42. D. Henderson and L. Blum, *J. Electroanal. Chem.*, **93**, 151 (1978).
43. G. M. Torrie and J. P. Valleau, *Chem. Phys. Lett.*, **65**, 343 (1979).
44. J. de Boer, in C. Prins, Ed., *Proceedings of the Van der Waals Centennial Conference on Statistical Mechanics*, North-Holland, Amsterdam, 1974, p. 17; see also *Physica*, **73**, 1 (1974).
45. M. J. Klein, in C. Prins, Ed., *Proceedings of the Van der Waals Centennial Conference on Statistical Mechanics*, North Holland, Amsterdam, 1974, p. 28; also in *Physica*, **73**, 28 (1974).
46. W. H. Keesom, *Proc. Akad. Sci. Amsterdam*, **18**, 636 (1915); **23**, 939 (1920).
47. P. Debye, *Phys. Z.*, **21**, 178 (1920); **22**, 302 (1921).
48. F. London, *Z. Phys.*, **63**, 245 (1930).
49. H. Kallmann and M. Willstätter, *Naturwissenschaften*, **20**, 952 (1932).
50. R. S. Bradley, *Phil. Mag.*, **13**, 853 (1932).
51. H. C. Hamaker, *Physica*, **4**, 1058 (1937).
52. J. H. de Boer, *Trans. Faraday Soc.*, **32**, 21 (1936).
53. H. B. G. Casimir, *Proc. K. Ned. Akad. Wet.*, **51**, 793 (1948).
54. E. M. Lifshitz, *Sov. Phys., J. Exp. Theor. Phys.*, **2**, 73 (1956).
55. I. E. Dzyaloshinskii, E. M. Lifshitz, and L. P. Pitaevskii, *Adv. Phys.*, **10**, 165 (1961).
56. J. Mahanty and B. W. Ninham, *Dispersion Forces*, Academic Press, London, 1976.
57. D. Langbein, *Theory of Van der Waals Attraction*, Springer-Verlag, Berlin, 174.
58. V. A. Parsegian and B. W. Ninham, *Nature* (*London*), **224** 1197 (1969); *J. Chem. Phys.*, **52**, 4578 (1970).
59. W. A. B. Donners, J. B. Rijnbout, and A. Vrij, *J. Colloid Interface Sci.*, **60**, 540 (1977).
60. J. Mahanty and B. W. Ninham, *Dispersion Forces*, Academic Press, London, 1976, Ch. 7.
61. B. V. Derjaguin and M. M. Kusakov, *Acta Phys. Chim. USSR*, **10**, 25, 153 (1939); B. V. Derjaguin and N. V. Churaev, *J. Colloid Interface Sci.*, **66**, 389 (1978).
62. J. N. Israelachvili and G. E. Adams, *J. Chem. Soc. Faraday Trans 1*, **74**, 975 (1978).
63. M. von Smoluchowski, *Ann. Phys.*, **25**, 205 (1908).
64. L. Mandelstamm, *Ann. Phys.*, **41**, 609 (1913).
65. A. Andronov and M. Leontowicz, *Z. Phys.*, **38**, 485 (1926).
66. R. Gans, *Ann. Phys.*, **74**, 231 (1924); **79**, 204 (1926).
67. C. V. Raman and L. A. Ramdas, *Proc. Soc. London, Ser. A*, **108**, 561 (1925); **109**, 150, 272 (1925); *Indian J. Phys.*, **1**, 29, 97, 199 (1927).
68. A. Vrij, *Adv. Colloid Interface Sci.*, **2**, 39 (1968).
69. F. P. Buff, R. Lovett, and F. H. Stillinger, Jr., *Phys. Rev. Lett.*, **15**, 621 (1965).
70. D. G. Triezenberg and R. Zwanzig, *Phys. Rev. Lett.*, **28**, 1183 (1972).
71. L. S. Ornstein and F. Zernike, *Proc. Akad. Sci. Amsterdam*, **17**, 793 (1914).
72. F. F. Abraham, *Chem. Phys. Lett.*, **58**, 259 (1978).
73. J. D. Weeks, *J. Chem. Phys.*, **67**, 3106 (1977).
74. M. S. Wertheim, *J. Chem. Phys.*, **65**, 2377 (1976).
75. V. Bongiorno, L. E. Scriven, and H. T. Davis, *J. Colloid Interface Sci.*, **57**, 462 (1976).
76. Lord Rayleigh, *Proc. R. Soc. London, Ser. A*, **79**, 399 (1907).
77. M. A. Bouchiat and D. Langevin, *J. Colloid Interface Sci.*, **63**, 193 (1978).
78. A. Vrij, *J. Colloid Sci.*, **19**, 1 (1964).
79. J. Adin Mann, Jr., K. Caufield, and G. Gulden, *J. Opt. Soc. Am.*, **61**, 76 (1971).
80. I. B. Ivanov and D. S. Dimitrov, *Colloid Polym. Sci.*, **252**, 982 (1974).
81. A. Vrij, *Discuss. Faraday Soc.*, **42**, 23 (1966).

82. A. Vrij and J. T. G. Overbeek, *J. Am. Chem. Soc.*, **90**, 3074 (1968).
83. J. W. Cahn, *J. Chem. Phys.*, **42**, 93 (1965).
84. E. Ruckenstein and R. K. Jain, *J. Chem. Soc. Faraday Trans. 2*, **70**, 132 (1974).
85. E. H. Lucassen-Reynders and J. Lucassen, *Adv. Colloid Interface Sci.*, **2**, 347 (1969).
86. P. G. de Gennes, *C. R. Acad. Sci., Ser. B*, **268**, 1207 (1969).
87. L. Onsager, *Phys. Rev.*, **37**, 405 (1931); **38**, 2265 (1931).
88. B. U. Felderhof, *J. Chem. Phys.*, **49**, 44 (1968).
89. S. Sche and H. M. Fijnaut, *Surf. Sci.*, **76**, 186 (1978).
90. J. Lucassen, M. van den Tempel, A. Vrij, and F. T. Hesselink, *Proc. K. Ned. Akad. Wet. B*, **73**, 109 (1970).
91. A. Vrij, F. T. Hesselink, J. Lucassen, and M. van den Tempel, *Proc. K. Ned. Akad. Wet. B*, **73**, 124 (1970).
92. I. B. Ivanov, "Physicochemical Hydrodynamics of Thin Liquid Films," thesis, Sofia, 1977.
93. T. T. Traykov and I. B. Ivanov, *Int. J. Multiphase Flow*, **3**, 471 (1977).
94. J. G. H. Joosten, A. Vrij, and H. M. Fijnaut, in *Proceedings of the International Conference on Physical Chemistry and Hydrodynamics, 1977*, Vol. 2, Advanced Publications, Guernsey, 1978, p. 639.
95. J. W. Gibbs, *Collected Works*, Vol. 1, Longmans, Green and Co. New York, 1931, p. 223.
96. V. G. Levich, *Physicochemical Hydrodynamics*, Prentice-Hall, Englewood Cliffs, NJ, 1962, Ch. XI.
97. R. S. Hansen and J. Ahmad, *Prog. Surf. Membrane Sci.*, **4**, 1 (1971).
98. D. Langevin and M. A. Bouchiat, *C. R. Acad. Sci., Ser. B*, **272**, 1422 (1971).
99. J. G. H. Joosten, to be published.
100. A. Prins, C. Arcuri, and M. van den Tempel, *J. Colloid Interface Sci.*, **24**, 84 (1967).
101. S. Sche, *J. Electrostat.*, **5**, 71 (1978).
102. B. U. Felderhof, *J. Chem. Phys.*, **48**, 1178 (1968).
103. F. T. Hesselink, unpublished calculations.
104. J. G. H. Joosten, unpublished calculations.
105. S. Hård and H. Löfgren, *J. Colloid Interface Sci.*, **60**, 529 (1977).
106. R. Dorrestein, *Proc. K. Ned. Akad. Wet. B*, **54**, 260, 350 (1951).
107. M. van den Tempel and R. P. van de Riet, *J. Chem. Phys.*, **42**, 2769 (1965).
108. W. A. B. Donners and A. Vrij, *Colloid Polym. Sci.*, **256**, 804 (1978).
109. F. Wagner, *J. Low Temp. Phys.*, **13**, 317 (1973).
110. J. Adin Mann, Jr., *J. Colloid Sci.*, **25**, 437 (1967).
111. J. B. Rijnbout, W. A. B. Donners, and A. Vrij, *Nature (London)*, **249**, 242 (1974); W. A. B. Donners, J. B. Rijnbout, and A. Vrij, *J. Colloid Interface Sci.*, **61**, 249 (1977).
112. W. A. B. Donners, J. B. Rijnbout, and A. Vrij, *J. Colloid Interface Sci.*, **61**, 535 (1977).
113. J. Lyklema, P. C. Scholten, and K. J. Mysels, *J. Phys. Chem.*, **69**, 116 (1965).
114. S. P. Frankel and K. J. Mysels, *J. Appl. Phys.*, **37**, 3725 (1966).
115. J. B. Rijnbout, *J. Phys. Chem.*, **74**, 2001 (1970).
116. P. Richmond, in D. H. Everett, Ed., *Specialist Periodical Reports: Colloid Science*, Vol. 2, Chemical Society, London, 1975, p. 130.
117. R. H. Katyl and U. Ingard, *Phys. Rev. Lett.*, **19**, 64 (1967).
118. A. T. Forrester, *J. Opt. Soc. Am.*, **51**, 253 (1961).
119. J. B. Lastovka and G. B. Benedek, *Phys. Rev. Lett.*, **17**, 1039 (1966).
120. R. H. Katyl and U. Ingard, *Phys. Rev. Lett.*, **20**, 248 (1968).
121. H. Z. Cummins and H. L. Swinney, in E. Wolf, Ed., *Progress in Optics*, Vol. VIII, North-Holland, Amsterdam, 1970, p. 133.
122. M. A. Bouchiat, J. Meunier, and J. Brossel, *C. R. Acad. Sci., Ser. B*, **266**, 255 (1968).

123. M. A. Bouchiat, J. Meunier, *C. R. Acad. Sci.*, *Ser. B*, **266**, 301 (1968).

124. B. Chu, *Laser Light Scattering*, Academic Press, New York, 1974.

125. B. J. Berne and R. Pecora, *Dynamic Light Scattering*, Wiley, New York, 1976.

126. H. Z. Cummins and E. R. Pike, Eds., *Photon Correlation and Light Beating Spectroscopy*, Plenum Press, New York, 1974.

127. H. Z. Cummins and E. R. Pike, Eds., *Photon Correlation Spectroscopy and Velocimetry*, Plenum Press, New York, 1977.

128. D. Langevin, *J. Chem. Soc. Faraday Trans. 1*, **70**, 95 (1974).

129. S. Hård, Y. Hammerius, and O. Nilsson, *J. Appl. Phys.*, **47**, 2433 (1976).

130. N. McQueen and I. Lundström, *J. Chem. Soc. Faraday Trans. 1*, **69**, 694 (1973).

131. M. Bird and G. Hills, in *Proceedings of the International Conference on Physical Chemistry and Hydrodynamics, 1977*, Vol. 2, Advanced Publications, Guernsey, 1978, p. 609.

132. M. A. Bouchiat and J. Meunier, *J. Phys. (Paris)*, **33**, C1-141 (1972).

133. E. S. Wu and W. W. Webb, *J. Phys. (Paris)*, **33**, C1-149 (1972).

134. E. S. Wu and W. W. Webb, *Phys. Rev. A*, **8**, 2077 (1973).

135. J. Zollweg, G. Hawkins, I. W. Smith, M. Giglio, and G. B. Benedek, *J. Phys. (Paris)*, **33**, C1-135 (1972).

136. J. A. Mann, J. F. Baret, F. J. Dechow, and R. A. Hansen, *J. Colloid Interface Sci.*, **37**, 14 (1971).

137. D. Langevin and M. A. Bouchiat, *J. Phys. (Paris)*, **33**, C1-77 (1972).

138. H. M. Fijnaut and J. G. H. Joosten, *J. Chem. Phys.*, **69**, 1022 (1978).

139. J. G. H. Joosten and H. M. Fijnaut, *Chem. Phys. Lett.*, **60**, 483 (1979).

140. J. G. H. Joosten and H. M. Fijnaut, *Proceedings: Workshop on Quasielastic Light Scattering Studies of Fluids and Macromolecular Solutions, 1979*, Plenum Press, New York (in press).

141. N. A. Clark and C. Young, Harvard University, personal communication at Verbier (Switzerland) light scattering conference, 1974; private communication, 1980.

142. S. Levine, J. R. Marriott, and K. Robinson, *J. Chem. Soc. Faraday Trans. 2*, **71**, 1 (1975).

143. E. Grabowski and J. A. Cowen, *Biophys. J.*, **18**, 23 (1977).

INTERACTION POTENTIALS AND GLASS FORMATION: A SURVEY OF COMPUTER EXPERIMENTS

C. A. ANGELL

Department of Chemistry
Purdue University
West Lafayette, Indiana

J. H. R. CLARKE

Department of Chemistry
University of Manchester Institute of Science and Technology
Manchester, England

AND

L. V. WOODCOCK

Laboratory for Physical Chemistry
University of Amsterdam
Amsterdam, The Netherlands

CONTENTS

Abstract

This chapter reviews recent applications of computer simulation techniques, in particular the method of molecular dynamics (MD) by us and others to supercooled liquids, glass transition phenomena, and amorphous solids.

We examine a variety of results for simple models to determine the features of glass formation that are quite general and those that are specifically dependent on the form of the interaction potential. We further consider the requirements in the pair potential for glass formation vis à vis homogeneous crystal nucleation, and for behavior reminiscent of more complex laboratory glasses.

The ultrafast effective cooling rate to which the liquid is subjected in computer simulation leads to a "glass transformation range," which is so broad that the usual methods of characterizing glass transitions in the laboratory are barely applicable. Nevertheless, the consequences of "freezing" the structure in amorphous phase space *can* be observed, and results for different potentials then lead to assignment of the pronounced change in heat capacity at the glass transition temperature, common to laboratory glasses, to the attractive part of the potential, while the loss of diffusivity and the resultant jamming of the structure is attributed to the repulsive part of the potential.

Provided care is exercised in the choice of models and questions asked, computer "experiments" will prove to be an increasingly valuable aid to the elucidation of glass transition phenomena and the study of amorphous structures, when carefully applied in conjunction with real experimental and theoretical studies.

I. INTRODUCTION

It is assumed at the outset that the reader is familiar with the basic Monte Carlo (MC) and molecular dynamics (MD) methods of computer simulation of liquids. This area is now the subject of a large body of literature, as well as review articles[1-3] and lecture series.[4] In recent years, a number of authors have taken advantage of the ability of the simulation methods to obtain detailed information on the properties of supercooled simple liquids in internal equilibrium. This is possible because the "measurements" are made on a time scale that is short with respect to the time scale on which the crystallization event occurs. These time-scale differences lead directly to the existence of the type of study on which this chapter focuses.

For many simple liquids, extensions of such studies to lower temperatures using simply programmed cooling schedules (which in fact correspond to enormous and experimentally inaccessible quenching rates) have resulted in the complete bypassing of crystallization and the consequent trapping of the system in a nondiffusing, configurationally arrested (i.e., glassy) state. This process, with its attendant characteristic thermodynamic manifestations, was first deliberately investigated for the case of a simple ionic liquid,[5] following earlier observations of "jammed" amorphous states for hard-sphere systems.[6] Since the initial work, several, mainly structural, studies of the simpler Lennard-Jones,[7-15] hard- and soft-sphere,[16-20] and repellent Gaussian "glasses"[21] have been reported, and attention has also been given to "real" glass-forming systems such as BeF_2 [22-24] and SiO_2 [23] and mixed oxides.[25] With the rapid growth of interest in quenched metallic glassy systems, applications of this initially academic use of computer "experiments" may be expected to grow rapidly.

In the above-referenced applications of molecular dynamics, a clear division of interests has emerged, which we summarize in the following paragraphs, together with the brief comments on the special utility of the simulation method.

The least emphasized interest, but the one that must be adequately explored before the full significance of the more popular (structural) applications can be properly assessed, concerns the investigation by molecular dynamics of the kinetically determined, hence history-dependent, nature of the glass transition itself. It is through this irreversible process that the configurationally frozen material—to which the word "glass" pertains—is obtained, and in laboratory studies it is well known that the structure and properties of the material obtained depend on the precise manner in which it was formed. Two questions need to be answered here before the usefulness of computer simulation methods to the study of the glassy state of common experience is properly established.

1. How does the behavior of a simple computer model system, in a highly irreversible quench, differ from that which would be observed, were it possible both to bypass crystallization and to examine the system on a laboratory time scale?

2. How does the behavior of each of these simple model systems, which are not obtainable as glasses in the laboratory, differ from that of the simplest type of laboratory glass-forming material, particularly in the description of transport behavior in the vicinity of the glass transition?

To answer the second question decisively, it may be necessary first to carry out a comparison of simulation and experiment for a known laboratory glass-forming liquid—for example, the simple four-atom chain molecule S_2Cl_2 in the temperature-pressure range of the simulation glass transition. (Any simple laboratory liquid that can in practice be maintained liquid into this region would be acceptable: it would not need to be "glass forming" in the normal sense.)

A related but somewhat more fundamental question concerns the underlying thermodynamic basis of the transition. Specifically, we need to know whether it is sufficient to understand the behavior of simplest atomic models to explain the thermodynamic aspects of the phenomenon (as is the case, e.g., for the equilibrium melting and freezing transitions.[26] If so, it would be possible for computer simulation experiments of sufficient accuracy to provide the basis for a molecular level statistical mechanical account of all aspects of the glass transition.

An alternative and distinct application of computer simulation methods to glass science problems lies in the examination of the equilibrium and dynamic structural properties of the amorphous *solid* phase. Such studies, which have been the focus of much of the simulation effort to date, yield "experimental" results with much greater information content, and also for more fundamental systems, than is at present possible by the common experimental techniques, X-ray and neutron scattering. They thereby permit the evaluation of the relative influence of the different components of the interaction potential. However, because the structures under study are produced by subjecting the liquid to a quenching process some orders of magnitude in cooling rate beyond the nearest experimental equivalent, the structures obtained are likely to be atypical and need to be evaluated in the light of considerations of relaxation time versus structure; that is, an understanding of the effects described in the preceding paragraphs is a prerequisite to the structure-oriented studies.

Despite the serious limitations imposed by the economic restriction to fast irreversible quenches for small systems, there is, in each of the objectives cited above, the distinct compensatory advantage that virtually no restrictions are placed on the choice of the intermolecular potential (except that for economic reasons only it must at present be pairwise additive). Thus computer simulation can be used to assess the requirements in the pair potential for particular modes of behavior in glass formation in:

1. The irreversible thermodynamics and dependence of transport coefficients.
2. The underlying equilibrium thermodynamics.
3. The structure-determining characteristics.

In simple liquids studied by computer simulation the problem of glass formation through structural arrest cannot be dissociated from the problem of nonequilibrium crystallization (homogeneous nucleation), since under many conditions the time scales for the two processes are comparable. A broad review emphasizing phase changes has recently been given by Frenkel and McTague,[27] to which we refer the interested reader for a more comprehensive reference list of this and related fields to December 1979. The primary purpose of this chapter is to assess the progress, or lack of it, toward a more comprehensive understanding of glass formation thanks to the molecular level output of computer simulations. In fulfilling this purpose, we find it necessary to deal at length with some of the limitations of computer experiments, not least by comparing results for a number of systems studied to date with the behavior of more complex laboratory glass-forming systems.

II. CONDITIONS FOR, AND LIMITATIONS ON, METASTABLE–STATE SIMULATIONS

It will be helpful to review first the conditions under which we can carry out metastable-state simulations for simple systems.

Although it is possible artificially to devise pair potentials that will energetically favor disordered over ordered states, in all the model systems studied to date the fluid state is actually metastable with respect to one or more ordered states under the conditions in which the structural arrest (glass formation) is observed. We should therefore discuss briefly the limitations on such metastable-state studies.

The focus of interest here must be the relation between the time for a metastable fluid system to evolve, following a thermodynamic perturbation (e.g., temperature decrease) toward a new state on the metastable liquid free energy surface on the one hand, and toward a state on the stable, crystal free energy surface on the other hand. Given sufficient time, the latter evolution must of course always be chosen; but if the configurational barriers along this path are large relative to those along the path to metastable equilibrium, the properties of the metastable state can always be observed. Operationally it is reasonable to require that there be at least an order of magnitude difference in these time scales for a discussion of metastable state properties to be meaningful: otherwise the system must always be considered as evolving continuously toward the crystal, and distinguishable physical properties for the metastable liquid cannot be determined.

For any liquid these two time scales exhibit quite different temperature dependences (Fig. 1a). Thus a metastable liquid branch may be investigable according to the criterion above in one temperature range (e.g., near the melting point) but not in another (e.g., at greater supercooling). Homogeneous nucleation theory[28—30] considers the independent influences of the height of the configurational barrier to crystal nucleation, and the growth rate of the fluctuations that permit passage over the barrier. In this theory there are two characteristic time scales: τ_{out}, the inverse of the nucleation rate, which characterizes the escape time from the metastable state, and τ_{in}, the characteristic relaxation time along the free energy surface. According to the theory, τ_{out} exhibits a minimum, or the probability of nucleation a maximum, as a function of tempreature (Fig. 1a), as is consistent with MD observations on the Gaussian core model.[21] This implies that the ratio τ_{in}/τ_{out} also has a maximum value (Fig. 1b), hence that a metastable free-energy surface may also be investigable at a temperature far below the melting point—for example, near but above the "ordinary" glass transition temperature, where the relaxation time along the liquid free energy surface is of the order of minutes, even if it is not investigable at intermediate temperatures. On the other hand, if at any temperature the two time scales become the same, then the system has, at that temperature, a mechanical instability or spinodal point and, given only the time for diffusion to occur, a phase change must occur. In view of the comparatively more specific rearrangement of particles required to generate a crystalline configuration, it is not clear that such an instability must exist, although it seems certain that in some systems it is closely approached.

Spontaneous nucleation of the type just described has been observed recently in computer simulation studies,[12, 21, 30–36] and details of the nucleating structures have been reported,[32–36] but the dependence of τ_{out} on T has not been studied (indeed this would be prohibitively costly at the moment). The nucleation rates in the simulation experiments can be approximately evaluated and are enormous, as is appropriate for argon and liquid metals. For instance, the observation of one nucleus forming in a sample of 500 atoms of a substance of atomic volume 40 cm³/mol in a run of 100 psec implies a nucleation rate[28] of $10^{30}/cm^3sec$.

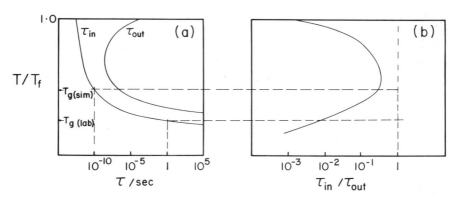

Fig. 1. (a) Variations with temperature (relative to fusion temperature, T_f) of the relaxation times within the amorphous phase, τ_{in}, and out of the amorphous phase into the crystalline phase (nucleation time), τ_{out}. (b) Variations with temperature of the relaxation time ratio τ_{in}/τ_{out}.

For a system in which τ_{in}/τ_{out} is always less than unity, a glassy state can be realized if the thermal energy can be removed on a time scale small with respect to τ_{out} at all temperatures. The structural arrest known as the glass transition will then occur (at a temperature denoted T_g) when τ_{in} becomes too long for the structure to follow (equilibrate with respect to) the temperature. For normal laboratory cooling this occurs when $\tau_{in} = 1-100$ sec, depending on how the glass transition temperature T_g is defined from the rather sudden but still continuous changes in derivative thermodynamic properties (typically, the heat capacity C_p) that accompany the structural arrest. (These thermodynamic manifestations and their relation to details of the pair potential are discussed in Section IV.) Naturally, this arrest occurs at higher temperatures, the faster the cooling rate. Much higher cooling rates are accessible in computer experiments than in laboratory experiments; hence simpler systems can be vitrified, and structures characteristic of very high temperature configurational states are frozen in, as indicated by T_g(sim) at $\tau_{in} = 10^{-10}$ sec in Fig. 1a. On the other hand, as we show in more detail in the next section, vitrification in computer simulations occurs under conditions in which the usually sharp "glass transition" is smeared out almost beyond recognition.

Before considering this issue, it is interesting and important to ask what factors predispose a system to rapid crystallization, and thereby to the exclusion of glass formation. A general but vague answer is, of course, the factors that tend to minimize the distinction between solidlike and liquidlike regions of phase space. A more quantitative guide is obtained from the magnitude of the changes of thermodynamic variables on fusion. Single-component atomic systems that exhibit large changes in volume (e.g., hard spheres) have large regions of low probability between solid and liquid regions of phase space and supercool readily, whereas systems with very weak first-order transitions behave conversely.[18] In more specific terms, the "strength" of the first-order transition for simple systems decreases as the hardness of the pair repulsion interaction decreases, basically because at a given thermal energy, particles in the amorphous state experience less difficulty in moving past one another (unjamming) to explore additional, otherwise excluded, configurations.

The latter comments are broadly consistent with experiment. Thus, one observes, in simulations of the one component plasma (a system with r^{-1} repulsive interactions) a frequent fluctuation in and out of the ordered state,[37] although in a 500-particle system, a supercooling to 0.95 T_f has recently been obtained.[38] Conversely, in the opposite extreme of the hard-sphere system, spontaneous ordering is not seen at all [in very small (32-particle) systems, the periodic boundaries seem to induce some artificial ordering at high effective pressures[6]]. NOTE ADDED IN PROOF: Crystallization of a 512 particle hard sphere system has now been observed after a total of $\sim 4 \times 10^6$ collisions in the metastable state during runs at progressively increasing densities. The nucleation event occurred suddenly during the run at $\rho\sigma^3 = 1.08$, $\sim 0.5 \times 10^6$ collisions after an increase of density from 1.07 (L.V. Woodcock, *Ann. N.Y. Acad. Sci.*, to be published).

Progressive tendencies to crystallize with softening of repulsion parameters are discernible in the tabulated results for soft-sphere fluids defined by

$$\phi(r) = \varepsilon\left(\frac{\sigma}{r}\right)^n \tag{1}$$

where ε is the potential at particle separation σ, for which the systems $n = 4$, 6, 9, and 12 have been studied.[18-20] In the laboratory, the influence of the repulsive potentials is seen in the general absence of the metals among easily vitrified liquids, and the prevalence of transition metals, that are best described as experiencing hard ($\sim r^{-12}$) repulsions, among the metal alloy systems that have been vitrified by splat quenching.[39, 40] The alkali metals for which the accepted pseudopotentials approximate very soft ($\sim r^{-6}$) repulsions have never been vitrified.

We may represent these ideas by the extension of Fig. 1 shown in Fig. 2, which takes note of the observation (referred to at the end of the next section) that relaxation times may show some system size dependence, and accordingly may have marked the horizontal axis to indicate relaxation times at constant k, where the k-value of interest is the inverse of the critical nucleus diameter. The lines represent the behavior of systems of different repulsive parameters for spherically symmetric pair potentials, (1). For small n, the lines approach arbitrarily close to the $\tau^* = 1.0$ line. The horizontal lines again represent temperatures at which the internal liquid (amorphous) state relaxation time reaches values of ~ 1 sec (which is the average relaxation time characteristic of laboratory systems when they enter the glass transition on cooling) and 100 psec (at which economic restrictions will impede most simulationists from pursuit of equilibrium properties, and therefore will serve as an indicator of T_g determined by computer studies). The spacing of these lines relative to T_m (at which $\tau_{in} \approx 10^{-12}$ sec) is a reminder that the relaxation time becomes a very strong function of temperature at low T/T_m, a property that is emphasized in the next section.

Figure 2 may be given a second and conceptually useful interpretation: as n increases, so does the space between the temperature-time ratio curve and the dotted area to the right of the $\tau_{in}/\tau_{out} = 1$ line. This gap may be regarded as a measure of the "gap" in phase space between regions of high probability for liquidlike states and those of high probability for crystallike

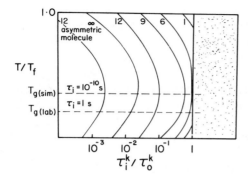

Fig. 2. Conjectured relation between the relaxation time ratio τ_{in}/τ_{out} and temperature (relative to the fusion temperature, T_f) for different repulsive interactions. Dashed lines represent temperatures at which the fluids have the internal relaxation times 10^{-10} sec characteristic of the system near the usual MD or MC glass transition, and 1 sec where departure from equilibrium becomes noticeable during normal cooling of laboratory glass-forming liquids.

states. The diagram then emphasizes that for systems with soft repulsions, the distinction between solidlike and liquidlike regions of phase space is reduced, and transitions between the two states are quite probable even at small supercoolings.

III. LOSS OF EQUILIBRIUM AND THE GLASS TRANSITION

A. General Considerations

A perfect crystal equilibrates very rapidly, following a perturbation from equilibrium, by a collisional energy exchange mechanism requiring of the order of 10^{-13} sec. Liquids and defective crystals, on the other hand, equilibrate slowly because of the need for the overall structure to change (to a new distribution of coordination numbers, and a new concentration of defects, respectively). At least at the lower temperatures, the latter processes require the system to explore phase space through an "activated" diffusion process (see below) for which the time requirements are $\gg 10^{-13}$ sec and are also strong functions of temperature.

To understand properly the relationship between the "glass transition" phenomenon observed in computer-simulated systems and that observed in laboratory systems, it is necessary to be familiar with the temperature dependence of the relaxation time. The point to be made is that the "transition," which is the thermodynamic manifestation of a failure to maintain equilibrium during cooling, occurs sharply in laboratory systems but diffusely in simulated systems, primarily because of a great difference in relaxation time temperature (or volume) dependence in the time-scale regimes in which the processes are observed in the two cases.

The essential diffuseness of the simulation transition is not and cannot be at all obvious from the plots of diffusion coefficient versus temperature. The latter have been the natural and frequently used criterion, $D \geq 0$, for deciding whether a system is in fluid or rigid ("configurationally arrested") state. The behavior of D versus T has led on several occasions to speculations that the glass transition observed in simple systems by simulation might be a singular point for the system. Because this is such an important matter, and because it has long been argued whether a thermodynamic basis for the laboratory glass transition indeed exists, we devote considerable attention to what may happen in a simulation in the region where D approaches zero, in the hope of clarifying an aspect of the liquid-glass transition that is consistently a source of interest and also, occasionally, of confusion.

We consider first the behavior of the diffusion coefficient (which is simply obtained in simulations, if not in the laboratory) in relation to temperature and density, and then look at what is implied for the behavior of the thermodynamic properties as D goes to zero in simulation "experiments". Comparisons with experimental observations are important in this connection.

B. Temperature and Density Dependence of Transport Properties

Before comparing the glass transition observed in simulations with that observed in the laboratory, it is necessary to review briefly the temperature and density dependence of transport properties. In some of the model systems studied (specifically, hard and soft spheres) there is only one system variable, and temperature- or density-dependent representations of the properties are a matter of choice only. With other systems and of course with all laboratory systems, the two types of plot display independent aspects of the system's behavior.

It has been common to display the results of computer simulation studies of systems approaching the glassy state as simple linear plots of D versus T, usually in dimensionless reduced units natural to the simulation: For example, for soft potentials,

$$D^* = D\sigma^{-1}\left(\frac{m}{kT}\right)^{1/2}\left(\frac{kT}{\varepsilon}\right)^{1/2} = D\left(\frac{m}{\varepsilon\sigma^2}\right)^{1/2}, \qquad T^* = \frac{Tk}{\varepsilon} \qquad (2)$$

where D is the normal diffusivity in units of square centimeters per second, m is the particle mass, σ the particle diameter determined, for attractive systems, either at $U(r)=0$ or at the potential minimum, and ε is the potential at $r=\sigma$, or the depth of the potential well for attractive systems. To permit some comparison with typical experimental results, we have, where necessary, converted the quoted simulation results to normal cm^2/sec diffusion coefficients for argonlike particles by introducing $\sigma = 3.4$ Å and $m=$

6.6×10^{-23} g and $\varepsilon/k = 120$ into expressions (2). These data are plotted against absolute temperature in Fig. 3, along with data for some of the simpler laboratory liquids for which data are available.[41-46] The latter are frequently represented in the high-D range by Arrhenius functions,

$$D = D_0 \exp\left(\frac{-E}{RT}\right) \tag{3}$$

where E is the "activation energy." However, in liquids of all varieties, systematic deviations from (3) are found when D falls much below 10^{-5} cm^2/sec.[41-48] The deviation is such that the activation energy increases exponentially with decreasing temperature. A comparable deviation has been described for the case of Lennard-Jones (LJ) argon,[12] though there are

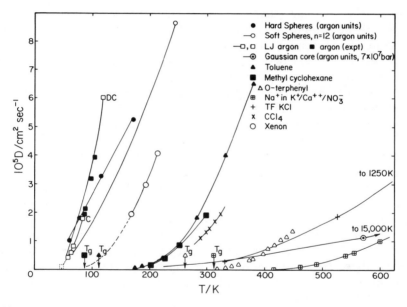

Fig. 3. Variation with temperature of the diffusion coefficients for various simulated fluids and actual laboratory fluids. Sources of data are, from left to right: LJ argon, simulated Refs. 7 (DC) and 12 (C) laboratory, Ref. 41; hard spheres (for which temperature axis corresponds to $pV/NkT \times 50$), Ref. 82; soft spheres, Ref. 20; xenon, Ref. 41; toluene, Ref. 42; methyl cyclohexane, Ref. 43; carbon tetrachloride, Ref. 44; o-terphenyl, Ref. 45; molten KCl, simulated using Tosi-Fumi (TF) potential parameters, Ref. 5; repellent Gaussian core particles, Ref. 21 (F. H. Stillinger kindly deduced the values his simulation results would infer for argonlike particles in familiar units); Na$^+$ ions diffusing in molten $6KNO_3 \cdot 4Ca(NO_3)_2$ solvent medium, Ref. 46. The dashed extension of lower temperature in the case of xenon is based on the Arrhenius parameters quoted for the data.[41]

considerable uncertainties in the lowest temperature points. It is important to note that the laboratory liquids may be observed in internally equilibrated, diffusing states to temperatures well below that of the lowest diffusivity recorded in Fig. 3, and data can in principle be obtained down to (and even somewhat below) the points marked T_g (to denote the laboratory glass transition temperature, see below) in each case. At the laboratory T_g, diffusivities have decreased to values of $\sim 10^{-18}$ cm^2/sec, which are almost immeasurably small (although the closely related viscosities, $\sim 10^{11}$ P, may be determined quite easily).

The simple atomic liquids studied by simulation methods seem to be approaching zero diffusivity more abruptly than the neighboring laboratory liquids, though it is not clear from Fig. 3 that this is more than a matter of temperature scaling. A difficulty in furthering the comparison in the interesting, even vital, region of vanishing diffusivity is raised by the economic problem of calculating reliable diffusion coefficients in a temperature region where the attainment of equilibrium liquid states requires inordinately long computation runs. This problem and its origin are considered in detail below. Suffice it to say here that any simulation value of D less than 0.5×10^{-5} cm^2/sec is to be suspected of being an overestimate for the temperature in question, unless the absence of continuing relaxation during the determination is specifically documented.

The possibility of inherent difficulties in the accurate MD simulation of slowly diffusing systems is also to be borne in mind following the finding of discrepancies between experiment and simulation observed in the nonliquid but still very relevant case of the crystalline "superionic" conductor CaF_2.[49] In this system, Rahman[50] has reported MD diffusivities of F^- that approach zero, on linear D scale plots of the Fig. 3 type, more rapidly than expected from the observed Arrhenius behavior of the measured diffusivities. (Rahman compared his results with extrapolations of D_{F^-} data measured below the weak lambda transition at 1200°C, but the electrical data[49] have validated the extrapolation to the accuracy of the comparison.) Rahman's high-temperature data have been accurately reproduced in one of the authors' laboratories using a rather different form of pair potential.[51]

In Fig. 4, we plot the volume dependences of the diffusivities of various model systems, all calculated for argonlike particles, and for the laboratory system methyl cyclohexane, which seems to be representative. Data for the latter are available for two temperatures from the accurate nuclear magnetic resonance (NMR) spin-echo measurements of Jonas et al.[43] To group the data for different systems for better comparison of their behavior, we have obtained a reducing parameter for each liquid by taking advantage of the observations of Batchinski[52] and Hildebrand[53-56] that fluidities ϕ, and diffusivities D of highly fluid molecular liquids vary linearly with volume.

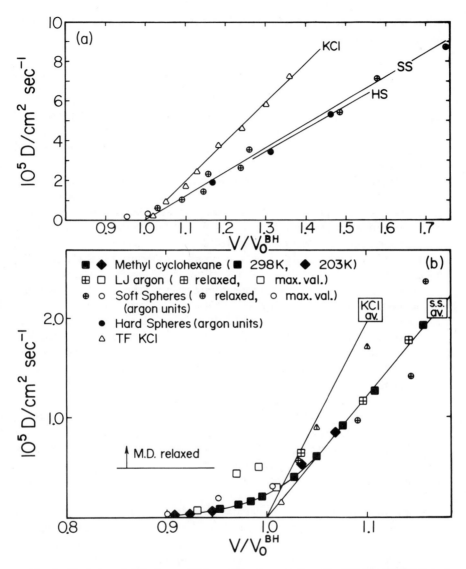

Fig. 4. Variation of diffusion coefficient with volume, reduced by Batchinski-Hildebrand V_0 in each case, for various simulated and actual fluids. (a) Extended diffusivity range (MD systems only). (b) Lower D range, (expanded scale), showing nonlinear volume dependence characteristic of glass-forming molecular liquids.

Hildebrand devoted much effort to establishing the general validity and utility of a modified form of Batchinski's original equation, namely,

$$\phi = \frac{1}{\eta} \propto D = A\left(\frac{V - V_0}{V_0}\right) \tag{4}$$

where A is a constant. We have accordingly presented the MD data and the methyl cylohexane data in Fig. 4 as a function of V_r, where $V_r = V/V_0^{BH}$ and V_0^{BH} is chosen for each liquid by extrapolation of the linear part of the data to $D = 0$.

Reduced in this manner, the various liquids considered show much similarity, though variations in slope are expected from Batchinski's and Hildebrand's findings. Again, the laboratory case shows a marked deviation from linearity at low diffusivity, which Hildebrand attributed, in the few cases of such deviations he considered, to entanglement or "cog-wheel" effects.[55] For the laboratory cases, it is possible to determine a V_g, at which the structural relaxation time reaches value of the order 1–100 sec and a glassy state forms during continuous compression. Like T_g, this V_g will fall far below the volume at which D in Fig. 4 tends to zero. Values of D below 0.5×10^{-5} cm^2/sec for LJ argon and soft spheres, which are represented by different symbols in Fig. 4, are maximum values because of incomplete equilibrations of the system, and at present the similarity in behavior to the (equilibrated) laboratory liquid should be treated as a coincidence, pending more accurate measurements.

To go from the behavior of the diffusion coefficient to the phenomenon of the glass transition and the relation between simulated and real systems, we need to consider (1) the observed diffusivity-temperature relation over the full temperature range, and (2) the relation between diffusivity and relaxation time, and between real and apparent thermodynamic properties. The first of these we consider now; the second is taken up in the Section III.C.

As mentioned above, all molecular liquids that have been studied at viscosities above ~ 1 cP ($D \lesssim 1 \times 10^{-5}$ cm^2/sec) have been found to depart the behavior described by (3) in a way that requires the "activation energy" to increase continuously and in an accelerating manner as the temperature decreases. Over much of the range, the behavior is described by a simple empirical modification of the Arrhenius relation,[47] often called the VTF or Fulcher[57] equation

$$D = A' \exp\left(\frac{-E}{R(T - T_0)}\right) = A' \exp\left(\frac{-B}{T - T_0}\right) \tag{5}$$

Here T_0 is a new nonzero "vanishing mobility" temperature that is characteristic of each liquid and lies far below the temperature at which each diffusivity plot in Fig. 3 tends to zero on the linear scale. The claim of T_0 to physical significance is that it falls below the observed "normal" T_g, such that $T_g \approx 1.1$ to 1.3 T_0 and is usually coincident with the temperature at which the supercooled liquid entropy, measured below the fusion temperature, extrapolates to the same value as for the crystal (though the free energy of the latter remains lower[58, 59]). Equation 5 implies an apparent activation energy that varies with temperature according to

$$E = R \frac{d \ln D}{d(1/T)} = B \left(\frac{T}{T - T_0} \right)^2 \qquad (6)$$

With $B = 500 \ K$, (6) implies that at $T = 2T_0$ (where $D \approx 10^{-5} \ cm^2/sec$) $E = 16.6 \ kJ/mol$, but rises to 149 kJ/mol at $T = 1.2T_0$ as at the experimental glass transition. This change is very important in understanding the origin of the "glass transition" as normally observed. We consider this matter next.

C. Relaxation Times and the Loss of Equilibrium During Cooling

In the laboratory process of glass formation, an imposed temperature gradient causes heat to flow from the interior of the initially liquid system, and the liquid vitrifies—that is, loses its equilibrium properties—first at the outer surface (frequently producing a pipe in the center, where contraction at liquidlike rates continues). In the computer simulation of vitrification, it is simpler to cool the system uniformly by rescaling all the particle velocities at the same moment. This may be done either by rescaling the velocities at each time step to produce a continuous cooling[7, 8, 21, 23] or by a succession of step processes[5, 12, 20] in which a larger change in temperature is imposed instantaneously followed by a constant-temperature "anneal" during which the system may or may not reestablish an equilibrium state (depending on the annealing time and the internal relaxation time).

Each suddenly imposed temperature change represents a perturbation of the equilibrium state from which the system attempts to recover by relaxation—an exploration of the phase space accessible under the new conditions. The relaxation process has a characteristic time, which is temperature dependent and about which something is known from various experimental studies.

As always, the relaxation kinetics depend on the nature of the perturbation being relaxed but, in dense single-component systems of symmetric molecules, the responses to different perturbations that involve rearrangement of molecules tend to occur with remarkably similar time constants.

For instance, dielectric relaxation times τ_D usually are close to, and scale simply with, the shear and bulk viscosity relaxation times, τ_s and τ_v. Since the Stokes-Einstein equation connecting diffusivity with shear viscosity is generally valid,[4] this implies that the structural relaxation time for a system could be estimated if the diffusivity were known. Indeed for a number of simple polar molecules the product $\tau_D D$ seems experimentally to be roughly constant at $\sim 2 \times 10^{-16}$ cm^2, implying that perturbations of a liquid of diffusivity 1×10^{-5} cm^2/sec would require of the order 10 psec to be relaxed to $1/e$ th of their initial value.

For atomic liquids, however, this time appears to be much shorter, presumably reflecting more modest requirements in the configuration space exploration needed to equilibrate. Although the relevant structural relaxation time can be measured only crudely (by neutron-scattering measurements of the decay of nonpropagating density fluctuations, $\tau_v \approx 0.2 \times 10^{-12}$ sec[60] for argon at T_f), it can be determined satisfactorily from analysis of MD results for LJ argon.[62, 63] Desai and Yip[61] find that their τ_0 (for an assumed "single relaxation time" process), which should closely approximate τ_v, is 0.22×10^{-12} sec at T_f. This accords well with the result obtained[64] from combining measured bulk viscosity η_v data with calculated bulk elastic modulus K_∞ calculations using the relation $\tau_v = K_\infty / \eta_v$.[65] For argon, then, the product $\tau_v D$ is much smaller than for molecular liquids. The root-mean-square center of mass displacement during the time $\tau_v (\bar{l} = (6 D\tau)^{1/2})$ is only 0.5 Å, compared with ~ 3.4 Å for a molecular system (calculated using τ_D). This is entirely reasonable, since configurational relaxation in a molecular liquid involves rotational as well as translational molecular motion.

Full relaxation requires the passage of some 10τ; thus internal equilibrium in a simulation of argon should be established in some 2 psec at the triple point, where $D = 1.7 \times 10^{-5}$ cm^2/sec.[12, 41, 62] The time necessary will increase rapidly with decreasing temperature, but structural equilibrium should be computationally accessible down to temperatures where $D = 1 \times 10^{-6}$ cm^2/sec. It is probably reasonable to assign similar relaxation times to other atomic systems.

With these ideas in mind, let us clarify the process of glass formation in computer simulations by considering a series of idealized MD simulations of a simple monatomic system conducted for fixed time periods at successively lower temperatures, such that at some point in the series the time allowed for equilibration becomes much less than the configurational relaxation time. Figure 5 displays as a function of time the behavior expected for a dynamic quantity, the mean squared particle displacement, and for a thermodynamic quantity, the enthalpy. Each successively lower temperature run is assumed to commence using the final configuration of the previous run, with an instantaneous decrease in volume and decrease of

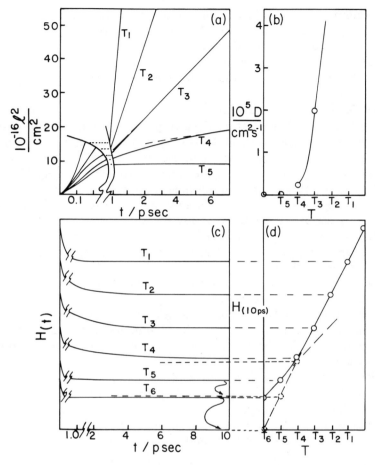

Fig. 5. Variations with time and temperature of dynamic (mean square displacement) (*a*), (*b*), and thermodynamic (enthalpy) (*c*), (*d*) properties of simple dense fluids evaluated by MD runs of limited time duration, showing partial failure to equilibrate at T_4, and "frozen" or glassy-state behavior at T_5 and T_6. Note the change in heat capacity from liquidlike to solid-like values implied by the change of slope of H versus T in (*d*).

T_g for this "experiment" would be located at T_4. The arrows indicate the values of H the system would reach if allowed sufficient time to reach complete structural equilibrium at T_5 and T_6.

temperature by appropriate scaling of the velocities, such that the final pressure is constant for the series. In practice it is common to reach the final temperature in a series of velocity rescalings to correct for the upward temperature drift, which results from decrease with time of the system's potential energy toward the value characteristic of the "new" temperature (see below).

The highest temperature T_1 in Fig. 5, corresponds to a condition where $D = 10^{-4}$ cm^2/sec. Temperature T_3 corresponds to a temperature where D has a "normal" liquid value of 2×10^{-5} cm^2/sec, and T_4 is a temperature at which D (for the equilibrated liquid) is only 2×10^{-6} cm^2/sec. The relaxation time is ~ 10 psec, and the time needed to reach equilibrium is therefore rather longer than the run time. Temperature T_5 shows a run under conditions where D at equilibrium would have a value 10^{-8} cm^2/sec or less. In this case $\tau \approx 2000$ psec, and only a loss in vibrational energy can occur; the configuration remains fixed. Clearly, for the purposes of the simulation experiment, the system at T_5 is in a "glassy" state, even though, in the laboratory, a liquid with diffusion coefficient of 10^{-8} cm^2/sec is described as a "viscous liquid," not a glass. The distinction is simply one of experiment time scale, the latter description being apt for "normal time scale" observations in which the response of the system to some perturbation (e.g., a shear stress) is judged after a time lapse of the order of seconds, compared with the 10-psec time scale of the above-described computer experiment.

Note that in all cases, two sources of energy loss exist, a "fast" loss of vibrational energy, and a "slow" loss of configurational, or potential, energy. At the highest temperature the two cannot be distinguished. The lowest temperature, T_6, is included to show how the glass heat capacity relates to that of the supercooled liquid.

Of the six temperatures, shown in Fig. 5, T_4 is perhaps the most interesting, because here the system is being observed in what glass scientists call the "transformation range," the usually narrow* range of temperature (relative to absolute temperature at midrange) in which the liquid properties are found to be time dependent *on the time scale of the experiment*. A laboratory example based on data from Ref. 66 and showing variations in the extensive property H and the corresponding intensive property C_p for experiments conducted at different heating and cooling rates, $q = dT/dt$, is given in Fig. 6. Inset 1 to Fig. 6a shows how cooling at a constant rate corresponds to a series of measurements of constant time scale Δt in which the relaxation time τ is changed progressively (with each ΔT). The transformation range is the change in T necessary to change the average relaxation time by at least 2 orders of magnitude, as can be understood by consider-

---

*For laboratory experiments.

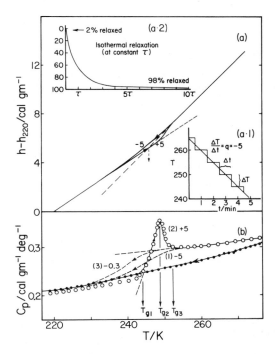

Fig. 6. ('a') Experimental enthalpy and (b) heat capacity changes during sequential [(1),(2),(3)] cooling and heating of a sample of a polyphenyl ether (5P4E) at different rates (rates marked in degrees per minute) to demonstrate characteristics of laboratory glass transition phenomena. The shapes of these plots are well understood.[66] The solid dotted C_p curve is a crude representation of the function expected for the same sample being cooled at the enormous effective cooling rates of the computer simulations MC and MD "experiments." (See also Fig. 1 in G. S. Grest and M. H. Cohen, *Phys. Rev. B*, **20**, 1077 (1980).)

Inset (a.1). Relations between the continuous cooling at rate $q = -5$ deg/min of the laboratory experiment, and the equivalent quench-hold sequence used in MD experiments for evaluating supercooled liquid properties and illustrated in Fig. 5. For the equivalent computer simulation experiment, each minute time interval would represent 10 psec. Inset (a.2). Relation between percentage of instantaneous perturbation relaxed and number of relaxation times elapsed, at constant τ (isothermal relaxation).

ing the range of t/τ values at constant temperature (hence constant τ) needed to pass from a 2%-relaxed to 98%-relaxed condition along a simple exponential decay curve (see Fig. 6a, inset 2). In practice, the range of times is usually increased by the existence of a "distribution" of relaxation times (i.e., the decay function is nonexponential[67]), which considerably lowers the base temperature needed to completely freeze the equilibrium. The temperature range is small, nevertheless, because of the large activation energy characteristic of glass-forming liquids near T_g (see below).

The time scale for a laboratory observation is normally of the order of minutes, corresponding to diffusion coefficients of 10^{-18} cm^2/sec, but for the computer experiment it is shifted enormously to smaller times—in fact, into a range accessible only to dielectric relaxation and inelastic neutron-scattering experiments. The shift to the short time-scale range also means a shift to the small activation energy range, as forewarned earlier. This has the important consequence that the temperature range in which the time-dependent behavior is observed inevitably is greatly extended, since the temperature interval that must be traversed to change the relaxation time by 2 orders of magnitude is greatly increased. We see this immediately by writing, for the range under consideration, the approximation

$$\tau = \tau_0 e^{E/RT} \tag{7}$$

from which

$$\ln\frac{\tau_1}{\tau_2} = \frac{E}{R}\left(\frac{1}{T_2} - \frac{1}{T_1}\right) \tag{8}$$

Rearranging, for the case where $\tau_1/\tau_2 = 100$, (8) yields, for the necessary temperature range,

$$\frac{1}{T_2} - \frac{1}{T_1} = \frac{4.606R}{E} \tag{9}$$

or, on further manipulation,

$$\Delta T = \frac{4.606RT_1^2}{E + 4.606RT_1} \tag{10}$$

where $\Delta T = T_1 - T_2$. For laboratory liquids, E typically has a value of ~ 200 RT_g,[48] thus for $T_g = 250°$K, $E \approx 400$ kJ/mol. In this case we can neglect the second term in the denominator and, taking $T_1 = 250°$K, ΔT for a 100-fold change in relaxation time is $\sim 6°$K (somewhat narrower, as expected, than the $\sim 11°$K observed in Fig. 6). However, if E is only ~ 12 kJ/mol, as it is for simple molecular liquids at room temperature ($D \sim 10^{-5}$ cm^2 sec), then ΔT is of the order $\underline{100°K}$ if T_1 is set at 250°K. Such a "transition" would have the form of the solid dotted line in Fig. 6, and would be almost unrecognizable.

For a substance like argon,[41] with T_1 set at 50°K and E only 3.0 kJ/mol, $\Delta T \approx 19°$K, which is essentially in agreement with the computer simulation results of Clarke.[12] A laboratory glass transition centered at 50°K would have $\Delta T \approx 2°$K. Some of the argon data are reproduced in Fig. 7, which

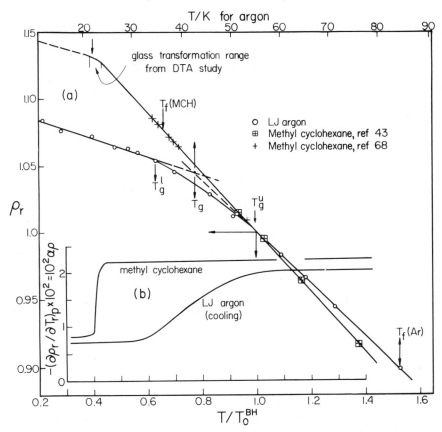

Fig. 7. (a) Comparison of the temperature dependence of the density for MD simulated LJ argon and laboratory-measured methyl cyclohexane (MCH), using temperatures reduced by the temperature T_0' at which diffusivity data in Fig. 3 extrapolate to zero, and densities reduced by the equilibrium density at the temperature T_0'. The reduced glass transformation range for LJ argon is much broader than for MCH, determined by a differential thermal analysis (DTA) experiment,[69] because of the much higher quenching rate of the computer experiment. (b) The derivative property from which T_g is normally defined (see Fig. 6); this plot better demonstrates the smeared-out transition. Note that the reduced melting point for MCH is far below that of argon.

shows the results of a series of densities (evaluated by a quench-compress-wait procedure, such that each point represents an approximately equal period of equilibration, like the sequence of Fig. 5), using reduced temperature and reduced density scales to permit a comparison with the laboratory results for methyl cylohexane.[43, 68] The basis for reduction is similar to that used for Fig. 4. Specifically, we reduce each temperature by the temperature T_0^{BH} at which $D = 0$ according to extrapolation of data in the region $D = 1 \times 10^{-5}$ to 2×10^{-5} cm²/sec for the substance in question. Likewise, ρ is reduced by its equilibrium value at the reference temperature above (since LJ argon is already falling out of equilibrium at this temperature, the reference density is obtained by short extrapolation of higher temperature data). The precise choice of reference temperature is not of great consequence provided it bears a straightforward relation to the diffusivity-temperature relation.

Given the empirical observation that simple liquids exhibit linear density versus temperature relations, and that methyl cyclohexane, in the laboratory studies, remains in internal equilibrium down to ~90°K,[69] the comparison in Fig. 7 implies that simulated LJ argon enters its "glass transformation" range at about 55°K (where $D \approx 5 \times 10^{-6}$ cm²/sec) and only emerges from it, fully arrested with crystallike expansivity, at 35°K—a span of 20°K compared with ~7°K for methyl cyclohexane (which, however, decreases to $\approx 2°K$ if scaled to the lower argon temperature range). The "smearing out" of the transition is best shown by the derivative property α, the thermal expansivity, which is compared with the methyl cyclohexane property in Fig. 7b. The systematic spreading out with decreasing experiment time scale can be seen at an intermediate stage in the behavior of the adiabatic compressibility of Arachlor, determined at several frequencies in the megahertz region by ultrasonic dispersion studies.[70]

Clearly the density of the structure frozen in will depend on the experiment time scale—that is, the cooling rate—as was noticed in the early MD study of Damgaard-Kristensen on LJ argon.[7] Figure 7 implies that the difference in density between the MD glass and the same glass formed by laboratory-time-scale cooling amounts to some 7% when the comparison is made at the same temperature, $T < T_g$ (lab). This difference needs to be borne in mind when comparing theoretical predictions based on computer simulation results with laboratory results. In systems with intrinsically low liquid-state expansion coefficients (e.g., liquid metals), the problem will be less acute.

The blurring of the transition in the region of the regime that is accessible to computer simulation raises the problem of definition of T_g for these cases. In laboratory studies it is most common to use the definition T_{g_1} in Fig. 6 obtained from *reheating* data. The latter have yet to be performed in simulation studies. The choice for MD systems corresponding most closely with the T_g of the sharper laboratory transitions is the temperature given by the intersection of the linear glass and liquid density plots shown in Fig. 7, and referred to as T_g.[16b] This corresponds to 43°K for argon, which is also the temperature at which the diffusivities of the incompletely relaxed system became immeasurably small[12] (see Fig. 3). The observation that some relaxation takes place at lower as well as higher temperatures leads to the definition of *lower* and *upper* glass transition temperatures as points where, *for the chosen experiment schedule* (e.g., Fig. 5), all detectable relaxation ceases (T_g^l) and the first departure from equilibrium is observed (T_g^u). These points were earlier designated T_g^r and T_g^k, respectively.[20] It is very important that the reheating experiment be simulated since the volume hysteresis and the expansivity (or heat capacity) "overshoots" during reheating, which are the trademarks of the glass transition, have yet to be seen for the simulated glasses. In fact some simulation studies,[5, 14, 23] though not all[7, 16b] have suggested the absence of history-dependent phenomena.

At any temperature within the transformation range, the thermodynamic properties, and probably also the mass transport properties, will be time dependent. Surprisingly, until the recent study by one of us on LJ argon,[12] the time dependence of properties of simulated liquids in the transformation range had not been specifically reported, but instead was either monitored and discarded in "equilibration" periods, or avoided by dropping the temperature to values so low that the system became "frozen" after a "fast" initial relaxation. Most data would, in any case, be unreliable because of the small system sizes involved. Some recent results illustrating these effects are shown in Fig. 8, where the total pressure in an 864-particle LJ system is plotted as a function of time following a succession of instantaneous quenches starting from near the normal melting temperature to a point on the zero-pressure isobar in the glass transformation region, $kT/\varepsilon = 0.33$, $\rho/\sigma^3 = 0.97$, corresponding to $T = 36°K$, $T/T_0' = 0.654$, $\rho_r = 1.054$ in Fig. 7). During these quenches the particle velocities were scaled every 50 time steps (0.5 psec) to correspond to the new temperature. This scaling is essential to simulate the temperature decrease and to control the annealing of the sample as potential energy is converted into kinetic energy.

We might expect these relaxation experiments to be highly nonlinear, such that no single relaxation time could be assigned to a given observation. It is well known in laboratory studies of the glass transformation region that the relaxation time depends not only on the temperature but also on the displacement from equilibrium (i.e., on the actual structure that is relaxing).[66, 71] If equilibrium is being approached from a higher temperature by a more open structure, the relaxation time increases as the more compact equilibrium structure is approached because the fundamental diffusion process itself slows down, as indeed Fig. 4 implies. In some systems the dependence on (instantaneous) "structure" is much more important than dependence on "pure" temperature, in which case nonequilibrium structures can continue to relax at quite low temperatures. This type of nonlinearity seems intuitively reasonable in cooperative systems, and it is of interest to determine whether there is direct evidence for

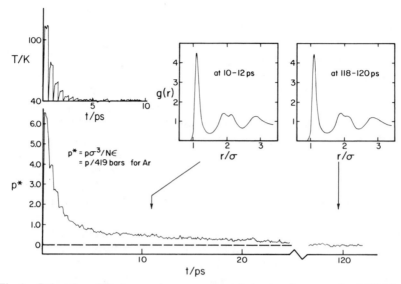

Fig. 8.   Relaxation of the temperature, pressure, and structure for a sample of 864 LJ particles quenched from near the triple point ($T^* = 0.75$) to a density ($\rho\sigma^3 = 0.97$) and temperature ($T^* = 0.33$, $T = 39°$K) in the glass transition region, see Fig. 7 (Ref. 12). To simulate the
quench, which is seen to take place over 3 to 4 psec, particle velocities are scaled to the required
temperature every 50 time steps (0.5 psec). Configurational relaxation—reflected in the pressure decay—takes more than 100 psec. (From J. H. R. Clarke, to be published.)

it in the structurally loose, short relaxation time, condition of the computer
simulations.

We have assumed the existence of the foregoing type of nonlinearity in
Fig. 5a for T_4, where the slope of the plot of l^2 in the initial 5 psec at T_4 is
drawn larger than that in the latter 5 psec, for which the structure is closer
to that of the equilibrium state. This implies that the glass transformation
in the simulated systems should be associated with a finding of diffusion
coefficients that are anomalously higher than anticipated from higher temperature measurements, as indeed has been the case.[7, 12, 20] The clearest evidence to date that "structure," as distinct from temperature, plays an important role in the behavior of these simple systems, is the recent observation by Cape and Woodcock[20] that a large (4000-particle) soft-sphere system quenched from high temperatures to a temperature below even the
"ideal glass transition temperature" for equilibrated structures [at which
$S(\text{liquid})_{(\text{extrapolated})} = S(\text{crystal})$[58, 59]—see Section V] continues to relax
slowly toward the smaller volumes required by the equilibrated fluid at
somewhat higher temperatures. The changes in the structure itself that

accompany such a relaxation are illustrated for the LJ case in the insets to Fig. 8, which are discussed further in Section IV. A.

At the moment the "noise" in the mean squared displacement versus time curves does not permit easy observation of this type of time-dependent behavior, and it remains a matter for future investigation. Also of interest is the question of whether, under the highly diffusive conditions of computer simulations, the systems very near equilibrium relax exponentially ("single relaxation time") or otherwise, as for laboratory glasses.[66, 67, 71, 72*] Much could be learned by comparing the approach to equilibrium after perturbations from both above and below the final temperature of equilibration.[71]

Finally, one other factor that may affect the approach to equilibrium for simulated systems should be mentioned. There are indications from a recent study of the LJ system[12] that supercooled samples approach constant energy much more rapidly when the sample size is small. Since macroscopic systems have a fixed (average) relaxation time for a given temperature and pressure, such a result is a matter of some concern. Extensive studies will be necessary to verify such effects, however, because statistical fluctuations obscure the relaxation of configurational properties close to equilibrium.

Since structural relaxation arises from the decay of density fluctuations, one might expect, from simple hydrodynamics,[73] that the relaxation time at a given self-diffusion coefficient, would show a k^{-2} dependence where k is the smallest wave vector allowed by the primary box dimension. A system size dependence for the relaxation time also is predicted by generalized hydrodynamics in which the transport coefficients are dependent on both k and w.[74]

IV. THE ROLE OF THE PAIR POTENTIAL

A. General Considerations

The relaxation phenomenology described above should apply rather generally to supercooling or supercompressing liquids irrespective of the interaction potential; but the changes in thermodynamic properties and the variations in structure that accompany the relaxation might be expected to

---

*Note Added In Proof: This question has now been answered. Stress relaxation functions for LJ argon obtained in recent simulations (S. M. Rekhson, D. M. Heyes, C. J. Montrose and T. A. Litovitz, *J. Non-Cryst. Solids* 38-39, 403 (1980); D. M. Heyes, J. J. Kim, C. J. Montrose and T. A. Litovitz, *J. Chem. Phys.* 73(8), 3987 (1980)) show that for argon also the approach to equilibrium is non-exponential. Remarkably enough, the decay function proves to be almost indistinguishable in form from that measured in the laboratory for SiO_2 glass.

show some specific dependence on potential. We now examine such dependences as have been found for the various pair potentials so far examined in computer simulations of supercooling liquids. The possible effects of many-body interactions on supercooling have not been investigated and are not considered here.

It is, of course, one of the major advantages of the simulation method that virtually no restrictions are placed on the choice of the intermolecular potential that defines the system. In the search for understanding of the changes of thermodynamic properties that define the glass-transition phenomenon, this advantage largely offsets the disadvantages, discussed above, of being restricted to observing a very "smeared-out" version of the transformation.

Let us first consider briefly the relationship between some of the simpler potentials currently being utilized in metastable-state simulations.

The simplest but also the least physical is the hard-sphere model first simulated by Alder and Wainwright.[6] The limiting ($T\rightarrow 0$, or $P\rightarrow\infty$) structure of the "glassy state" of this model, now evaluated from larger systems[16] than the original 32-particle case,[6] is essentially the same as that obtained from the classical steel ball packing experiments of Bernal and Finney[75, 76] (see Section IV. A). A step closer to real systems are the "soft-sphere" models studied now by several authors.[18–20] The soft-sphere model, (1), stands at the limit for high temperatures or pressures, of realistic models involving attractive forces, of which the Lennard-Jones 6-12 potential, (2), is the classic case

$$\phi(r)=4\varepsilon\left[\left(\frac{\sigma}{r}\right)^{12}-\left(\frac{\sigma}{r}\right)^{6}\right] \tag{11}$$

More work has been done on the properties of argonlike LJ than on any other fluid.

Systematic variations in pair potential between these extremes hold promise of revealing (via the glass transition—no matter how smeared out) how the features that distinguish between completely equilibrated and completely frozen states are related to the different parts of the pair potential. For instance, the finding that both hard-sphere and soft-sphere systems "jam" into amorphous nondiffusing structures at high densities associates the onset of shear rigidity, or "glassiness," with the repulsive potential. On the other hand, the absence of the typical C_p discontinuity at the "transition" from these cases, but its presence in the case of LJ argon, associates this phenomenon, most characteristic of laboratory glasses, with the attractive part of the potential.

These second-order-like characteristics of the transition in systems with realistic potentials might therefore be expected to weaken and vanish at very high temperatures and pressures, where the attractive part of the pair potential becomes ineffective, yielding soft-sphere behavior. The latter systems, defined by (1), exhibit scaling properties (due to the particular simple form for the partition function[77]), which lead to correspondences between the temperature and density dependences of all the thermodynamic and transport properties.[18] It has been found[78] that many condensed-phase properties of real atomic substances, such as the transport coefficients of diffusion and viscosity, structure factors, and melting points, closely obey the exact scaling laws of the soft-sphere model with n approximately 12. The reason for such simplicity of behavior is that all these properties depend primarily on the repulsive part of the pair potential, and the attractive tail is only a small perturbation. The soft-sphere model can be shown, for example, to obey the Lindemann law of melting exactly, and to follow the empirical Simon equation for the melting line.[26] Thus we might expect that studies of the glass-forming properties of soft-sphere models might lead to valuable generalizations, besides forming a bridge between the idealized hard-sphere model and the more realistic (at low temperatures) models with attractive potentials.

In the sections that follow we compare the thermodynamic, transport, and structural properties obtained from the above-mentioned studies of hard and soft spheres and Lennard-Jones atoms in an attempt to assess the present level of our understanding of the effects of the pair potential on these aspects of glass formation.

B. Thermodynamic Properties

The foremost thermodynamic property associated with any phase boundary is the location of its surface in the p-V-T phase diagram. Most laboratory experiments of glass formation are carried out in a particular V-T plane, usually for atmospheric pressure, and the temperature dependence of volume through the transition is determined. If the glass transition is indeed dictated by the repulsive part of the potential, we expect, at least for simple steeply repulsive systems, that it will occur at the same molecular-reduced volume for many real and model systems and that this will be largely insensitive to the strength of the attractive component of the pair potential relative to kT.

Thus, if we take the packing fraction of the glass transition in the hard-sphere model as a reference, a straightforward prediction of the glass transition volume in real systems based on a complete neglect of temperature effects can be regarded as a perturbation treatment of zeroth order. This question has been examined by Andersen[13] and Abraham[14] for the glass transition in amorphous metals using the hard-sphere reference, with favorable comparisons; they conclude that provided the effective hard-sphere diameter is carefully defined, the hard-sphere model offers an excellent reference state for the location of the glass transition in the p-V-T plane. With the available data for soft-sphere and Lennard-Jones models, we can now examine these postulates in more detail, bearing in mind the difficulties due to the diffuseness of the transition in simulation and the insufficient accuracy of the present data.

The existing p-V-T data on the glass transition for hard spheres, soft spheres, and Lennard-Jones molecules are summarized in Fig. 9. Here we have followed a common practice among glass scientists in relating glass properties to those of the corresponding crystalline phases, and reducing the temperature relative to that of the equilibrium freezing transition. A more fathomable set of curves is obtained if one reduces the volume by the crystal volume at $0°K$ (or at $NkT/\rho\sigma^3 = 0$) instead of at T_f, as shown elsewhere.[10] Both these comparisons can be misleading, however, and they should be viewed with caution, since the variations are almost certainly due to a sensitive dependence of both the freezing transition and the properties of the crystalline phase on the details of the pair potential, rather than a reflection of any aspects of the behavior of the amorphous phases in the supercooled liquid and glassy states. The excess volumes are of the same order as most experimental molecular glasses, and the glass transitions range from

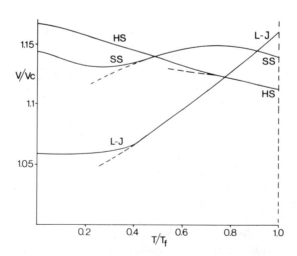

Fig. 9. Volumes of the supercooled fluid and vitreous phases of hard spheres, soft spheres, and Lennard-Jones molecules (at $p\sim0$) relative to their respective crystalline phases as a function of temperature reduced according to the equilibrium freezing temperatures.

around 0.3 T_f for the Lennard-Jones model and 0.4 T_f for the soft-sphere model, to 0.8 T_f for hard spheres, as defined by the extrapolation of diffusion data. It is interesting to note that the density of the amorphous close-packed, hard-sphere MD glass is the same as that found for steel balls.[76, 79]

Hudson and Andersen[13] have argued that the glass transition should occur at approximately the same "effective" hard-sphere packing fraction, based on the Weeks, Chandler, and Andersen (WCA) expansion,[80] in several different simple systems and that this is close to the packing fraction found for the hard-sphere glass transition. Abraham[14] showed that this criterion applied irrespective of whether an LJ glass was formed by cooling, pressurizing, or a combination of both. Although this is, or could be, a useful generalization, which emphasizes the dominant role of the repulsive part of the potential in glass formation, there are, at present, difficulties in testing the postulate accurately. The first arises in the definition of the glass transition point in computer experiments, discussed in Section III. C, and the second difficulty rests in the definition of the effective hard-sphere diameter for realistic potentials. Thus in the zeroth order of perturbation it is always possible to vary this parameter to fit the data for a single system. What we seek is a universal criterion that is approximately applicable for many systems.

Clarke,[12] for example, reports a value of the effective packing fraction for the low-temperature, Lennard-Jones system of 0.6, somewhat higher than the value 0.516 obtained by Hudson and Andersen using the MD data of Rahman et al.[8] The latter value is close to the packing fraction at the hard-sphere glass transition. In this case the discrepancy is due not to differing definitions of the effective packing fraction but to the definitions of the glass transition point. If the same criterion is applied to both sets of data for T_g, the same result is obtained. We are at present at liberty to choose these definitions to reproduce a variety of effective packings in the high-density range; much more accurate data are required.

For the soft-sphere model the same WCA criterion for defining the effective hard-sphere diameter[80] gives totally unrealistic values for the glass transition packing fraction. This illustrates the limitations of the zeroth-order perturbation treatment and suggests that agreement with the hard-sphere model is being forced by the single adjustable parameter in the low-temperature range. An alternative procedure is to fix the effective diameter by the density of the hard-sphere system when the height of the first peaks in the static structure factor are the same. This criterion gives then a value of the packing fraction for soft spheres at the glass transition of 0.52, in excellent agreement with the hard-sphere model.[20]

Turning again to Fig. 9, then, we see that neither the degree of under-cooling, nor the excess volume relative to the crystal, is a determining feature in glass formation. These are largely manifestations of crystal properties. The important criterion is the volume of the system, expressed in molecular units, in excess of that for a system of hard spheres having the same effective diameter at the *onset* of glass formation. An interesting property of the soft-sphere model is that as a consequence of its scaling properties, the pressure dependence of T_g is fixed. It follows the relation

$$T_g^* = Cp^*, \qquad \text{that is,} \qquad \frac{kT_g}{\varepsilon} = C\left(\frac{p\sigma^3}{\varepsilon}\right)^{4/5} \qquad (12)$$

where the constant C has been determined[20] to be 0.0421. It will be interesting to see to what extent this law describes real systems, and also the Lennard-Jones model at low temperatures, when more accurate data have been obtained. Deviations from this law would be a reflection of the effect of the attractive part of the pair potential, which is expected to be small for the reasons discussed above.

Having established that the glass transition in simple liquids occurs in a narrow density range and is only weakly temperature dependent, it is instructive to proceed to compare the behavior of second-order thermodynamic properties, in particular, the isobaric heat capacity through the transition as a function of density. A pronounced discontinuity in this property is the most general manifestation of the laboratory glass experimental transition. Unfortunately, there is still a dearth of reliable simulation data for derived properties of these simple models; the meagre results to date for hard-sphere, soft-sphere, and Lennard-Jones molecules are collected in Fig. 10. These data, however, serve to emphasize some important points in our understanding of computer glasses in relation to laboratory glasses.

Figure 10 compares the isobaric heat capacity as a function of density, scanning the entire amorphous range from the ideal gas to the high-density glass. To simplify the comparison, we choose the Lennard-Jones effective hard-sphere diameter to be the potential minimum; this is somewhat higher than given by the WCA criterion. The soft-sphere density is then the appropriate high-temperature limit, and it is seen that when $\varepsilon = kT$ in the soft-sphere model, the density at the glass transition, 1.47, is more or less the same as that of the Lennard-Jones model at approximately zero pressure. This shows that the soft-sphere model (with $\varepsilon = kT$) is a better zeroth-order perturbation representation of the LJ glass transition than the hard-sphere model with $\sigma = r_0(\text{LJ})$, which occurs at a slightly lower density.

It is interesting to observe in Fig. 10 the manner in which the change in C_P on glass formation develops on going from the hard-sphere model to the soft-sphere model and then down the temperature scale to the zero pressure LJ model. In the hard-sphere model the only contribution to C_p arises from density fluctuations (i.e., the pV derivative with respect to T), and

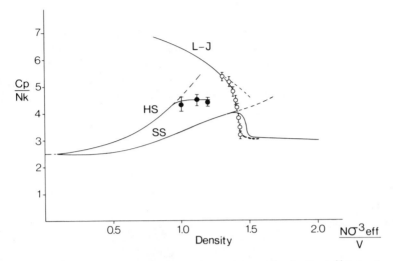

Fig. 10. Isobaric heat capacities of the amorphous phases of hard spheres,[16] soft spheres,[20] and Lennard-Jones[12] molecules as a function of density.

these can be regarded, above the glass transition, as being of vibrational and diffusional contributions. When the density is increased through the transition, only the vibrational modes remain; thus there is a discontinuity in the derivative of C_p but no pronounced discontinuity in C_p itself. Therefore the absence of potential energy fluctuations in the hard-sphere system leads to a glass transition without the usual pronounced second-order-like discontinuity. In the soft-sphere system the potential energy fluctuations, which can only be repulsive in character, are weak, and the C_p discontinuity is thus weak also.

In the case of the Lennard-Jones system, on the other hand, there is a large contribution from the attractive interactions to potential fluctuations that determine the C_v component of C_p, which is absent in hard spheres. On traversing the glass transformation range, a single configuration becomes frozen in (assuming no crystallization) and associated with this, C_p decreases by about 40% (though not as abruptly as in laboratory experiments for the reasons discussed previously). This percentage ΔC_p is typical of laboratory observations (see, e.g., Fig. 6).

Since the soft-sphere system represents the high-pressure, high-temperature limit of the LJ system, it is predicted from the data above that the thermal manifestation of the laboratory glass transition will weaken with increasing pressure as it is displaced to higher temperatures. Indeed, increasing difficulties in T_g detection have been noted in high-pressure studies.[81]

C. Transport Properties

Owing to the difficulties arising from the limitations in relaxation times in computer experiments, the existing data cannot say very much about the form of the mass transport coefficient behavior in the low-diffusivity region approaching the glass transition, apart from the obvious and necessary statement that the diffusion and fluidity coefficients are approaching zero (Figs. 3 and 4). For hard and soft spheres, the diffusion coefficient and the fluidity appear to be linear functions of volume and to extrapolate to zero at or around the densities at which the breaks in C_p were observed in Fig. 10.

In Fig. 11 we have plotted the fluidity of the hard-sphere fluid (from Alder, et al.[82]), together with some very recent data of Michels[83] on the fluidity of the square-well liquid. The square-well model has a uniform attractive potential between σ and 1.5σ of depth ε. When we extrapolate linearly the fluidity of the square-well system, an interesting result is obtained that vindicates the inferences of the preceding section.

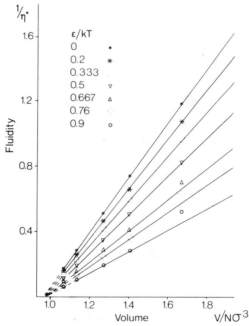

Fig. 11.   Fluidity of the square-well fluid (width 1.5σ) as a function of volume over a wide temperature range. These data were supplied prior to publication by Dr. J. Michels of the Van der Waals laboratory, University of Amsterdam.

From Fig. 11 we see that in the high-temperature limit the hard-sphere model has a glass transition volume predicted by this linear extrapolation to occur at $N\sigma^3/V = 1$, that is, at just the same density as the break in the derivative of C_p shown in Fig. 10. Yet more interesting is the effect of the attractive well as the temperature (kT/ε) is lowered. Figure 11 shows that although the reduced fluidity at constant T and V is affected by the attractive well at large V, the glass transition density seems as though it will be entirely determined by the hard core. Thus, in the simple square-well model, we expect the zeroth-order perturbation prediction for the glass transition density to be almost exact, the thermal effects associated with the attractive well having negligible effect. Figure 11 bears a striking resemblance to Hildebrand's collection of liquid metal fluidity data.[55]

We return now to the important question introduced in Section III. B dealing with p, V, and T dependence of transport properties in the vicinity of the glass transition. The most accurate data so far seem to be those obtained for the soft-sphere model. These have in part been seen already in Fig. 4 in normal units based on argonlike soft spheres. It is more natural, for internal comparisons, to use reduced soft-sphere $(n=12)$, self-diffusion coefficients defined by (2). Fig. 12 plots results for several authors[19, 20, 84] against the reduced volume, which is defined as follows:

$$V^* = \frac{V}{N\sigma^3}\left(\frac{kT}{\varepsilon}\right)^{1/4}$$

The data are accurately described by the Batchinski-Hildebrand form of

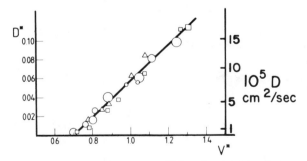

Fig. 12. Values of the reduced diffusion coefficient for the soft-sphere model as a function of the reduced volume from molecular dynamics simulations: circles, from Cape and Woodcock[20]; squares, from Hiwatari et al.[19]; triangles, from Ross and Schofield.[84] The scale at the right shows the equivalent diffusivities for argon-like soft-spheres.

reduced volume dependence

$$D^* = C(V^* - V_0^{BH^*})$$ (13)

for reduced diffusivities greater than $D^* = 6.5 \times 10^{-3}$ (corresponding to argonlike normal diffusivities of $> 0.6 \times 10^{-5}$ cm^2/sec). It is in the same fluidity-diffusivity range that the Batchinski-Hildebrand relation was established as an experimental law. The limiting reduced density implied by this relation is 1.43, which is in the middle of the glass transition range of the heat capacity decrease seen in Fig. 10. This is quite consistent with a "break" (however diffuse) in thermodynamic properties being associated with the "freezing in" of a single configuration.

Much interest remains, however, in the question of the long time limiting density because Fig. 4 indicated that the deviation from the Batchinski equation seen in all laboratory studies at low D becomes greater than the uncertainty in the simulation diffusion coefficients (only) at $D < 5 \times 10^{-6}$ cm^2 sec. At this diffusivity, relations discussed in Section II. B imply that the time needed for full structural equilibration would exceed 10 psec. The dependence of D on the degree of equilibration is not known at this time, but it seems probable that a major computing effort will be needed to determine whether the volume dependence of the *dense* atomic liquid diffusivity is as simple and significant as (13) implies or is more complex, as for the known behavior of low-diffusivity laboratory liquids (Fig. 4).

In this region the laboratory liquids follow a three-parameter exponential relation first suggested by Doolittle[85]:

$$\phi T \propto D = A \exp - \left(\frac{B}{V - V_0^D} \right)$$ (14)

The point of interest is to decide whether this behavior, and the denser ultimate packing it implies, reflects the asymmetry of the molecules,[55] which seems to be a prerequisite for experimental accessibility to this diffusivity region, or reflects some more fundamental aspect of the relation between motion and packing of particles, which could embrace simple atomic liquids as well, as has been theorized.[86, 87] It is probable that (well-funded) MD experiments will be able to decide this issue in the near future, unless the increasing probability of nucleation in the same temperature region should frustrate the inquiry.

NOTE ADDED IN PROOF: This issue has now been resolved. A system of 500 hard spheres has been studied at successively decreasing volumes between 1.08 and 0.94 on the reduced scale of Fig. 4 (equivalent to a reduction from 150% to 131% of the close-packed volume) allowing up to 10^6 collisions to establish a linear l^2 vs t plot at each density. Before the sequence was finally terminated by crystallization at the highest density, a distinct departure from the linear volume dependence of diffusivity was observed, closely following the behavior of the laboratory fluids. This is discussed elsewhere (L. V. Woodcock and C. A. Angell, *Phys. Rev. Lett.*, to be published; L. V. Woodcock, *Ann. N.Y. Acad. Sci.*, to be published).

D. Structural Effects

All the time-dependent phenomena discussed in the foregoing sections are consequences of increasingly retarded structural relaxation in the supercooled state as the number densities approach conditions of amorphous close packing. It is of some interest to consider structural effects in these systems and to examine any dependence on the form of the interaction potential.

Structure in liquids and amorphous solids often has been discussed in terms of the radial (pair) distribution function $g(r)$. Although this function is important as an experimentally accessible quantity and therefore is a convenient starting point in simulation studies, the attention it has received is out of all proportion to its value in revealing fine details of structure. That the glass transition is seen more clearly in second-order thermodynamic properties suggests that an examination of structural fluctuations might lead to more insight into its structural origin, but we postpone further discussion on this point.

1. Structural Characterization of the Glass Transition

Since the process of glass formation is a central interest of this chapter, we consider first some results of dynamic simulation experiments similar to those referred to schematically in Fig. 5. As the density of a system of spherical particles is increased from the fusion points, three significant, continuous, effects occur in $g(r)$ as shown in Figs. 14 and 16 (Section IV.D.2) for LJ particles and soft spheres. First, the peaks become narrower and more sharply defined as a result of restricted particle displacements. Second, discernible structure begins to extend to larger distances. Third, the second-nearest-neighbor peak begins to split into two components.

In regard to the nearest-neighbor geometry, the narrowing of the first peak in $g(r)$ is partly due to decreased amplitudes of vibration, but also it is indicative of a narrower range of coordination numbers. This is demonstrated in Fig. 13 for the LJ system, where the distribution of coordination numbers, defined here as the number of particle neighbors within a distance corresponding to the first minimum in $g(r)$, is shown at a range of densities through the zero pressure glass-transformation region.[12] There is nothing unpredictable about these data. The most probable coordination number is 13 in each case and, as should be expected, there is no distinction between the metastable liquid and the glass. An empirical criterion for the glass transition using $g(r)$ has been suggested by Wendt and Abraham,[71] who, following Raveché et al.[93] measured the ratio of the first minimum to the first maximum for the supercooled LJ liquid. They found that the ratio showed a characteristic break in the glass transformation region, similar to that observed for the density.[7, 12, 14] Since the height of the first maximum must reflect the bulk density, such a result may not be unexpected. However, the gradual increase in the ratio above the transition[11] may be partly associated with the onset of diffusive motions,[94] which necessarily increase the height of the first minimum.

The doublet structure of the second peak in $g(r)$ has been discussed at length in connection with the geometry of amorphous close packing of spheres.[87, 88] The splitting is not universal in dense systems, however. It is not observable in cases of very soft interaction potentials, as has been noted in static simulations[88] and also in a recent MD study of the Gaussian core model.[21] It should be noted, however, that contrary to occasional suggestions, the splitting can in no way be said to *distinguish* between the glassy and metastable liquid states. If it occurs at all, the feature appears gradually as a liquid is cooled and/or compressed. It is often discernible for liquids close to the normal freezing temperatures. The longer of the two preferred lengths corresponds to collinear arrangements of atom triplets, whereas the shorter one corresponds to the distance of closest approach of a second-nearest neighbor across an intervening pair of atoms in the nearest-neighbor shell. Even in the case of the liquid KCl model, where the second-nearest neighbors are like ions, the doublet structure is clearly visible at high densities.[23] This is perhaps remarkable because close packing is constrained by the strong tendency for charge ordering in this system.

Since the glass transition is characterized by discontinuous changes in second- (and higher) order thermodynamic properties, it would seem relevant to give some attention to what might be termed "second-order" structural properties. Just as $n(r)$, from which $g(r)$ is derived, is the average number of particles within a sphere of radius r around a reference particle, so we can define moments about this mean distribution. Here we focus on the second moment, termed the radial fluctuation function $W(r)$, and defined as follows:

$$W(r) = \left[\langle n(r)^2 \rangle - \langle n(r) \rangle^2 \right] \left[\frac{4}{3} \pi r^3 \rho \right]^{-1} \qquad (15)$$

One significance of $W(r)$ is that its limiting value gives the isothermal

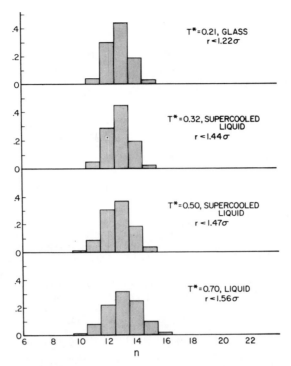

Fig. 13.   Probability distributions $P_{(n,r)}$ of nearest-neighbor "coordination number" (n) for a sample of 216 LJ particles quenched to various states on the zero-pressure isobar from near the triple point. In each case the data were obtained after allowing 20 psec for "equilibration." Nearest neighbors are defined (arbitrarily) by separations up to the first minimum in $g(r)$ as indicated; $T^* = kT/\varepsilon$.

compressibility, K_T

$$K_T = \left[\frac{\partial (\ln V)}{\partial p} \right]_T = W(r \to \infty) \frac{V}{NkT} \qquad (16)$$

A small discontinuous change in K_T at the isobaric glass transition has been observed[12] for the LJ liquid; thus it is perhaps appropriate to examine the behavior of $W(r)$ in this system, although of course only the *short-range* behavior can be determined. Figure 14 displays $W(r)$ for the LJ liquid and glass at various densities along the $P = 0$ isobar. The form of this function follows very closely the form of $g(r)$, indicating quite sensibly that the fluctuations are largest where the local density of particles is largest. The

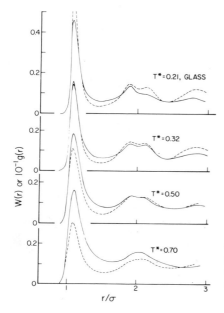

Fig. 14. Radial fluctuation functions $W(r)$ (solid curves) and radial distribution functions $g(r)$ (dashed curves) for the LJ states discussed in Fig. 15. Note the behavior of the outer of the second-nearest-neighbor peaks in $W(r)$ through the transition.

significant narrowing of the first peak with increasing density is again indicative of a consequent decrease in the distribution of coordination numbers, but there is no apparent discontinuity through the glass transition. As for $g(r)$, a doublet structure appears in the peak corresponding to second-nearest-neighbor distances as the density is increased. One interesting point to note, however, is the change in relative intensities of these two components between the metastable liquid and the glass. The small r component in fact changes little, but the larger r component decreases significantly in height. Fluctuations at distances corresponding to this peak appear to be suppressed in the glass.

One can distinguish two contributions to $W(r)$ in a dense liquid. The first arises from purely vibrational motions and would be observed even for a finite temperature crystal. A second contribution, however, can be envisaged as arising from motions giving rise to diffusion. The latter may not of course be simple translation motions but may represent highly anharmonic vibrations of particles between neighboring potential energy wells. In any case, any description in terms of single-particle motions is likely to be inadequate. Since a glass is characterized by the absence of diffusion on the experimental time scale, $W(r)$ here will be dominated by purely vibrational effects. Perhaps the intensity increase in the large r component of the second peak can therefore be associated with the onset of diffusional motions. The "local compressibility" for spheres defined by this radial distance (which depends on the available distribution of configurations) increases in going from the glass to liquid. This has more to do with the dynamics of second-nearest-neighbor packing than nearest-neighbor effects, emphasizing the cooperative nature of any particle motions at high densities.

An advantage of dynamic simulations experiments is that they permit evaluation of the time dependence of structural relaxation for a particular system. The importance of such studies in the present context is demonstrated in Fig. 8, which shows some radial distribution functions for the metastable LJ liquid at $\rho\sigma^3 = 0.97$ and $kT/\varepsilon = 0.33$, which is in the glass transformation region. The only difference between the functions is the allowed relaxation time for structural rearrangement. As this time is increased, the shorter distance component of the doublet grows at the expense of the other. The second-nearest-neighbor distribution achieves a more compact arrangement, which generates a low pressure at the same density (or a higher density state at constant pressure). It is not now known whether a similar effect can occur in the hard-sphere fluid. The effect of this densification on macroscopic properties such as energy or density is apparently rather small (amounting to 1% or so at most) and has been difficult to detect within the limits of statistical error.[7, 12] Nevertheless the result described emphasizes the entirely operational definition of the glass transition. Some account must clearly be taken of any differences in cooling and/or compression schedules when comparing structural features in different glasses.

2. Structural Characterization of the Glass

Considerable effort has been expended in characterizing the instantaneous structures of close-packed static amorphous assemblies of spherical particles, and this work has been extensively reviewed.[87, 88] High-density, simple model liquids also have been examined.[7-14] Some attention has been directed toward dynamic structural properties,[8] an area of great potential interest.

First, it is of interest to compare (Fig. 15) the radial distribution functions for an amorphous close-packed assembly of macroscopic (steel) spheres[76, 78] and for the amorphous hard-sphere solid obtained by gradual densification by MD of a 500-sphere system with the usual periodic boundary conditions.[10] Although there are some minor differences for the third peak in $g(r)$, there is little doubt that the mechanical shaking, and the more thermodynamic shaking in the MD experiment, are leading the system toward the same structure. The mechanical shaking experiment can give

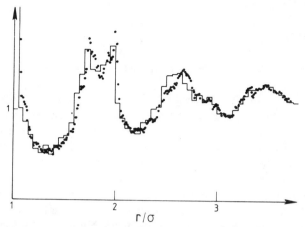

Fig. 15. Comparison of radial distribution functions for an MD-generated amorphous hard-sphere solid at a density ($\rho\sigma^3 = 1.2$) very slightly less than the amorphous close-packed density (circles), and a random closed-packed assembly of macroscopic spheres at $\rho\sigma^3 = 1.22$ (histogram). From Refs. 10 and 78.

information on the structural changes occurring as the close-packed limit is approached only by stopping the process after different periods of time. This is roughly equivalent to the MD experiments carried out recently to determine the zero-temperature packing density using different quenching rates.[16] Although the results of both types of experiment suggest that the system jams at lower densities in the "faster" experiments, neither has been analyzed to provide the $g(r)$ needed to make comparisons.

It is interesting to note that it has so far apparently proved impossible to grow the high-density, amorphous close-packed structures characterized in Fig. 15 by successive attachment of single hard spheres to an amorphous heap,[90] although the reverse process of dismantling such an assembly must clearly be possible. This emphasizes that the densification process involves increasingly narrow pathways in phase space, with cooperative reorganization possibly being an essential ingredient. This "annealing" process can result from the phase-space exploration implicit in computer simulation or from the shaking operations on macroscopic assemblies.

Turning to the results of dynamic simulation experiments, Fig. 16 compares radial distribution functions for hard-sphere, soft-sphere, and LJ glasses, all in comparable states. The compression rates ($\rho^{-1}d\rho/dT$) for the preparation of the soft-sphere sample,[91] the LJ sample,[12] and the hard-sphere case[16] were 0.024, 0.010, and 0.003 psec^{-1}, respectively, all on a time scale appropriate for LJ argon. Also the densities are all comparable as judged by the reference scales discussed in Section IV.C. If one uses the

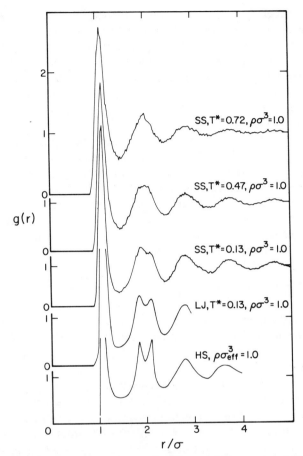

Fig. 16.   Comparison of radial distribution functions $g(r)$ for samples of 4000 soft spheres (SS, Ref. 91), 500 hard spheres (HS, Ref. 10), and 216 LJ particles (Ref. 12) in the glass transformation region. Three soft-sphere states have been chosen to show structural changes on "cooling" from above the freezing temperature (0.57 at $\rho\sigma^3 = 1.0$) to below the glass transition temperature (0.24 at $\rho\sigma^3 = 1.0$). The hard-sphere glass is at $\rho\sigma_{HS}^3 = 1.20$; on the basis of perturbation theory this is a reference state for the LJ liquid at $T^* = 0.13$, $\rho\sigma^3 = 1.0$ (see Ref. 12). Using (1) and (11), σ is as defined by the LJ or SS potentials. For the hard spheres, an *effective* value is used with $\sigma = 0.94\sigma_{HS}$. Vertical scales are not identical.

perturbation theory criterion (e.g., Ref. 12), the LJ glass should have a structurally equivalent hard-sphere reference state with a density $\rho\sigma^3 = 1.2$. It is seen from Fig. 16 that the relative intensities of the doublet components of the second-nearest-neighbor peak appear to vary according to the interaction potential. In particular, there is a striking reversal of the relative intensities in going from the hard spheres to the soft systems.

A similar effect was noted previously for experiments on static assemblies in which the structure is allowed to relax along a potential energy surface by substituting a soft LJ potential for a hard-sphere interaction.[95, 96] The comparison for the hard- and soft-sphere glasses is particularly convincing because the latter sample was produced at a higher compression rate and for a much larger sample. Both these effects should decrease the extent of structural relaxation, but a much more compact local structure is nevertheless achieved by the soft spheres. "Softness" of the interaction potential appears to have an important influence on relaxational behavior in the glass transition region. Densification relative to hard spheres is to be expected, and this can be largely associated with structural relaxation involving second-nearest neighbors.

Very striking effects of the pair potential parameters can be observed if we admit for consideration the glasses of single-component ionic substances for which a simple and successful[23] form of effective pair potential is the simplified Born-Meyer-Huggins potential.[97]

$$U(r)_{ij} = z_i z_j \frac{e^2}{r} + A \exp\left(\frac{\sigma_i + \sigma_j - r}{\rho} \right) \tag{17}$$

In these cases the packing problem must be solved subject to the constraint of local electroneutrality, under which conditions the change of a single ion size parameter by a few percent can completely alter the preferred coordination scheme.

It is in the simulation of ionic systems that some of the closest contact with experiment has been made. Almost quantitative agreement, for instance, has been found between simulated and X-ray diffraction scattering patterns for vitreous BeF_2.[22] The agreement of simulated and X-ray-based pair distribution functions for the classical oxide glass SiO_2 has also proved to be surprisingly good.[23] In each of these cases it has been therefore shown that the celebrated anion-bridged tetrahedral network structure can originate from simple spherically symmetric interactions alone.

It is not computationally expensive to explore the geometric consequences of changes in pair potential parameter in these cases. Figure 17 illustrates the time step requirements to establish the basic coordination groupings following a change in the cation repulsion parameter, σ_i of (17), of an initial BeF_2 simulation to a value more appropriate to Ba^{2+}.[101] The figure shows the radial displacement of F^- neighbors around a randomly chosen doubly charged cation as a function of time. What were initially only anharmonic oscillations characteristic of the network glass or viscous liquid become violent flights on the instantaneous change of cation size. However, in a little more than 200 time steps (<1 psec), the essential geometry

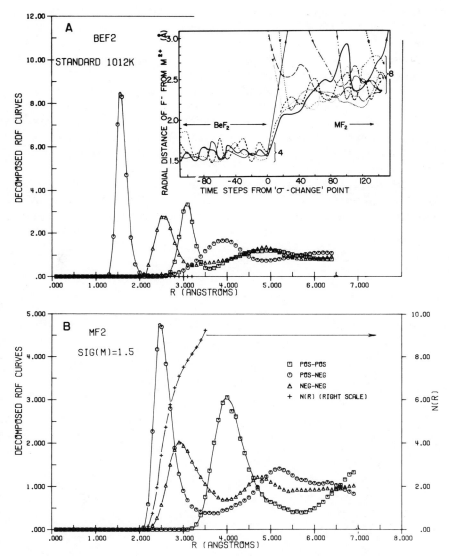

Fig. 17. Demonstration of effect of cation repulsion parameter σ_{z+} [see (17)] on structure of MX_2 ionic systems. (a) Radial distribution pair functions for BeF_2 at the simulation glass transition temperature ($D = 1 \times 10^{-6}$ cm²/sec). (b) Radial distribution pair functions for MF_2 ($M \approx Ba^{2+}$) near simulation glass transition temperature: cation coordination number (right-hand ordinate) has increased to ~8 from 4.0 characteristic of BeF_2. Inset illustrates rapidity with which the structural change occurs after instantaneous change of σ_{z+} at time step zero. (Data from ref. 101.)

of the new state is established. Of course, subtle changes will continue for several relaxation times, but generally these are not the changes of major interest.

An interesting polymeric structure, based on octahedrally coordinated cations corner-linked to each other by shared anions, results from a similar increase of cation size in an initially molecular MX_4 liquid.[101] Such structures, and their more cross-linked relatives with still higher coordination numbers consequent on smaller anion-cation radius ratios, are no doubt fundamental to the new class of fluoride glasses based on ZrF_4 and HfF_4, which are currently a focus of attention in laser and infrared transmission line technology.[98, 99]

Finally, particular and very practical advantages accrue from the simulation of multicomponent ionic glasses for which individual pair distribution functions of great physical importance in understanding—for example,

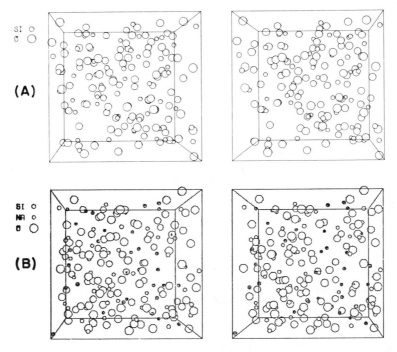

Fig. 18. Stereoscopic projections of (a) SiO_2 tetrahedral network structure at 300°K and (b) $Na_2O \cdot 3SiO_2$ "glass" at 1500°K. System is below the simulation glass transition temperature, since there is no Si^{4+} or O^{2-} diffusion occurring, though Na^+ diffuses rapidly through the disrupted network. Note breaking of Si—O—Si bridges to give nonbridging oxygens where Na^+ ions tend to cluster. (Data from ref. 51.)

electrical conductivities—can be obtained while they are currently inaccessible to laboratory determination. The basic structures of alkali-silicate, alkali-borate, and borosilicate glasses have recently been explored with new insight in the light of MD studies.[25, 100] Figure 18 shows stereoscopic projections of the structures of the SiO_2 tetrahedral network and the clustered alkali ion structure of the $Na_2O \cdot 3SiO_2$ melt.[25, 51] Again these structures are high "fictive" temperature structures, though they relate closely to what is known of the room-temperature structures of these systems. For another example, detailed information on the coordination, hence spectroscopic characteristics, of rare earth ions in BeF_2 glass (including relation to different "cooling" schedules) has recently been obtained by MC calculations.[102] Carried out with an awareness of the likely effects of dealing with ultra-high "fictive" (i.e., freezing-in) temperatures, such studies can be very informative and will doubtless receive much attention in the future. Most recently, Brawer has observed ions which are locally mobile below T_g and which are associated with network defects. The concentration of these is a function of cooling rate. He suggests these may be the origin of the ultra-low temperature anomalies (two-level systems) which have been the focus of much attention in recent years.

3. Characterization of Crystal Nucleation

It is of interest to inquire further into the structure of the metastable state, and particularly its relationship with the stable crystal into which it will eventually transform. It has been suggested that the temporary stability of supercooled states of spherical particles is related to the predominance of structures with fivefold symmetry, such as the icosohedron.[76] Such structures cannot fill space uniformly; nevertheless, quite large units of this kind (Fig. 19) have been identified in recent computer simulation experiments.[36, 91]

The spectrum of structural fluctuations within a sample contains both those characteristic of the liquid and those leading to homogeneous nucleation. As mentioned in Section II, the relative time scales and magnitudes of the respective fluctuations determine the lifetime of the metastable state. The study of such fluctuations constitutes one area in the investigation of supercooled liquids to which the MD method can make a unique contribution because of its ability to capture all nuances of the structural changes in time and space for tiny particle groupings beyond the range of observation by laboratory methods. Spontaneous crystallization via such fluctuations has now been observed in several prolonged simulation experiments on metastable liquids.[12, 18, 19, 21, 30–36] For small systems ($\sim$100 particles) the situation is rather confused, since crystallization appears to be promoted (if not initiated) by the periodic boundary conditions. For larger samples these

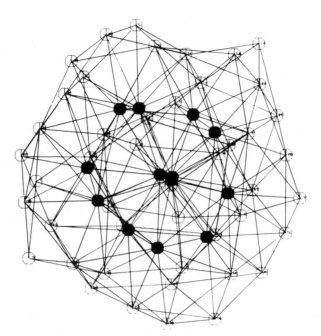

Fig. 19. An "amorphon"—a cluster with fivefold symmetry containing 61 atoms from a configuration of 4000 soft spheres compressed to a density $\rho/\sigma^3 = 1.0$ at $kT/\varepsilon = 0.13$. The central icosohedral atoms are shaded. (From Ref. 91).

spurious effects are probably absent, so that homogeneous nucleation is observed. It is interesting to note that there was apparently no sign of crystal nucleation in a recent extensive MC study of metastable LJ states using only 108 particles in an (N, P, T) ensemble with up to 24×10^5 moves at each point.[14] Under comparable conditions, crystal nucleation would certainly have been expected in an (N, V, E) MD ensemble.[12] It remains a possibility that the relevant structural fluctuations have different characteristics in the two cases.

One of the first indications of nucleation of ordered structure within a metastable liquid is the appearance and growth of a shoulder on the second-nearest-neighbor peak in $g(r)$ at about 1.54 r_0, which is a characteristic distance of the body-centered cubic lattice. It is at first sight surprising that the amorphous structure does not nucleate face-centered cubic, since both the second-nearest-neighbor distances are already very close to those of successive shells in the face-centered-cubic lattice, whereas only the outer peak corresponds to a body-centered cubic distance. However, Alexander and McTague[103] have argued that, other things being equal, the bcc fluctuations are more probable because, in essence, they may arise by co-operation of three density waves whereas FCC and HCP each require at least four.

For a more sensitive detection and classification of structural changes within the metastable liquid, the radial distribution function proves quite inadequate, and this had led to the development of alternative approaches. For instance, analyses in terms of the structure factor[31, 32, 34, 35] $S(k)$ and Voronoi polyhedra[33–36] have been useful in characterizing the nucleation event. Triplet correlation functions also have been discussed.[35, 93] In metastable liquid rubidium it has been shown by these methods[34, 35] that, as might be expected in this case, the initial growing nuclei have a body-centered-cubic structure. For the supercooled LJ liquid there have been separatereports of the formation of nuclei with body-centered-cubic[30, 32] and face-centered-cubic structures.[34] These may not be entirely conflicting results, however, but a manifestation of a general point made some years ago by Wood concerning simulation experiments on dense fluids.[104]

In small samples the character and course of the structural fluctuations that lead eventually to a single nucleation event are likely to be strongly dependent on the initial conditions. It is nuclei with the fastest local growth rate (or lowest local free-energy barrier) that will be observed. Rearrangement to some other macroscopically stable structure could occur at a later stage. In this regard it is interesting to note the results of a recent study of a very large sample (4000 particles) of the metastable soft-sphere fluid, in which *both* body-centered-cubic and face-centered cubic nuclei were identified.[36, 91] The results to date do seem to support the notion that structural fluctuations must yield a critical size of nucleus before crystallization occurs.[32, 35] It is largely unknown, however, how the growth of these fluctuations is related to the structural evolution of the metastable phase.

V. ON THE EXISTENCE OF AN UNDERLYING SINGULARITY

In discussion of the thermodynamics of supercooled liquids, considerable attention has been given to the observation by Kauzmann[105] in 1948 that as the glass transition temperature is approached on cooling, the entropy difference between the metastable liquid and the stable crystal is approaching a negative value. Below the experimental glass transition point, internal equilibrium no longer prevails; this leaves a puzzle concerning the thermodynamic description and physical nature of the low-temperature limit to the liquid state.

From the cases reviewed by Kauzmann, and many studied since,[58, 106, 107] the phenomenon seems to be universal, and it includes even the case of argon,[12] which is illustrated, and compared with a laboratory example, in Fig. 20b. The difference in heat capacity shows no indication of decreasing with temperature until the onset of glass formation; in fact it generally increases, as in Fig. 20a for the case of 2-methyl pentane. It therefore appears that the two entropy functions would inevitably intersect at a temperature (T_K) not too far below the glass transition temperature (T_g), were it not for the loss of internal equilibrium occasioned by the increasingly long relaxation time of the viscous liquid.

In the absence of any interceding phenomena, it seems necessary only to allow the supercooled liquid longer and longer periods of equilibration to ensure that its entropy will fall below that of the crystal at the temperature T_K (denoted T_2 and $T_{0(cal)}$ in Refs. 58 and 106, respectively). Such an occurrence, though not actually in violation of the third law at finite temperatures, would imply a contradiction of the Nernst heat theorem on the approach to $0°K$. This seems unlikely and raises the question of the existence of a thermodynamic singularity underlying the glass transition at or above the Kauzmann temperature T_K.

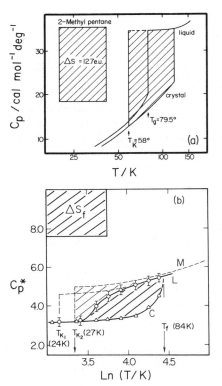

Fig. 20.   Two examples of the vanishing excess entropy paradox.[105,106] The area representing the entropy of fusion on the C_p versus T (log scale) plot has been matched to the area between the crystal and supercooled liquid heat capacity curves (the latter being extrapolated naturally below the T_g to indicate the supercooled liquid C_p for very slow (equilibrium) measurements. At the temperature T_K, S (supercooled liquid) would equal $S_{crystal}$, as can be seen from the relation $\Delta S = \int \Delta C_p \, d\ln T = \Delta S_f$ for limits T_f and T_K). (a) Data for 2-methyl pentane. [From D. R. Douslin and H. M. Huffman, J. Am. Chem. Soc., 68, 1704 (1946).] (b) Data for LJ argon from Ref. 12. Extrapolations using both MD heat capacities extrapolated below lowest internally equilibrated temperature, and using the equation of state of I. R. McDonald and K. Singer [Mol. Phys., 23, 29 (1972)] marked M, yield estimates of T_K differing by only $3°K$.

Kauzmann did not accept this implication, however. He argued instead that in the temperature interval between T_g and T_K, the probability of crystallization was increasing so that before the isoentropic point could be reached, the time scale for crystallization would become the same as that for configurational relaxation in the amorphous phase, as discussed in Section II above. This would precipitate a first-order phase transition by the spontaneous growth of fluctuations in the appropriate direction. Such an event would necessarily terminate the liquid-state metastable free-energy surface and make the apparent entropy-crossing problem metaphysical and, Kauzmann therefore reasoned, of no consequence.

Since the homogeneous nucleation probability cannot be measured for glass-forming liquids, it has not been possible to either prove or disprove this denial of an in-principle ground state for the liquid state of simple substances. The plausibility of Kauzmann's resolution, however, has suffered from the identity of behavior of crystallizable and atactic (noncrystallizable) polymers,[58] and by the experimental contrasts in the composition dependencies of homogeneous nucleation temperatures (T_H) and glass transition temperatures $(T_g$ and $T_K)$ observed in binary solutions.[108]

It is possible but not easy to imagine conditions in which two phases of the same laboratory substance could have identical entropies and also maintain the identity over a range of temperatures; it is not possible, however, in the case of classical hard and soft sphere systems since, at constant pressure, equal entropy in these cases implies equal volume, hence the same phase. Since, at constant pressure, there is only one point in temperature—the fusion point—where the free energies of the fluid and crystal phases of the same substance can be equal, a T_K cannot exist for hard spheres. This raises the question of whether there are other occurrences that might terminate the supercooled fluid state above T_K. Two have been suggested.

The first is a theoretical prediction by Cohen and Grest[109] that for a system of spherical particles, the decrease in volume available during cooling would result in arrival, between T_g and T_K, at a free-volume percolation threshold. The percolation transition is predicted to be first order in character, leading to an amorphous ground state of fixed configuration. However, as long as such a transition is asserted to be first order, the original problem remains, because it is always possible in principle to supercool through a first-order transition. The theory leaves the fate of the now doubly metastable supercooled liquid unresolved.

The second is a conjecture, based on evidence from MD simulations for simple systems, that the liquid state is terminated above T_K as a result of the relaxation time for the equilibrium structure diverging at positive excess entropy.[20] This would imply a finite zero-frequency shear modulus,[65] hence indefinite mechanical stability. With the associated vanishing of nonpropagating density fluctuations would be found a vanishing of communal contributions to the entropy and a second-order thermodynamic transition yielding a glass with a fixed and characteristic residual entropy.[16,]

A primary application of computer experiments is to provide unambiguous empirical routes to the testing and development of theoretical descriptions. It is therefore pertinent to inquire whether computer simulation studies such as have been described in this chapter have or could shed any light on this question of the existence of the underlying thermodynamic transition, or its nature, if it exists as such. As a consequence of the limitations to fast irreversible quenches imposed in computer studies, simulation seems to be less favorably placed to deal with this limit problem than laboratory experiments that approximate much more closely the metastable thermodynamic limit in the vicinity of the glass transition. In the case of simple spherical repulsive models, however, the equations of state through the supercooled fluid ranges are exactly represented by convergent virial series. Thus a complete knowledge of the equation of state in the equilibrium fluid range determines the metastable branch and, in principle therefore, contains the answer to the question of whether an underlying thermodynamic transition exists. We therefore discuss briefly the present state of knowledge of the virial series for hard and soft spheres, and compare the predictions with the MD observations.

In the case of the hard-sphere model, the situation at present is that only the first seven virial coefficients have been determined[111] (the first four are obtained exactly by analytic methods), and unfortunately this is insufficient to predict accurately the path of the metastable branch beyond the freezing transition or to say anything definite about the singularities in the virial series itself. Experimentally, there are four singular points that have been or could be associated with possible singularities in the virial series. These are the freezing transition, the glass transition, and the first-order poles associated with the amorphous and crystalline close-packed states at infinite pressure or zero temperature. The general consensus at present is that virial series has nothing to do with either the freezing transition or the MD glass transition, but there seem to be two schools of thought as to whether the first singularity in the virial series should coincide with an amorphous close-packed density or with the crystal close-packed density (or some other density in that vicinity).

Those of the former school—namely Turnbull and Bagley,[112] LeFevre,[113] Kratky,[114] Gordon et al.,[17] Hoare,[115] Hiwatari,[92] and Frenkel and McTague[27]—either imply or explicitly assert that the virial series could be a continuous representation of the supercooled fluid and the amorphous solid, with its first singularity at the zero-temperature point of the ground-state glass. This would require that the essentially Arrhenius behavior of self-diffusion and fluidity though the "liquid" density range would be maintained up to $pV/NkT = \infty$ and that any observed glass transition behavior is purely a kinetic phenomenon. The alternative point of view (Woodcock,[16] Cape and Woodcock[20]) is that since the system has undergone a glass transition instigated by an underlying thermodynamic singularity, there is no justification for associating the virial series with the amorphous solid and any singularities it may have.

The latter interpretation gains some "experimental" support from MD behavior of second-order thermodynamic properties in both the hard- and soft-sphere models[16-20] around the MD glass transition temperatures. Although Gordon et al.[17] interpreted Alder's results on 32-particle systems as showing a break in expansivity at "T_g," both these authors[17] and Hiwatari[92] imply the absence of any discontinuities in their respective parameterizations of the metastable amorphous phases of the hard- and soft-sphere models, when they used a continuous free-volume equation over the whole amorphous solid-supercooled fluid range. The higher order anomalies that were bypassed in these parameterizations are, however, weak and somewhat diffuse for hard spheres, so there is room for dispute. At this time it is not possible to prove or disprove the possibility that the higher virial coefficients in the hard-sphere model expression

$$\frac{pV}{NkT} = 1 + \sum_{n=2}^{\infty} B_n \left(\frac{\rho}{\rho_0} \right)^{n-1}$$

(where ρ_0 is the pole density), instead of following an appealing asymptotic approach[116] to the value D^2 (where D is the dimensionality of the system), undergo a second increase, for example, terminating at infinity and implying a continuous nonsingular thermodynamic transformation from supercooled fluid to a solid, with a pole at Bernal's random close-packing volume.

Both Padé approximant[111] and the more powerful Tova approximant[117] predictions of the tails of the virial series are consistent with the location of the first-order pole at the crystal close-packed density, as required by the D^2 closure for the virial series. In fact, Baram and Luban[117] give this as a conclusion of their work. The known virial coefficients in the soft-sphere, inverse twelfth power models[17] also imply that the virial series contains information on the crystalline phase at very high pressure, but is unrelated to the freezing transition, the glass transition, or the amorphous solid equation of state.[20]

With these possibilities in mind, we plot the courses in temperature of the fluid- and solid-phase entropies for hard- and soft-sphere systems, and compare them with the MD results in Figs. 21 and 22. It should be noted that in classical mechanical models the entropy is usually defined relative

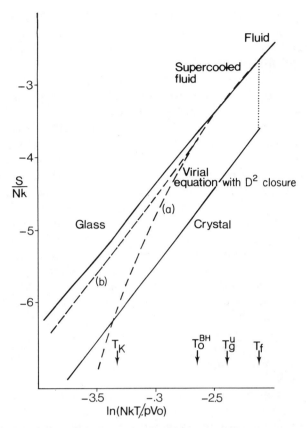

Fig. 21. Entropy versus log-temperature diagram for the hard-sphere model. The solid curves give the computer simulation values for the supercooled fluid, glass, and crystal. The dashed curves have the following bases: (a) a calculation from the virial equation using the known first seven coefficients and higher coefficients obtained from the conjectured D^2 closure[116] (the plot corresponds quite closely[116] with that calculated from the Carnahan-Starling equation[120]); and (b) an extrapolation of higher temperature behavior such as that used by Gordon et al.,[17] which implies a maximum in the series of virial coefficients. The entropy is defined in excess of that for the ideal gas at the same temperature and pressure. Some characteristic temperatures are identified: T_f, fusion point; T_g^u, upper glass transition temperature; T_K, Kauzmann isoentropic point according to D^2 closure virial equation.

to the ideal gas at the same temperature and pressure and that at low temperature the heat capacities approach the constant values of $4.5Nk$ and $3Nk$ for the anharmonic and harmonic oscillator limits, corresponding to the hard- and soft-sphere models, respectively. In consequence the entropy approaches minus infinity as T approaches absolute zero. Curve a in Fig. 21 is that obtained for the case of the first-order pole occurring at the crystal

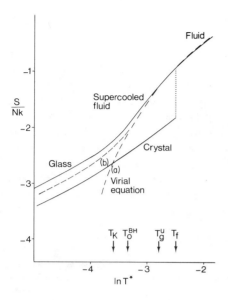

Fig. 22. Entropy versus log-temperature diagram for the soft-sphere model. The solid curves give the computer simulation values for the supercooled fluid, glass, and crystal. The dashed curves correspond to the internally equilibrated fluid behaving in accord with (a) a six-term virial series[20] and (b) an alternative hypothetical equation of state giving a continuous transition to an amorphous ground state with residual entropy $0.15Nk$ virial coefficient series with maximum. The entropy is defined in excess of that for the ideal gas at the same temperature and pressure.

density, whereas curve b gives approximate behavior for the alternative case of the pole occurring at some limiting density characteristic of random close packing.

This limit is commonly believed to be that obtained by Bernal,[75] Finney,[76] and Scott and Kilgour[118] from ball-packing experiments and closely reproduced by finite compression rate MD experiments.[6, 16a, 113] particularly for large (500-particle) systems.[16b] However, it is not known whether for larger systems, this is a true or kinetics-determined limit, notwithstanding the attractive agreement (to 0.2%) of the packing fraction ($\eta = 1/6\pi d^3 N/V$) with the number $2/\pi = 0.6366$, which frequently arises in random statistics problems.[119] Gordon et al.,[17] for instance, assume that this packing density would be exceeded for slower compression cooling rates and estimate the limiting $V_{\text{glass}}/V_{\text{crystal}}$ ($= 0.7405/\eta$) to be 1.1328 ($\eta = 0.6537$; cf. crystal, 0.7405). One of us[16b] has actually observed higher limiting densities for very slowly compressed 32-particle systems, but these are suspect because small systems tend to crystallize—a periodic boundary condition artifact.

In Fig. 22, the two dashed curves have a similar significance. Curve a, as in Fig. 21, requires the existence of a thermodynamic transition associated with a divergence of the relaxation time at some temperature $T > T_K$, the most natural choice being the temperature (marked T_0^{BH}), which corresponds to the Batchinski-Hildebrand V_0^{BH} for the system (see Fig. 4). Curve b represents an alternative resolution of the Kauzmann paradox for this system and implies that the heat capacity must have a maximum, *under conditions of internal equilibrium*, somewhat similar to that shown in Fig. 2 of Ref. 20.

It is a matter for some enthusiasm that the resolution of some of these fundamentally challenging problems in the thermodynamics of very simple systems is probably within reach. Simulations into the low-temperature (or high-density) long relaxation time region in which the resolution can be found are now becoming computationally feasible, and will soon be attempted. However, at the same time it must be recognized that it is for the simple atomic systems of greatest theoretical interest that the probability of nucleation of the stable phase is the greatest. That is, the problems of low-temperature amorphous phase thermodynamics may finally be resolved by an irrevocable escape of the system from amorphous phase space, as Kauzmann originally proposed for complex liquids.

Acknowledgments

This collaborative work could not have been performed without the assistance of a grant from the Scientific Affairs Division of NATO (Grant No. 892), which we gratefully acknowledge. We are grateful also to Dr. T. A. Litovitz for many helpful discussions. C. A. Angell acknowledges the support of the National Science Foundation under Grant No. DMR 77-04318.

References

1a. J. P. Valleau and S. G. Wittingham, in B. J. Berne, Ed., *Modern Theoretical Chemistry*, Vol. 5A, Plenum Press, New York, 1977.

1b. J. P. Valleau and G. M. Torrie, in B. J. Berne, Ed., *Modern Theoretical Chemistry*, Vol. 5A, Plenum Press, New York, 1977.

2. M. J. L. Sangster and M. Dixon, *Adv. Phys.*, **25**, 247 (1976).

3. L. V. Woodcock, in J. Braunstein, G. Mamantov, and G. P. Smith, Eds., *Advances in Molten Salt Chemistry*, Plenum Press, New York, 1975, Ch. 1.

4. I. R. McDonald, in J. Dupuy and A. J. Dianoux, Eds., *Microscopic Structure and Dynamics of Liquids*, (NATO Advanced Study Institute), Plenum Press, New York, 1977.

5. L. V. Woodcock, in *Proceedings of the Conference on Computer Simulation*, San Diego, 1972, Section 4, Physical Science, Vol. 1, p. 847.

6. B. S. Alder and T. E. Wainwright, *J. Chem. Phys.*, **33**, 1439 (1960).

7. W. Damgaard Kristensen, *J. Non-Cryst. Solids*, **21**, 303 (1976).

8. A. Rahman, M. Mandel, and J. McTague, *J. Chem. Phys.*, **64**, 1564 (1976).

9. H. J. Ravaché, *Ann. N.Y. Acad. Sci.*, **279**, 36 (1976).

10. C. A. Angell, P. Cheeseman, J. H. R. Clarke, and L. V. Woodcock, in P. H. Gaskell, Ed., *Structure of Non-crystalline Materials*, Taylor and Francis, London, 1977.

11. H. R. Wendt and F. F. Abraham, *Phys. Rev. Lett.*, **41**, 1244 (1978).

12. J. H. R. Clarke, *J. Chem. Soc. Faraday Trans. 2*, **75**, 1371 (1979).

13. S. Hudson and H. C. Andersen, *J. Chem. Phys.*, **69**, 2323 (1978).

14. F. F. Abraham, *J. Chem. Phys.*, **72**, 359 (1980).

15. J. H. R. Clarke, J. F. Maguire, and L. V. Woodcock, *Discuss. Faraday Soc.*, No. 69, Exeter, 1980.

16a. L. V. Woodcock, *J. Chem. Soc. Faraday Trans. 2*, **72**, 1667 (1976).

16b. L. V. Woodcock, *J. Chem. Soc., Faraday Trans. 2*, **74**, 11 (1978).

17. J. M. Gordon, J. H. Gibbs, and P. D. Fleming, *J. Chem. Phys.*, **65**, 2771 (1976).
18. W. G. Hoover, S. G. Gray, and K. W. Johnson, *J. Chem. Phys.*, **55**, 1128 (1971).
19. Y. Hiwatari, H. Matsuda, T. Ogawa, N. Ogita, and A. Veda, *Prog. Theor. Phys.*, **52**, 1105 (1974); Y. Hiwatari, *ibid.*, **59**, 1401 (1978).
20. J. N. Cape and L. V. Woodcock, *J. Chem. Phys.*, **72**, 976 (1979).
21. F. H. Stillinger and T. A. Weber, *J. Chem. Phys.*, **68**, 3837 (1978); **70**, 4879 (1979).
22. A. Rahman, R. H. Fowler, and A. H. Narten, *J. Chem. Phys.*, **57**, 3010 (1972), and unpublished work.
23. L. V. Woodcock, C. A. Angell, and P. Cheeseman, *J. Chem. Phys.*, **65**, 1565 (1976).
24. S.A. Brawer, *J. Chem. Phys.* **72**, 4264 (1980).
25. T. F. Soules, *J. Chem. Phys.*, **71**, 4570 (1979).
26. W. G. Hoover and M. Ross, *Contemp. Phys.*, **12**, 339 (1971).
27. D. Frenkel and J. P. McTague, *Annu. Rev. Phys. Chem.*, 1981 (in press).
28. D. Turnbull and J. C. Fisher, *J. Chem. Phys.*, **17**, 71 (1949).
29. D. R. Uhlmann, *J. Non-Cryst. Solids*, **7**, 337 (1972).
30. M..J. Mandel, J. P. McTague, and A. Rahman, *J. Chem. Phys.*, **64**, 3699 (1976).
31. H. J. Ravaché and W. B. Streett, *J. Res. Natl. Bur. Stand. A*, Section 80, 59, 1976.
32. M. J. Mandel, J. P. McTague, and A. Rahman, *J. Chem. Phys.*, **66**, 3070 (1977).
33. M. Tanemura, Y. Hiwatari, H. Matsuda, T. Ogawa, N. Ogita, and A. Veda, *Prog. Theor. Phys.*, **58**, 1079 (1977).
34. C. S. Hsu and A. Rahman, *J. Chem. Phys.*, **70**, 5234 (1979).
35. C. S. Hsu and A. Rahman, *J. Chem. Phys.*, **71**, 4974 (1979).
36. J. N. Cape, J. L. Finney, and L. V. Woodcock, to be submitted to *J. Chem. Phys.*
37. S. G. Brush, H. L. Sahlin, and E. Teller, *J. Chem. Phys.*, **45**, 2102 (1966).
38. W. L. Slattery, G. D. Doolen, and H. E. DeWitt, *Phys. Rev. A*, **21**, 2087 (1980).
39. H. S. Chen, *Rep. Prog. Phys.*, **43**, 353 (1980).
40. D. E. Polk and W. C. Giessen in J. J. Gilman and H. J. Leamy, Eds., *Metallic Glasses*, American Society for Metals, Metals Park, Ohio, 1978, Ch. 1.
41. J. Nahgizadeh and S. A. Rice, *J. Chem. Phys.*, **36**, 2710 (1962).
42. D. E. O'Reilly and E. M. Peterson, *J. Chem. Phys.*, **56**, 2262 (1972).
43. J. Jonas, D. Hasha, and S. G. Guang, *J. Chem. Phys.*, **71**, 3996 (1979).
44. A. F. Collings and R. Mills, *Trans. Faraday Soc.*, **66**, 2761 (1970).
45. D. W. McCall, D. C. Douglass, and E. W. Anderson, *J. Chem. Phys.*, **31**, 1555 (1959).
46. B. J. Welch and C. A. Angell, *Aust. J. Chem.*, **25**, 1613 (1972).
47. G. Tammann and W. Hesse, *Z. Anorg. Allg. Chem.*, **156**, 245 (1926).
48. C. A. Angell and W. Sichina, *Ann. N.Y. Acad. Sci.*, **279**, 53 (1976).
49. M. O'Keefe and B. G. Hyde, *Phil. Mag.*, **33**, 219–224 (1976); M. Dixon and M. J. Gillan, *J. Phys. C, Solid State Phys.*, **11**, L165 (1978).
50. A. Rahman, in P. Vashishta, J. N. Mundy, and G. K. Shenoy, Eds., *Fast Ion Transport in Solids*, North-Holland, Amsterdam, 1979.
51. P. A. Cheeseman and C. A. Angell, unpublished work.
52. A. J. Batchinski, *Z. Phys. Chem.*, **84**, 643 (1913).
53. J. H. Hildebrand, *Viscosity and Diffusion*, Wiley, New York, 1977.
54. J. H. Hildebrand, *Science*, **174**, 490 (1971).
55. J. H. Hildebrand and R. H. Lamoreaux, *Proc. Natl. Acad. Sci. U.S.A.*, **69**, 3248 (1972).
56. J. H. Hildebrand and R. H. Lamoreaux, *Proc. Natl. Acad. Sci. U.S.A.*, **73**, 998 (1976).
57. G. S. Fulcher, *J. Am. Ceram. Soc.*, **8**, 339 (1925).
58. J. H. Gibbs, in J. D. McKenzie, Ed., *Modern Aspects of the Vitreous State*, Butterworths, London, 1960, Ch. 7.

59. C. A. Angell, *J. Chem. Educ.*, **47**, 583 (1970).
60. B. A. Dasannacharya and K. R. Rao, *Phys. Rev.*, **137**, A417 (1965).
61. R. C. Desai and S. Yip, *Phys. Rev.*, **166**, 129 (1968).
62. A. Rahman, *Phys. Rev.*, **136**, A405 (1964).
63. B. R. A. Nijboer and A. Rahman, *Physica*, **32**, 415 (1966).
64. A. R. Dexter and A. J. Matheson, *J. Chem. Phys.*, **54**, 203 (1971).
65. K. F. Hertzfeld and T. A. Litovitz, *Absorption and Dispersion of Ultrasonic Waves*, Academic Press, New York, 1959.
66. C. T. Moynihan, P. B. Macedo, C. J. Montrose, P. K. Gupta, M. A. DeBolt, J. F. Dill, B. E. Dom, P. W. Drake, A. J. Easteal, P. B. Elterman, R. P. Moeller, H. Sasabe, and J. A. Wilder, *Ann. N.Y. Acad. Sci.*, **279** (1976).
67. J. Wong and C. A. Angell, *Glass: Structure by Spectroscopy*, Dekker, New York, 1976, Ch. 11.
68. C. A. Angell, unpublished work.
69. J. M. Sare, E. J. Sare, and C. A. Angell, *J. Phys. Chem.*, **82**, 2622 (1978).
70. T. A. Litovitz and J. Lyon, *J. Acoust. Soc. Am.*, **30**, 856 (1950).
71. M. Goldstein, in J. D. McKenzie, Ed., *Modern Aspects of the Vitreous State*, Butterworths, London, 1960, Ch. 4.
72. A. Barkatt and C. A. Angell, *J. Chem. Phys.*, **70**, 901 (1979).
73. P. A. Egelstaff, *An Introduction to the Liquid State*, Academic Press, New York, 1967, Chs. 10 and 12.
74. R. W. Zwanzig and M. Bixon, *Phys. Rev. A*, **2**, 2005 (1970); C. H. Chung and S. Yip, *Phys. Rev.*, **182**, 323 (1969).
75. J. D. Bernal, *Nature (London)*, **188**, 910 (1960).
76. J. L. Finney, *Proc. Soc. London, Ser. A*, **319**, 479, 495 (1970).
77. W. G. Hoover, M. Ross, K. W. Johnson, D. Henderson, J. A. Barker, and B. C. Brown, *J. Chem. Phys.*, **52**, 4931 (1970).
78. J. A. Barker, M. H. Hoare, and J. L. Finney, *Nature (London)*, **257**, 120 (1975).
79. J. D. Scott, *Nature (London)*, **194**, 956 (1962).
80. R. A. Weeks, D. Chandler, and H. Andersen, *J. Chem. Phys.*, **54**, 5237 (1971).
81. T. Atake and C. A. Angell, *J. Phys. Chem.*, **83**, 3218 (1979).
82. B. J. Alder, D. M. Gass, and T. E. Wainwright, *J. Chem. Phys.*, **53**, 3813 (1970).
83. J. P. J. Michels and N. J. Trappeniers, *Chem. Phys. Lett.*, **66**, 20 (1979).
84. M. Ross and P. Schofield, *J. Phys. C*, **4**, 305 (1971).
85. A. K. Doolittle, *J. Appl. Phys.*, **22**, 1471 (1951).
86. M. H. Cohen and D. Turnbull, *J. Chem. Phys.*, **31**, 1164 (1959); D. Turnbull and M. H. Cohen, *ibid.*, **52**, 3038 (1970).
87. G. S. Cargill, *Solid State Phys.*, **30**, 227 (1975).
88. M. R. Hoare, *J. Non-Cryst. Solids*, **31**, 227 (1975).
89. M. J. Mandel, J. P. McTague, and A. Rahman, *J. Chem. Phys.*, **64**, 3699 (1976).
90. C. H. Bennett, *J. Appl. Phys.*, **43**, 2727 (1972).
91. J. N. Cape, Ph.D. thesis, University of Cambridge, 1979.
92. Y. Hiwatari, *J. Phys. Soc. Jpn.*, **47**, 733 (1979).
93. H. J. Raveché, R. D. Mountain, and W. B. Streett, *J. Chem. Phys.*, **61**, 1970 (1974).
94. S. Basak, R. Clarke, and S. R. Nagel, private communication—preprint.
95. J. A. Barker, M. R. Hoare, and J. L. Finney, *Nature (London)*, **257**, 120 (1975).
96. L. von Heimendahl, *J. Phys. F*, **5**, L141 (1975).
97. M. P. Tosi and F. G. Fumi, *J. Phys. Chem. Solids*, **25**, 31 (1964).
98. M. Poulain, M. Poulain, J. Lucas, and P. Brun, *Mater. Res. Bull.*, **10**, 243 (1975).
99. M. Poulain, M. Chanthanasinh, and J. Lucas, *Mater. Res. Bull.*, **12**, 151 (1977).

100. T. F. Soules, *J. Chem. Phys.*, in press (1981).
101. C. A. Angell, I. M. Hodge, and P. A. Cheeseman, in J. P. Pemsler, Ed., *Molten Salts, Proceedings of the International Conference on Molten Salts*, Electrochemical Society, London, 1976, p. 195.
102. S. A. Brawer and M. J. Weber, (a) *Phys. Rev. Lett.* **45**, 460, (1980) (b) J. Non-Cryst. Sol. **38, 39**, 9 (1980).
103. S. A. Alexander and J. P. McTague, *Phys. Rev. Lett.* **41**, 702 (1978).
104. W. W. Wood in H. N. V. Temperley, J. S. Rowlinson, and G. S. Rushbrooke, Eds., *The Physics of Simple Liquids*, North Holland, Amsterdam, 1968.
105. W. Kauzmann, *Chem. Rev.*, **43**, 219 (1948).
106. C. A. Angell and K. J. Rao, *J. Chem. Phys.*, **57**, 470 (1972).
107. A. A. Miller, *J. Polym. Sci. A*, **6**, 249 (1968).
108. C. A. Angell and J. Donnella, *J. Chem. Phys.*, **67**, 4560 (1977).
109. M. H. Cohen and G. S. Grest, *Phys. Rev. B*, **20**, 1077 (1979).
110. L. V. Woodcock, in P. H. Gaskell, Ed., *Structure of Non-crystalline Materials*, Taylor and Francis, London, 1976, pp. 187–189.
111. F. H. Ree and W. G. Hoover, *J. Chem. Phys.*, **46**, 4181 (1967).
112. D. Turnbull and B. G. Bagley, Ch. 10, 1975, pp. 513–554.
113. E. J. LeFevre, *Nature (Phys. Soc.)*, **20**, 235 (1972).
114. K. W. Kratky, *Physica*, **87A**, 584 (1977).
115. M. R. Hoare, *J. Non-Cryst. Solids*, **31**, 157 (1978).
116. L. V. Woodcock, *J. Chem. Soc. Faraday Trans. 2*, **72**, 731 (1976).
117. A. Baram and M. Luban, *J. Phys. C*, **12**, L659 (1979).
118. G. D. Scott and D. M. Kilgour, *J. Phys. D*, **2**, 963 (1969).
119. A. Gamba, *Nature (London)*, **256**, 521 (1975).
120. N. F. Carnahan and K. E. Starling, *J. Chem. Phys.*, **51**, 635 (1969).

LIQUIDS, GLASSES, AND THE GLASS TRANSITION: A FREE-VOLUME APPROACH

GARY S. GREST

Department of Physics
Purdue University
West Lafayette, Indiana

AND

MORREL H. COHEN

James Franck Institute and
Department of Physics
University of Chicago
Chicago, Illinois

CONTENTS

I. INTRODUCTION

The liquid-glass transition has been intensively studied for many years. Despite the many papers on the subject, both experimental and theoretical, there is still no clear understanding of this transition. However a few relatively simple phenomenological theories have been developed to explain an extensive body of observations, especially those of viscosity, heat capacity, and volume. The focus of one set of these theories is on the temperature dependence of the diffusion in dense liquids.

Two complementary attempts to interpret such transport phenomena in a variety of molecular systems are the free-volume model and the entropy theory. Both have been useful in describing diffusion in liquids in a physically appealing manner. A second set of theories, largely phenomenological, has involved the temperature dependence of the heat capacity and the volume near the liquid-glass transition temperature. Nevertheless, in spite of all the work done, there are still no rigorous theoretical results. To begin with, our understanding of the liquid state, particularly at high densities, is inadequate. In addition, the liquid-glass transition is a nonequilibrium phenomenon in which the time scale for relaxation becomes comparable to experimental times. The observed glass is not even a metastable state of the system, let alone the lowest free-energy state, so that equilibrium thermodynamics cannot always be applied.

In the absence of a rigorous or at least microscopically sound theory, one is forced back to the modelistic theories, which are in a fragmented state. Accordingly, the aim of this chapter is to tie together many of these theoretical ideas on the liquid-glass transition by use of a single model, something not previously accomplished. We concentrate on the free-volume model because of its past success in describing the behavior of the viscosity. However, our results are much broader and transcend any one model.

After describing and justifying the underlying assumptions of the free-volume model, we apply it to calculate all the thermodynamic quantities of interest. We also include in our model the effects of slow relaxation rates, so that our results can be compared directly to experiment. We still do not have many rigorous results. Instead, we have a simple picture of a quite complex set of phenomena. We have built a single, unified theory from which one can continue to study all aspects of the transition. Although the theory has its weaknesses, which principally result from the necessity to make approximations along the way, it has the desired robustness from which to proceed further.

II. BASIC PHENOMENA AND EARLY THEORIES

The liquid state of matter[1-3] is characterized by zero shear modulus, a finite fluidity ϕ, and time-average translational invariance. We know that when a liquid is cooled slowly, the system usually crystallizes via a first-order phase transition (finite latent heat) at the melting temperature T_m. At T_m, the fluidity vanishes abruptly, and the system finds itself in a new, stable, translationally ordered, solid state. However, in many systems, it is possible to supercool the liquid to temperatures below T_m by rapid cooling.[4-9] In systems in which crystallization can be avoided, several interesting phenomena occur. Most important, the fluidity is observed to decrease continuously, finally reaching a limiting value so small that it becomes unobservable as relaxation rates become on the order of days or more. When this has occurred, the system is referred to as a glass.[10-15] The glassy state differs from the crystalline state in that it is disordered and only metastable. Put most simply,[16] the glassy state is an extension of the liquid state in which the viscosity $\eta = \phi^{-1}$ increases above about 10^{15} P: thus we have quantified the statement, "A solid does not flow."

This raises the possibility that the liquid-to-glass transition is in fact a real phase transition, given that in the glass the time-average translational symmetry of the liquid is broken, at least on the time scale above. However, there is no experimental evidence for a sharp phase transition. In fact, the exact thermodynamic nature of the glassy state is not known. The glass and the supercooled liquid could form a single, metastable thermodynamic phase that is in local equilibrium. However, it is also possible that the observed glassy state is only a stationary frozen configuration that is not in local equilibrium and occurs only because the relaxation rates are very long. In either case, there may exist an underlying phase transition obscured by kinetic effects, or there could be no phase transition at all. Indeed, whatever the case, it is convenient to refer to this new state as the amorphous phase, even if the liquid-glass transition is not a true phase transition in the limit of slow temperature variations.

It is commonly believed that the glassy state is never the most stable state of any material.[5] That is, a solid is always more stable in its crystalline phase than as a glass. Although there is no rigorous theoretical proof for this statement, it is consistent with the experimental observations that the glass transition temperature T_g is always lower (usually by a significant amount) than the crystallization temperature T_m and that the viscosities of most liquids are of the order of only 10^{-2} P at T_m.[13] However, it is also probably true that the glass transition can occur in all classical liquids, including monatomic ones, provided crystallization can be bypassed. This idea is based on the hypothesis that the glassy state is truly metastable and not unstable and leads one to the concept of an "ideal" glassy state first suggested by Gibbs and DiMarzio[17] for polymeric systems and by Cohen and Turnbull[18, 19] for simpler, even monatomic glasses.

An "ideal" glass is a solid in internal equilibrium in which there exists a definite set of equilibrium positions about which each atom oscillates.

However, unlike the crystalline solid, these positions are randomly distributed and do not exhibit translational symmetry. If one such structure exists, there must be a large number of similar random structures of nearly equal energy. Nevertheless, the Gibbsian entropy of each would be zero, because these structures are mutually inaccessible for classical systems.[19] One particular structure would be picked out during cooling, and the system would remain in that structure. A large increase of entropy would occur only on heating back into the liquid state, when the equivalent random structures again become mutually accessible. Since there exist a large number of such mutually accessible, nearly equivalent configurations, the system has a finite residual entropy even at zero temperature. When quantum mechanical effects are included, these configurations are no longer mutually inaccessible. The system can tunnel between them, as has been observed at low temperatures.

To produce a glass, we know that the melt must be cooled fast enough to avoid crystallization.[4-8] What determines the rate of crystallization and what are the best conditions for glass formation are continuing subjects of research activity.[13] We do know that the crystallization rate of a supercooled liquid is dependent both on the rate of crystal formation or nucleation and the speed with which the crystal-liquid interface advances. Both the rates are strongly dependent on the reduced temperature $T_r = T/T_m$. It is not difficult to see that a melt can more easily be undercooled through its metastable liquid regime from T_m to T_g, the higher the cooling rate and the higher the reduced glass transition temperature $T_{rg} = T_g/T_m$. For many glass formers, in particular the metallic glasses, crystallization can be avoided only by using small sample volumes, since this helps to increase the cooling rate and decreases the nucleation probability. Simple nucleation theory indicates that melts with $T_{rg} \gtrsim \frac{2}{3}$ should readily form glasses.[5, 8] This condition has been observed to hold for many simple molecular substances that easily form glasses in bulk. However very large quench rates, on the order of 10^6 °K/sec, are necessary to form most metallic glasses for which T_{rg} is lower.[9] For these systems, the glassy state has been formed only in very narrow compositional ranges, usually near a eutectic composition, for which T_m' is reduced. All the experimental results are consistent with the statement that the higher T_{rg}, the greater the glass-forming tendency.

Solids can be classified according to bonding type and by their constituents. Glasses are found to occur among all such classes of solids. The simplest ones have spherical or nearly spherical constituents and include the metallic glasses[9, 13, 21-27] and some molecular substances. In this group one also includes the dense random packing (DRP) of hard spheres[28-31] and systems of Lennard-Jones particles[32-34] studied by molecular dynamics.[35] The more complex glasses include the ionic,[36-41] polymeric,[42-47] and organic materials.[48-56] There are also the network glasses,[57-72] which include systems covalently bonded in two and three dimensions, both insulators and semiconductors. These distinctions are not sharp and the groupings overlap, but the important point is that the occurrence of glasses in nature seems to be a universal phenomenon.

In the liquid↔glass transition region, a homogeneous and continuously increasing resistance to flow develops, as measured by the shear viscosity η. The choice of a limiting viscosity to mark the transition from fluid to a configurationally frozen solid is somewhat arbitrary. A value of $\eta \sim 10^{15}$ P is often used, since at this viscosity a body would substantially maintain its shape against small shearing forces for periods up to 1 day. This is to be compared with the viscosity of simple liquids such as water or alcohol, which at room temperature have $\eta \sim 10^{-2}$ P. For $\eta \sim 10^{15}$ P, the time constant for relaxation becomes so long that the system is frozen into a single configurational state, the amorphous state, neglecting tunneling.

The dependence of the shear viscosity η in both the equilibrium liquid and supercooled liquid phases is shown in Fig. 1. Vogel[73] and Fulcher[74] first observed that in the high-temperature, low-viscosity regime, η could be fitted to the form[73-75]

$$\eta = \eta_0 \exp\left(\frac{b}{T - T_{0H}}\right) \qquad (2.1)$$

where η_0, b, and T_{0H} are constants. This form works well in the low-viscosity region ($\eta \lesssim 10^4 - 10^6$ P) for most glasses, including the ionic, polymeric, and organic materials, as shown by curves b and c in Fig. 1 with $T_\eta \equiv T_{0H} \neq 0$. A few simple organic glasses[48] and some network glasses,[57-59] including SiO_2 and GeO_2, follow an Arrhenius behavior,

$$\eta = \eta_0 \exp\left(\frac{b}{T}\right) \qquad (2.2)$$

over almost the entire temperature range studied, as shown by curve a of

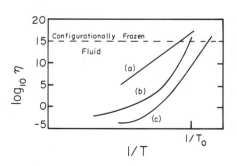

Fig. 1. Sketch of the logarithm of viscosity η (in poise) with reciprocal temperature (when the liquid is cooled from the liquid to the glassy state). Curve a corresponds to Arrhenius behavior, $T \rightarrow 0$. Curves b and c show the typical form for simple molecular glass formers. Curves b and c correspond to the Doolittle equation, where the free volume $\bar{v}_f \propto T - T_{0H}$, $T_{0H} \neq 0$ at the high temperature and $\bar{v}_f \propto T - T_{0L}$ at low temperatures. In curve b, $T_{0L} \neq 0$, and in curve c, $T_{0L} \rightarrow 0$.

Fig. 1. In Table I, we present a partial list of glass formers and their characteristic temperatures T_η and T_g. However, most of the experimental results do not extend to lower temperatures near T_g, where η becomes $\gtrsim 10^8$ P. In systems that have been measured at higher viscosities ($10^8 < \eta < 10^{14}$ P), it is found that most glasses become Arrehenius, $T_{0H} \to 0$ in (2.1), curve c of Fig. 1, whereas others are described by a value of $T_{0H} \to T_{0L} < T_{0H}$ in (2.1), curve b of Fig. 1. Examples of the former include the organic liquids salol,[54] α-phenyl-o-cresol, n-butylbenzene,[48] and di-n-butyl phthalate,[48] and the ionic glass $0.60KNO_3 - 0.40Ca(NO_3)_2$.[60] The organic liquids o-terphenyl[54, 55] and tri-α-naphthylbenzene[53] are examples of the latter. Results for salol and $0.60KNO_3 - 0.40Cu(NO_3)_2$ as examples of the former and o-terphenyl and tri-α-naphthylbenzene of the latter are shown in Fig. 2. We have found only one system, the metallic glass[20] $Au_{0.77}Ge_{0.136}Si_{0.094}$, that can be fitted in both regimes with a value of $T_{0H} \approx T_{0L}$ (curve b in Fig. 1). Unfortunately, the metallic glasses can be studied only in the liquid phase,

TABLE I

Characteristic Temperatures for Glass Formers

Glass former	T_g (°K)	T_η (°K)	T_s (°K)
$Au_{0.77}Ge_{0.136}Si_{0.094}$ [a]	295	241	—
Tri-α-naphthylbenzene[b]	342	342	—
o-Terphenyl[c]	240	248 / 231	200
Salol[d]	230	226	—
α-Phenyl-o-cresol[d]	230	210	—
2-Methyl pentane[e]	79.5	59	58
Glycerol[e]	180	138	134
Methanol[e]	103	60	63
Ethylene glycol[e]	152	107	112
Sorbitol[e]	266	236	236
B_2O_3 [e]	539	402	335
$ZnCl_2$ [e]	375	260	250
$H_2SO_4 \cdot 3H_2O$ [f]	158	128	135
$Ca(NO_3)_2 \cdot 4H_2O$ [e]	217	205	202
$0.62KNO_3 \cdot 0.38Ca(NO_3)_2$ [g]	345	276	—

[a] Ref. 20.
[b] Refs. 53 and 54.
[c] Refs. 14, 50, 51, and 54.
[d] Ref. 54.
[e] Ref. 39.
[f] Ref. 36.
[g] Refs. 38 and 78.

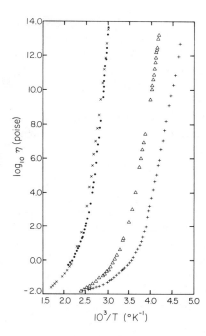

Fig. 2. Logarithm of the viscosity η (in poise) as a function of temperature for four glass formers. Data for o-terphenyl (triangles) from Refs. 54 and 55, tri-α-naphthylbenzene (crosses) from Ref. 53, salol (pluses) from Ref. 54, and 0.60KNO$_3$·0.40Ca(NO$_3$)$_2$ (circles) from Ref. 40.

with $T \gtrsim T_m$, and in the glass phases $T \lesssim T_g$, and not in the intermediate supercooled liquid region ($10^{-2} \lesssim \eta \lesssim 10^9$ P), since they crystallize too readily. Therefore a large extrapolation is necessary to interpret results for these systems. Tweer et al.[76] have shown that many of these systems, which cannot be fitted by the three-parameter Vogel-Fulcher equation, (2.1), over the entire range ($10^0 < \eta < 10^{14}$ P) with a single value of T_0, can be fitted with a more general five-parameter equation for η that diverges only as $T \rightarrow 0$. Included in this fit is the organic liquid tri-α-naphthylbenzene, which has a finite T_{0L} when fitted to the usual equation for η. This leaves open the possibility that T_{0L} always goes to zero, and all viscosities become Arrhenius at a low enough temperature.

The deviation of η from its high-temperature behavior, (2.1), could be regarded as a nonequilibrium phenomenon. However, the deviation from this form usually begins at a temperature that is greater than the temperature at which the system is considered to fall out of thermal equilibrium (i.e., the temperature at which the anomaly in the specific heat occurs). For the organic glasses,[54, 55] this deviation occurs at a viscosity $\eta \sim 10^4$ P, whereas $\eta \geq 10^8$ P at the glass transition. In the measurements of these organic glasses, care was taken to assure that the equilibrium value of η was obtained.[54, 55] For the metallic glass Au-Ge-Si, the viscosity when measured at its highest accessible values[20] begins to deviate sharply from the simple exponential behavior, presumably because of nonequilibrium effects. We expect that nonequilibrium phenomena will lead to a quite different pattern of temperature variation for η than occurs when (2.1) no longer holds. Except for Au-Ge-Si, we consider the observed deviation of η from its high temperature behavior to represent typical behavior.

During the configurational freezing of the supercooled liquid, the heat capacity C_p usually changes markedly within a narrow temperature range around a temperature T_g, conventionally referred to as the glass transition temperature. This decrease reflects the loss of configurational freedom. The magnitude of the drop in C_p in going from the equilibrium liquid to the glass is usually very large, approximately 50% for most glass formers.[13] However, it is hardly detectable in the tetrahedrally coordinated network glasses[70] SiO_2 and GeO_2. More important, the value of T_g and the shape of C_p are found to be very dependent on the heating and cooling rates of the measurement and on the thermal history of the sample.[15, 42, 77–80] It is known that the *observed* changes in heat capacity are the consequences of the falling out of complete thermodynamic equilibrium of the system under observation as the time of measurement becomes comparable to the relaxation times of the system.

In Fig. 3, we show schematically the behavior of the enthalpy H and heat capacity C_p when the liquid is cooled isobarically through the glass transition region at a constant rate

$$q = \frac{dT}{dt} \tag{2.3}$$

and subsequentially reheated at the same rate.[78] The two sketches for H and C_p correspond to two different cooling rates $|q_A| > |q_B|$. The rate of heating or cooling may be thought of as a series of incremental temperature steps ΔT during a time interval Δt. At high temperatures above T_g, the cooling rate is sufficiently high that the time required for structural relaxation is short compared to the experimental time scale, and the enthalpy H can follow its equilibrium liquid value. At lower temperatures near T_g, the time required for structural modes to equilibrate becomes longer, so that these modes may not completely relax before the next time increment Δt, and the experimental H begins to depart from its equilibrium value. For $T \ll T_g$, these structural relaxation times become so long that no further changes can take place in the configurational state of the system, and H follows its value for a glass. As seen in Fig. 3, the higher the cooling rate, the higher the temperature at which H departs from its liquid value. As pointed out by

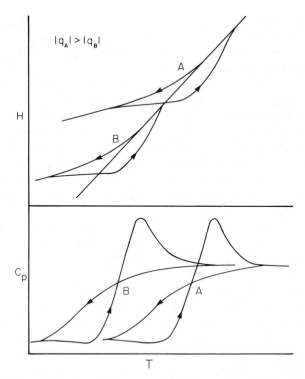

Fig. 3. Enthalpy and heat capacity versus temperatures for a glass cooled and then reheated through the transition region at different rates. The cooling rate $|q_A| > |q_B|$. [After Moynihan et al., *J. Phys. Chem.*, **78**, 2673 (1974).]

Moynihan et al.,[15, 78] irrespective of the direction of the temperature change, the direction of the structural relaxation process is always toward equilibrium. That is, when the glass is heated or cooled, the enthalpy H changes in such a fashion as to move toward the equilibrium value. Consequently, on reheating, H follows a path different from that found on cooling, as indicated in Fig. 3.

Since the heat capacity $C_p = \partial H / \partial T$, one expects that at a constant cooling rate in materials that do not crystallize, C_p will be a monotonic,[78] decreasing function as shown in Fig. 3. However, the majority of the measurements are done at a constant heating rate, starting from the glassy state, and the observed anomaly is shown in Fig. 3: C_p departs from its value in the glass rather sharply, passes through a maximum, then decreases at a high temperature, where it merges smoothly with the equilibrium liquid value for C_p.

G. S. GREST AND M. H. COHEN

It is found experimentally that when the heating or cooling rate q increases in magnitude, the value of T_g also increases. This follows from the behavior of the enthalpy shown in Fig. 3. The higher the cooling rate, the earlier the enthalpy departs from its equilibrium value, and the higher T_g will be. Moynihan et al.[78] have carried out a systematic analysis of the dependence of the glass transition temperature on the heating and cooling rate q for several glass formers. They find, as shown in Fig. 4, for B_2O_3 and $0.6KNO_3 \cdot 0.40Ca(NO_3)_2$, that the logarithm of the cooling rate $|q|$ varies linearly with the glass transition temperature for all the glass formers studied. The observed dependence of the shape of C_p and the value of T_g on the rate q has been used to study the temperature-dependent kinetic processes occurring during the glass transition, as we discuss in greater detail in Section IX.

The specific volume $\bar{v}$ also shows an anomalous temperature dependence[79] near T_g. The behavior of $\bar{v}$ is universal among all systems having a finite T_g. When measured at a constant cooling rate q, $\bar{v}$ follows the behavior of the enthalpy, shown in Fig. 3, as it decreases linearly with T and changes shape at a temperature dependent on q but close to the T_g observed for C_p. Below this breakaway temperature, the system is not in equilibrium. If the system is annealed, not far below T_g, the behavior shown in Fig. 5 is observed. Volume $\bar{v}$ decays toward a lower asymptotic value $\bar{v}_0$, which can lie either on the extrapolated volume-temperature curve for the liquid or above it, if the annealing temperature is low enough. The latter observation suggests the existence for metastable equilibrium of a $\bar{v}_0$ versus T curve that breaks away from the extrapolated liquid curve, but no information is yet available on where or how it breaks away. There is also

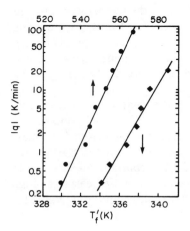

Fig. 4. Logarithm of cooling rate $|q|$ versus limiting fictive temperature, T_f'. (The fictive temperature is used to give a precise measure of the glass transition temperature T_g). Data for $0.60KNO_3 \cdot 0.40Ca(NO_3)_2$ (diamonds) and B_2O_3 (circles) from Moynihan et al. [*J. Am. Cera. Soc.*, **59**, 12 (1976)]. The slope of T_f' versus $\log_{10}|q|$ is approximately $3.8°K$ for $0.60KNO_3 \cdot 0.40Ca(NO_3)_2$ and $14.5°K$ for B_2O_3.

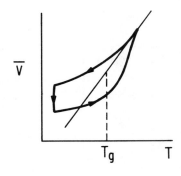

Fig. 5. Isobaric volume $\bar{v}$ versus T, illustrating volume hysteresis effect. The equilibrium curve in the liquid well above T_g is unique. At a constant cooling rate q, the $\bar{v}$ falls out of equilibrium below T_g. When the sample is then annealed at constant T, the volume becomes densified and may reach a relaxed glass state, depending on the temperature. The lower portion of the solid curve represents the behavior after heating at constant q. Note $\bar{v}$ remains under its liquid value to $T > T_g$. This plot is similar to that found for amorphous Selenium in Ref. 71 and for a polymeric system in Ref. 79.

hysteresis upon heating at a constant rate q. The $\bar{v}$ remains under the extrapolated liquid volume, then rises sharply to a final equilibrium value at temperature exceeding T_g. The amount of hysteresis is dependent on the value of q.

These experimental results indicate that as studied in the laboratory, the glass transition is a kinetic phenomenon that depends on the time scale of the measurements. The question still remains whether, as may be suggested by the dielectric annealing data on selenium, there exists a real thermodynamic glass transition in the limit of infinitely slow heating or cooling. That such a thermodynamic transition should occur has been inferred by extrapolating the heat capacity versus temperature for liquids with large values of $C_{pr} = C_{p(\text{liquid})} / C_{p(\text{glass})}$ and observing the manner of change of the difference in entropy between the liquid and crystalline states (which was the entropy of melting ΔS_m at the melting point).

Data for $H_2SO_4 \cdot 3H_2O$ by Kunzler and Giauque[36] and for $Ca(NO_3)_2 \cdot 4H_2O$ by Angell and Tucker[39] are shown in Fig. 6. These plots suggest that at only about 20°K below the normal value of T_g, the difference in entropy between the liquid and crystalline phases would vanish at a temperature T_s and become negative below T_s. A partial list of values of T_s compiled by Angell and co-workers is included in Table I. The existence of an amorphous phase with lower total entropy than the stable crystal phase at the same temperature is unreasonable. Moreover, the heat capacity of the glass is known to be higher than that of the crystal at low temperatures. To avoid this apparent paradox, C_p has to decrease precipitously at a temperature not far below the observed T_g. From Fig. 6, we see that this must occur at a temperature no lower than 130°K for $H_2SO_4 \cdot 3H_2O$ and 200°K for $Ca(NO_3)_2 \cdot 4H_2O$.[14] As pointed out by Angell and Sichina,[14] the most gradual possible decrease of C_p below T_g that remains consistent with the requirement that $S_{(\text{liquid})} > S_{(\text{crystal})}$ still amounts to a very sharp change in C_p,

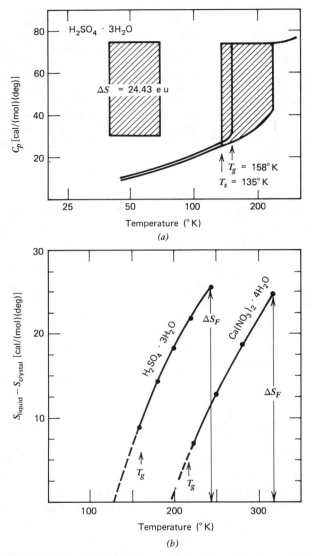

Fig. 6. Graphical estimates of the limits on the temperature range below T_g over which the observed equilibrium liquid heat capacity could be maintained in "slow" (equilibrium) experiments (a) for $H_2SO_4 \cdot 3H_2O$. The condition $(S_{\text{liquid}} - S_{\text{crystal}})_T = 0$ would appear to define a temperature $T = T_s$ to which, if not before, rapid decrease in the equilibrium liquid heat capacity must occur. [From C. A. Angell and W. Sichina, *Ann. N.Y. Acad. Sci.*, **279**, 53 (1976).] This is shown in (b) for two glass formers.

comparable to the high-temperature end of a common λ transition. The existence of the vanishing excess entropy was first recognized by Kauzmann; it points to the possible existence of an "ideal" glass transition T_p at a temperature $T_s \lesssim T_p < T_g$. In the limit of slow heating or cooling, T_g would approach T_p. It is also interesting to note that $T_\eta \simeq T_s$ for many glass formers, as recognized by Angell and co-workers,[39] (Table I).

The existence of a temperature of vanishing excess entropy T_s suggests the notion that a real phase transition, either second or first order, may occur at T_g in the limit of slow heating and cooling. The amorphous system would undergo a transition to a solid with negligible configuration entropy. However, the configuration of this new solid state is still unresolved. Kauzmann[81] predicted that it must be crystalline, which means that a first-order phase transition, a low-temperature crystallization ($T_p < T_m$), must occur before the excess entropy is exhausted at T_s. However, recent work by Angell and Donnella[82] on the homogeneous nucleation temperature T_H for the binary solution $Ca(NO_3)_2 \cdot H_2O$ provides evidence that this cannot be true in all cases. They find that there is a composition range in which by extrapolation T_H lies below T_s. This is consistent with the view that an "ideal" amorphous solid state, characterized by a short-range order, is topologically unique and distinct from that of the crystal. The structural basis for this concept was first provided by Zachariasen's random network model for amorphous solids.[83] The statistical basis for such a state has been provided by Gibbs and DiMarzio,[17] who suggested that an underlying equilibrium transition of second order occurs at T_g in the limit of infinitely slow cooling.

The entropy theory of the glass transition was developed by Gibbs and DiMarzio[17] and by Adams and Gibbs[84] to describe polymeric systems. By mixing the polymer links with holes or missing sites on a lattice to account for thermal expansion as in a lattice gas model, they could determine the entropy of mixing and the configurational entropy of the polymer. They found a second-order transition at a temperature $T_2 \neq 0$. They then pointed out that this temperature would correspond to T_g if the experiments could be done so that system were always in equilibrium. Below T_2, the configurational entropy would remain zero instead of going to a meaningless negative number. They were able to derive a very useful result for the viscosity,[84]

$$\eta = \eta_0 \exp\left(\frac{C}{TS_{ex}}\right) \qquad (2.4)$$

where C is constant and the excess entropy S_{ex} is defined as $S_{(liquid)} - S_{(glass)}$.

This result also suggests that the sluggish relaxation behavior governing T_g is itself just a manifestation of the smallness of S_{ex} in the region immediately above T_2. The expression (2.4) can be used to rationalize the Vogel-Fulcher result for the viscosity. In systems in which $C_{p,r}$ is very close to unity, such as GeO_2 and SiO_2, the relaxation properties will show Arrhenius behavior because S_{ex} is only weakly temperature dependent.[14] However for systems in which $\Delta C_p = C_p(\text{liquid}) - C_p(\text{glass})$ is large, S_{ex} should change rapidly with temperature, and non-Arrhenius behavior should be encountered. In fact, in many cases one can approximate ΔC by constant$/T$. Combining this with the observation that S_{ex} vanishes at T_s, we obtain ($T \cong T_s$)

$$S_{ex} = \int_{T_s}^{T} \frac{\Delta C_p}{T} dT = \Delta C_p(T_s) \frac{(T - T_s)}{T} \tag{2.5}$$

and the Vogel-Fulcher result[14]

$$\eta = \eta_0 \exp\left(\frac{b'}{T - T_s}\right) \tag{2.6}$$

where b' is another constant. Equation 2.6 is consistent with the experimental observation that $T_\eta \simeq T_s$; that is, the temperature of vanishing excess entropy should agree with the high-temperature estimate for the temperature at which η diverges. However, since the data for η taken near T_g show an Arrhenius temperature dependence of the viscosity in most glasses, the results in (2.5) and (2.6) fail near T_g. The crossover to Arrhenius behavior can be understood in terms of the free-volume model, which we develop in a later section.

Although the entropy theory of Gibbs and DiMarzio provides a justification for the Vogel-Fulcher result for polymers, it is difficult to apply it to monatomic or rigid molecules. For those glasses, their theory reduces to Frenkel's hole theory of glasses, which is known not to be very accurate. Before the development of the entropy theory, other explanations of the viscosity data had been proposed. The most successful of these had been the free-volume model first applied by Fox and Flory[85] and developed subsequently by a large number of authors.[86-95] This model was motivated by the interesting experimental observation that the fluidity of many liquids decreases markedly with increasing pressure. From this pressure dependence alone, one can infer that fluidity must be closely connected with the average free volume v_f, defined by

$$v_f = \bar{v} - v_0 \tag{2.7}$$

where $\bar{v}$ is the average volume per molecule in the liquid and v_0 is the van der Waals volume of the molecule. Batschinski,[96] as early as 1913, suggested that the fluidity increases linearly with free volume. Fox and Flory[85] made a critical contribution with their suggestion that the glass transition can be attributed to the falling of the free volume below some critical value.

Doolittle[97] found that the fluidity of many simple hydrocarbon liquids could be represented by the simple form

$$\phi = \phi_0 \exp\left(\frac{-bv_0}{v_f}\right) \qquad (2.8)$$

where b is a constant of order unity. The Doolittle equation correctly predicted the abrupt decrease in $\phi = \eta^{-1}$ in a narrow temperature range as v_f becomes small. Williams et al.[86] then showed that this result is valid for a large number of glass formers. They proposed a description the free volume by the simple form

$$v_f = v_g\left[0.025 + \alpha(T - T_g)\right] \qquad (2.9)$$

where v_g is the volume at T_g and $\alpha = \alpha_l - \alpha_g$ is the difference between the thermal expansion coefficients of the liquid and the glass. Cohen and Turnbull later developed a simple theory of molecular transport in liquids in a series of papers.[18, 87, 88] The principal idea of their work, which we develop in greater detail in a later section, is that molecular transport occurs by movement of molecules into voids of a size greater than some critical value formed by the redistribution of the free volume. The idea is analogous to that of molecular cooperation, which underlies the earlier treatments of Bueche[98] and Barrer.[99]

Up to this point, we have discussed in general terms the behavior of a system near the liquid↔glass transition. What happens in the low-temperature range for $T \ll T_g$ is also a very intriguing question. Near T_g, we have seen that the system falls out of complete thermodynamic equilibrium and is trapped in a region of configuration space far removed from its crystalline ground state. There are many other energetically equivalent regions into which the glass could have been trapped. The residual entropy is a measure of that number. Most of those states are mutually inaccessible because they are distant in configuration space; it is possible, however, that a few mutually accessible states of essentially the same ground-state energy exist. These nearly equivalent states could then effect the low-temperature properties of the system in a remarkable way, since at low temperature any additional degrees of freedom would play an important role.

That such nearly mutually accessible states do in fact exist was first postulated by Anderson et al.[100] and by Phillips[101] to explain the linear temperature dependence of the specific heat at low temperatures found in glasses. They proposed that what they called *tunneling centers*, which consist of a certain number of atoms or groups of atoms, have accessible two nearly equivalent equilibrium configurations (corresponding to the minima of asymmetric double-well potentials), and tunnel between them. With that model they could explain several other interesting experimental results.[102-105] These include a T^2 dependence of the thermal conductivity and an ultrasonic attenuation that decreases at low temperature and saturates at large strain amplitude. It also has been very successful in explaining many other experimental observations, including the existence of phonon and electric echoes. Nevertheless, there has been no microscopic analysis of what a tunneling level really is. Instead, tunneling is treated via a model Hamiltonian for two-level systems describing the ground states in the two local energy wells.

The tunneling states are commonly held to consist of a small group of atoms undergoing a local rearrangement. The number of atoms involved is assumed to be reasonably small, to minimize the distance between states in configuration space. However, the larger the number of atoms, the easier it is to find two ground states of roughly equivalent energies. It is believed that this competition between accessibility and degeneracy determines the size of the tunneling states.[105] Just as we have seen that the existence of the glass transition and the behavior of the diffusion and viscosity are universal, it is also known that the excitations just mentioned are universal among all types of glass former.[106-108] With regard to universality, the behavior at the two temperatures $T \approx 0$ and $T \approx T_g$ are thus very similar. How the glassy state is formed directly affects the tunneling states, and both should be describable within the same theory and model.

Here we show how one particular model, the free-volume model, can account for the glass transition, the behavior of η, C_p, and $\bar{v}$ near T_g, and the existence of low-temperature tunneling states.

III. PHYSICAL BASIS OF THE FREE–VOLUME MODEL

Fox and Flory[85] first postulated that the liquid-glass transition resulted from the decrease of the free volume of the amorphous phase below some critical value. The subsequent derivation of Doolittle's fluidity equation[97] within the free-volume model is based on four simple assumptions,[87, 88, 94] which can be worded as follows:

1.   It is possible to associate a local volume v of molecular scale with each molecule (or motile segment of a flexible molecule).

2. When v reaches some critical value v_c, the excess can be regarded as free.

3. Molecular transport occurs only when voids having a volume greater than some critical value v^* approximately equal to the molecular volume v_m form by the redistribution of the free volume.

4. No local free energy is required for free-volume redistribution. The resulting equation for the diffusion constant is of the form given by (2.8).

Assumption 1 is valid if each molecule is restricted to movement within a cell or cage defined by its nearest neighbors. Kirkwood[109] supposed such cells to exist in a liquid, but for simplicity supposed further that the cells were all identical, forming a regular lattice.[110] He took for the free energy of N interacting molecules

$$F = Nf - TS_c \qquad (3.1)$$

where f is the free energy of an atom or molecule moving within its cell in the mean potential of its neighbors, properly corrected for double counting of the interactions, and S_c is the communal entropy. We may regard f as the local free energy referred to in assumption 4. The S_c arises in a liquid from the additional freedom each molecule possesses because it can diffuse throughout the entire volume. It may be written as

$$S_c = k \ln\left[(N-1)v_s \right]$$

$$k \ln v_s = s = -\frac{\partial f}{\partial T} \qquad (3.2)$$

which expresses the fact that each molecule can move in any other cell in addition to the one arbitrarily assigned to it. Clearly, the glass transition might profitably be examined within the framework of a thermodynamic theory such as Kirkwood's. Equally clearly, justification of assumptions 1 through 4 would justify also such a thermodynamic theory.

Hsu and Rahman[111] have shown by MD calculations that the diffusion coefficient of a dense, supercooled rubidiumlike liquid at constant volume extrapolates to zero at a temperature that is essentially zero. Earlier, a Lennard-Jones liquid had been quenched into what appeared to be a glassy state by rapid cooling to a temperature that turned out to be below T_{0H}. Subsequent simulations in a variety of systems showed, however, that the diffusion coefficient remained finite though small ($< T_{0H}$) and that crystallization ultimately occurred. Thus the crystal is the lowest energy state of the system, but there is a disordered state of low atomic mobility that is metastable, persisting for a time that is very long on the molecular scale. We identify this state as a glass, as did Rahman et al.[32]

Clarke[34] has examined the thermodynamic equation of state and the specific heat for a Lennard-Jones liquid cooled through T_g at zero pressure. He found that C_p drops with decreasing temperature near where the self-diffusion becomes very small. Wendt and Abraham[112] have found that the ratio of the values of the radial distribution function at the first peak and first valley shows behavior on cooling much like that observed for the volume of real glasses (Fig. 6), with a clearly defined T_g. Stillinger and Weber[113] have studied a Gaussian core model and find a self-diffusion constant that drops essentially to zero at a finite temperature. They also find that the ratio of the first peak to the first valley in the radial distribution function showed behavior similar to that found by Wendt and Abraham[112] for Lennard-Jones liquids. However, the first such evidence for a nonequilibrium (i.e. kinetic) nature of the transition in a numerical simulation was obtained by Gordon et al.,[30] who observed breakaways in the equation of state and the entropy of a hard-sphere fluid similar to those in real materials.

The radial distribution function found for the putative Lennard-Jones glass at finite T was closely similar to that found for the dense random packing of hard spheres in a calculation effectively carried out at $T=0$.[114] The differences of detail suggest that the structure be considered the DRP of soft spheres. In the DRP of hard spheres, all spheres are completely constrained by their nearest neighbors.[18, 115] It is thus possible to construct around each atom Voronoi polyhedra that differ in shape and volume, and to take these as the cells or cages in a cellular model. The polyhedra are constructed by passing perpendicularly bisecting planes through each line connecting the center a molecule to the center of every other molecule in the system. The innermost polyhedron formed by this method is known as the Voronoi polyhedron. The construction is an extension of the Wigner-Seitz construction for crystals. The results of Rahman et al.[32] strongly suggest that the same construction can be carried out for the Lennard-Jones glass, and presumably for all simple molecular glasses, thus validating the use of a cellular model and thereby justifying assumption 1.

Jacucci[33] has examined the structure of a dense, argonlike Lennard-Jones liquid in a novel and instructive way. MD calculations are typically done in about 10^4 time steps of 10^{-14} sec each. The positions are retained and the radial distribution function constructed from sets of position pairs taken at equal time. Jacucci first averaged the molecular positions over an intermediate time interval of 10^2 steps or 10^{-12} sec and then constructed the radial distribution function. He found a radial distribution function very close to that found by Rahman et al.[32] apart from an expansion of the distance scale. Thus a well-defined cage or cellular structure persists even in the liquid for a time scale longer than 10^{-12} sec. It would be of interest to determine on what time scale and at what temperature this structure begins to break down.

Further MD calculations by Jacucci[33] provide both detailed justification of assumption 3 and additional support for the cage picture. Jacucci examined the separation of a pair as a function of time in his dense, supercooled, argonlike Lennard-Jones liquid. He found that the separation underwent occasional large changes of order of the particle spacing. In between these, the separation showed rapid small fluctuations around a nearly constant value. In a cage picture, the former correspond to the diffusive steps of Cohen and Turnbull,[18, 87, 88] which lead to an occasional reorganization of the cages, whereas the latter correspond to oscillatory motion within a cage structure that remains roughly static in between diffusive steps.

Cohen and Turnbull's critical free-volume fluctuations picture of self-diffusion in dense liquids is similar to the vacancy model of self-diffusion in crystals. However, in crystals individual vacancies exist and retain their identity over long periods of time, whereas in liquids the corresponding voids are ephemeral. The free volume is distributed statistically so that at any given instance there is a certain concentration of molecule-sized voids in the liquid. However, each such void is short-lived, being created and dying in continual free-volume fluctuations. The Frenkel hole theory of liquids ignores this ephemeral, statistical character of the free volume.

These results of Rahman et al.,[32] Hsu and Rahman,[111] and Jacucci[33] strongly suggest that there is a cage structure in a sufficiently dense liquid that, as the liquid is cooled, persists for longer and longer times until finally it is frozen. Assumption 1 is thus justified, and (3.1) can be made the basis for a simple theory of the glass transition. The computer quenching experiments and the structural results demonstrate as well the validity for the Lennard-Jones fluid of the concept of a single, condensed, amorphous phase comprised of the vitreous and liquid states; the results of Wendt and Abraham[112] suggest that there can be a T_g below which the system goes out of equilibrium.

In a cellular model of a glass or liquid derived from the Voronoi construction, each cell i differs in shape, size, and location within it of the molecule. We suppose, following Kirkwood, that each cell i contributes an additive term f_i, the cellular or local free energy, to the total free energy. We suppose further that the cellular free energy f_i depends only on the cell volume v_i, $f_i = f(v_i)$. The total free energy F of (3.1) then becomes

$$F = N \int P(v) \left[f(v) + kT \ln P(v) \right] dv - TS_c \qquad (3.3)$$

where we have added the configuration entropy of $P(v)$, the probability that a cell has a volume v, which is analogous to an entropy of mixing of the cells.

The local free energy function $f(v)$ contains two contributions, the negative of the work to remove a molecule from the interior of a cage of volume v, $f_0(v)$, and the work to expand the cage to the volume v from some suitable average value, $f_1(v)$. The work to remove a molecule from a cage at $T = 0°\mathrm{K}$ has the same general shape as a function of cage volume that the intermolecular pair potential has as a function of pair separation. Thermal effects will not modify this, and we use for $f_0(v)$ the shape shown in Fig. 7a. The essential features of $f_0(v)$ are a minimum at v_0 and a point of inflection at v_1. The discussion of $f_1(v)$ is more intricate because it depends on the state of the neighborhood of the particular cell in question. One can, however, argue that for smaller v, the total $f(v)$ must be quadratic in v. Moreover, as v increases beyond the minimum in $f_1(v)$, $f_1'(v)$ decreases because of the reduced effect of interaction across the void. The asymptotic behavior of $f_1(v)$, however, depends on whether v is in a liquidlike or solidlike environment. In the latter case one has both a surface energy and an elastic component, in the former case only surface energy. We are not interested in such wide variation of v, and for us the significant thing in both $f_0(v)$ and $f_1(v)$ is the decrease in slope below that of quadratic dependence on v as v increases away from the minimum. Thus a shape for $f(v)$ like that of $f_0(v)$ is good enough for our purpose. Accordingly, we suppose $f(v)$ to have the model form shown in Fig. 7. For T near or above T_g, the effects of curvature in $f(v)$ are small compared to kT_g near its point inflection and

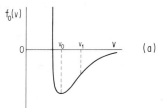

(a)

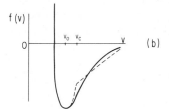

(b)

Fig. 7. (a) The negative of the work to remove a molecule from the center of a cell $f_0(v)$ versus cell volume v; v_0 marks the minimum and v_1 the point of inflection of $f_0(v)$. (b) Local free energy $f(v) = f_0(v) + f_1(v)$, where $f_1(v)$ is the work to expand the cage to the volume v from its average value. For $v > v_c$, $f(v)$ can be approximated as quadratic, and for $v > v_c$, linear in its dependence on v as shown by the dashed curve.

we can approximate $f(v)$ by the following simple form,[94]

$$
\begin{aligned}
f(v) &= f_0 + \tfrac{1}{2}\kappa(v - v_0)^2 && \text{for} \quad v < v_c \\
&= f_0 + \tfrac{1}{2}\kappa(v_c - v_0)^2 + \zeta(v - v_c) && \text{for} \quad v > v_c
\end{aligned}
\tag{3.4}
$$

The parameters f_0, v_0, κ, $v_0 < v_c < v_1$, and ζ can be determined on a best-fit basis. All are functions of T and particle density. The f_0 and v_0 turn out to be irrelevant. Thermal smearing effects and thermal expansion at constant pressure give $\partial v_c / \partial T, \partial \kappa / \partial T < 0$.

The segmenting of $f(v)$ in (3.4) enables us to divide the cells into two classes. Those with $v > v_c$ we call liquidlike, and those with $v < v_c$ we call solidlike. Following Turnbull and Cohen,[87, 88] we say that only liquidlike cells with $v > v_c$ have a free volume, which we take as[94]

$$
v_f = v - v_c \quad \text{for} \quad v > v_c
\tag{3.5}
$$

It immediately follows that the part of the sum of local free energy $\sum_i f_i(v)$ contributed by liquidlike cells, $v_i > v_c$, depends only on the average value of v_i among the liquidlike cells. That is, according to (2.5) it depends only on their average free volume $\bar{v}_f$ and is unchanged by any repartition of the free volume among the liquidlike cells. It is in this sense that a free volume can be defined, as in assumption 2, and can be redistributed without change of free energy, as in assumption 4. This is as far as Turnbull and Cohen got in formulating their version of the free-volume model; they were unable to proceed to a thermodynamic calculation.

As stated above, $f(v)$ has two contributions $f_0(v)$ and $f_1(v)$, and the latter depends sensitively on the nature of the cell's immediate environment. This dependence is not so crucial for smaller expansions, $v < v_c$ in the quadratic range, but in the linear range $v > v_c$ it must be taken into account. We therefore decompose ζ into two corresponding parts ζ_0 and ζ_1, leave ζ_0 as a constant, and introduce the environment dependence into ζ_1. The system clearly becomes more rigid as the volume decreases; ζ_1 is maximal when the system is entirely solidlike. We can characterize the deviation from solidlike behavior through the mean free volume within the liquidlike fraction of the material:

$$
\bar{v}_f = \frac{\displaystyle\int_{v_c}^{\infty} (v - v_c) P(v)\, dv}{\displaystyle\int_{v_c}^{\infty} P(v)\, dv}
\tag{3.6}
$$

As will emerge later, we are primarily interested in liquidlike cells with a substantial number of liquidlike neighbors. Accordingly, as $\bar{v}_f$ increases, the environment becomes progressively less rigid and ζ_1 decreases monotonically. A very simple function of $\bar{v}_f$, which has that monotonic behavior over a large range of $\bar{v}_f$ and approaches a finite limit when $\bar{v}_f \to 0$ (solidlike), is

$$\zeta_1 = \frac{kT_1}{v_a + \bar{v}_f} \tag{3.7}$$

where v_a and T_1 are constants of the dimension of volume and temperature, respectively. This form for ζ_1 will have important implications for the final forms of the average free volume $\bar{v}_f$ and viscosity η. We show below that this term is not an entropic contribution to the free energy but must come from effects of environment on the free energy of each cell. We arrive therefore at

$$\zeta = \zeta_0 + \frac{kT_1}{v_a + \bar{v}_f} \tag{3.8}$$

for ζ in (3.4).

The entire cage picture on which the free-volume model and the mobility theory are based is valid only for a sufficiently dense material. As the material expands, the time scale over which the cage structure persists becomes comparable to the time scale of motion within the cages, and the picture loses its meaning. We emphasize that it should be used primarily for discussion of supercooled liquids where the Doolittle equation holds, of the glass transition, and of certain aspects of the glassy state. Our model excludes the possibility of a channel to crystallization. Since most glasses are stable for periods greater than years, this does not pose a problem.

IV. FREE EXCHANGE OF FREE VOLUME; CLUSTERS AND PERCOLATION

The essence of the free-volume theory described in Section III is that the only change in free energy associated with a redistribution of free volume is in the entropy of the probability distribution of the free volume. This arises from the decomposition of the free energy into a sum of terms depending only on the volume of a single cell, the local free energy $f(v_i)$, and from the linearity of $f(v_i)$ in v_i for liquidlike cells. Of the two, the former is the more serious approximation.

Consider two liquidlike cells that are not nearest neighbors and are individually surrounded by solidlike cells. From the construction of the Voronoi polyhedra defining the cell volume, it is clear that the cell volumes are not all independent variables. It is not possible to change the volume of an isolated liquidlike cell without also changing the volumes of the neighboring solidlike cells. Thus a change in the local free energy of an isolated liquidlike cell that is linear entails quadratic changes in those of the neighboring solidlike cells. An exchange of free volume between isolated liquidlike cells therefore entails a change in the sum of the local free energies of all the cells and will be an activated process.

A free exchange of free volume can take place only between liquidlike cells that (1) are nearest neighbors and (2) have enough other liquidlike nearest neighbor cells ($\geq z$) to ensure that the volumes of any neighboring solidlike cells are not constrained to change simultaneously. This defines a type of percolation problem.[116-121]

The fraction of liquidlike cells is

$$p = \int_{v_c}^{\infty} P(v) \, dv \qquad (4.1)$$

When p is nonzero, there are clusters of liquidlike cells, each one of which has at least z liquidlike neighbors. It is well known that in such situations there is a critical concentration p_c above which there exists an infinite cluster. Thus for $p > p_c$, there is an infinite, connected liquidlike cluster, and we can consider the material within it to be liquid. For $p < p_c$, only finite liquidlike clusters exist, which might imply a glass phase because the fluidity would be reduced. However, percolation theory tells us that just above p_c the infinite cluster is very stringy or ramified so that bulk liquid properties are not fully developed.

We have defined a liquidlike cell to be in a cluster if it has at least z neighbors that are also liquidlike.[94] Within such a liquidlike cluster, cells can exchange their free volume freely without restriction by neighboring solidlike cells. The usual percolation problem has $z = 1$, so that all isolated liquidlike cells would be clusters of size one. Thus we have introduced a new percolation problem, which we call environmental percolation.[94, 122, 123] In addition to its present applications for the glass transition, this generalization of ordinary percolation has interesting applications for the study of many magnetic alloys[124-126] including Ni-Cu as well as the properties of supercooled water.[127] We do know that a system with $z = 2$ has the same value of p_c and is essentially the same as the $z = 1$ case, with only a few dangling cells excluded from the clusters. For $z > 2$, the percolation threshold p_{cz} becomes z-dependent. The various critical exponents associated with the percolation problem are expected to be z-dependent as well, and we have no knowledge of their values. However, the general structure of the theory must remain unchanged, and it is only this which we use in the following. The value of z appropriate to the free-volume theory is uncertain; a reasonable estimate is $z \simeq \frac{1}{2} n$, where n is the average number of nearest neighbors.

At this point one is tempted to anticipate the results of the quantitative analysis and suppose that the liquid phase has $p > p_{cz}$, the glass phase has $p < p_{cz}$, and $p = p_{cz}$ at the glass transition temperature. If so, the transition would be second order because the infinite cluster is formed sharply.[116–121] Calculations based on the model show that it cannot be second order in most circumstances, but is first order, with a range of values of p around p_c excluded. Elimination of the simplifications we have introduced wipes out the second-order phase transition, but the first-order phase transition persists in the circumstances we believe to hold experimentally, as we shall discuss after presenting the calculations.

We note also that atomic mobility occurs within finite liquidlike clusters that exist below the transition. Thus the fluidity of the system would in principle persist below T_p.

In the usual percolation problem with $z = 1$, all pN liquidlike cells are in clusters if one counts all isolated liquidlike cells as clusters of size one. That is no longer true when $z \neq 1$. Only a fraction $a_z(p)$ of the pN liquidlike cells are now in the cluster $[a_1(p) \equiv 1]$. The cluster distribution $C_{\nu z}(p)$, $\nu = 1, 2, \ldots$, is normalized so that[119]

$$\sum_{\nu=1} \nu C_{\nu z}(p) = 1 \qquad \text{for} \quad p \leq p_{cz}$$

$$= 1 - P_z(p) \qquad \text{for} \quad p > p_{cz} \qquad (4.2)$$

where $P_z(p)$ is the percolation probability, the probability of being in the infinite cluster $[P_z(p) = 0, p < p_{cz}]$. Thus the number of cells in finite clusters is $Npa_z(p)\sum_\nu C_{\nu z}(p)$, that in the infinite cluster is $Npa_z(p)P_z(p)$, and the total in clusters is $Npa_z(p)$.

The reader is referred to recent reviews[116, 117] of percolation theory for $z = 1$ for a more complete study. Here we summarize some important results, which we expect to carry over to the environmental percolation problem with $z \neq 1$. For $p > p_{cz}$ and $|p - p_{cz}| \ll 1$, $P_z(p)$ is assumed to have a scaling form,

$$P_z(p) = B_z(p - p_{cz})^{\beta_z} \qquad (4.3)$$

which defines the critical exponent β_z. For $z = 1$, $\beta_1 \simeq 0.39$ in three dimensions, and $p_{c1} \approx 0.15$ for the continuum percolation problem and $p_{c1} \approx 0.18$ for the site percolation problem on a face-centered cubic lattice. Numerical results are not available for $z > 2$. However, we do expect a larger value of p_{cz} for $z > 2$ and a slower growth of the finite cluster, that is, a larger value of β_z for $z > 1$. Stauffer has shown that in mean field theory for $z = 1$, the cluster distribution function $C_{\nu z}(p)$ has a particularly simple form. Generalizing his results for all z, we find[119]

$$C_{\nu z}(p) = q_0 \nu^{-\tau_z} \exp(\varepsilon \nu^{\sigma_z}) \tag{4.4a}$$

$$\varepsilon = q_1(p_{cz} - p) \tag{4.4b}$$

$$\sigma_z = \beta_z^{-1} \delta_z^{-1} = (\gamma_z + \beta_z)^{-1}, \qquad \tau_z = 2 + \delta_z^{-1} \tag{4.4c}$$

where the q_0 and q_1 are positive quantities that may depend analytically on p and depend on the specific percolation problem. The terms τ_z and σ_z are the critical indices appropriate to the cluster distribution function $C_{\nu z}(p)$. They are related to other critical indices, β_z, γ_z, and δ_z through (4.4c). Equation 4.4a can be generalized beyond the mean field approximation to[119]

$$C_{\nu z}(p) = q_0 \nu^{-\tau_z} g(\varepsilon \nu^{\sigma_z}) \tag{4.5}$$

where g is a universal function of its argument, depending only on dimension and z.

In the absence of theoretical results for the exponent β_z, which we need in later sections, we turn to some experiments on magnetic alloys for useful information. There are several systems—for example, Ni-Cu, in which a magnetic atom (Ni) must have a minimum number of like neighbors before it can have a magnetic moment.[124-126] For Ni-Cu, $n = 12$ and $z = 8$. The percolation threshold appears at $p_{c8} = 0.44$, and our estimates give $0.5 < \beta_8 < 0.7$ for β_8. These estimates were made by analyzing the p-dependence of the low-temperature magnetization above but near p_{c8}. They are necessarily very crude because of the limited accuracy of the data, contributions to the magnetization from finite clusters (superparamagnetism plus weak ferromagnetic coupling of the finite clusters), and a probable dependence of the Ni moment on the number of Ni neighbors above the critical value of $z = 8$. A more refined description must await further investigation (e.g., by Monte Carlo techniques). The essential point is that there exists a well-defined p_{cz} and a characteristic exponent β_z whose values we do not yet know.

V. COMMUNAL ENTROPY AND THE FREE–ENERGY FUNCTIONAL

Atoms can diffuse when a fluctuation in cellular volume of atomic size v_m or greater occurs. Such large fluctuations can arise with significant probability only from redistribution of the free volume (free exchange of free volume) within a liquidlike cluster in the present picture; otherwise an activation free energy is required. The total free volume within a cluster of size ν must therefore be greater than v_m for diffusion to occur.

$$\sum_{i=1}^{\nu} (v_i - v_c) > v_m \qquad \text{for} \quad v_i > v_c \tag{5.1}$$

The average free volume within a liquidlike cluster is $\bar{v}_f$, given by (3.6). Thus for diffusive motion to take place within a given cluster, its size ν must be at least $v_m/\bar{v}_f$,

$$\nu \geq \nu_m = \frac{v_m}{\bar{v}_f} \tag{5.2}$$

We call all clusters liquidlike. However, a cluster for which (5.2) holds is liquid, rather than liquidlike, in the sense that each atom or molecule within it moves in time through the entire cluster. That is, each molecule finds accessible the configuration space of every other molecule in the cluster. We now suppose that exchange of free volume between solidlike and liquidlike cells is so slow compared to exchange between liquidlike cells that we can ignore it in the computation of equilibrium properties. We return to this point later. As we shall see in Section X, the two time scales differ by much more than 2 orders of magnitude.

In a liquid cluster, an atom or molecule is not confined to a particular cell or cage but can wander over the entire volume of the cluster. The communal entropy[110, 128] of a single cluster of size $\nu > \nu_m$ is then given by $k\nu \ln(\nu - 1)\bar{v}_s$, where $\bar{v}_s$ is the average configurational volume of a liquidlike cell,

$$\bar{v}_s = p^{-1} \int_{v_c}^{\infty} v_s(v) P(v) \, dv \tag{5.3}$$

where

$$s(v) = -\frac{\partial f(v)}{\partial T} = k \ln v_s(v) \tag{5.4}$$

When the infinite cluster is present, the communal entropy is greatly enhanced, since many atoms have the possibility of extending their movement over the entire system. Therefore,[94] *the communal entropy is that entropy associated with the accessibility of all the configurational volume within the finite liquid clusters and within the infinite cluster when present.* That is,[94]

$$S_c = Nkpa_z(p) \left\{ \sum_{\nu=\nu_m}^{\infty} \nu C_{\nu z}(p)\ln[(\nu-1)\bar{v}_s] \right.$$

$$\left. + P_z(p)\ln[Npa_z(p)P_z(p)\bar{v}_s] \right\} \qquad (5.5)$$

where Np is the total number of liquidlike cells, $Npa_z(p)$ is the number of liquidlike cells in clusters, and k is the Boltzmann constant. The first term in (5.5) arises from finite clusters and the second from the infinite cluster. Note that an extra $N\ln N$ in (5.5) arises from treating the molecules as distinguishable. Although this term could be eliminated by treating the molecules as indistinguishable, it is very convenient to keep it and simply regard the $N\ln N$ correction as understood. Using (4.2), we can bring (5.5) into a more convenient form,

$$S_c = Nkpa_z(p) \left\{ A_z(p)\ln \bar{v}_s + \sum_{\nu=\nu_m}^{\infty} C_{\nu z}(p)\ln(\nu-1) \right.$$

$$\left. + P_z(p)\ln[Npa_z(p)P_z(p)] \right\} \qquad (5.6)$$

where

$$A_z(p) = 1 - \sum_{\nu=\nu_m}^{\infty} \nu C_{\nu z}(p) \qquad (5.7)$$

We have thus expressed S_c in terms of p, $\bar{v}_s$, ν_m, $a_z(p)$, $C_{\nu z}(p)$, and $P_z(p)$. The first two quantities depend only on $P(v)$, and in Section VII we develop a method of determining $P(v)$. The $a_z(p)$ and $P_z(p)$ depend on $C_{\nu z}(p)$. Because $C_{\nu z}(p)$ enters the free energy through S_c, (5.6), it should be determined by minimization of the free energy. As we see below, the results are the same as those of a percolation problem somewhat different from the environmental percolation problem described in Section IV. The essential point is that clusters with $\nu \geq \nu_m$ are favored because they contribute to the communal entropy, while the formal structure of the problem remains that of a percolation problem.

We now consider the effects of the communal entropy on the features of the percolation problem. Clusters of sizes less than ν_m tend to be suppressed; $a_z(p)$ and $A_z(p)$ move toward unity. The problem moves away from environmental percolation based on an uncorrelated distribution of cells toward the continuum percolation problem studied by Scher and Zallen[129] and Webman et al.[130] These, together with the considerations in Section V on Ni-Cu alloys, give us very rough bounds on the values of p_{cz} and β_z appropriate to the present percolation problem,[94]

$$0.15 \leq p_{cz} \leq 0.45 \quad \text{for} \quad 0.4 \leq \beta_z \leq 0.7$$

Qualitatively speaking, as p increases toward p_{cz}, the mean cluster size grows; that is, the scale of ν's important in (5.6), grows until at p_{cz} an infinite cluster emerges. Above p_{cz}, the percolation probability $P_z(p)$ increases. Thus S_c shows a monotonic increase with p, with maximum and possibly divergent slope as $p \rightarrow p_{cz} \pm 0^+$.

We can go no further toward the explicit evaluation of S_c without introducing more information about $C_{\nu z}(p)$. Since the results we obtained for the cluster distribution function in Section IV are valid only in the pure percolation problem, they do not apply here. The cluster distribution function can be obtained in a mean field approximation by minimizing the free energy with $P(\nu)$, therefore p fixed. The $C_{\nu z}(p)$ enters only through the communal entropy and a cluster surface free energy discussed below, which has not yet been explicitly introduced into the free energy. The procedure developed by Fisher[131] in his study of the droplet model of condensation can then be followed with only slight modification.[132] The entropy now contains an additional $\nu \ln \nu$ contribution for "droplets" with $\nu > \nu_m$, which must be included.

We consider the energy E_ν of a cluster of ν molecules. For all but the very small clusters, we can decompose E_ν into a bulk term proportional to the number of molecules in the cluster and a cluster surface energy contribution,[131]

$$E_\nu = \nu E_0 + W \tag{5.8a}$$

where E_0 is the average energy per molecule. A positive surface energy contribution W is given by

$$W \simeq ws \tag{5.8b}$$

where s is the surface area of the cluster. Similarly, we expect the entropy of the cluster to have the form[131]

$$S_\nu = \nu S_0(\nu) + \omega s \tag{5.9a}$$

Here $S_0(\nu)$ is the average entropy per molecule

$$S_0(\nu) = \bar{s}_0 + ka_z(p)\ln\nu \qquad \text{for} \quad \nu > \nu_m$$
$$= \bar{s}_0 \qquad \text{for} \quad \nu < \nu_m \qquad (5.9b)$$

where $\bar{s}_0$ is independent of ν and ω is the surface entropy density. We have added the $\nu\ln\nu$ contribution for $\nu > \nu_m$, which follows from (5.5) and (5.6). This is where our result differs from the usual droplet model. The surface entropy density ω is a measure of the number of different configurations of the cluster that have the same surface area. Fisher constructed the classical configurational partition function,[131]

$$q_\nu(T) = \frac{1}{\nu!} \int dr_1 \cdots \int dr_\nu \exp\left(-\frac{E_\nu}{kT}\right) \qquad (5.10)$$

where the integrations are restricted to configurations in which ν molecules form a drop. Assuming further that the pair interaction potential has an infinite hard core and an attractive square well, Fisher showed that

$$q_\nu(T) = V\sum_s g(\nu, s)\exp\left[\frac{\nu E_0 - ws}{kT}\right] \qquad (5.11)$$

where the combinatorial $g(\nu, s)$ is the number of configurations of ν indistinguishable molecules that form a drop of surface area s. Although our problem differs from that of Fisher in that we have clusters within a condensed phase instead of drops in a gas, we can reinterpret what he does in terms of a Landau-Ginzburg-Wilson type of theory and carry (5.11) over into the present problem. The derivation of the Landau-Ginzburg-Wilson theory is discussed in the next section.

The crux of the analysis is the factor $g(\nu, s)$ entering (5.11). It is convenient to define the additional quantity[131]

$$G_\nu(T) = \sum_s g(\nu, s)\exp\left(-\frac{ws}{kT}\right)$$
$$= \exp\left(-\frac{\nu E_0}{kT}\right)\frac{q_\nu(T)}{V} \qquad (5.12)$$

where $G_\nu(T)$ is related to the bulk entropy per particle $S_0(\nu)$ in a larger cluster by the relation

$$S_0(\nu) = k\lim_{\nu\to\infty}\frac{1}{\nu}\ln G_\nu(T) \qquad (5.13)$$

Fisher points out further that the surface of a cluster cannot exceed some constant multiple of the number of the particles, nor can it be less than some minimum surface attained by some approximately spherical cluster; thus

$$a_1 \nu^{1-1/d} \leq s \leq a_2 \nu \tag{5.14}$$

where a_1 and a_2 are constants. Since the terms in (5.12) are all positive, the standard maximum term argument leads one to the conclusion that for ν large, $G_\nu(T)$ must have the form[131]

$$\ln G_\nu(T) = \ln \bar{g}[\nu, \bar{s}(\nu, T)] - \beta w \bar{s}(\nu, T)$$
$$+ \tilde{\tau} \ln \nu + \mathcal{O}(\nu^0) \tag{5.15}$$

where $\bar{s} = \bar{s}(\nu, T)$ is the most probable or mean surface area and $\tilde{\tau}$ is a positive constant. For the surface to be well defined, we must have $\bar{s}/\nu \to 0$ as $\nu \to \infty$, in which case it is convenient to write simply

$$\bar{s} = a_0 \nu^\sigma \quad \text{for} \quad 0 < \sigma < 1 \tag{5.16}$$

where σ could be a function of T. From (5.13) to (5.16), we see that $k \ln g(\nu, \bar{s})$ varies as $S_0(\nu)$ for large ν, and the difference for smaller ν is a measure of the surface entropy $\omega \bar{s}$

$$k \ln g(\nu, \bar{s}) - \nu S_0(\nu) = \omega \bar{s} \tag{5.17}$$

as $\nu \to \infty$.

Combining these results one finds

$$\ln G_\nu(T) \approx \frac{\nu S_0(\nu)}{k} - \frac{\bar{s}(w - \omega T)}{kT} - \tilde{\tau} \ln \nu + \ln g_0 \tag{5.18}$$

This expression is different for $\nu > \nu_m$ and for $\nu < \nu_m$ because only clusters with $\nu > \nu_m$ contribute to the communal entropy $S_0(\nu)$ and are therefore favored. This difference in the dependence of $S_0(\nu)$ and the resulting discontinuity at $\nu = \nu_m$ is the only difference from the usual percolation theory. The desired cluster distribution function $C_{\nu z}(p)$ is porportional to $q_\nu(T)/V$, obtained by substituting (5.18) into (5.12) and treating p as a function of temperature T. Thus $C_{\nu z}(p)$ is the same as that already given by (4.4) with $\tau = \tilde{\tau} - a_z(p)$ for $\nu > \nu_m$. For $\nu < \nu_m$, τ is replaced by $\tilde{\tau} > \tau$, so that clusters with $\nu < \nu_m$ are suppressed. Beyond mean field theory, we expect $C_{\nu z}(p)$ to have the scaling form[119, 132] given by (4.4) with comparable changes, which will be sufficient for the present discussions.

We can now evaluate S_c near p_c. Inserting (4.5) into (5.6) and replacing the sum over ν by an integral leads for p near p_c and $\beta < 1$ to: (we drop the subscript z for convenience):

$$S_c = Nkpa_z(p)\{A_z(p)\ln \bar{v}_s + B_1^{\pm}|p-p_c|^{\beta}$$
$$+ B|p-p_c|^{\beta}u(p-p_c)\ln[Npa_z(p)B|p-p_c|^{\beta}]\} + S_c' \quad (5.19)$$

for the communal entropy, where $u(x)$ is the unit step function, S_c' stands for analytic contributions to S_c and

$$B_1^{\pm} = (\beta\delta)^2 q_0 q_1 \int_0^{\infty} dx\, x^{-1-\beta}[g(\pm x)-g(0)]\ln\left(\frac{x}{|\varepsilon|}\right) \quad (5.20)$$

the plus corresponding to $p < p_c$ and minus to $p > p_c$.

We know that as $p \to 0$, $S_c \to 0$, while as $p \to 1$, $S_c \to Nk\ln(N\bar{v})$. From (5.9), S_c is a monotonic function of p and is sketched in Fig. 8. The slope of S_c,

$$\frac{\partial S_c}{\partial p} \underset{p \to p_c}{\to} A_{\pm}|p-p_c|^{\beta-1} \quad (5.21)$$

diverges as $p \to p_c$ for $\beta < 1$. The behavior of the slope of S_c and, therefore, the value of β, turns out to be critical in determining the order of the phase transition at T_p, as discussed in Section VI.

There is also the possibility that β will equal or exceed unity. Such a situation may occur in more complex glasses (e.g., polymer glasses). In such systems, the moving units are molecular segments constrained by their connections to the rest of the molecule. Such constraints imposed by the complexity of the material increase z and otherwise decrease the growth rate of the clusters with increasing p, increasing the value of β. If this were the case, then

$$\frac{\partial S_c}{\partial p} \underset{p \to p_c}{\to} A_0 \quad (5.23)$$

where A_0 is a constant. Thus the slope of S_c is constant as $p \to p_c$ for $\beta \geq 1$.

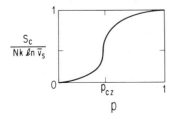

Fig. 8. Sketch of the communal entropy S_c as a function of the fraction of liquidlike cells p.

VI. MEAN FIELD THEORY

A. Statistical Mechanics

Now that we have an expression for the communal entropy, we can derive the probability distribution $P(v)$. We start from the configurational free energy $\mathcal{F}$ given by

$$e^{-\beta\mathcal{F}} = \int D[N] e^{-\beta\mathcal{V}[N]} \tag{6.1}$$

Here $D[N]$ is the volume element in the N-particle configuration space, and $\mathcal{V}[N]$ is the sum of the pairwise interaction potentials among the N molecules. We deal with the case of spherical molecules for simplicity; the results are general. We must convert (6.1) into an integral over a set of N independent dynamic variables that have a monatomic relationship to the cell volume variable v we have used thus far. Consider the configurations $[N-i]$ of the $N-1$ molecules other than the ith that contribute significantly to (6.1). These produce an interaction potential with the ith molecule,

$$\mathcal{V}[\vec{r}_i | N-i] = \sum_j{}' V(|\vec{r}_i - \vec{r}_j|) \tag{6.2}$$

where $V(|\vec{r}_i - \vec{r}_j|)$ is the pairwise interaction, which for the densities we are dealing with has closed contours (defining a cell for the ith molecule) up to values of $\mathcal{V}$ so large that they are unimportant in (6.1), a consequence of the steepness of the repulsive part of V. For fixed $[N-i]$, the volume τ_i accessible to the center of the ith molecule is the volume inside the contour of constant $\mathcal{V}[\vec{r}_i | N-i]$ corresponding to the largest value of $\mathcal{V}$ of significant probability $\mathcal{V}^*$. Fortunately, we do not need to know $\mathcal{V}^*$ because of the steepness of the repulsive part V; using the contour $\mathcal{V}[\vec{r}_i | N-i] = 0$ gives us adequate accuracy. We therefore define, formally,

$$\tau_i = \int d^3x\, u(\mathcal{V}[\vec{x} | N-i]) \tag{6.3}$$

where $u = 1$ for $\mathcal{V} < 0$ (i.e. for x inside the contour $\mathcal{V} = 0$) and zero otherwise; τ_i is the volume inside this contour.

We now transform (6.1) into an integral over the τ's

$$e^{-\beta\mathcal{F}} = \int D[\tau] e^{-\beta F'[\tau]} \tag{6.4a}$$

where

$$e^{-\beta F'[\tau]} = \int D[N]\Pi_i \delta\left(\tau_i - \int d^3x\, u(\mathcal{V}[\vec{x}|N-i])e^{-\beta\mathcal{V}[N]}\right) \quad (6.4b)$$

There is a monotonic relationship between τ_i, the volume accessible to the center of the molecule within its cell, and v_i, the volume of the cell. We can thus transform (6.4) into an integral over the v_i, exponentiating the Jacobian and absorbing it into the free energy $\bar{F}[v]$,

$$e^{-\beta\mathcal{F}} = \int D[v]e^{-\beta\bar{F}[v]} \quad (6.5)$$

The results embodied in (6.4) and (6.5) are obviously too formal to be directly useful, so we assume further that the probability $P[v] = \Pi_i P(v_i)$. We can therefore take for $F[v]$ form we have developed thus far, (3.3) as modified in Section V to include cluster surface energies. The $F(v)$ depends only on the probability distribution $P(v)$, the cluster size distribution $C_{vz}(p)$, and p. Thus (6.5) can be converted into a functional integral over $P(v)$, $C_{vz}(p)$, and p, and F is replaced by $F[P, C, p]$, a Landau-Ginzburg-Wilson free-energy functional

$$e^{-\beta\mathcal{F}} = \int D[P, C, p]e^{-\beta F[P, C, p]} \quad (6.6)$$

The transformation from $[N]$ to $[\tau]$ has not been shown to exist. One requirement for its existence is that τ_i be finite for any configuration $[N-i]$ of significant probability. Assign each molecule $j \neq i$ a diameter equal to the separation at which $V(|\vec{r}_i - \vec{r}_j|)$ goes through zero. Define a continuous percolation problem with p' the fraction of the space outside the molecules. As long as p' is below a percolation threshold p'_c, all the τ_i will be finite. We can expect p'_c to be substantially larger than the value for the purely random case, 0.15, because large voids are suppressed by the large free energy they require and because the minimum linear dimension of a continuous void must exceed a molecular diameter. For hard spheres, the density of melting is two-thirds of the density of the ideally close-packed crystal. The density for dense random packing (the glass form) is 15% larger than that of the crystal. Thus the thermal expansion on melting is about 18%, close to the random percolation threshold and presumably much less than p'_c. We expect therefore that the τ_i are finite and that the transformation $[N] \rightarrow [\tau]$ exists.

To find $C_{vz}(p)$, $P(v)$, and p itself, we do a mean field calculation. We have in fact already found $C_{vz}(p)$ by such a mean field calculation in Section V. We require that $\delta F/\delta X = 0$ subject to normalization conditions—for example,

$$\int_0^\infty P(v)\,dv = 1$$

where X is $C_{vz}(p)$, $P(v)$, and p in turn. In as much as $C_{vz}(p)$ has already been studied, we find $P(v)$ and p in the present section. We choose $F[P, C, p]$ to be of the form given by (3.3), insert $f(v)$ from (3.4) and (3.8), and S_c from (5.6). We note that $f(v)$ in (3.3) is regarded as independent of $P(v)$ and yet that in (3.4) and (3.8) $f(v)$ has in it an implicit dependence on $P(v)$ through the presence of $\bar{v}_f$. The most convenient way to resolve this inconsistency is to treat $f(v)$ as though it were a self-consistent field, itself the results of a first functional derivative of the total local free energy with respect to $P(v)$, and not differentiate it further in the variation of $F[P, C, p]$. Otherwise, to avoid multiple counting, we would have to deal more explicitly with a complex nonlinear functional of $P(v)$ with no direct physical interpretation. Moreover, the introduction of $\bar{v}_f$ into $f(v)$ forces it into the role of a self-consistent field.

The term $P(v)$ enters both explicitly in $F[P, C, p]$ [cf. (3.3)] and implicitly through the presence of ν_m, $\bar{v}_s$, and p in S_c [cf. (5.2), (5.3), and (4.1)]. A convenient simplification of (5.3) is

$$\bar{v}_s = v_s(v_c) + v_s'(v_c)\bar{v}_f \tag{6.7}$$

We now write the communal entropy in the form

$$S_c = Nk\mathcal{S}(p, \nu_m, \bar{v}_s) \tag{6.8}$$

Its variation is, from (6.7), (3.6), (5.2), and (4.1),

$$\delta S_c = Nk\int_0^\infty \left[R_1(v - v_c) - R_2 \right] u(v - v_c)\delta P(v)\,dv \tag{6.9a}$$

$$R_1 = \frac{1}{p}\left(\frac{-\nu_m}{\bar{v}_f}\frac{\partial \mathcal{S}}{\partial \nu_m} + v_s'\frac{\partial \mathcal{S}}{\partial \bar{v}_s} \right) \tag{6.9b}$$

$$R_1 = a_z(p)\left[\frac{\nu_m^2}{\bar{v}_F}C_{\nu_m}(p)\ln(\nu_m - 1) + A_z(p)\frac{v_s'}{\bar{v}_s} \right] \tag{6.9c}$$

$$R_2 = R_1\bar{v}_f - \frac{\partial \mathcal{S}}{\partial p} \tag{6.9d}$$

where (6.9c) is obtained from (6.9b) and (5.6). Inserting this into the variation of $F[P,C,p]$ gives us the desired expression for $P(v)$,

$$P(v)=(1-p)v_g^{-1}\exp\left[\frac{-\kappa(v-v_0)^2}{2kT}\right] \qquad \text{for} \quad v<v_c \qquad (6.10a)$$

$$P(v)=p\bar{v}_f^{-1}\exp\left(-\frac{v-v_c}{\bar{v}_f}\right) \qquad \text{for} \quad v\geq v_c \qquad (6.10b)$$

where

$$v_g=\int_0^{v_c}\exp\left[-\frac{\kappa(v-v_0)^2}{2kT}\right]dv \qquad (6.11)$$

$$\bar{v}_f^{-1}=\Gamma=\frac{\varsigma}{kT}-R_1 \qquad (6.12)$$

$$p=\bar{v}_f\left[\bar{v}_f+v_gQ\right]^{-1}=\left[1+v_g\Gamma Q\right]^{-1} \qquad (6.13)$$

$$Q=e^{\mu_c/kT} \qquad \mu_c=\frac{\kappa(v_c-v_0)^2}{2}+kTR_2 \qquad (6.14)$$

Our result for $v>v_c$ is essentially identical to that derived earlier by Cohen and Turnbull[87, 88] for the most probable distribution of free volume x, $P(x)=\gamma/v_f\exp(-\gamma x/v_f)$, where v_f is the free volume averaged over all cells, $v_f=p\bar{v}_f$, and γ is a numerical factor between $\frac{1}{2}$ and 1 introduced to correct for overlap of the volume between neighboring cells. Comparing the exponent $\gamma x/v_f=(\gamma/p)x/\bar{v}_f$ with the exponent in (6.10b), $(v-v_c)/v_f$, we see that the two distributions are identical if x is taken as $(v-v_c)$ and γ is taken as p, which would be close to $\frac{1}{2}$ in the temperature region considered in Ref. 88.

Equation 6.13 is a self-consistency condition for p, since its right-hand side contains p through the presence of R_1 in Γ of (6.12), and R_2 in Q of (6.14). We must therefore solve (6.13) before we can calculate the heat capacity or thermal expansion and characterize the glass transition. Indeed, (6.13) is precisely the mean field equation for p obtained by the variation of F directly, as shown below.

B. p and the Order of the Transition

The self-consistency condition (6.13) is the key to the glass transition. To make its content and meaning clearer, we derive it by an alternative procedure. Instead of considering all possible variations of $P(v)$ that leave it normalized, we consider only those that leave p invariant. Variation of $F[P]$ subject to these two constraints leads again to (6.10) for $P(v)$ but with p unspecified, that is, without (6.13). Insertion of $P(v)$ in the form (6.10) back into (3.3) gives

$$F[P] = N \left\{ f_0 + \frac{p\kappa(v_c - v_0)^2}{2} \right\} + NkT \left\{ (1-p)\ln\left(\frac{1-p}{v_g}\right) \right.$$

$$\left. + p\ln\left(\frac{p}{\bar{v}_F}\right) + pR_1\bar{v}_f - \mathcal{S} \right\}$$

$$= \mathcal{F}(p) + Nf_0 \tag{6.15}$$

for the free energy as a function of p. In (6.15), $-NkT(1-p)\ln v_g$ is the part of the free energy $\mathcal{F}(p)$ associated with the solidlike cells,

$$Np\left[\frac{\kappa(v_c - v_0)^2}{2} + kTR_1\bar{v}_f - kT\ln\bar{v}_f \right]$$

that associated with the liquidlike cells,

$$Nk\left[(1-p)\ln(1-p) + p\ln p \right]$$

the entropy of mixing of solid- and liquidlike cells, and $Nk\mathcal{S}$ the communal entropy.

The values of p that make $F[P, C, p]$ in (6.10) stationary are obtained by differentiating (6.15) with respect to p and setting the result equal to zero. Carrying out the differentiation and simplifying the result by using the stationarity of $F[P, C, p]$ with respect to $P(v)$ at constant p gives (6.10) as the condition of stationarity of the free energy with respect to p. We now study its solution in each of the three cases $\beta = 1$, $\beta < 1$, and $\beta > 1$.

Returning then to (6.13), we see that it has the form $p = h(p) = [1 + Qv_g\Gamma]^{-1}$, where v_g is independent of p and Γ is a smoothly decreasing function of p. Both R_1 and $\bar{v}_f$ are essentially independent of p as shown below, but R_2 is nonmonotonic, going to $-\infty$ at p_c because of the divergence to $+\infty$ of $\partial\mathcal{S}/\partial p$ for $\beta < 1$. Thus μ_c goes to $-\infty$ at p_c, Q to zero, and $h(p)$ to unity. Therefore (6.13) does not possess a solution for p in the vicinity of p_c, and p cannot increase continuously through p_c with increasing temperature. The graphical solution of (6.13) is sketched in Fig. 9 for several temperatures. There is a bifurcation from one solution $p_1 > p_c$ at high temperatures to three solutions $p_1 > p_c$, $p_2 < p_c$, $p_3 < p_c$ at low temperatures.

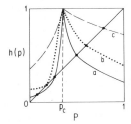

Fig. 9. Graphical solution of (6.13). Solid (a), dotted (b), and dashed (c) curves correspond to $h(p)$ (see text) for $\beta < 1$ at three different and increasing temperatures. The circles give the values of p that satisfy (6.13). Case a has three solutions, and case c has one. Case b corresponds to the bifurcation point between a and c.

Recall that these p_i are stationary points for the free energy $\mathscr{F}(p)$. We therefore pass from a single minimum in $\mathscr{F}(p)$ at high temperatures with $p > p_c$, corresponding to the liquid, to three extrema in $\mathscr{F}(p)$ at lower temperatures. Figure 10 sketches these features of $\mathscr{F}(p)$. There are two minima at $p_1 > p_c$ and $p_3 < p_c$, separated by a maximum at p_2 at lower temperatures. At some temperature T_p, $\mathscr{F}(p_1) = \mathscr{F}(p_3)$, and there is a first-order phase transition. Below T_p, $\mathscr{F}(p_1) > \mathscr{F}(p_3)$, and the minimum free-energy state has $p = p_3 < p_c$ corresponding to a solid, the glass. Above T_p, $\mathscr{F}(p_1) < \mathscr{F}(p_3)$, and the minimum free-energy state has $p = p_1 > p_c$, corresponding to the liquid. At a temperature $T_3 > T_p$ (curves b in Figs. 9 and 10), the minimum at p_3 disappears, and the glassy state is no longer locally stable. The minimum at p_1 persists down to $T = 0$, so that there is no critical end point for the liquid state. This results only because $(\partial \mathscr{S}/\partial p)|_{p_c} = \infty$. The expected dependence of p on T in the vicinity of T_p is shown in Fig. 11. The persistence of the minimum at p_1 down below T_p can give rise to hysteresis and relaxation effects associated with cooling, and the persistence of the minimum at p_3 up to T_3 can give rise to similar effects on heating. These are not the relaxation effects commonly observed at $T_g \gtrsim T_p$ that are associated with non-equilibrium liquid states.

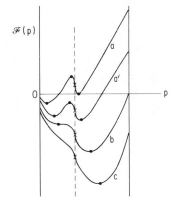

Fig. 10. Sketches of the free energy $\mathscr{F}(p)$ as a function of the liquidlike cell fraction p. Curves a, b, and c correspond to those so labeled in Fig. 9. The positions of the solutions of Fig. 9 are indicated by dots. Curve a' corresponds to $\mathscr{F}(p_1) = \mathscr{F}(p_3)$ at the temperature T_p, where the first-order phase transition occur. Crosses emphasize the infinite negative slope of $\mathscr{F}$ at p_c, the percolation threshold.

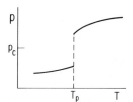

Fig. 11.   Sketch of the probability of liquidlike cells p versus T near T_p for $\beta < 1$.

The first-order transition is a direct result of the divergence of $\partial S/\partial p$ as $p \to p_c$ for $\beta < 1$. The latter is caused by the rigid and arbitrary manner in which the cells were divided into liquid- and solidlike cells. We discuss below the consequences of eliminating this unphysical feature of the model.

For $\beta > 1$, $\partial S/\partial p$ is finite at p_c, though nonanalytic, and therefore R_2 is always finite. The order of the transition now depends critically on the value of $(\partial S/\partial p)|_{p_c}$. When this quantity is large, the function $h(p)$ is similar to that found for $\beta < 1$. That is, $h(p_c)$ approaches 1 and there are three solutions for p for some range of temperatures. In our search of the parameter space, as discussed below, this occurred most of the time. For smaller values of $(\partial S/\partial p)|_{p_c}$ coupled with a slow variation in $\partial S/\partial p$ versus p, $h(p)$ will vary more slowly as a function of p and can give rise to only one solution p for each T that satisfies (6.13). Nevertheless, S changes its functional dependence on p at p_c, and there is a phase transition. Since p is a continuous function of T, the transition is second order, with T_p corresponding to $p = p_c$. Thus the transition for $\beta \geq 1$ may be either first or second order, depending on the value of $(\partial S/\partial p)|_{p_c}$ and the magnitude of the variation in $\partial S/\partial p$ versus p. We find, as discussed in Section VIII, that the former is most often the case. This situation with $\beta \geq 1$ may correspond to the case of the complex organic and polymeric glasses, where the infinite cluster has difficulty developing because of the additional constraints. Therefore β increases and may become greater than or equal to 1. Even if β remained less than unity in the complex glasses, the interplay between the configurational entropy (present in them but not in the simple glasses) and the communal entropy could lead to a variation in the total entropy, with p characterized by an exponent $\tilde{\beta}$ near p_c, with $\tilde{\beta} = \beta + x > 1$. Thus x describes the retardation in the growth of the internal configurational entropy imposed by the constraints on the free volume. If either β or $\tilde{\beta} > 1$, our thermodynamic results for some values of the parameters may be compared to those for polymers of Gibbs and Dimarzio,[17] who predict that the glass transition is a second-order phase transition in the limit of slow cooling. We make this comparison in Section VIII.

VII. FREE VOLUME AND VISCOSITY

In the preceding section, we found that p is usually a discontinuous function of T. Here, we show that both the free volume $\bar{v}_f$ and the viscosity η are only weakly dependent on p and therefore do not reflect strongly the existence of the transition.

The free volume $\bar{v}_f$ is given by the self-consistency condition, (6.13). The contribution to R_2 from $\partial S/\partial v_m$ can be neglected because $C_{vz}(p)$, the number of clusters of size v_m, is exponentially small compared to the total number of clusters present. We have, after combining (3.8), (6.9c) and (6.12), a simple equation for $\bar{v}_f$,

$$\frac{\bar{v}_f a_z(p) A_z(p)}{\bar{v}_f + v_s(v_c)/v_s'(v_c)} = \frac{\zeta_0 \bar{v}_f}{kT} + \frac{T_1 \bar{v}_f}{T(\bar{v}_f + v_a)} - 1 \qquad (7.1)$$

Here $A_z(p)$ measures the number of clusters that are larger than some minimum value v_m and $a_z(p)$ measures the number of liquidlike cells that belong to liquid clusters. As we found in Section VI, clusters of size $v < v_m$ are less favored because they do not contribute to S_c. We expect the number of these clusters to be reduced compared to an ordinary percolation problem; therefore $a_z(p)$ and $A_z(p)$ should be close to unity for p near p_c and nearly independent of p. In the extreme limit that no clusters of size $v < v_m$ are allowed, $a_z(p)$ and $A_z(p)$ would be identically equal to one. Thus (7.1) gives a cubic equation for $\bar{v}_f$ that depends on T and only very weakly on the probability p. Neglecting this dependence on p by setting $a_z(p)$ and $A_z(p)=1$, we see $\bar{v}_f$ is a smooth function of T near T_g. The contribution on the left-hand side of (7.1) arises from the communal entropy and would be absent if we had neglected the communal entropy in our deviation of the free energy. The first two terms on the right-hand side follow from our choice of ζ in (3.8). The second term, which depends on ζ_1, is similar to the entropic contribution R_1, except for the temperature dependence. The latter is important at low temperature, where one can neglect the contribution on the left-hand side. This important difference in the temperature coefficients arises because S_c enters the free energy only as TS_c. Thus the dependence of ζ_1 on $\bar{v}_f$ in (3.7) cannot be an entropic effect, since it has both the wrong T dependence and sign. For higher temperatures, $R_1\bar{v}_f$ is constant, and including it in (7.1) does not affect the general form of $\bar{v}_f$. Thus including R_1 in the expression for $\bar{v}_f$ will make only a quantitative, not qualitative, modification at high T and no change at low T, and we can absorb its effect in the remaining three parameters. This helps to reduce the proliferation of free parameters and gives a simple result for the free volume that is valid at all temperatures,

$$\bar{v}_f = \frac{k}{2\zeta_0}\left[T - T_0 + \left((T-T_0)^2 + \frac{4v_a\zeta_0 T}{k}\right)^{1/2}\right] \qquad (7.2)$$

where $kT_0 = kT_1 + v_a\zeta_0$. Equation 7.2 is noteworthy because $\bar{v}_f$ vanishes only as T goes to zero, in contrast to the earlier view that the free volume vanishes at some finite temperature. This limiting behavior is independent of the exact form of R_1 or ζ and follows from (6.12), since R_1 is finite as T goes to zero. Because of the explicit free-volume dependence in ζ in (3.8), $\bar{v}_f$ has a characteristic dependence proportional to $(T-T_0)$ at high temperatures. This introduction of $\bar{v}_f$ into ζ, (3.7), as a phenomenological way to account for the effect of neighboring cells, does not affect the low-temperature behavior of $\bar{v}_f$, only the high-temperature region.

The original work on the free-volume model by Cohen and Turnbull[87, 88] showed that the fluidity obeyed the Doolittle equation (2.8). We show that the percolation ideas developed in this paper give rise to the same equation for the fluidity.

Let D_ν be the diffusion coefficient for a particle moving in a cluster of size ν. The total diffusion coefficient D is then given by

$$D = \sum_{\nu = \nu_m} D_\nu \nu C_{\nu z}(p) \qquad (7.3)$$

From Ref. 88, we have that

$$D_\nu = \frac{1}{3}\bar{u}\int_0^\infty P_\nu(v)\tilde{f}(v)a(v)\,dv \qquad (7.4)$$

where $\bar{u}$ is the average gas kinetic velocity, $\tilde{f}(v)$ is a correlation factor that may be associated with the magnitude of each displacement within the cluster, and $a(v)$ is a mean net displacement, which depends on the immediate neighborhood of the moving particles and incorporates cooperative effects. The $\tilde{f}(v)$ must go from 0 at $v=0$ to 1 at $v=\infty$. The basic assumption is that $\tilde{f}(v)$ is a step function such that it is zero for $v < v^*$ and unity for $v > v^*$; that is, for $v < v^*$ the only motion that takes place is oscillation within a cage, oscillation having a vanishing correlation factor. Here $v^* = v_c + v_m - b\bar{v}_f$, where b is a constant and v^* is the minimum size of the cell necessary to accommodate two atoms. The correction $b\bar{v}_f$ in v^* is present because the second atom does not have to fit entirely into the cell but can use some of the free volume of the neighboring cells. The value of $a(v)$ can be chosen to be proportional to v in the small v limit. Then since $P_\nu(v)$

$=p\Gamma\exp[-\Gamma(v-v_c)]$, we have

$$D=D_0 pa_z(p)A_z(p)e^{-v_m/\bar{v}_f} \tag{7.5}$$

where we have added the contribution from the infinite cluster for $p>p_c$ and D_0 is a constant. We have included in D only the contributions from the liquid clusters. Equation 7.5 can be rewritten in the form

$$D=D_0 pe^{-v_m/\bar{v}_f} \tag{7.6}$$

when $v_m/\bar{v}_f \ll \bar{v}$, the average cluster size, then

$$\sum_{\nu=1}^{\nu_m} \nu C_\nu(p) \ll 1$$

and one obtains the Doolittle equation. The latter is valid for p near and greater than p_c. For $p\ll p_c$, (7.6) is no longer valid because $a_z(p)$ and especially $A_z(p)<1$. However, $p\ll p_c$ probably occurs only at lower temperatures in the region that is not experimentally accessible. Since D depends on p, we could expect a discontinuity at T_p that is not seen experimentally because the material goes out of equilibrium at T_g and $T_p<T_g$. The value of the viscosity at T_g is usually between 10^8 and 10^{11} P. Thus a value of T_p only 20 to 30 K below T_g would usually give a value of η too large to measure, even if an equilibrium measurement could be made. In any event, the jump in η would be difficult to detect.

In fitting the experimental data for η to the Doolittle equation, $\bar{v}_f$ is successfully approximated as proportional to $T-T_{0H}$ for the high-temperature, low-viscosity regime, and to $T-T_{0L}$ for the lower temperature regime, where T_{0L} may vanish. A viscosity in the range $10^4 \to 10^6$ P is typical of the crossover region separating these two regimes. Our theory gives a formula for $\bar{v}_f$, (7.2), which is more general and can be expected to fit in both regimes. We have fitted the viscosity of several organic glasses, B_2O_3 and $0.60KNO_3\cdot 0.40Ca(NO_3)_2$, for which data are available over a large range, to

$$\log_{10}\eta=A+\frac{2B}{T-T_0+\left[(T-T_0)^2+4v_a\zeta_0 T\right]^{1/2}} \tag{7.7}$$

where $B=v_m\zeta_0\log_{10} e$. The fit was excellent for all available data, and the parameters of best fit are shown in Table II. The goodness-of-fit parameter $\chi^2=\Sigma[(\log_{10}\eta^{cal}-\log_{10}\eta^{exp})^2]/(N-4)$, where N is the number of data points, was typically between 0.001 and 0.007 for the glasses considered. The error in the fit is comparable to the experimental error. For all the glasses in Table II, η was measured over at least 12 orders of magnitude. We find from the fits that $T_0 > T_\eta$ for each of these glasses. Included in the fit were two glasses, o-terphenyl and tri-α-naphthylbenzene for which $T_{0L} \neq 0$, whereas the other four showed Arrhenius behavior at low temperature, $T_{0L} \to 0$. The fits for tri-α-naphthylbenzene and for o-terphenyl are shown in Figs. 12 and 13, respectively. The best fits to the Doolittle equation with $\bar{v}_f \propto (T - T_{0H})$ are shown by the dashed curves in Figs. 12 and 13. We also tried fits where the preexponential A was temperature dependent, since $\eta \sim T^{1/2} D^{-1}$. However this dependence is so weak compared to the exponential dependence on $v_m / \bar{v}_f$ that the fits are comparable to those with A constant.

The generalization of these results to include the effects of the pressure P is straightforward. The free energy $f(v)$ then contains an additional term Pv. The effect of the term on $P(v)$, $v > v_c$, is simply to change $\zeta_0 \to \zeta_0 + P$. Thus the free volume has the same form as in (7.2) with $\zeta_0 \to \zeta_0 + P$. The characteristic temperature T_0 then has the form

$$T_0(P) = T_0 + \frac{v_a P}{k} \qquad (7.8)$$

High-temperature viscosity measurements[38] at elevated pressure for $0.62 KNO_3 \cdot 0.38 Ca(NO_3)_2$ were fitted to the usual three-parameter Doolittle equation with a T_0 that had a linear dependence on pressure as in (7.8).

TABLE II
Parameters Used in the Fit for the Viscosity η (7.7)

Glass former	A	$2B$ (°K)	$4v_a\zeta_0$ (°K)	T_0 (°K)
Tri-α-naphthylbenzene[a]	-2.44	345.3	10.6	401.8
o-Terphenyl[b]	-2.65	253.0	6.4	278.7
α-Phenyl-o-cresol[c]	-1.11	92.4	2.9	252.8
Salol[c]	-0.52	25.3	0.94	264.6
B_2O_3[d]	0.63	1825.4	72.6	609.5
$0.60 KNO_3 \cdot 0.40 Ca(NO_3)_2$[e]	-1.79	362.0	6.4	365.4

[a] Ref. 53.
[b] Refs. 54 and 55.
[c] Ref. 54.
[d] Ref. 60.
[e] Ref. 40.

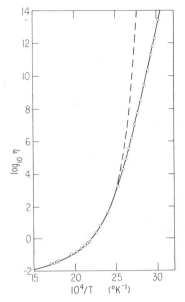

Fig. 12. Viscosity versus temperature for tri-α-naphthylbenzene, based on data of Ref. 53. The solid curve is the best fit to (7.2) and (7.6), for the parameters given in Table II. The dashed curve is the best fit (Ref. 53) with $\bar{v}_f \propto T - T_{0H}$, $T_{0H} = 342°K$.

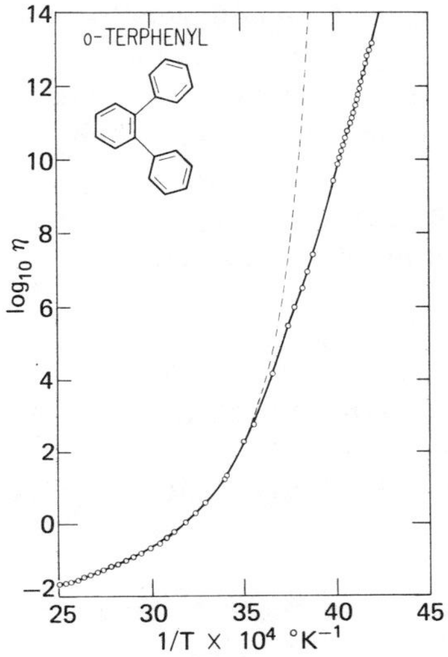

Fig. 13. Viscosity versus temperature for o-terphenyl based on data of Refs. 54 and 55. The solid curve is the best fit to (7.2) and (7.6), for the parameters given in Table II. The dashed curve is the best fit with $\bar{v}_f \propto T - T_{0H}$, $T_{0H} = 248°K$.

The fit gave a value of $v_a = 1.1$ Å^3. Unfortunately, we know of no high-pressure, low-temperature studies of η that can be used to check fully the temperature and pressure dependence of (7.2) and (7.6). Recently there have been some high-pressure studies ($\eta \lesssim 10^8$ P) for a number of inorganic glasses at room temperature.[133] Further studies as a function of both T and P would be very useful for testing the present model.

From Table II and (7.7), we obtain the value of 260.5 for v_m/v_a in $0.60KNO_3 \cdot 0.40Ca(NO_3)_2$. The two compositions ($0.62 - 0.38$ and $0.60 - 0.40$) are close enough for the difference to be ignored. These values of v_m/v_a and v_a lead to a value of v_m of 237 Å^3. This corresponds to a radius r_m of 3.8 Å, which is to be compared to the ionic radii of 1.33 Å for K^+, 0.99 Å for Ca^{2+}, and approximately 3.0 Å for NO_3^-. The comparison shows clearly that the diffusive units are most probably not the individual ionic species but complexes instead.

Angell and co-workers[14] have pointed out that $T_\eta = T_{0H}$ and T_s are nearly equal in many materials (e.g., see Table I). This can be understood in our model as follows. From (7.6) and the Stokes-Einstein relation $\bar{v}_f$ extrapolates to zero at T_η in our theory. However, if $\bar{v}_f \to 0$, then $p \to 0$ from (6.13), so that the communal entropy and entropy of mixing also vanish. The only nonvanishing contribution to S is the entropy of the solidlike cells, which is essentially the same as the entropy of the crystal. This is true because no liquidlike cells remain when p goes to zero. Thus according to our theory, extrapolating the entropy to the crystal values using C_p data from essentially the same range of temperature as that from which the extrapolation of η to infinity was made, must yield

$$T_s \simeq T_\eta \qquad (7.9)$$

because $p = 0$ both at T_η and at T_s and p is a single-valued function of T.

VIII. EQUILIBRIUM THERMODYNAMIC PROPERTIES

A. Specific Heat

The specific heat can be calculated in the vicinity of T_g from $C_p = T(\partial S/\partial T)$, where from (6.15),

$$S = Nk\left\{ -(1-p)\ln\left(\frac{1-p}{v_g}\right) - p\ln\left(\frac{p}{\bar{v}_f}\right) - pR_1\bar{v}_f + \mathcal{S} \right\} \qquad (8.1)$$

The heat capacity is the sum of contributions from the configuration and communal entropy,

$$C_p = C_p^{\text{conf}} + C_p^{\text{comm}} \qquad (8.2)$$

$$C_p^{\text{conf}} = T\left\{ \frac{p}{\bar{v}_f}\left(1 - R_1\bar{v}_f\right)\frac{\partial \bar{v}_f}{\partial T} + \frac{1-p}{v_g}\frac{\partial v_g}{\partial T} \right. \qquad (8.3)$$

$$\left. + \ln\left[\frac{\bar{v}_f(1-p)}{v_g p}\right]\frac{\partial p}{\partial T} - \bar{v}_f\frac{\partial(pR_1)}{\partial T} \right\}$$

and

$$C_p^{\text{comm}} = T\frac{\partial \mathcal{S}}{\partial p}\frac{\partial p}{\partial T} \qquad (8.4)$$

where we assume that p is a smooth function of T away from T_p so that we can use the chain rule. We know little about $\tilde{C}_p = \partial \mathcal{S}/\partial P$ except its scaling form near p_c. From (5.21), we know that $\tilde{C}_p \propto |p - p_c|^{-\alpha}$ where $\alpha = 1 - \beta$.

With our earlier estimates of β, we crudely estimate α to be positive and in the range 0.3–0.6. Note that the usual exponent α for the percolation problem is negative, since the entropy is related to the configuration probability of being in a cluster. Here the communal entropy is present and results in a different expression for α, dependent only on β. For the critical contribution to $\tilde{C}_p$, which is dominant in the vicinity of p_c, we take

$$\tilde{C}_p = A \left| \frac{p - p_c}{p_c} \right|^{-\alpha} - D + \frac{E(p_c - p)}{p_c} \qquad (8.5)$$

for $p < p_c$ and the same function with primed parameters for $p > p_c$. We set $D = D'$ and $E = E'$ because these contributions arise from analytic terms in S_c that are continuous through p_c. We also set $A' > A$, because the presence of the infinite cluster makes a large additional contribution to S_c and therefore $\tilde{C}_p$ above p_c.

To complete the calculation of C_p, we must know the temperature dependence of $\bar{v}_f$, v_g, and most important, p. This can only be done by solving the self-consistency condition $p = h(p)$, (6.13), in more detail. To reduce the number of free parameters, we use the viscosity data fitted by the parameters in Table II for the temperature dependence of $\bar{v}_f$. For this reason, we limit our discussion to systems in which η has been measured over a wide temperature range. The remaining parameters are then v_0, v_c, and κ to describe $f(v)$ and p_c, α, A, A', D, and E in $\tilde{C}_p$. If we scale all volumes by v_0 taken equal to v_m, that leaves only v_c/v_0 and $\tilde{\kappa} = \kappa v_0^2$ as unknowns in $f(v)$. The latter is constrained, since we know

$$\left. \frac{df(v)}{dT} \right|_{v_c^-} > \left. \frac{df(v)}{dT} \right|_{v_c^+}$$

that is, the derivative of the free energy is rising more rapidly in the solid than in the liquidlike regimes. This gives

$$\kappa(v_c - v_0) \gtrsim \zeta_0 + \frac{T_1}{\bar{v}_f + v_a}$$

where we can take $\bar{v}_f$ equal to its value at T_p. The value of v_c/v_0 is taken to be between 1.05 and 1.15. This leaves only the parameters in $\tilde{C}_p$ to be determined. However, these are constrained by the requirement that for three solutions of p in the vicinity of T_g and $\mathcal{F}(p_1) = \mathcal{F}(p_3)$ at T_p, which should satisfy the condition $T_s \leq T_p \leq T_g$. Here T_g is the temperature at which the system falls out of thermodynamic equilibrium for the experimental heat capacity measurements. We also require that $\Delta C_p = C_{p(\text{liquid})} - C_{p(\text{glass})} > 0$. This reduces the range of parameter space that is allowed. In practice, choosing values for p_c, α, $\tilde{\kappa}$ and v_c/v_0 effectively reduces the freedom in choosing the remaining parameters in $\tilde{C}_p$. We have not tried to span the complete parameter space but show here typical results for C_p versus T. In the examples below, we choose $\alpha = 0.3$. The results are not critically dependent on α because we have found similar results for both larger and smaller values of α. The important effect of the parameter α is on the size of the latent heat at T_p. In general, the larger the value of α, the larger the latent heat.

In our first paper,[94] we presented the equilibrium results for C_p and $\bar{v}$ for the organic glass tri-α-naphthylbenzene. For comparison, we present here results for the ionic glass $0.60\text{KNO}_3 \cdot 0.40\text{Ca}(\text{NO}_3)_2$. We find similar results for o-terphenyl and B_2O_3 glasses. In Fig. 14 we show the result for the probability distribution function $P(v)$ versus v for $0.60\text{KNO}_3 \cdot 0.40\text{Ca}(\text{NO}_3)_2$ for the parameters given in the figure legend. Note the bimodal distribution in $P(v)$, which is discontinuous at $v = v_c$. In Fig. 15 we show a result for C_p versus T near T_p for a set of parameters given in the figure legend. Since T_s is unknown for this system, we have chosen parameters to give $T_p = 335°$ K. The rise in C_p in the equilibrium liquid phase is characteristic of several of the organic glasses. Since the transition is first order, there is a latent heat at T_p, denoted by a delta-function spike. There may also be other contributions to C_p that we have not included, but they should be smooth near T_p.

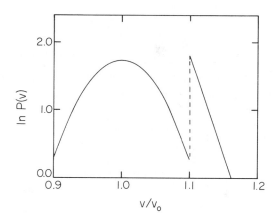

Fig. 14. Plot of the logarithm of the probability distribution function $P(v)$ for $0.60\text{KNO}_3 \cdot 0.40 \text{ Ca}(\text{NO}_3)_2$ at $T = 340°$K and $p = 0.20$. The free-volume parameters from Table II and (7.2) are used for $\bar{v}_f$, and we have chosen $\tilde{\kappa} = \kappa v_0^2 = 10^{5°}$K an $v_c/v_0 = 1.1$.

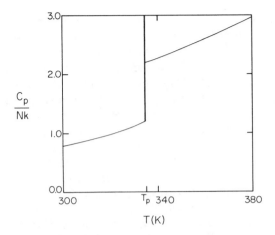

Fig. 15. Equilibrium heat capacity C_p versus T near T_p. The transition is first order and the latent heat is represented by a spike at T_p in C_p. We have used the free-volume parameters from Table II for $0.60KNO_3 \cdot 0.40Ca(NO_3)_2$, $T_0 = 365.4°K$, $v_m T_0 = 416.8°K$, and $v_a/v_m = 0.0038$. The other parameters chosen are $\bar{\kappa} = 10^{5}°K$, $v_c/v_0 = 1.1$, $R_1 = 0$, $p_c = 0.15$, $\alpha = 0.3$, $A = 0.6$, $A' = 2.0$, $D = 1.0$, and $E = 0.5$, which give $T_p = 335°K$.

This result cannot be compared directly to the experimentally measured C_p, since that is greatly affected by kinetic phenomena occurring around T_g. Those measurements do not give an equilibrium result for C_p, which is what we have calculated.

As a second example, we have calculated the heat capacity for the metallic glass $Au_{0.77}Ge_{0.136}Si_{0.094}$. Using the viscosity and heat capacity measured by Chen and Turnbull,[20] we learn that $\bar{v}_F/v_m = (T - 241.3)/540.6$ and $T_g = 295° K$. Results for C_p versus T are shown in Fig. 16 for parameters given in the legend. Here we have chosen parameters to give $T_p = 280°$ K, but the results are similar for other values of T_p. Note that the heat

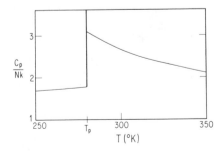

Fig. 16. Equilibrium heat capacity C_p versus T near T_p using the free-volume parameters for the metallic glass $Au_{0.77}Ge_{0.136}Si_{0.094}$: $T_0 = 241.3°K$, $v_m \zeta_0 = 590.6°K$, and $v_a = 0$. The latent heat is represented by a spike at T_p. The other parameters chosen that give $T_p = 280°K$ are $\bar{\kappa} = 10^{5}°K$, $v_c/v_0 = 1.1$, $R_1 = 0$, $p_c = 0.2$, $\alpha = 0.3$, $A = 0.7$, $A' = 2.0$, $D = 0.5$, and $E = 0.30$.

capacity decreases with increasing temperature above T_p, as observed experimentally. As discussed above, it is difficult to compare our results directly with the measurements, which are not carried out at equilibrium.

For $\beta \geq 1$, the temperature dependence of p relates to the value of $(\partial S / \partial p)|_{p_c} = -D > 0$ in (8.5). For most values of D, the transiton is similar to that found for $\beta < 1$. That is, Q is very small near p_c and $h(p_c) \lesssim 1$. This gives three solutions of the equation $p = h(p)$ for some range of temperatures. The only differences are a reduction in the latent heat and the introduction of a critical end point for the liquid. The latter is the temperature below which the liquid state does not exist. There is a much smaller range of parameter space for which p is a continuous function of T through T_p, and a second-order phase transition occurs at $T = T_p$ when $p = p_c$. This requires values of D, A, and A' in (8.5) so small that C_p becomes a very smooth function of T, with no anamolous rise. The decrease in C_p with decreasing T is then too gradual to avoid the entropy crisis. We therefore conclude that although a second-order transition is possible in our model, it is very improbable and does not correspond to the type previously proposed by Gibbs and DiMarzio.[17] In our model, internal configurational entropy is omitted, but entropy of mixing and communal entropy are included explicitly. In the entropy model of Gibbs and DiMarzio,[17] the communal entropy is omitted, but entropy of mixing and internal configurational entropy are included explicitly. By increasing the exponent β above 1, we have attempted to subsume the internal configurational entropy missing from our model into the communal entropy. The loss of the entropy catastrophe thereby indicates that this is an inadequate way to deal with the internal configurational entropy. The problem of how to incorporate communal entropy into the entropy theory or internal configurational entropy into the free-volume theory remains unsolved.

B. Thermal Expansion

Experimentally, the volume depends on the cooling and heating rates during the measurement and shows hysteresis. Here we calculate the average volume $\bar{v}$ as a function of T; it is given by

$$\bar{v} = \int_0^\infty v f(v)\, dv \qquad (8.6)$$

Inserting (6.10a) and (6.10b) into (8.6), we have

$$\bar{v} = p(v_c + \bar{v}_f) + (1-p)v_0$$
$$+ (1-p)\frac{kT}{\kappa v_g}\left\{ \exp\left(\frac{-\kappa v_0^2}{2kT}\right) - \exp\left[\frac{-\kappa(v_c - v_0)^2}{2kT}\right] \right\} \qquad (8.7)$$

Thus with the exception of p, the parameters in (8.7) for $\bar{v}$ are smooth functions of temperature and show no anomaly at T_p. However, p is a discontinuous function of T at T_p and, as a result, this model predicts a jump discontinuity in $\bar{v}$ at T. The volume will show a change of slope near T_p, as the contributions from the infinite cluster dominate for $p > p_c$, while the solidlike and finite-size clusters dominate for $p < p_c$. In Fig. 17 we show the result for $\bar{v}$ versus T using the parameters for $\bar{v}_f / v_m$ for $0.60KNO_3 \cdot 0.40Ca(NO_3)_2$ and those in the legend of Fig. 15.

Above T_g, where the material is in equilibrium, a direct comparison can be made between our theory and experiment. The observed values of $\tilde{\alpha} = d\ln(v/dT)$ are constant there. Our theoretical values are not constant as $\tilde{\alpha}$ continues to increase with T. For instance, for tri-α-naphthylbenzene, the observed value of $\tilde{\alpha}$ is $5.2 \times 10^{-4}/°K$. This difference is probably related to our mean field theory, which neglects interaction between neighboring cells. These interactions should retard the growth of $\bar{v}$ in the liquid. Note that we have omitted from consideration the temperature dependence of v_0 and v_c. These quantities would largely compensate for each other in the liquid domain. Below T_p, the calculated value of $\tilde{\alpha}$ for the solidlike domain is far smaller than the value observed below T_g. Since the variation of p with T is largely frozen out below T_g, and the temperature dependence of $\bar{v}_f$ makes a small contribution below T_p, the value of $\tilde{\alpha}$ observed below T_g should in fact correspond to our calculated value below T_p. The discrepancy arises from our neglect of dv_0/dT and can yield an estimate of it. These results indicate that the volume $\bar{v}$ is very sensitive to the simplifications we have introduced in our model. However, we expect that η and C_p are less sensitive and should be more characteristic of a dense liquid and glass.

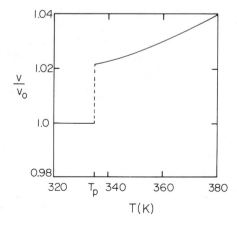

Fig. 17. Equilibrium volume $\bar{v}$ versus T near T_p for $0.60KNO_3 \cdot 0.40Ca(NO_3)_2$. The parameters are the same as in Fig. 15.

C. Robustness of the Theory

Now that we have described in some detail the analysis of the free-volume model, it is worthwhile to pause and consider the effect of the numerous simplifications introduced. The important question is, of course, how do they affect the physics of the equilibrium phase transitions as developed in the preceding sections? Fortunately, we do not expect that any of our assumptions will affect qualitatively the nature of the liquid-glass transition.

To see this explicitly, let us review the most important simplifications introduced. These include the artificially sharp distinction between liquid and solidlike cells, which allowed us to approximate the local free energy by quadratic and linear regions. From this assumption, we were led immediately to the conclusion that only liquidlike cells exchange volume freely. However, there surely exist activated processes by which solid- and liquidlike cells can exchange volume, and these will be particularly important near the surface of liquidlike clusters. We have ignored all thermally activated motion, which should be important at low temperatures where the number and sizes of the liquidlike cells are reduced. There are also time-dependent effects, in which solidlike cells become liquidlike, and vice versa. We have assumed that the time scale on which this occurs is substantially longer than that for atomic motion with a liquid cluster. For T near and just above T_p, this should not be important, since the relaxation rates are long and the cage structure should be long-lived. However, for $T \gg T_p$, this effect will be important. Finally, we have factored the joint probability distributions of cell volumes and treated the total free energy in a mean field theory. We assumed that $f(v)$ and $P(v)$ depend only on the cellular volume v and neglected any interaction with neighboring atoms. The only local environment effect we included was in the dependence of ζ_1 on $\bar{v}_f$, (3.7), and this we treated as a mean field correction.

However, we expect none of these effects to change the qualitative nature of the transition. This can most easily be seen by considering how one might begin to generalize the solution of the model within mean field theory, to eliminate these artificial discontinuities. First, one would introduce the probability $u(v)$ that a cell of volume v will be liquidlike. If a cell's local environment were such that it could participate in a free exchange of free volume, we would call that cell liquidlike. This requires only that in a localized region of configuration space (i.e., in cell v-space) the consequences of finite curvature in the free energy F on a scale given by kT can be neglected. That is, $(\Delta X)^2 \delta^2 F / \delta^2 X$ is small in comparison to kT, where X is a generalized configuration coordinate and ΔX is its probable range of variation. In the local regions in which this condition is not satisfied, the cells are considered to be solidlike. One would expect that $u(v)$ would be an S-shaped function such that $u(v)=0$ for v much less than a critical value v_c and $u(v)=1$ for $v>v_c$. There would be a narrow range of v about v_c in which the transition from 0 to 1 is made. Most important, the probability p survives elimination of the discontinuities, since one can define

$$p = \int_0^\infty P(v)u(v)\,dv \qquad (8.8)$$

The theory now proceeds as developed in Sections V and VI, essentially unchanged. For example, $P(v)$ will have the same bimodal structure as shown in Fig. 14, but will now be continuous. Similar smoothing of all artificially introduced discontinuities will not affect the theory in any essential way. The loss of a sharp distinction between liquid- and solidlike cells could vitiate use of the percolation theory. The nonanalyticity in S will certainly be lost, leading to a communal entropy for which $\partial S/\partial p$ is always less than infinity. However, the first-order phase transition should be preserved, just as it was for most of the parameter space even when $\beta>1$. The discontinuity in p and $\bar{v}$ would be reduced, as would be the latent heat. One important effect of this smearing will be the appearance of a critical end point for the liquid, a temperature below which the liquid phase is no longer even metastable. The second-order transition, which is only a small region of parameter space for $\beta \geq 1$, is now wiped out completely by the restoration of analyticity. Our theory thus leads to a first-order phase transition or no transition at all. However, the entropy catastrophe can be resolved within our theory only if a transition occurs.

We have now completed our discussion of the equilibrium liquid↔glass transition. However, as discussed above, the results for C_p and $\bar{v}$ obtained in this section are not seen experimentally. This leads us to consider nonequilibrium phenomena in the next section.

IX. STRUCTURAL RELAXATION AND THE FREE-VOLUME MODEL

The marked changes characteristic of a first-order transition shown in Figs. 15 and 17 are not seen experimentally. The observed anomalies in C_p and $\bar{v}$ are the consequences of the falling out of complete thermodynamic equilibrium of the system as the time of measurement becomes comparable to the relaxation time of the system. The results just described cannot therefore be compared directly with experiment. First we must develop a relaxation theory of the kinetic effects observed around and below T_g.

The changes of the volume $\bar{v}$ and enthalpy H consequent to stepwise or continuous changes in temperature have been investigated by a number of workers, following the early work of Tool.[15] The essential features that relate the dependence of T_g on q to the temperature dependence of the structural relaxation time $\tau(T)$ determined from cooling the liquid was first presented by Ritland.[134] Results for heating are more difficult to interpret, since the temperature dependences of $\bar{v}$ and H are affected by the detailed thermal history of the glass during cooling and annealing. To account for these effects quantitatively, it is necessary to develop a multiparameter approach, which has been done by Narayanaswamy,[135] Moynihan and coworkers,[15, 78] and Kovacs and co-workers.[79] These theories all involve the same basic assumptions but use different techniques and approximations. However, they are all built on the assumption that the volume $\bar{v}$ and enthalpy H of the nonequilibrium glass would always relax to its metastable equilibrium value for the liquid, in the limit of infinitely slow cooling. That is, none of these theories has allowed for the possibility of an "ideal" glass phase in which the long time values of $\bar{v}$ and H differ from their extrapolated liquid value, as suggested by Ref. 71. The results of the recent work by Moynihan et al.[15, 78] and Kovacs et al.[79] have shown that one can understand quantitively the experimental results for both C_p and $\bar{v}$ by introducing only a limited number of free parameters. Both theories stress the importance of including both a distribution of relaxation times $\tau_i(T)$ for the structural modes and keeping track of the thermal history of the sample. However, no microscopic specification of the structural modes is made. One crucial result obtained by Moynihan et al.[78] is the dependence of $\ln|q|$ on temperature. They find

$$\frac{d\ln|q|}{d(1/T_g)} \approx -\frac{\Delta h^*}{R} \tag{9.1}$$

where Δh^* is an activation enthalpy and R is the ideal gas constant. This result is in agreement with the experimental observation that $\ln|q|$ varies linearly with $1/T_g$ for all the glass formers that have been systematically studied. Since T is large, a plot of $\ln|q|$ over a limited temperature range would be approximately linear with respect to T_g as shown in Fig. 14 for B_2O_3 and $0.60KNO_3 \cdot 0.40Ca(NO_3)_2$. The reader is referred to the recent reviews by Moynihan et al.[78] and Kovacs et al.[79] for further details of their analysis.

Here we wish to describe how the effect of these strongly temperature-dependent structural relaxation rates near T_g can be incorporated into the free-volume model.[95] Our analysis differs significantly from that described above because we have at our disposal an equilibrium theory of the heat capacity and volume to begin our discussion. By starting from a concrete model of a glass, the free-volume model, we can identify the dominant structural mode and establish its relaxation mechanism. From our discussions above, we have that the free energy is a function of the form

$$F = F[P(v); C_{vz}(p); p] \tag{9.2}$$

A characterization of the free energy and therefore the structure of the system is obtained by a specification of $P(v), C_{vz}(p)$ and p, which we can denote by the set $\{X_i\}$, and which we refer to as the structural variables. The equilibrium state was obtained from the variational condition

$$\frac{\delta F}{\delta X_i} = 0 \tag{9.3}$$

which determines the equilibrium variables of X_i^{eq} found in Section VI. A study of relaxation near equilibrium can be made by examining the generalized equation of motion which we write in the form,

$$\dot{X}_i = \lambda_{ij} \frac{\delta F}{\delta X_j} \tag{9.4}$$

where the set of parameters λ_{ij} describes all possible relaxation processes and a sum over repeated indices has been assumed. By considering the possible processes that contribute to the structural relaxation, we can separate out the rapidly relaxing quantities from those that are more slowly relaxing. The rapidly relaxing quantities can then be assumed to be in secular equilibrium during the variation of the slowly relaxing quantities.

The variable $P(v)$ should reach its equilibrium value very rapidly for fixed p, since both the vibrations of the atom or molecule in the solidlike cells and the free exchange of free volume associated with the oscillations in cages for liquidlike cells occur readily. Since the cluster size distribution $C_{vz}(p)$ is fixed for fixed p, this leaves p as the one fundamental variable that determines structural relaxation. The value of p can change only by exchange of volume between solid and liquidlike cells, that is, by a reorganization of the cell structure. Therefore it can change only by self-diffusion, which is *not* localized within a cluster. In more detail, this exchange of volume between solidlike and liquidlike cells occurs at a cluster boundary and requires a cage rearrangement. This requires a diffusive step and becomes more difficult as T approaches T_g, since the diffusion rates are decreasing exponentially. This same process that changes p allows $C_{vz}(p)$ to equilibrate with a new value of p. Thus p is the controlling variable. Everything else is in secular equilibrium with p.

In the free-volume model, p becomes the dominant "structural mode" of the relaxation theories.[95] The rearrangement of the cage structure requires diffusion, which is slowly being frozen out. As T approaches T_g, p can no longer follow its equilibrium value, but becomes frozen at a value $p > p_c$. Since the variation in p no longer contributes to the heat capacity, C_p decreases. The relevant contribution to C_p arises from the communal entropy, which depends primarily on p and ceases to change. However, at temperatures above T_g, the structure can equilibrate rapidly compared to the measurement time and both p and C_p approach their equilibrium value.

Since the heat capacity is measured using a differential scanning calorimeter, the dependence of C_p on time and temperatures is complicated because C_p is measured continuously while the sample is being heated or cooled at a constant rate. Therefore, the time for the heat capacity measurement is not well defined, and thermal history effects complicate the shape of the step at T_g. To overcome this difficulty, one can approximate the real situation by changing the temperature in small discrete steps ΔT at time intervals Δt. Stephens[72] and Moynihan et al.[78] follow this procedure and calculate the change in the sample's enthalpy H and its heat capacity from $C_p = \Delta H / \Delta T$. However, we already have an explicit expression for the equilibrium enthalpy and heat capacity. Since the result for C_p, (8.2), depends partially on the rate of change of p with respect to T, we need only include the effect of p falling out of equilibrium (i.e., $dp/dT \to 0$) for $T < T_g$ in that result.[95]

The exact dependence of the probability p on the detailed microscopic mechanisms that determine the interchange of volume between liquidlike and solidlike cells is unknown. In the development of the equilibrium results, we have assumed the local free energy f_i of each cell depends only on its volume v_i, $f = f(v)$ as shown in Fig. 7. Consistent with this assumption, we assume a single average relaxation time $\tau(T)$ for which the structural modes (i.e., p) equilibrate. We can then allow in an average way for the fact that these modes may not completely relax before the next time increment Δt. Although we do not expect results for C_p that are quantitatively correct, we do expect the results to be qualitatively correct. The goal of the present analysis is to develop a clear, consistent picture of the freezing process, devoid of any unnecessary complications. We leave open at this point the more involved study that would be essential to obtain quantitative comparisons with experiment. We can then write an expression for $p(T)$, the probability of having a fraction of liquidlike cells at temperature T,[95]

$$p(T) = p_{\text{eq}}(T) + \left[p'(T) - p_{\text{eq}}(T) \right] e^{-\Delta t / \tau(T)} \tag{9.5}$$

where $p'(T)$ is the value of p at the end of the previous step and $p_{\text{eq}}(T)$ is the equilibrium probability for that temperature. For high temperatures $\Delta t/\tau(T)$ is large and $p(T)=p_{eq}(T)$. However as T approaches T_g, $p(T)$ can no longer follow $p_{eq}(T)$ and will finally freeze out a value p_{froz}, which will depend on the cooling rate and $\tau(T)$, for all $T<T_g$. That is,[95]

$$p(T\rightarrow0)\approx p_{froz}\neq0 \tag{9.6}$$

even though $p_{eq}(T\rightarrow0)=0$.

The program is then simply to start at a high temperature, where $p(T)=p_{eq}(T)$ and lower the temperature at a fixed $q<0$. The result for $p(T)$ can then be used in (8.2) to (8.4) for C_p to obtain results that can be compared directly with experiment. The only quantity that we must specify in addition to those in the equilibrium theory is the relaxation time $\tau(T)$. Since $\tau(T)$ is to describe the relaxation by diffusion of structural modes represented by the variation of p, it should have the same temperature dependence as the shear viscosity η. That is, we suppose that the same microscopic movement processes underlie self-diffusion, viscosity, and structural relaxation. This supposition is consistent with existing theories and with a number of experimental results indicating that the activation enthalpy Δh^* for volume or enthalpy relaxation is generally the same as the activation enthalpy for the viscosity η.[78] We therefore assume that $\tau(T)$ can be expressed by the Doolittle equation,

$$\tau(T)=\tau_0 e^{v_m/\bar{v}_f} \tag{9.7}$$

in analogy with the result for η. The only additional free parameter so introduced will be the preexponential factor τ_0, since the three parameters for $v_m/\bar{v}_f$ can be taken from a best fit of (7.7) to the viscosity data.

Using the values for $p_{eq}(T)$ obtained for the glass formers, tri-α-naphthylbenzene, $0.60KNO_3 \cdot 0.40Ca(NO_3)_2$, and B_2O_3, we have calculated the dependence of C_p on heating and cooling rates and history of the sample. Our results convince us that the effects are universal, as found experimentally for all glass formers. The value of τ_0 was chosen so that T_g is consistent with the experimentally observed values. We found[95] results that were insensitive to the choice of increments ΔT and Δt for fixed q, provided $\Delta T<1°K$. In the present analysis, we choose $\Delta T=0.05°K$. The final results are dependent only on the product $q\tau_0$, and not q and τ_0 independently, since they enter only through this product in (9.5), where Δt is replaced by $\Delta T/q$. The results for tri-α-naphthylbenzene have been described elsewhere.[95] Here we show results for $0.60KNO_3 \cdot 0.40Ca(NO_3)_2$ and B_2O_3.

In Fig. 18, we show results for C_p versus T for $0.60KNO_3 \cdot 0.40Ca(NO_3)_2$ for three values of $q\tau_0$ obtaining on cooling the system from high temperature. The parameters used to determine $p_{eq}(T)$ are given in the legend of Fig. 15. If we consider these curves to have τ_0 fixed, say $\tau_0 = 10^{-10}$ sec, curves a to c correspond to $q = -0.1$, -10.0, and $-1000.0°K/min$, respectively. Another way to think of these curves is to have q fixed, say $q = -10.0°K/min$, then curves a to c correspond to $\tau_0 = 10^{-12}$, 10^{-10}, and 10^{-8} sec, respectively. As expected, C_p is a monotonic, decreasing function. For faster cooling rates, C_p departs from its equilibrium value at a higher temperature, since the system falls out of equilibrium sooner. The value of p_{froz} is larger for larger cooling rates. In most materials, C_p cannot be measured on cooling, since crystallization occurs before the glass phase is reached. However, in our calculations, we have ignored the channel to crystallization and are able to study C_p under cooling conditions.

Most experimental measurements are made upon heating the sample from its glassy state. To consider this case, we start at low temperatures where $p'(T)$ equals its frozen-in value and $\tau(T)$ is very large. Then we take $\Delta T > 0$ and follow the same procedure described above. In Figs. 19 and 20, we show results for $0.60KNO_3 \cdot 0.40Ca(NO_3)_2$ obtained by heating.

Figure 19 gives results for five values of $q\tau_0$, in which the initial frozen-in state is obtained by cooling at the same respective rates. If we consider $q = 10.0°K/min$, then curves a to e correspond to $\tau_0 = 10^{-13}$, 10^{-12}, 10^{-11}, 10^{-10}, and 10^{-9} sec, respectively. Another way to think of these curves is to have τ_0 fixed, say $\tau_0 = 10^{-10}$ sec, then curves a to e correspond to $q = 0.01$, 0.1, 1.0, 10.0, and $100.0°K/min$, respectively. As expected, the larger the value of $q\tau_0$, the larger the transition temperature T_g. If we consider T_g to be the midpoint of the rise in C_p (an arbitrary, but useful definition), we

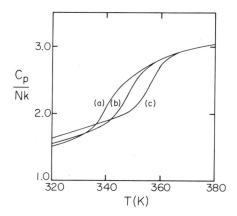

Fig. 18. Heat capacity C_p versus T for $0.60KNO_3 \cdot 0.40Ca(NO_3)_2$ cooled through the transition for three values of $q\tau_0$. For $\tau_0 = 10^{-10}$ sec, these curves correspond to (a) $q = -0.1$, (b) -10.0, and (c) $-1000.0°K/min$. If, instead we consider q fixed, say equal to $-10.0°K/min$, then (a) $\tau_0 = 10^{-12}$, (b) 10^{-10}, and (c) 10^{-8} sec.

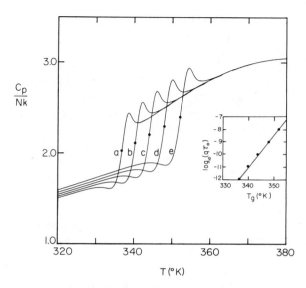

Fig. 19. Heat capacity C_p versus T for $0.60KNO_3 \cdot 0.40Ca(NO_3)_2$ heated through the transition region at various rates after cooling the glass at the same respective rate. For $\tau_0 = 10^{-10}$ sec, these curves correspond to (a) $q = 0.01$, (b) 0.1, (c) 1.0, (d) 10.0, and (e) 100.0°K/min. If instead we consider q fixed, say equal to 10°K/min, then (a) $\tau_0 = 10^{-13}$ (b) 10^{-12}, (c) 10^{-11}, (d) 10^{-10}, and (e) 10^{-9} sec. The inset shows our results for T_g (determined by the midpoint in the rise in C_p) as a function of $\log_{10}(q\tau_0)$. The slope of the line is in excellent agreement with experiment.

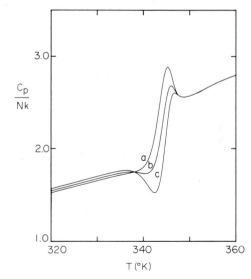

Fig. 20. Heat capacity C_p versus T for $0.60KNO_3 \cdot 0.40Ca(NO_3)_2$ heated through the transition region at a fixed rate, after cooling at different rates. For $\tau_0 = 10^{-11}$ sec and a heating rate $q = 10$°K/min, these curves correspond to (a) cooling at -1.0°K/min, $p_{froz} = 0.203$; (b) -10.0°K/min, $p_{froz} = 0.207$; and (c) -100.0°K/min, $p_{froz} = 0.211$.

find that $\log_{10}(q\tau_0)$ depends linearly on T_g, as shown in the inset. This linear dependence of $\log_{10}(q\tau_0)$ also occurs if one uses the temperature T_{max}, for the maximum in C_p, or T_{min}, for the minimum in C_p to describe the transition temperature. We find the slope of T_g versus $\log_{10}(q\tau_0)$ equals 3.9°K, in remarkable agreement with the experimental result 3.8°K found by DeBolt et al.[78] and shown in Fig. 4. This result is remarkable because we make no reference in our equilibrium calculation for $p_{eq}(T)$ and C_p to this slope. Once $p_{eq}(T)$ is calculated, the only free parameter is τ_0, which shifts the position of T_g, not the slope of T_g versus $\log_{10}(q\tau_0)$. The important conclusion to be drawn from this agreement is the correctness of the choice of $\tau(T)$, (9.7), to have the same temperature dependence as the viscosity. This conclusion is further supported by the results for B_2O_3 shown below (Fig. 21), for which the slope of T_g versus $\log_{10}(q\tau_0)$ also is remarkably close to its experimental value. From the published data for $0.60KNO_3 \cdot 0.40Ca(NO_3)_2$ at $q = 10°K/min$, we can estimate that $\tau_0 \sim 10^{-10}$ sec.

In Fig. 20 we plot results for heating at a constant rate $q\tau_0 = 10^{-10}/60°K$ starting from different frozen-in states. These frozen-in states could be obtained by cooling at different rates or annealing at low temperatures. If we take $\tau_0 = 10^{-11}$ sec and $q = 10.0°K/min$, the curves correspond to cooling as shown in the legend; the large dip in C_p for case c results because p_{froz} is larger than it would have been for a cooling rate $q = -10.0°K/min$, and p decreases to approach its equilibrium value when Δt first becomes comparable to $\tau(T)$, instead of increasing. In this region $dp/dT < 0$ and C_p decreases. Although this dip in C_p is often seen experimentally, the depth in case c is too large. Stephens[72] has suggested that a spectrum of relaxation times would reduce the depth of this minimum in C_p. The dip before the T_g step would be washed out by more rapidly relaxing processes and the relaxation peak would be considerably broadened. For case a, starting from the frozen state, p_{froz} is lower than it would have been if cooled at the same rate used in heating, and as seen, the peak in C_p is greatly enhanced. Above T_g, $\Delta t/\tau(T)$ is large and C_p equals its equilibrium value shown in Fig. 20.

We should point out that C_p continues to rise above T_g, while experimentally C_p is essentially constant for $0.60KNO_3 \cdot 0.40Ca(NO_3)_2$. This, we believe, is caused by the contributions to C_p proportional to $\partial \bar{v}_f/\partial T$. Although the phenomenological form for $\bar{v}_f$ when used in (7.2) fits the viscosity data very well, it does not necessarily predict the correct curvature of $\bar{v}_f$. From fitting the viscosity data to (7.2) and (7.7), $\partial \bar{v}_F/\partial T$ appears to have its maximum far above T_g, at $T \simeq 400°K$. Studies of the thermal expansion, however, indicate that it should be at a lower temperature, near T_g. This spuriously high temperature for the maximum in $\partial \bar{v}_f/\partial T$ produces the spurious rise in C_p above T_g. We suspect that fitting a more accurate form for $\bar{v}_f$ (more free parameters) to η would eliminate it. However, this is a small point and should not detract from the more important results near T_g.

To check that these effects are universal, we have now calculated C_p for a total of three glass formers. Results for tri-α-naphthylbenzene were shown in Ref. 95. We choose a third example, the network glass B_2O_3, for two reasons. First, it should be very different from both the inorganic tri-α-naphthylbenzene and the ionic $0.60KNO_3 \cdot 0.40Ca(NO_3)_2$, and second, data for η are available over a wide range, which we can use to fit the parameters in $\bar{v}_f$. The dependence of T_g and C_p on scanning rates q have also been studied in detail and are shown in Fig. 4. First we calculated $p_{eq}(T)$ following the procedure outlined in Section VI. We chose the unknown parameters to give an equilibrium liquid-glass transition at a temperature T_p below the observed T_g. The parameters used in our calculation for $p_{eq}(T)$ are $\tilde{\kappa} = 2 \times 10^5$°K, $v_c/v_0 = 1.1$, $R_1 = 0$, $p_c = 0.15$, $\alpha = 0.3$, $A = 0.6$, $A' = 3.0$, $D = 1.0$, and $E = 0.5$, which give $T_p = 480$°K. Results for equilibrium C_p are similar to those shown in Fig. 18 for $0.60KNO_3 \cdot 0.40Ca(NO_3)_2$. We then calculated C_p following the procedure outlined above. Results for cooling at fixed q were similar to those shown in Fig. 21, which shows results for heating through the transition region at various rates after cooling the glass at the

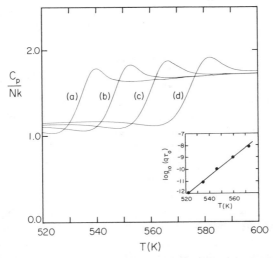

Fig. 21. Heat capacity C_p versus T for B_2O_3 heated through the transition at various rates after cooling the glass at the same respective rate. For $\tau_0 = 10^{-10}$ sec, these curves correspond to (a) $q = 0.1$, (b) 1.0, (c) 10.0, and (d) 100.0°K/min. If instead we consider q fixed, say equal to 10.0°K/min, then (a) $\tau_0 = 10^{-12}$, (b) 10^{-11}, (c) 10^{-10}, and (d) 10^{-9} sec. The inset shows our results for T_g as a function of $\log_{10}(q\tau_0)$.

same respective rates for B_2O_3. As seen in the inset, T_g was found to depend linearly on $\log_{10}(q\tau_0)$ over the temperature range studied. We find that the slope of T_g versus $\log_{10}(q\tau_0)$ equals 13.1°K, also in good agreement with the experimental result 14.5°K shown in Fig. 4. From the results of these three glass formers, we believe the results for the nonequilibrium heat capacity can easily be explained in terms of the free-volume model. Although we have made several simplifying assumptions, particularly in regard to the existence of a single relaxation time, we believe that we can describe the essential features of the transition without introducing any additional free parameters beyond those used in the equilibrium calculation, apart from τ_0.

It is interesting to attempt a simple theory of the parameter τ_0. Relaxation of p occurs at the cluster surfaces via a diffusive jump across the interface. Using the free-volume model of self-diffusion (Section VII), we obtain the following for τ_0,

$$\frac{1}{\tau_0} = \frac{1}{16} \frac{S}{V} \sqrt{\frac{3kT}{M}} \tag{9.8}$$

where S/V is the surface-to-volume ratio of the liquid clusters. Since we are near the percolation threshold and the minimum-sized liquid cluster contains $v_m/\bar{v}_f$ molecules, S/V is well approximated by $5\,(v_m^2/\bar{v}_f)^{-1/3}$. The result for τ_0 is 10^{-10} sec for $0.60KNO_3 \cdot 0.40Ca(NO_3)_2$, in excellent agreement with the value found from the $\ln q$ versus T_g data.

X. TUNNELING MODES

We saw in Section IX that the system's falling out of equilibrium at T_g has a profound effect on p, the fraction of liquidlike clusters. Since the decrease of p requires a structural rearrangement of molecules in the dense liquid, it becomes more difficult as T approaches T_g, where the value of p is frozen at $p_{froz} > p_{cz}$. This freezing in of liquid clusters affects the low-temperature properties of glasses by giving rise to the tunneling levels, described in Section II.

A number of liquid clusters of size $v > v_m$ is frozen in at $T \leq T_g$ because of the finite cooling rate used experimentally to avoid crystallization. Within those clusters, there remains the possibility that large ($\sim$atomic size) displacements can occur for $T \leq T_g$ by the coupling of translational motion and density fluctuations that lies at the heart of the free-volume model of diffusion. For temperatures near T_g, the curvature in $f(v)$ is negligible, and the local free energy is independent of the relevant configuration coordinates. However, when the glass is cooled down, $T \ll T_g$, this curvature becomes significant and gives rise to local energy minima separated by saddlepoint barriers on the order of kT_g in size. The energy scale is set by kT_g because it is near T_g that the system falls out of equilibrium, freezing in potential variations of at most this size. This last result is quite general and does not depend on the free-volume model.

At high temperatures ($\geq T_g$), the curvature in the free energy of the liquid clusters can be ignored, diffusion can occur, and, ultimately, extended motion can occur. At low T, the system freezes into one of its many total free-energy minima. The most probable pattern is a freezing inward into the cluster from the interface with the solidlike or glassy regions. However, there is no significant relaxation between the clusters and the solidlike environment.[95] Whatever excess volume Δv_ν was in the cluster stays within the cluster. Within each cluster of size ν, this excess volume Δv_ν is $\nu[(v_c - v_0) + \bar{v}_f]$. The $\nu\bar{v}_f$ comes from the free volume that is still freely redistributed at high temperatures ($\leq T_g$). The $\nu(v_c - v_0)$ is the thermal expansion of each cell required to bring it into the liquidlike range. Some fraction of the $\nu(v_c - v_0)$ will become available for redistribution as the cluster freezes into the minima.

The freezing of a liquid into a glass requires atomic movement, which occurs by the concentration the free volume into ephemeral voids of the size v_m.[94] This process can be regarded as a diffusion of the void away from the interface into the interior, where it becomes trapped as the freezing is completed. Thus within the originally liquid clusters there forms one void of volume somewhat greater than v_m for each $\nu_m = v_m / \bar{v}_f$ atoms.

The free-volume model leads us directly to an estimate of the number N_T of these voids present at low T. There is a void of size v_m for each group of approximately ν_m atoms in a cluster. We find

$$N_T \sim \tilde{N} \frac{P_z(p_{\text{froz}})}{\nu_m} + \tilde{N}\left[\sum_{\nu=\nu_m}^{\nu=2\nu_m - 1} C_{\nu z}(p_{\text{froz}}) \right.$$

$$\left. + 2 \sum_{\nu=2\nu_m}^{\nu=3\nu_m - 1} C_{\nu z}(p_{\text{froz}}) + \cdots \right] \qquad (10.1)$$

where $\tilde{N} = pa_z(p)N$ is the total number of atoms in clusters, $P_z(p)$ is the probability of being on the infinite cluster, and $C_{\nu z}(p)$ is the cluster distribution. In general $a_z(p) < 1$, since we require each atom in a cluster to have at least z nearest neighbors that are also in the cluster. The quantities $a_z(p)$, $P_z(p)$, and $C_{\nu z}(p)$ have not yet been determined theoretically, but we can estimate useful upper bounds on ν_m and N_T.

From Section IX, our best fits to the measured viscosity data for six glass formers give $v_m = v_m / \bar{v}_f \approx 30$ at T_g. That is, for each 20 to 40 atoms within a cluster, we expect to find a void of the size $\sim v_m$. Having chosen the percolation threshold value of $p_c = 0.15$, we found that $p_{\text{froz}} \sim 0.2$ was a typical result from our relaxation studies of the specific heat near T_g. Then if we consider

$$a_z(p_{\text{froz}}) \sim 0.1 \quad \text{and} \quad P_z(p_{\text{froz}}) \sim 0.1$$

we find

$$\frac{N_T}{N} \lesssim 10^{-4}$$

A realistic range would be $10^{-3} - 10^{-5}$. A more precise estimate must await a study of the cluster distribution function $C_{yz}(p)$.

It is important to note that N_T depends only on p_{froz}, which in turn depends on the product $q\tau(T)$. In our simulations of several glass formers we find that p_{froz} changes by only a small amount for very different cooling rates q or during annealing at temperatures near T_g. There should be little variation of the number of tunneling levels N_T with the cooling rates used to produce the glass. To our knowledge, there has not yet been any systematic study of the dependence of N_T on q.

We assert that these voids are the tunneling centers. Any one of the neighboring atoms can move into the void, but in contrast to the crystal, the void itself and the configuration of the surrounding atoms will be irregular, so that such motion can be expected to be substantially easier for one particular atom. That atom then tunnels into the void by *multiparticle tunneling* along a suitable one-dimensional path. Ignoring path curvature, we have for the effective Hamiltonian,[136]

$$H = \frac{p^2}{2\mu} + V_0(x) + \sum_\lambda \frac{1}{2} \hbar \omega_\lambda(x)$$

$$+ \left(\Phi_0(x), \frac{p^2}{2\mu} \Phi_0(x) \right)$$

$$= \frac{p^2}{2\mu} + V(x) \tag{10.2}$$

provided the tunneling energy separations are much smaller than the mean phonon frequency $\omega_\lambda(x)$. Here μ is the mass of the atom, $V_0(x)$ is the equilibrium potential energy, $\Phi_0(x)$ is the ground-state eigenfunction for all the remaining degrees of freedom λ, and the integration in the last term is over those degrees of freedom. The $V(x)$ will be of the general form shown in Fig. 22, assumed by Anderson et al.[100] for the voids as large as those we consider here. It is possible but highly unlikely that more than two minima exist. There will be tunneling between the two minima, with a tunneling energy

$$\Delta_0 = \omega_0 e^{-\Lambda} \tag{10.3}$$

and energy

$$E = \pm \left(\Delta^2 + \Delta_0^2\right)^{1/2} \tag{10.4}$$

relative to the mean energy $\frac{1}{2}\omega_0$ above the lower minimum. Here Δ is the usual splitting between the lowest two energy levels. An estimate of Λ for the symmetric-well case can be obtained by using the triangular barrier approximation, which gives

$$\Lambda = \frac{d}{\hbar}\left[2\mu\left(V - \frac{1}{2}\omega_0\right)\right]^{1/2} \tag{10.5}$$

where d is the separation in x of the two minima. Because the liquid clusters freeze in at $T \leq T_g$, the energy scale of Δ is set by T_g. Either of the two can be the lower minimum, so that Δ has a symmetric, roughly Gaussian distribution around zero with a root-mean-square value for which T_g sets the scale. The $|\Delta|$ can be represented as distributed over a rectangular distribution of width $\leq T_g$. Since ω_0 is a typical vibrational frequency, it will be on the same scale; V is also governed by the same scale. Therefore values of V so close to $\frac{1}{2}\omega_0$ that according to (10.5) Λ is small enough for $\Delta_0^2 \simeq \Delta^2$ are improbable. Consequently, we almost always have $\Delta_0 \ll |\Delta|$, and the density of states for the tunneling levels is given by the distribution of $|\Delta|$.

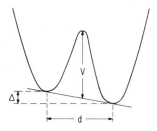

Fig. 22. Potential energy surface producing a tunneling state characterized by a barrier height V, asymmetry Δ, and generalized distance d.

At low T, the important values of $|\Delta|$ for the specific heat are $\sim kT \ll kT_g$. We can restrict ourselves to the case $\Delta = 0$ in estimating Λ and use (10.5). We are thus dealing with a void containing two energetically equivalent positions for a given atom in estimating Λ. If it were a vacancy in a crystal, the value of Λ so calculated would be sharply defined, having a delta function distribution. The disorder in the glass smears this distribution, roughly as a Gaussian, as shown in Fig. 23. For there to be a two-level system, there is a Λ_{min} for each barrier shape ~ 5–10. Since ultrasonic experiments are insensitive to large Λ,[103, 105] there is an effective cutoff $\Lambda^{ua} \simeq \Lambda_{min} + 1$. The $P(\Lambda)$ is found to be essentially constant[103] between Λ_{min} and Λ^{ua}_{max}. From the fast and slow specific heat measurements,[103, 104, 137–140] we know that $P(\Lambda) \simeq 0$ for $\Lambda > \Lambda_{max}$ and $\Lambda_{max} < \Lambda_c = 15$–$20$. This tells us that all the tunneling levels contribute to C_p, and we can expect the distribution in curve b of Fig. 23 to be appropriate.

Loponen et al[139] set limits on the relaxation times T_1 for the coupling of the phonon system to the tunneling center systems by faster heat capacity measurements. From their data we[141] infer that 5μ sec $\leq T_1 \leq 5$m sec, which corresponds to $\Lambda_{min} \simeq 12$ and $\Lambda_{max} \simeq 17$, consistent with the previous estimates and Fig. 23. Since all Λ contribute to C_p, we can calculate the fraction of the centers $\overline{N}_T$ contributing to C_p at $1°$K to be $1°$K$/|\Delta|_{max}$ of the total N_T. Since potential variations in Δ of at most T_g in size are frozen in at the glass transition, one expects $|\Delta|_{max}$ to scale with T_g, which would lead to a value of $\overline{N}_T/N \sim 10^{-6}$, in good agreement with experiment. This T_g^{-1} dependence of the density of states has recently been observed experimentally by Reynolds[142] and Raychaudhuri and Pohl[143] for a number of glass formers. However, we are unable to estimate the value for the peak in $P(\Lambda)$ or Λ_{max}. This involves solving a multiparticle tunneling problem, which cannot even be started until more details of the configuration are known.

The essential point of the free-volume picture on which all else depends is the notion that the free energy as a function of the cellular volumes taken as configuration coordinates is flat, on the scale of kT for $T \geq T_g$, in certain regions of the configuration space. It is this feature, and this feature only,

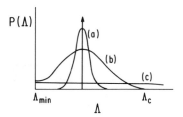

Fig. 23.   Probability distribution $P(\Lambda)$ versus Λ.

that we are invoking to explain the existence of tunneling levels. A free-energy flat in localized regions of the configuration space on a scale of kT_g simply cannot be regarded as flat any longer at temperatures much lower. Instead kT_g becomes the bound to the magnitudes of the variations, that is, maxima, minima, and saddle points (enhanced somewhat by thermal contraction) of importance for tunneling at low temperatures. In this view, tunneling is a continuation of diffusion. This view is independent of the particular set of configuration coordinates convenient for any particular material and thus portrays tunneling as a general phenomenon, universal in the same way that diffusion, viscosity, and the glass transition are universal.

XI. CONCLUSIONS

In this chapter we have tried to give a thorough description of the transition from a liquid to a glass within the context of a single model, the free-volume model. The major results of the theory are more general, however, and are not specific to one model. The model, in spite of its phenomenological nature, can provide a useful understanding of the complex phenomena occurring as a dense liquid solidifies without crystallizing.

The free-volume model was originally derived to explain the temperature dependence of the viscosity. We have shown that it has a much broader application and can explain many of the outstanding experimental observations. This includes the existence of an entropy catastrophe at T_s and the approximate equality of T_η and T_s, first observed by Angell and co-workers.[14] The relation between $\ln|q|$ and T_g, measured by Moynihan et al.,[78] also follows naturally and quantitatively from the notion that the liquidlike cell fraction p is the important variable that ceases to reach equilibrium when the relaxation rates become longer than the time scale for the measurement.

We also find that a simple estimate of τ_0 within the model also produces remarkable agreement with estimates from the experimental observations. The connection between the low-temperature properties of the glass and its behavior at the transition region follows as a natural consequence of our nonequilibrium theory. The estimate of the number of tunneling levels is very close to the usual predictions. The remarkable agreement of these predictions with the experimental results, of course, cannot be taken too seriously, but it does support our belief that the model correctly describes many of the essential features of the transition.

We should point out explicitly that our theory is not a theory of melting, though at first sight it has many similar features. Since this is an important point, let us make it very explicit. In our theory we derived a free energy $F[X] = F[P(v); C_{vz}(p); p]$. The equilibrium values for $P(v)$, $C_{vz}(p)$ and p

are obtained from $\delta F/\delta X = 0$. This free energy and, therefore, the equilibrium values of $P(v)$, $C_{vz}(p)$ and p, depend parametrically on the interfacial tension σ_{sl} between the liquidlike and solidlike regions. If one were to develop a theory for melting, a similar argument could be made in which one introduces a free energy $F_c[X]$ of the crystalline state containing clusters of liquidlike cells. This $F_c[X]$ depends on the interfacial tension σ_{cl} between crystal- and liquidlike regions. However, the structure of the solidlike regions is very close to that of the liquidlike regions; for example, for simple substances, the cage structure is based on dense random packing in both cases. Therefore, very little mismatch occurs across the interface. However, the mismatch is much greater at the crystal-liquid interface, so that $\sigma_{cl} \gg \sigma_{sl}$. Following (5.8) to (5.18), one sees that as σ increases, p and $C_v(p)$ for $v > v_m$ are rapidly suppressed. Thus, although the same theory that yields a first-order phase transition between the glass and the liquid at T_p yields also a first-order phase transition between the crystal and the liquid at T_m, it follows from $\sigma_{cl} \gg \sigma_{sl}$ that $T_m \gg T_p$ and that there are no premelting phenomena.

The conceptual structure of our theory emphasizes the importance of entropy. There are in general three possible contributions to the entropy: communal, internal configurational, and mixing. The movement of the molecules or molecular segments within a cluster gives rise to the communal entropy and is fundamental to it; it also gives rise to the configurational entropy, though less directly. Although the entropy of mixing was included in our model, we did not include the internal configurational entropy, as is done in the complementary "entropy" theory. Remember that entropy is central to both theories. Both theories yield what is essentially a vacancy model for molecular motion, the free-volume model theory giving a more realistic description by including the fluctuations of the voids. The free-volume theory, however, admits in addition a description of the evolution of the communal entropy, which is the most fundamental aspect of the passage from the solid to the liquid state.

The essential features of the glass transition do *not* require the free-volume concepts. On a more abstract level, the theory is based only on the notion that there exists a configuration space X in which $F[X]$ is locally flat on the scale of kT_g or greater and that these configuration coordinates control movement and, therefore, the buildup of configurational and communal entropy. At low temperatures, far below T_g, one can no longer assume that $F[X]$ is locally flat. The finite curvature gives rise to tunneling levels because the atoms or molecules can no longer move freely within the configuration space. In this general view, the localized tunneling motions are the low-temperature residuum of the free exchange of coordinate values and of the attendant classical diffusion at higher temperatures.

Although the free-volume theory has been successful, there are, of course, places where the model fails or additional work is needed. The model does not produce a satisfactory dependence of volume v on temperature. This is due in large part to our approximate form for $f(v)$: this local function neglects, except within mean field theory, the interaction of atoms with their neighbors. The form used for $f(v)$ gives rise to a temperature dependence for v that is not linear, as found experimentally. We have also not included in any explicit manner the fact that in many glass formers the motile segment is a flexible molecule with internal configuration coordinates. This effect should be included for the polymer systems in which neither the internal configurational entropy nor the communal entropy dominates. The explicit elimination of the artificially introduced simplifications should be pursued, as discussed in Section VIII. We do not expect it to make qualitative changes in the nature of transition, but it is essential if any quantitative agreement with experiment is to be expected. Moreover, a reduction in the number of sensitive free parameters in the theory is needed.

There has been no direct verification of the conceptual structure of the theory. That is, a microscopic determination of the cluster distribution function has not been made, and the effects of percolation have not been seen. Assuming that the structure of the glass is well-defined liquidlike clusters in a denser solidlike background, one might expect to be able to see these clusters by either neutron or X-ray scattering. Since v_m is probably between 100 and 400 Å^3 and $\nu_m \simeq 30$ at $T \lesssim T_g$, one would expect that a probe compatible with scattered wave vectors on the order of 0.1 Å^{-1} could be used.

Acknowledgments

This work supported in part by the Materials Research Laboratory Program of the National Science Foundation at the University of Chicago by Grant No. DMR78-24698 and at Purdue University by Grant No. DMR77-23798.

References

1. J. P. Hansen and I. R. McDonald, *Theory of Simple Liquids*, Academic Press, New York, 1976.
2. P. A. Egelstaff, *An Introduction to the Liquid State*, Academic Press, New York, 1967.
3. C. A. Croxton, *Introduction to Liquid State Physics*, Wiley, New York, 1975.
4. D. Turnbull, *Contemp. Phys.*, **10**, 473 (1969).
5. D. Turnbull, in J. W. Prins, Ed., *Physics of Non-Crystalline Solids*, North-Holland, Amsterdam, 1964, p. 41.
6. D. R. Uhlmann, *J. Non-Cryst. Solids*, **7**, 737 (1972).
7. H. A. Davies and B. G. Lewis, in N. J. Grant and B. C. Giessen, Eds., *Proceedings of the Second International Conference on Rapidly Quenched Metals*, MIT Press, Cambridge, MA, 1976, p. 259.

8. F. Spaepen and D. Turnbull, in N. J. Grant and B. C. Giessen, Eds., *Proceedings of the Second International Conference on Rapidly Quenched Metals*, MIT Press, Cambridge, MA, 1976, p. 205.

9. P. Duwez, *Trans. Am. Soc. Met.*, **60**, 607 (1967); *Ann. Rev. Mater. Sci.*, **6**, 83 (1976).

10. D. Turnbull and B. G. Bagley, in N. B. Hannay, Ed., *Treatise on Solid State Chemistry*, Vol. 5, Plenum Press, New York, 1975, p. 513.

11. G. C. Berry and T. G. Fox, *Adv. Polym. Sci.*, **5**, 261 (1968), and references therein.

12. G. S. Cargill, III, in H. Ehrenreich, F. Seitz, and D. Turnbull, Eds., *Solid State Physics*, Vol. 30, Academic Press, New York, 1975.

13. P. Chaudhari and D. Turnbull, *Science*, **199**, 11 (1978).

14. C. A. Angell and W. Sichina, *Ann. N.Y. Acad. Sci.*, **279**, 53 (1976).

15. C. T. Moynihan et al., *Ann. N.Y. Acad. Sci.*, **279**, 15 (1976).

16. E. U. Condon, *Am. J. Phys.*, **22**, 43 (1954).

17. J. W. Gibbs and E. A. DiMarzio, *J. Chem. Phys.*, **28**, 373 (1958); E. A. DiMarzio and J. W. Gibbs, *ibid.*, **28**, 807 (1958); *J. Polym. Sci.*, A, **1**, 1417 (1963); *J. Polym. Sci.* **40**, 121 (1959); E. A. DiMarzio and F. Dowell, *J. Appl. Phys.* **50**, 6061 (1979).

18. M. H. Cohen and D. Turnbull, *J. Chem. Phys.*, **31**, 1164 (1959).

19. M. H. Cohen and D. Turnbull, *Nature* (*London*), **203**, 964 (1964).

20. H. S. Chen and D. Turnbull, *J. Chem. Phys.*, **48**, 2560 (1968); *J. Appl. Phys.*, **38**, 3646 (1967); *Acta Metal.*, **17**, 1021 (1969).

21. D. E. Polk and D. Turnbull, *Acta Metal.*, **20**, 493 (1972).

22. H. S. Chen and M. Goldstein, *J. Appl. Phys.*, **43**, 1642 (1972).

23. H. S. Chen, *J. Non-Cryst. Solids*, **12**, 333 (1974); *Acta Metal.*, **22**, 897 (1974); 1505 (1974); **24**, 153 (1976); *Mater. Sci. Eng.*, **23**, 151 (1976).

24. S. Takayama, *J. Mater. Sci.*, **11**, 164 (1976).

25. H. A. Davies, *Phys. Chem. Glasses*, **17**, 159 (1976).

26. H. Jones, *Rep. Prog. Phys.*, **36**, 1425 (1973).

27. J. J. Gilman, *Science*, **208**, 856 (1980).

28. B. J. Alder and T. E. Wainwright, *Phys. Rev. Lett.*, **18**, 988 (1967).

29. B. J. Alder, W. G. Hoover, and D. A. Young, *J. Chem. Phys.*, **49**, 3688 (1968).

30. J. M. Gordon, J. H. Gibbs, and P. D. Fleming, *J. Chem. Phys.*, **65**, 2771 (1976).

31. L. V. Woodcock, *J. Chem. Soc. Faraday Trans. 2*, **72**, 1667 (1976); **73**, 11 (1977).

32. A. Rahman, M. J. Mandell, and J. P. McTague, *J. Chem. Phys.*, **64**, 1564 (1976).

33. G. Jacucci, unpublished.

34. J. H. R. Clarke, *J. Chem. Soc. Faraday Trans. 2*, **75**, 1371 (1979); C. A. Angell, J. H. R. Clarke, and L. V. Woodcock, in S. A. Rice and E. Prigogine, Eds., *Advances in Chemical Physics*, Vol. 48, Wiley, New York, 1981.

35. L. V. Woodcock, in J. Braunsteen, G. Mamantov, and G. P. Smith, Eds., *Advances in Molten Salt Chemistry*, Vol. 3, Plenum Press, New York, 1975.

36. J. E. Kunzler and W. F. Giaque, *J. Am. Chem. Soc.*, **74**, 797 (1954).

37. C. A. Angell, *J. Phys. Chem.*, **70**, 2793 (1966); *J. Chem. Phys.*, **46**, 4673 (1967).

38. C. A. Angell, L. J. Pollard, and W. Strauss, *J. Chem. Phys.*, **50**, 2964 (1969).

39. C. A. Angell and K. J. Rao, *J. Chem. Phys.*, **57**, 470 (1972); C. A. Angell and J. C. Tucker, *J. Phys. Chem.*, **78**, 278 (1974); C. A. Angell and D. Smith (private communication).

40. R. Weiler, S. Blaser, and P. B. Macedo, *J. Phys. Chem.*, **73**, 4147 (1969); H. Tweer, N. Laberge, and P. B. Macedo, *J. Am. Ceram. Soc.*, **54**, 121 (1971).

41. F. S. Howell, R. A. Bose, P. B. Macedo, and C. T. Moynihan, *J. Phys. Chem.*, **78**, 639 (1974).

42. A. A. Miller, *J. Chem. Phys.*, **49**, 1393 (1968); *Macromolecules*, **2**, 355 (1969); **3**, 674 (1970).

42. S. M. Wolpert, A. Weitz, and B. Wunderlich, *J. Polym. Sci.*, *A-2*, **9**, 1887 (1971); A. Weitz and B. Wunderlich, *J. Polym. Sci*, *Polym. Phys. Ed.*, **12**, 2473 (1974).
43. J. E. McKinney and M. Goldstein, *J. Res. Natl. Bur. Stand. A*, **78**, 331 (1974).
44. C. Lacabanne and D. Chatain, *J. Phys. Chem.*, **79**, 283 (1975).
45. S. Rogers and L. Mandelkern, *J. Phys. Chem.*, **61**, 985 (1957).
46. L. A. Woods, *J. Polym. Sci.*, **28**, 319 (1958).
47. A. J. Barlow, J. Lamb, and A. J. Matheson, *Proc. R. Soc. London, Ser. A*, **292**, 322 (1966).
48. A. C. Ling and J. E. Williard, *J. Phys. Chem.*, **72**, 1918 (1968); 3349 (1968).
49. R. J. Greet and J. H. Magill, *J. Phys. Chem.*, **71**, 1746 (1967).
50. R. J. Greet and D. Turnbull, *J. Chem. Phys.*, **46**, 1243 (1967); **47**, 2185 (1967).
51. S. S. Chang, J. A. Horman, and A. B. Bestul, *J. Res. Natl. Bur. Stand. A*, **71**, 293 (1967); S. S. Chang and A. B. Bestul, *ibid.*, **75**, 113 (1971); *J. Chem. Phys.*, **56**, 503 (1972).
52. M. R. Carpenter, D. B. Davies, and A. J. Matheson, *J. Chem. Phys.*, **46**, 2451 (1967).
53. D. J. Plazek and J. H. Magill, *J. Chem. Phys.*, **45**, 3038 (1966); **46**, 3757 (1967); J. H. Magill, *ibid.*, **47**, 2802 (1967).
54. W. T. Laughlin and D. R. Uhlmann, *J. Phys. Chem.*, **76**, 2317 (1972).
55. M. Cukierman, J. W. Lane, and D. R. Uhlmann, *J. Chem. Phys.*, **59**, 3639 (1973).
56. A. C. Wright and A. J. Leadbetter, *Phys. Chem. Glasses*, **17**, 122 (1976).
57. G. Hetherington, K. H. Jack, and J. C. Kennedy, *Phys. Chem. Glasses*, **5**, 130 (1964).
58. E. H. Fontana and W. A. Plummer, *Phys. Chem. Glasses*, **7**, 139 (1966).
59. J. P. DeNeufville, C. H. Drummond, III, and D. Turnbull, *Phys. Chem. Glasses*, **11**, 186 (1970).
60. P. B. Macedo and A. Napolitano, *J. Chem. Phys.*, **49**, 1887 (1968).
61. D. R. Uhlmann, A. G. Kolbeck, and D. L. DeWitte, *J. Non-Cryst. Solids*, **5**, 426 (1971).
62. K. Arai and S. Saito, *Jpn. J. Appl. Phys.*, **10**, 1669 (1971).
63. J. P. DeNeufville, *J. Non-Cryst. Solids*, **8–10**, 85 (1972).
64. A. Feltz, H. J. Büttner, F. J. Lippmann, and W. Maul, *J. Non-Cryst. Solids*, **8–10**, 64 (1972).
65. J. P. DeNeufville and H. K. Rockstad, in J. Stuke and W. Brenig, Eds., *Proceedings of the Fifth International Conference on Amorphous and Liquid Semiconductors*, Taylor and Francis, London, 1974, p. 419.
66. S. S. Chang and A. B. Bestul, *J. Chem. Thermodyn.*, **6**, 325 (1974).
67. D. D. Thornburg and R. I. Johnson, *J. Non-Cryst. Solids*, **17**, 2 (1975).
68. B. A. Joiner and J. C. Thompson, *J. Non-Cryst. Solids*, **21**, 215 (1976).
69. D. J. Sarrach, J. P. DeNeufville, and W. L. Haworth, *J. Non-Cryst. Solids*, **22**, 245 (1976).
70. R. Bruckner, *J. Non-Cryst. Solids*, **5**, 123 (1970).
71. M. Abkowitz and D. M. Pai, *Phys. Rev. Lett.*, **38**, 1412 (1979); *Phys. Rev. B*, **18**, 1741 (1978).
72. R. B. Stephens, *J. Appl. Phys.*, **49**, 5855 (1978).
73. H. Vogel, *Phys. Z.*, **22**, 645 (1921).
74. G. S. Fulcher, *J. Am. Ceram. Soc.*, **6**, 339 (1925).
75. G. Tammann and G. Hesse, *Z. Anorg. Alleg. Chem.*, **156**, 245 (1926).
76. H. Tweer, J. H. Simmons, and P. B. Macedo, *J. Chem. Phys.*, **54**, 1952 (1971).
77. C. T. Moynihan, A. J. Easteal, J. Wilder, and J. Tucker, *J. Phys. Chem.*, **78**, 2673 (1974).
78. C. T. Moynihan, A. J. Easteal, M. A. DeBolt, and J. Tucker, *J. Am. Ceram. Soc.*, **59**, 12 (1976); M. A. DeBolt, A. J. Easteal, P. B. Macedo, and C. T. Moynihan, *ibid.*, **59**, 16 (1976).
79. J. M. Hutchinson and A. J. Kovacs, *J. Polym. Sci.*, *Polym. Phys. Ed.*, **14**, 1575 (1960); A. J. Kovacs, J. M. Hutchinson, and J. J. Aklonis, in P. H. Gaskell, Ed., *The Structure of Non-Crystalline Materials*, Taylor and Francis, London, 1977, p. 157; A. J. Kovacs et al., *J. Polym. Sci.*, **17**, 1097 (1979).

80. M. Lasocka, *Mater. Sci. Eng.*, **23**, 173 (1976).
81. W. Kauzmann, *Chem. Rev.*, **43**, 219 (1948).
82. C. A. Angell and J. Donnella, *J. Chem. Phys.*, **67**, 4560 (1977).
83. W. H. Zachariasen, *J. Am. Chem. Soc.*, **58**, 3841 (1932).
84. G. Adams and J. H. Gibbs, *J. Chem. Phys.*, **28**, 139 (1965).
85. T. G. Fox and P. J. Flory, *J. Appl. Phys.*, **21**, 581 (1950); *J. Phys. Chem.*, **55**, 221 (1951); *J. Polym. Sci.*, **14**, 315 (1954).
86. M. L. William, R. F. Landel, and J. D. Ferry, *J. Am. Chem. Soc.*, **77**, 3701 (1955).
87. D. Turnbull and M. H. Cohen, *J. Chem. Phys.*, **34**, 120 (1961).
88. D. Turnbull and M. H. Cohen, *J. Chem. Phys.*, **52**, 3038 (1970).
89. J. Naghizadeh, *J. Appl. Phys.*, **35**, 1162 (1964).
90. P. B. Macedo and T. A. Litovitz, *J. Chem. Phys.*, **42**, 245 (1965).
91. A. J. Matheson, *J. Chem. Phys.*, **44**, 695 (1966).
92. H. S. Chung, *J. Chem. Phys.*, **44**, 1362 (1966).
93. Y. Lipatov, *Adv. Polym. Sci.*, **26**, 63 (1978), and references to other polymer literature therein.
94. M. H. Cohen and G. S. Grest, *Phys. Rev. B*, **20**, 1077 (1979).
95. G. S. Grest and M. H. Cohen, *Phys. Rev. B*, **21**, 4113 (1980).
96. A. J. Batschinski, *Z. Phys. Chem.*, **84**, 644 (1913).
97. A. K. Doolittle, *J. Appl. Phys.*, **22**, 1471 (1951).
98. F. Beuche, *J. Chem. Phys.*, **21**, 1850 (1953); **24**, 418 (1956).
99. R. M. Barrer, *Trans. Faraday Soc.*, **38**, 322 (1942).
100. P. W. Anderson, B. I. Halperin, and C. M. Varma, *Phil. Mag.*, **25**, 1 (1972).
101. W. A. Phillips, *J. Low Temp. Phys.*, **7**, 351 (1972).
102. J. Jäckle, *Z. Phys.*, **257**, 212 (1972).
103. B. I. Halperin, *Ann. N.Y. Acad. Sci.*, **279**, 173 (1976).
104. J. L. Black, *Phys. Rev. B*, **17**, 2740 (1978).
105. J. L. Black, in H. J. Guntherodt, Ed., *Metallic Glasses*, Springer-Verlag, New York, 1980.
106. R. C. Zeller and R. O. Pohl, *Phys. Rev. B*, **4**, 2029 (1971).
107. R. B. Stephens, *Phys. Rev. B*, **8**, 2896 (1973); *Phys. Rev. B*, **13**, 852 (1976).
108. W. A. Phillips, *J. Non-Cryst. Solids*, **31**, 267 (1978).
109. J. G. Kirkwood, *J. Chem. Phys.*, **18**, 380 (1950).
110. For a review, see T. L. Hill, *Statistical Mechanics*, McGraw-Hill, New York, 1956.
111. C. S. Hsu and A. Rahman, *J. Chem. Phys.*, **70**, 5234 (1979); **71**, 4974 (1979).
112. H. R. Wendt and F. F. Abraham, *Phys. Rev. Lett.*, **41**, 1244 (1978).
113. F. H. Stillinger and T. A. Weber, *J. Chem. Phys.*, **68**, 3837 (1978); *ibid.*, **70**, 4879 (1979).
114. J. L. Finney, *Proc. R. Soc., London, Ser. A*, **319**, 479, 495 (1970).
115. J. D. Bernal, *Nature (London)*, **183**, 141 (1959); **185**, 68 (1960); J. D. Bernal and J. Mason, *ibid.*, **188**, 910 (1960).
116. V. K. S. Shante and S. Kirkpatrick, *Adv. Phys.*, **20**, 325 (1971); S. Kirkpatrick, *Rev. Mod. Phys.*, **45**, 574 (1973).
117. J. W. Essam, in C. Domb and M. S. Green, Eds., *Phase Transition and Critical Phenomena*, Vol. II, Academic Press, New York, 1972, p. 197.
118. C. Domb, *J. Phys. C*, **7**, 2677 (1974).
119. D. Stauffer, *Z. Phys. B*, **25**, 391 (1976).
120. P. L. Leath, *Phys. Rev. B*, **14**, 5046 (1976); P. L. Leath and G. R. Reich, *J. Phys. C*, **11**, 4017 (1978).
121. S. Kirkpatrick, in D. P. Tanner, Ed., *Electrical Transport and Optical Properties of Inhomogeneous Media*, American Institute of Physics Conference Proceedings, 1977, AIP, New York.

122. G. R. Reich and P. L. Leath, *J. Stat. Phys.*, **19**, 611 (1978).
123. L. Turban and P. Guilmin, *J. Phys. C*, **12**, 961 (1979); L. Turban, *J. Phys. C.*, **12**, 5009 (1979).
124. V. Jaccarino and L. R. Walker, *Phys. Rev. Lett.*, **15**, 258 (1965).
125. T. J. Hicks, B. Rainford, J. S. Kouvel, G. G. Low, and J. B. Comly, *Phys. Rev. Lett.*, **22**, 531 (1969).
126. J. P. Perrier, B. Tissier, and R. Tournier, *Phys. Rev. Lett.*, **24**, 313 (1970).
127. H. E. Stanley, *J. Phys. A*, **12**, 961 (1979); H. E. Stanley and J. Teixeira, *J. Chem. Phys.*, **73**, 3404 (1980).
128. L. V. Woodcock, in P. H. Gaskell, Ed., *The Structure of Non-Crystalline Materials*, Taylor and Francis, London, 1977, p. 187.
129. H. Scher and R. Zallen, *J. Chem. Phys.*, **53**, 3759 (1970); R. Zallen and H. Scher, *Phys. Rev. B*, **4**, 4471 (1971).
130. I. Webman, J. Jortner, and M. H. Cohen, *Phys. Rev. B*, **11**, 2885 (1975); **14**, 4737 (1976).
131. M. E. Fisher, *Physics*, **3**, 255 (1967).
132. D. Stauffer, *Phys. Rev. Lett.*, **35**, 394 (1975).
133. R. G. Munro, S. Block and G. J. Piermarini, *J. Appl. Phys.*, **50**, 6779 (1979).
134. H. N. Ritland, *J. Amer. Ceram. Soc.*, **37**, 370 (1954).
135. O. S. Narayanaswamy, *J. Amer. Ceram. Soc.*, **54**, 121 (1971).
136. K. Freed (private communication).
137. W. M. Goubau and R. H. Tait, *Phys. Rev. Lett.*, **34**, 1220 (1975).
138. R. B. Kummer, R. C. Dynes, and V. Narayanamurti, *Phys. Rev. Lett.*, **40**, 1187 (1978).
139. M. T. Loponen, R. C. Dynes, V. Narayanamurti, and J. P. Garno, *Phys. Rev. Lett.*, **45**, 457 (1980).
140. M. Meissner and K. Spitzmann, *Phys. Rev. Lett.*, **46**, 265 (1981).
141. M. H. Cohen and G. S. Grest, *Phys. Rev. Lett.*, **45**, 1271 (1980); *Solid State Commun.* (to be published, 1981).
142. C. L. Reynolds, Jr., *J. Non-Cryst. Solids*, **30**, 371 (1979); **37**, 125 (1980).
143. A. K. Raychaudhuri and P.O. Pohl, *Solid State Commun.* (to be published, 1981).

AUTHOR INDEX

Numbers in parentheses are reference numbers and indicate that the author's work is referred to although his name is not mentioned in the text. Numbers in italics show the pages on which the complete references are listed.

SUBJECT INDEX